THE CONTEMPORARY CARIBBEAN

PEARSON EDUCATION

We work with leading authors to develop the strongest educational materials in Geography, bringing cutting-edge thinking and best learning practice to a global market.

Under a range of well-known imprints, including Prentice Hall, we craft high quality print and electronic publications which help readers to understand and apply their content, whether studying or at work.

To find out more about the complete range of our publishing, please visit us on the World Wide Web at: www.pearsoned.co.uk

THE CONTEMPORARY CARIBBEAN

ROBERT B. POTTER,
DAVID BARKER,
DENNIS CONWAY AND THOMAS KLAK

Harlow, England • London • New York • Boston • San Francisco • Toronto
Sydney • Tokyo • Singapore • Hong Kong • Seoul • Taipei • New Delhi
Cape Town • Madrid • Mexico City • Amsterdam • Munich • Paris • Milan

Pearson Education Limited
Edinburgh Gate
Harlow
Essex CM20 2JE
England

and Associated Companies throughout the world

Visit us on the World Wide Web at:
www.pearsoned.co.uk

First published 2004

ISBN 0–582–41853–4

British Library Cataloguing-in-Publication Data
A catalogue record for this book is available from the British Library

Library of Congress Cataloging-in-Publication Data
Potter, Robert B.
The contemporary Caribbean / by Robert B. Potter . . . [et al.]
p. cm.
Includes bibliographical references and index.
ISBN 0–582–41853–4 (pbk.)
1. Caribbean Area—Economic conditions—1945– 2. Caribbean Area—Social conditions—1945– 3. Caribbean Area—Politics and government—1945– 4. Agriculture—Economic aspects—Caribbean Area. 5. Caribbean Area—Rural condition. I. Title.

HC151.P68 2004
330.9729—dc22

2004044420

10 9 8 7 6 5 4 3 2 1
08 07 06 05 04

Typeset in 10½/12½ Ehrhardt MT by 35
Printed in Great Britain by Henry Ling Ltd at the Dorset Press, Dorchester, Dorset

The publisher's policy is to use paper manufactured from sustainable forests.

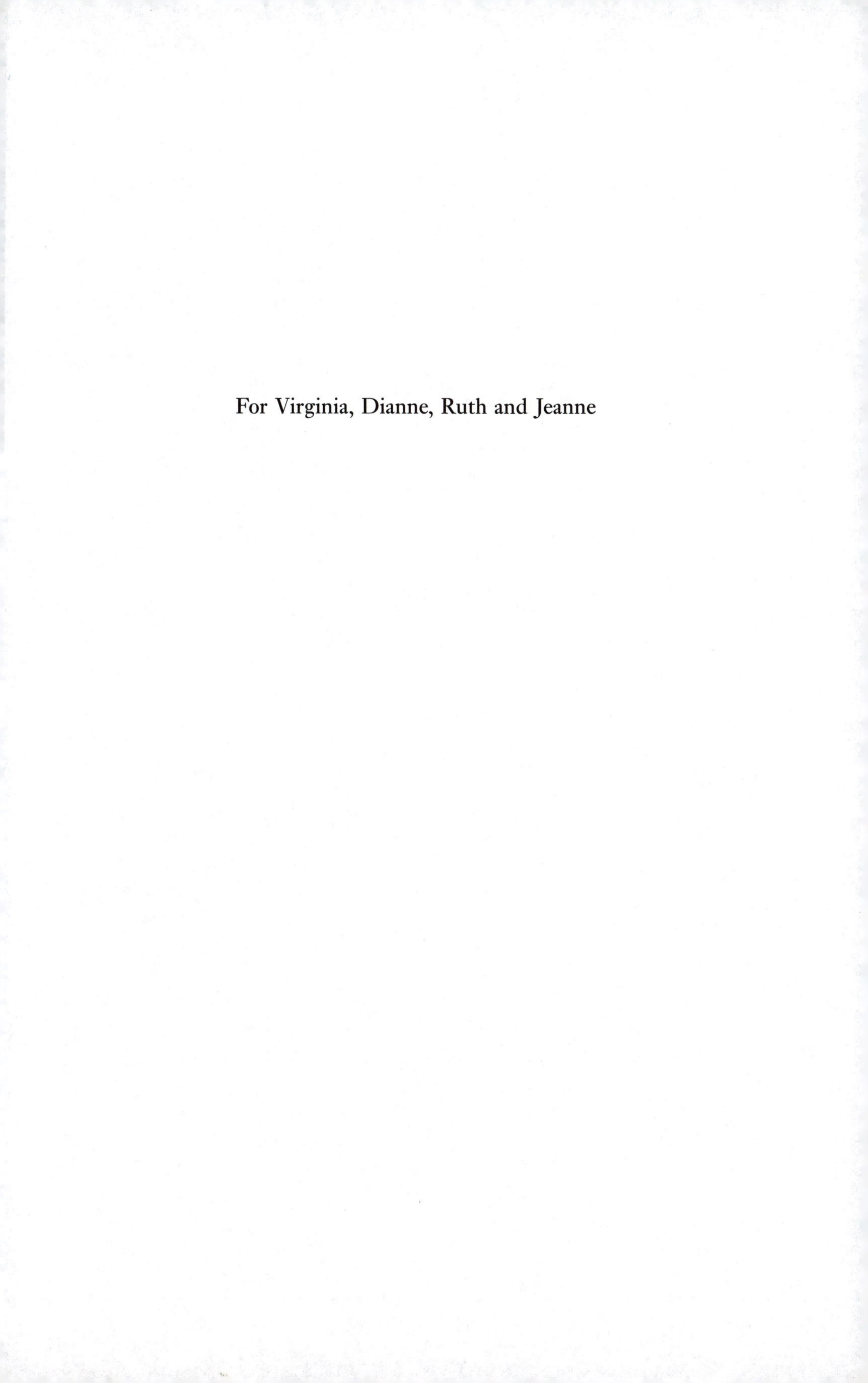

For Virginia, Dianne, Ruth and Jeanne

CONTENTS

Part II
Rural and urban bases of the contemporary Caribbean

LIST OF FIGURES

LIST OF PLATES

LIST OF TABLES

PREFACE

The aim of *The Contemporary Caribbean* is to focus attention on the present-day social, economic, political, cultural and environmental realities of the Caribbean region. Important historical aspects of the Caribbean, such as slavery, the plantation system, plantocracy and colonialism, are dealt with in order to explain the contemporary nature of, and challenges faced by, the Caribbean region and its peoples. As authors, we started from the realisation that there are many good books on the Caribbean, but that the vast majority deal with particular topics, be it the history, society, environment, politics, business, medicine, flora or fauna of the region. Few up-to-date texts are available that present a holistic view of the extant nature of this diverse, vibrant and complex world region. This is the remit that we adopted for *The Contemporary Caribbean*, and we wanted to provide an overview of the region for starting students and those more generally interested in this distinctive area of the world.

The resultant book is divided into three parts, dealing respectively with the foundations of the Caribbean, the rural and urban bases of the contemporary Caribbean, and global restructuring and the Caribbean: industry, globalisation, tourism and politics. The first part of the book serves to provide an introduction to the physical landscapes of the region and the character of its populations. Migration from, back to, and within Caribbean territories has always been such a salient part of the region that this vital topic receives extended treatment in Chapter 2. The nature of agriculture and rural occupancy are reviewed in Part II, as is the impact of natural events and disasters, before social conditions, housing, urbanisation and townscapes are considered in turn. These first two parts of the volume act as gateways to the consideration of global restructuring and the contemporary Caribbean region in Part III. The focus here is principally, although by no means exclusively, on economic activities, particularly manufacturing, offshore services and tourism. Part III also covers the pressing realities of globalisation and political change as they are affecting the present-day Caribbean region.

Thus, *The Contemporary Caribbean* is a comprehensive tertiary-level textbook aimed at students taking courses that involve consideration of the extant character of the Caribbean region. The book should, therefore, be of value to undergraduates, and indeed, postgraduates who are taking courses in the fields of geography, development studies, black studies, sociology, social policy, economics and politics, as well as more generally in the social sciences.

Among the distinctive features of this book, as authors we would specifically pinpoint the following:

- *The Contemporary Caribbean* stresses the present-day nature of the Caribbean region as its primary focus, but in so doing, covers its foundations in deeply rooted historical processes.
- Reflecting the topics dealt with, the text is well illustrated by means of plentiful figures, maps and photographs. Indeed, over 100 such illustrations are included in the text.
- In order to assist students and readers more generally, boxed case studies are presented within the chapters, often so as to provide country- or region-wide specific case studies of the thematic topics under consideration in the main text. Over 30 boxed case studies are presented.
- All of the authors have considerable experience of working in the Caribbean, and of undertaking first-hand research in the region.

As already noted, the treatment afforded the Caribbean in this book is thematic, but in dealing with a given topic, every effort has been made to ensure that relevant examples and instances are drawn from all parts of the region (see Preface Figures 1 and 2). This is shown by the content of the boxed case studies, which extend from the consideration of Haiti's revolutionary beginnings, to urbanisation and urban structure in Puerto Rico, protected areas in Cuba and patterns of settlement and economic activity in Guyana, to Montserrat's volcanic eruption and Grenada's experiment with socialism. Although our primary focus is on the insular Caribbean, attention is also given to the wider Caribbean Basin, including territories like Guyana and Belize, which historically and culturally are regarded as part of the Caribbean (see Preface Table 1 and see also Figure 1.1).

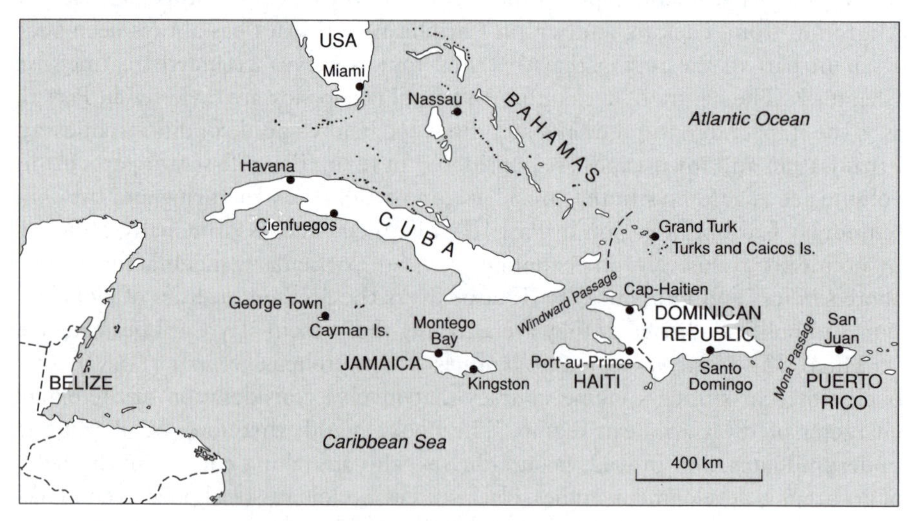

Preface Figure 1 Detailed map of the western Caribbean Basin
Source: Klak (1998)

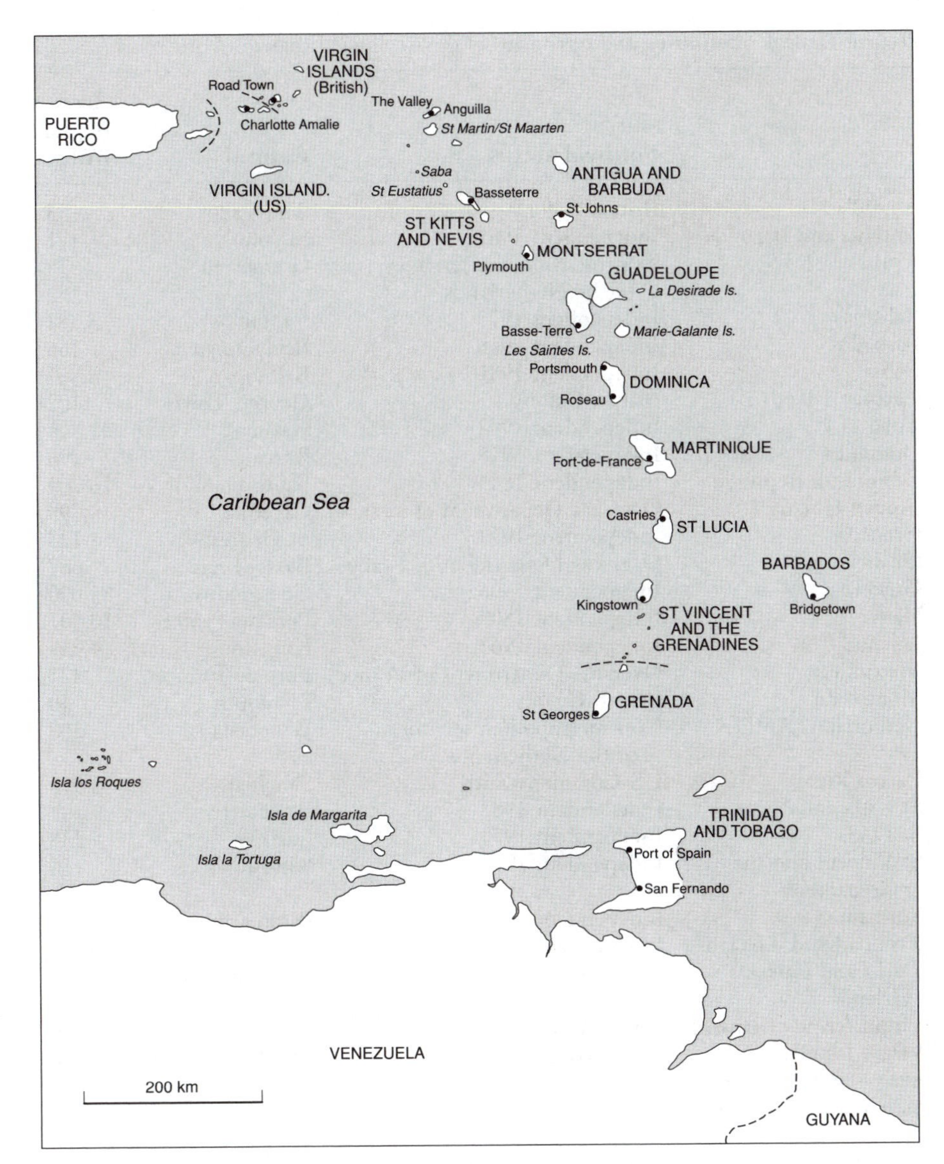

Preface Figure 2 Detailed map of the eastern Caribbean Basin
Source: Klak (1998)

The Caribbean is a fascinating world region, as attested by the broad coverage of topics provided in this book. The Caribbean may well be a sea rather than a landmass, and heterogeneity may be one of the distinctive characteristics of the territories making up the region, but the text also seeks to exemplify the major commonalities that serve to characterise the region.

Considering the fact that the respective authors were resident in Kingston, Jamaica, Indiana and Ohio USA, Luxembourg and Surrey, England, during the time the book was conceived and subsequently written, the process of collaborative

Preface Table 1 Countries and territories of the Caribbean region

	Political status	Capital	Area (square miles)
Anguilla	British Colony	The Valley	35
Antigua and Barbuda	Independent 1981	St John's	171
Aruba	Non-metropolitan territory of the Netherlands	Oranjestad	75
Bahamas	Independent 1973	Nassau	5,382
Barbados	Independent 1966	Bridgetown	166
Belize	Independent 1981	Belmopan	8,867
Cayman Islands	British Colony	George Town	102
Cuba	Independent 1902	Havana	42,804
Dominica	Independent 1978	Roseau	290
Dominican Republic	Independent 1844	Santo Domingo	18,704
French Guiana	Overseas Department of France	Cayenne	33,399
Grenada	Independent 1974	St George's	133
Guadeloupe	Overseas Department of France	Basse-Terre	687
Guyana	Independent 1966	Georgetown	83,000
Haiti	Independent 1804	Port-au-Prince	10,597
Jamaica	Independent 1962	Kingston	4,244
Martinique	Overseas Department of France	Fort-de-France	421
Montserrat	British Colony	Plymouth	40
Netherlands Antilles	Non-metropolitan territory of the Netherlands	Willemstad	308
Puerto Rico	US Commonwealth	San Juan	3,515
St Kitts and Nevis	Independent 1983	Basseterre	104
St Lucia	Independent 1979	Castries	238
St Vincent and the Grenadines	Independent 1979	Kingstown	150
Suriname	Independent 1975	Paramaribo	63,251
Trinidad and Tobago	Independent 1962	Port of Spain	1978
Turks and Caicos Islands	British Colony	Cockburn Town	193
Virgin Islands (British)	British Colony	Road Town	59
Virgin Islands (US)	US Territory	Charlotte Amalie	136

Source: Richardson (1992)

writing proved to be productive and challenging, but at the same time extremely collaborative and convivial. This was undoubtedly assisted by our meeting up in twos or threes as part of our respective travels within the Caribbean region and elsewhere, as well as at various international conferences and meetings. It was perhaps fitting that we finally all managed to be in the same place (and indeed, purely by chance, at the same budget-price hotel – although this obviously says something about academic salaries!), just a few months before the final completion of the manuscript. This was occasioned by our attendance at the 99th Annual Conference of the Association of American Geographers, held in New Orleans in March 2003. On all other occasions, e-mail communication along with a measure of good humour kept us working towards our self-imposed goal of producing this pan-Caribbean text.

Rob originally floated the idea of the book with Matthew Smith, who saw it through the contract stage with his usual measure of enthusiasm. We were delighted that when Matthew Taylor took over the reins he showed an equal measure of enthusiasm for the project, and more than a modicum of patience in dealing with four such wide-flung authors. Sarah Wild ably assisted us through the editorial process. Closer to home, we would like to thank Sophie Bowlby and Joan Phillips for reading a draft of Chapter 5 and David Miller for reading Chapters 1 and 4. Chris Holland helped with the typing of the preliminaries toward the end of the project. Erika Meller reproduced several of the photographs included in Chapter 6. Special thanks go to Allison Chapman-Andrews for allowing us to reproduce one of her fabulous paintings as part of the cover design. Reflecting the initiation of his interest in Caribbean affairs in general, and Caribbean geography in particular, Dennis would like to acknowledge the many friends he and Ruth made while he was teaching geography at Harrison College back in the mid-1960s, and thank them for stimulating his long-held interest in Caribbean small island development. Maintaining long distance, life-long ties with these friends, along with memories of the remarkable boys he taught, provided the basis for a deeply felt commitment to do something worthwhile for the people of this insular region. If this project helps, then these boys, now grown men, played their part. Dennis and Ruth's West Indian friends established the 'ties that bind'. Some still live in the region – Jean and Bruce, Allison and Stan, Hamish and Phyllis, Ben, Diana and Douglas, Irma, Gavin, Victor, Richard, Lady Grace Adams (now deceased) – others beyond it – Colin and Paula, Alex and Moira, Adrian, John, Mike and Mike. Their friendship was, and still is, a warm and treasured link to the region that Dennis and Ruth think of as their third 'home'.

Why should anybody write a student textbook in this day and age? Certainly not in the hope of getting rich! We are clear that our own nascent interests in the Caribbean were promoted by one or two texts. This might have been Helmut Blume's *Caribbean Lands* (Longman, 1974) or David Lowenthal's *West Indian Societies* (Oxford University Press, 1972). As all who have done so know only too well, setting the parameters of a text like this in a fashion that seeks to ensure that all the relevant topics are included, and that the various issues covered represent a genuine synthesis of knowledge in the field, presents major scholarly challenges. Any downgrading of such a task is misguided in the extreme, as it fails to recognise that the main ambition of such a text is to enthuse the next generation of scholars and researchers, who may in turn be drawn to work in the Caribbean region. If this book promotes enough interest to launch even a handful of scholars onto the path of studying and researching within this fascinating world region, then this will have proved a worthwhile project. We certainly hope that this will prove to be the case over the coming years.

Rob Potter, David Barker, Dennis Conway and Tom Klak
August 2003

ACKNOWLEDGEMENTS

We are grateful to the following for permission to reproduce copyright material:

Preface Figures 1 and 2 from adaptations of The Lesser Antilles (Map 2) and The Greater Antilles (Map 3), on pp. xx and xxi, from *Globalization and Neoliberalism: the Caribbean Context*, Rowman & Littlefield, (Klak, T. (ed.), 1998); Preface Table 1 from table from *The Caribbean in the Wider World, 1492–1992,* Cambridge University Press, (Richardson, B. C., 1992); Figure 1.3 adapted from figure from 'Geologic provinces of the Caribbean region', (Draper, Jackson, T. and Donovan, S., 1994), in *Caribbean Geology: An Introduction*, University of the West Indies Press, (Donovan, S. & Jackson, T., (eds.), 1994); Figure 1.7 adapted from a figure from *The Science and Wonders of the Atmosphere*, by Gedzelman, S. D., John Wiley & Sons, Inc., (1980), Copyright © 1980 John Wiley & Sons, Inc., This material is used by permission of John Wiley & Sons, Inc.; Figure 1.10 adapted from a table from 'The classification of tropical American vegetation types', in *Ecology*, Vol. 36, 1955, pp. 89–100, Ecological Society of America, (Beard, J. S., 1955); Figure 1.12 from NEPA and photos of turtles from WIDECAST; Figure 3.11 from an adaptation of a figure from *'Introduction: Household Gardens and Small-scale Food Production'*, in *Food and Nutrition Bulletin,* 1985, United Nations University Press, (Niñez, V., 1985); Figure 3.12 from 'West Indian Kitchen Gardens: a Historical Perspective with current Insights from Grenada', in *Food and Nutrition Bulletin,* 1985, United Nations University Press, (Brierley, J., 1985); Figures 4.7a and 4.7b from volcano risk maps of September 1997 and April 1999 produced by the Montserrat Volcano Observatory, Copyright © 1997 and 1999 Montserrat Volcano Observatory and the Government of Montserrat; Figures 4.9 and 4.17 from adaptations of figures from *Fundamentals of Physical Geography*, Routledge (Briggs, D. J., and Smithson, P., 1985); Figure 5.1 from figure in Ch. 7, pp. 271–310, from 'Power, politics and society', (Potter, R. B., and Binns, J., 1988), in *The Geography of the third world: Progress and prospects*, Routledge, (Pacione, M., 1988); Figure 5.3 and Table 5.6 from adaptations of a figure and table from *Urbanization and Urban Growth in the Caribbean*, Cambridge University Press, (Cross, M., 1979). Figures 5.4 and 5.5 from figures on pp. 25–50 from 'Social conditions in St Lucia: aggregate analysis of the 1991 Census at Quarter level', in *Journal of the Eastern Caribbean Studies,* Vol. 24, 1996, Sir Arthur Lewis Institute of Social and Economic Studies (SALISES), University of the West Indies, (Potter, R. B., and Jacyno, J., 1996); Table 5.5 from an adaptation of a table from 'Race, ethnicity and social stratification in three Windward Islands', in *Journal of the Eastern Caribbean Studies,*

Vol. 24, 1999, pp. 1–28, Sir Arthur Lewis Institute of Social and Economic Studies (SALISES), University of the West Indies, (De Albuquerque, K. and McElroy, J., 1999); Table 5.7 from an adaptation of a wider table from 'The geography of relative affluence and poverty in Barbados', in *Caribbean Geography*, Vol. 10, 1999, pp. 79–88, Department of Geography and Geology, University of the West Indies, (Potter, R. B., 1999); Figures 6.1, 6.4, 6.8, 6.9 from figures from *Low-income Housing and State Policy in the Eastern Caribbean* , University of the West Indies Press, (Potter, R. B., 1994); Figures 6.2 and 6.10 from figures from *Low-cost Housing in Barbados: evolution or social revolution?*, University of the West Indies Press, (Watson, M. R. and Potter, R. B., 2001); Figures 6.3, 6.6, 6.7, 6.11 and Table 6.1 from figures and a table from *Housing Conditions in Barbados: A Geographical Analysis*, Institute of Social and Economic Research, Sir Arthur Lewis Institute of Social and Economic Studies (SALISES), University of the West Indies, (Potter, R. B., 1992); Table 6.2 from table from Potter, R. B., and Watson, M. R. (1999), 'Current housing policy issues in Barbados: with particular reference to vacant sub-divisions', in *Third World Planning Review*, Vol. 21, pp. 237–260, Liverpool University Press; Figure 6.5 from Ch. 11, 'Housing and the state in the French Caribbean', (Condon, A. and Ogden, P. E., 1997), in *Self-help Housing, the Poor, and the State in the Caribbean*, pp. 217–242, University of the West Indies Press, (Potter, R. B. and Conway, D., 1997); Tables 7.1 and 7.2, Figure 7.6 from a figure and adaptations of two tables reprinted from *Cities: the international quarterly on urban policy*, Vol. 6, Rojas, E., (1989), 'Human settlement of the Eastern Caribbean: development problems and policy options', pp. 243–258, Copyright ©1989 Butterworth Scientific, (Elsevier), with permission from Elsevier; Figure 7.2 from figure from 'Urbanisation and development in the Caribbean', in *Geography*, Vol. 80, 1995, pp. 334–341, The Geographical Association, (Potter, R. B., 1995); Figures 7.3, 7.8, 7.15 and Table 7.3 adapted from figures in chapters by Strachan, Hall, and Clarke and table on p. 121 in Chapter 5 by Potter, R. B. and Wilson, M., 'Barbados', in *Urbanisation, Planning and Development in the Caribbean*, Mansell, (Potter, R. B. (ed.), 1989), reproduced by kind permission of Thomson Publishing; Tables 7.4 and 7.5 and Figures 7.7, 7.9 and 7.11 from adaptations of figures and tables from *The Urban Caribbean in an Era of Global Change*, Ashgate Publishing, (Potter, R. B., 2000); Figures 7.10 and 7.17 from figures from 'Urban Castries revisited: global forces and local responses', in *Geography*, Vol. 86, 2001, pp. 329–336, The Geographical Association, (Potter, R. B., 2001); Figure 8.1b from graph from *100 ways of Seeing an Unequal World*, by Sutcliffe, Robert, B., Zed Books, (2001), by Permission of Zed Books; Figure 11.3 from figure from McElroy and de Albuquerque, (1996), 'Sustainable alternatives to insular mass tourism: recent theory and practice', pp. 47–60, in *Sustainable tourism in islands and small states: issues and policies*, Pinter, (Briguglio, L., Archer, J., Jafari, J. and Wall, G. (eds.), 1996), reproduced by kind permission of Thomson Publishing.

In some instances we have been unable to trace the owners of copyright material, and we would appreciate any information that would enable us to do so.

Part I

FOUNDATIONS OF THE CARIBBEAN

Chapter 1

CARIBBEAN NATURAL LANDSCAPES

Introduction

At the dawn of the new millennium, the landscapes of the Caribbean region are quite different in appearance from those a thousand years ago. Then, indigenous Amerindian peoples lived throughout the islands and a recent estimate suggests a pre-1492 population of three million (Denevan 1992). Less is known about the pre-Colombian landscapes of the Caribbean islands than those of mainland America, but Sauer's (1966) discussion of Taino (Island Arawak) raised fields or *montones*, based on the writing of contemporary Spanish chroniclers, suggests that humanised landscapes, at least in lowland areas of the larger islands like Hispaniola and Cuba, were extensive. Watts (1987) concludes that much of the native mature forests (especially in upland regions) remained intact until the arrival of the Europeans. Over the last 500 years, however, the natural landscapes familiar to Amerindian peoples have largely disappeared and only remnants survive. The impacts of colonisation, commerce, economic development and urbanisation have created new and unique Caribbean cultural landscapes. The transformations have been dramatic, far-reaching and irreversible. Many of these changes are considered in the chapters that follow.

In the new millennium, however, the vestigial natural landscapes have acquired a vibrant significance for Caribbean people. Their importance is epitomised by the marketing of forests, wetlands, flora and fauna in concerted efforts to broaden tourist attractions beyond the traditional nexus of sun, sand and sea (see Chapter 11). But the focus goes far beyond the desire to improve the prospects for tourism. There has been a growing concern for environmental conservation, resource management and cultural heritage, reflected in increasingly active public sector planning agencies and community-based NGOs. The Internet enables planners, environmentalists, educators, media journalists and the general public to exchange ideas and gain access to environmental information in ways inconceivable a generation ago. In recent years there has been a proliferation of useful environmental websites and networking of information resources throughout the region, for individuals and organisations alike (McGregor and Barker 2003).

Two themes underpin the heightened attention given to environmental issues in Caribbean policy debates. The first concerns the omnipresent concept of sustainable development, which has become the main paradigm linking environment, development and human welfare (Potter *et al.* 2004; Lloyd Evans *et al.* 1998). Though it is proving an elusive goal for developing countries, sustainable development tries to balance economic growth and social equity with the sustainable use of natural resources. An integral aspect of sustainable development is the wise management of the natural environment. This is particularly relevant to Caribbean islands, which have fragile tropical ecosystems and limited natural resource bases (McGregor and Barker 1995). Agenda 21 forcefully presents the problems of small island developing states [SIDS] as a special case:

> Small island developing states, and islands supporting small communities are a special case both for environment and development. They are ecologically fragile and vulnerable. Their small size, limited resources, geographical dispersion and isolation from markets, place them at a disadvantage economically and prevent economies of scale (*United Nations Conference on Environment and Development*, 1992, AGENDA 21:17.123).

A second theme that links environmental policy issues across the region is the important role of international institutions and conventions, or global governance (Potter *et al.* 2004). Caribbean nations increasingly operate in this policy arena. When governments become signatories to such international conventions, they assume obligations and commitments to formulate policies and legal mechanisms to implement these written agreements. In return, countries gain access to international financial resources, and scientific and technical advice and cooperation. Figure 1.1(a) lists some international conventions concerning environmental policy that a number of Caribbean governments have signed. One of these, the *Convention on Biological Diversity*, unveiled at the 1992 Rio Earth Summit, had been ratified by most Caribbean nations by the year 2000 (Figure 1.1(b)).

The *Convention on Biological Diversity* is a good example of how the global community tries to foster sustainable development through environmental conservation. Biodiversity is important because we depend on the huge variety of plants and animals in the world for food, medicine and other raw materials (WRI/IUCN/UNEP 1992). Common sense and the precautionary principle argue for the preservation of all species, so that actual or potentially useful species are not allowed to become extinct, even if we do not know their present value to society or role in natural ecosystems (WCMC 1992). However, the *Convention on Biological Diversity* clearly acknowledges that conserving biodiversity goes beyond protecting natural landscapes and saving endangered species. It can play a role in achieving sustainable development, and its three objectives are: the conservation of biological diversity; the sustainable use of its components; and fair and equitable sharing of the benefits. A coral reef ecosystem exemplifies the idea that biodiversity (short for biological diversity) has an inherent aesthetic value that people want to see and experience. Healthy coral reefs contribute to species conservation and protect coastal environments, but also provide significant economic

Figure 1.1(a) Selected international conventions relating to environmental policy to which some Caribbean nations are signatories

Convention on Biological Diversity
Convention to Combat Desertification
Convention on International Trade in Endangered Species (CITES)
Convention for the Protection and Development of the Marine Environment of the Wider Caribbean
Convention on the Trans-Boundary Movement of Hazardous Waste (Basel Convention)
Convention on Wetlands of International Importance (RAMSAR)
International Convention on the Prevention of Pollution from Ships (MARPOL)
UN Framework Convention on Climate Change and the Kyoto Protocol
Vienna Convention and Montreal Protocol on Substances that Deplete the Ozone Layer

Figure 1.1(b) Ratification of Convention on Biological Diversity

1993	St Kitts and Nevis, Antigua and Barbuda, St Lucia, Bahamas, Barbados, Belize
1994	Cuba, Dominica, Grenada, Costa Rica, Guyana, El Salvador
1995	Jamaica, Panama, Guatemala, Honduras, Nicaragua
1996	Suriname, St Vincent and Grenadines, Trinidad and Tobago, Haiti, Dominican Republic

Source: www.biodiv.org/conv

and social benefits by promoting tourist activities like diving, snorkelling and trips in glass-bottomed or semi-submersible boats.

The purpose of this chapter is to explain the formation of Caribbean natural landscapes in the context of their diversity of landforms, ecosystem and species biodiversity, and the conservation of natural landscapes through the establishment national parks and protected areas. We highlight too, in this and subsequent chapters, the environmental impacts of the transformation of natural landscapes through human settlement and economic development. The emphasis is on the insular Caribbean, though we will refer also to the wider Caribbean Basin and territories like Guyana and Belize, which are historically and culturally part of the West Indies. Specifically, the chapter aims to:

- explain how earth-building processes, especially tectonic activity, contribute to the formation and diversity of island landforms;
- explain how denudation processes, such as weathering and erosion, influence island landforms especially with respect to the region's ubiquitous karst landscapes;

- show how Caribbean climate and terrain influence patterns of natural vegetation;
- explain the significance of Caribbean biodiversity and highlight the need for species and ecosystem conservation;
- explain the linkages between different components of the coastal environment, highlighting the importance of coral reef ecosystems;
- illustrate how national parks and protected areas can be used to protect natural landscapes and contribute to sustainable development.

The Caribbean Basin and plate tectonics

The islands of the Caribbean Basin comprise three main geographic groupings; the Greater Antilles, the Lesser Antilles, and the islands of the Bahamas and Turks and Caicos archipelagos (Figure 1.2). A line of islands also fringes the north coast of South America, and the Cayman Islands lie to the west of the Greater Antilles. The Caribbean islands and their landforms have evolved as a result of the simultaneous interaction of endogenic (internal) processes from within the earth's interior and exogenic (external) forces at work on the earth's surface. Endogenic processes such as volcanic activity and earthquakes produce new landforms through uplift and island-building, whereas exogenic forces such as weathering and erosion remove surface materials and transport them elsewhere. In this section we explain how endogenic forces have shaped the geographical configuration of islands shown in Figure 1.2, and produced some of their distinctive landforms.

The Caribbean Basin was formed over an immense period of geological time as a result of tectonic forces from deep inside the earth's interior which shifted, warped, folded and buckled sections of the earth's surface. The theory of plate tectonics provides a basic framework for understanding these processes of deformation and recycling of rocks and the tectonic reconstruction of landforms. The theory was formulated in the 1960s and incorporates earlier theories of continental drift proposed in 1912 by Alfred Wegener, and sea-floor spreading proposed in the 1950s following deep-ocean drilling. The plate tectonics of the Caribbean Basin are a small but complex part of the planetary system of plate movements, and geologists have made great strides in unlocking the regional puzzle in recent years (Draper *et al.* 1994).

The earth's lithosphere is segmented into rigid plates composed of either ocean crust or continental crust. Ocean crust is denser and heavier, but is only about 5–10 km in thickness (average = 7 km) compared with the continental crust, which may be as thick as 20–75 km (average = 33 km). Plates are of unequal size and are in constant relative motion. They move at infinitesimally slow rates of only a few centimetres per year (about the speed at which human fingernails grow). However, over the huge span of geological time, these minuscule annual plate movements are translated into distances of thousands of miles. Plates travel in different directions and at slightly different speeds, and over the millennia they collide and

Figure 1.2 The islands of the Caribbean Basin
Source: Cartographic service, Geography Department, Indiana University

fuse together, or break up and fragment, altering the shape and geometry of the continents.

A significant aspect of the theory deals with the points of contact between plates because these narrow boundary zones correlate closely with global patterns of volcanic and earthquake activity and so constitute natural hazards for human populations. There are four types of plate boundary zone:

- *Constructive boundaries* – zones where plates move apart and new material emerges from the earth's mantle. The African Rift Valley is a site of aborted spreading on land, and the mid-Atlantic ocean ridge is an example of the process occurring at the bottom of the ocean. The process is also termed 'plate divergence'.

- *Destructive boundaries* – zones where plates collide, a process called 'plate convergence'. When continental plates collide, fold mountains (e.g. the Himalayas) are thrust upwards. When two ocean plates collide, one is pushed beneath the other in a process known as 'subduction'. As a plate is subducted, it sinks back into the asthenosphere under great pressures and temperatures, which cause it to melt as it descends. Melting releases basaltic and andesitic magma, which rises through the overriding plate to the earth's surface, creating a linear pattern of volcanoes called a volcanic island arc. Destructive boundaries are associated with ocean trenches, the deepest parts of oceans.

- *Transform boundaries* – zones where plates slide past each other in opposite directions in a process called 'transform movement'. These are also called 'conservative margins'. Crust is neither being produced nor destroyed at these boundaries, which are most common in ocean settings. An example of a transform boundary on land is the San Andreas fault.

- *Plate boundary zones* – broad belts where boundaries are poorly defined and the effects of plate interactions are unclear.

Figure 1.3 shows the Caribbean is moving eastwards, at a rate of about 1–2 cm per year (Mann *et al.* 1990), relative to the North and South American ocean plates, which are moving westwards. The northern and southern boundaries of the Caribbean plate are associated with transform boundary movements, and are zones of earthquake activity. Their boundaries are difficult to pinpoint (Draper *et al.* 1994) so they are best characterised as plate boundary zones. On the western boundary, the Cocos and Pacific plates are moving eastward and are being subducted under the western margins of the Caribbean plate. This has created a chain of active volcanoes in Costa Rica, El Salvador, Nicaragua and Guatemala. On the eastern boundary, the westward-moving Atlantic Ocean crust is being subducted under the Caribbean plate, forming the Puerto Rican ocean trench, which, at its maximum depth (9,220 m), is further below sea level than the summit of Mt Everest is above. The violent tectonic forces associated with Atlantic Ocean

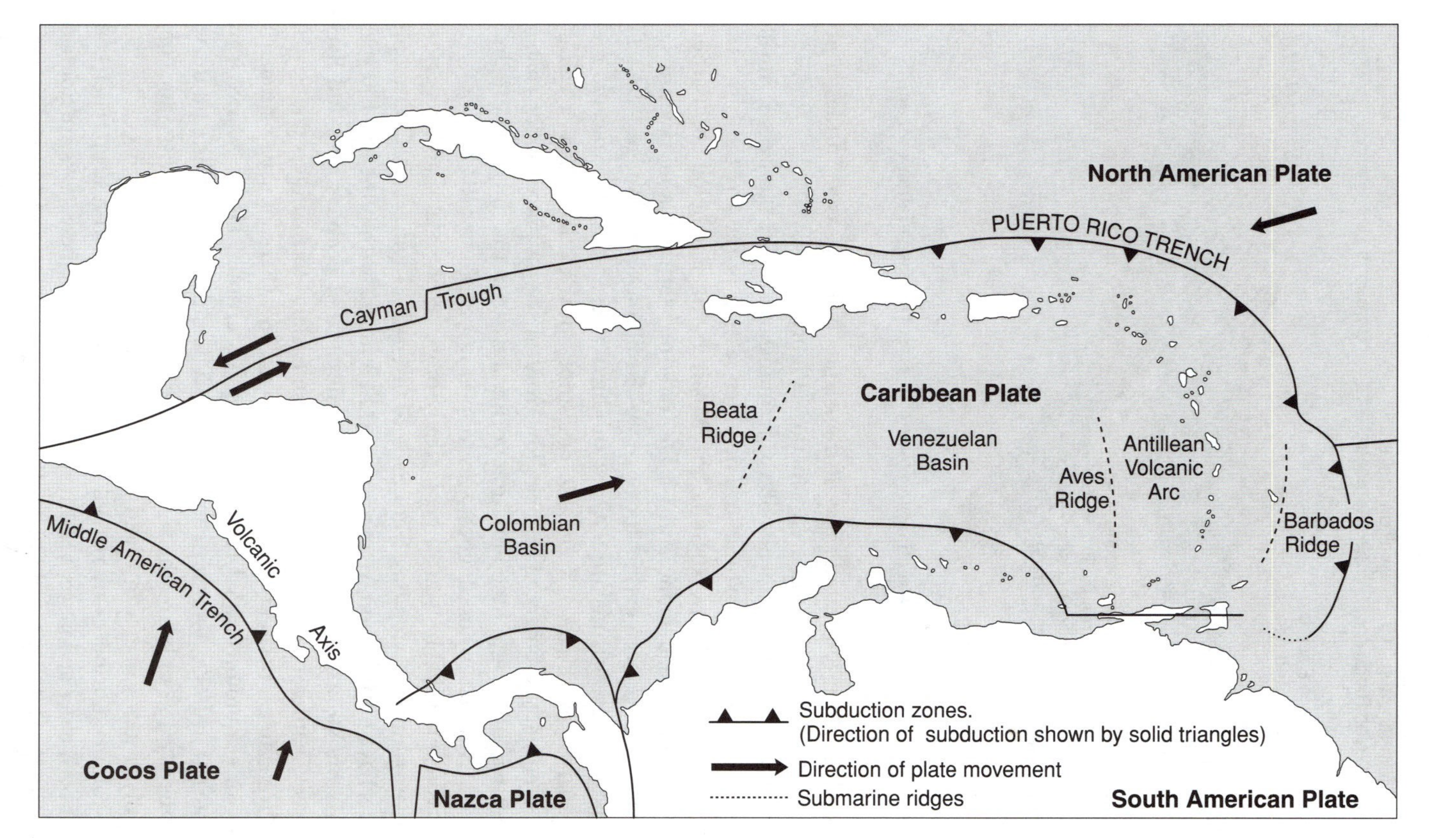

Figure 1.3 The position and movement of the Caribbean plate
Source: After Draper *et al.* (1994)

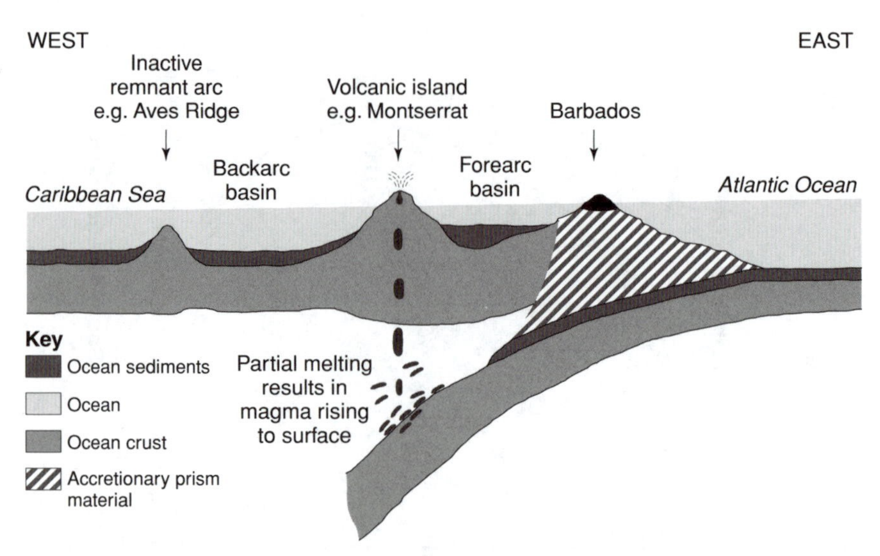

Figure 1.4 Generalised cross-section across the Lesser Antilles subduction zone

crust subduction have created most of the islands of the Lesser Antilles, a process known as volcanic island arc formation (Figure 1.4).

Scientists have found evidence of a volcanic island arc in the Lesser Antilles that bifurcates north of Martinique. The older section of the arc, the Limestone Caribbees, stretches from Anguilla, St Maarten, St Bartholomew, Barbuda, Antigua, eastern Guadeloupe and La Desirade to Marie Galante (Figure 1.5). The volcanoes have been extinct for possibly 28 million years (Wadge 1994). The landscapes of these islands are flat and subdued where the older volcanic rocks are covered by more recent Cenozoic limestone, clear evidence that a volcanic arc was eventually submerged by a shallow tropical sea. In places, older volcanic rocks protrude through these limestone formations as in the hilly topography of south-west Antigua, whose highest elevation is Boggy Point (402 m).

The Volcanic Caribbees are a geologically more recent island arc of dormant and active volcanoes extending from Saba to Grenada (Figure 1.5). Catastrophic eruptions have occurred on Martinique and St Vincent within the last 100 years, and those on Montserrat in the 1990s have devastated people's lives and livelihoods (see Chapter 4). An early stage in volcanic island formation is taking place in the Grenadines, between Grenada and Carriacou, where an active submarine volcano (or sea mount) called Kick 'em Jenny has erupted 11 times since it was first reported in 1939. A survey conducted in 2002 indicated that its summit was approximately 183 m below the surface of the ocean (see www.uwiseismic.com). Eventually, a new Caribbean island is likely to be formed.

The Volcanic Caribbees are rugged, mountainous islands with steep relief, composed of both older and newer volcanic structures. The highest peaks are Soufrière on Guadeloupe (1,467 m), and Morne Trois Pitons (1,423 m) and Morne Diablotins (1,421 m) on Dominica, all active or dormant volcanoes. Dominica, in fact, has

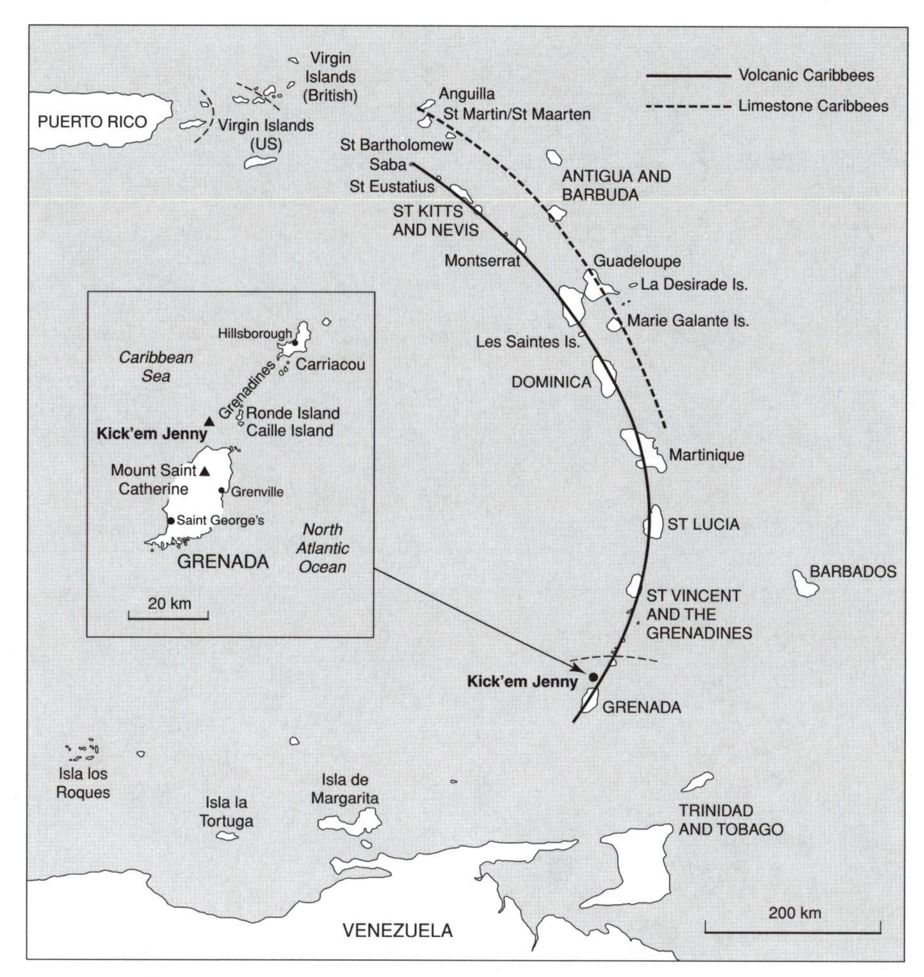

Figure 1.5 The volcanic island arcs of the Lesser Antilles
Source: Compiled from various sources at the Seismic Research Unit, UWI, Trinidad

seven distinct zones of volcanic activity (see www.uwiseismic.com). St Lucia is famous for its beautiful and impressive Pitons (examples of extinct volcanic plugs), and the island's equally famous sulphur springs are located in the caldera of the Qualibou volcano (marketed for tourism as the world's only drive-in volcano) and have potential as a source of geothermal power.

Barbados and Trinidad lie outside the volcanic arc systems. Barbados was formed as the abyssal ocean sediments of the Atlantic plate crumpled during subduction, and eventually an island emerged above sea level (Speed 1994). In plate tectonic theory such a feature is known as a forearc ridge, or an accretionary prism (Figure 1.4). The northern range of Trinidad once formed a separate island (Donovan 1994), but other sections of the island were connected to the South American mainland in recent geological times.

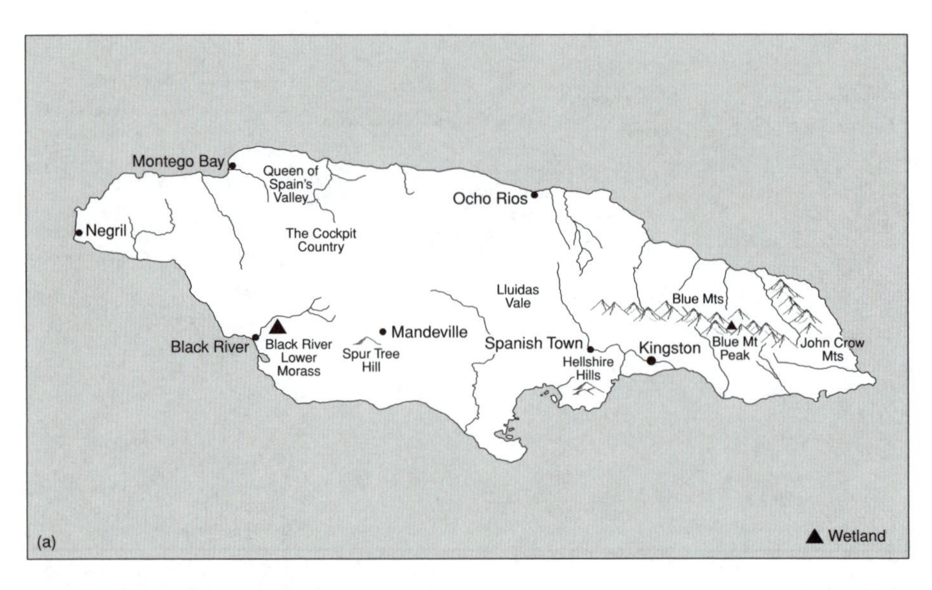

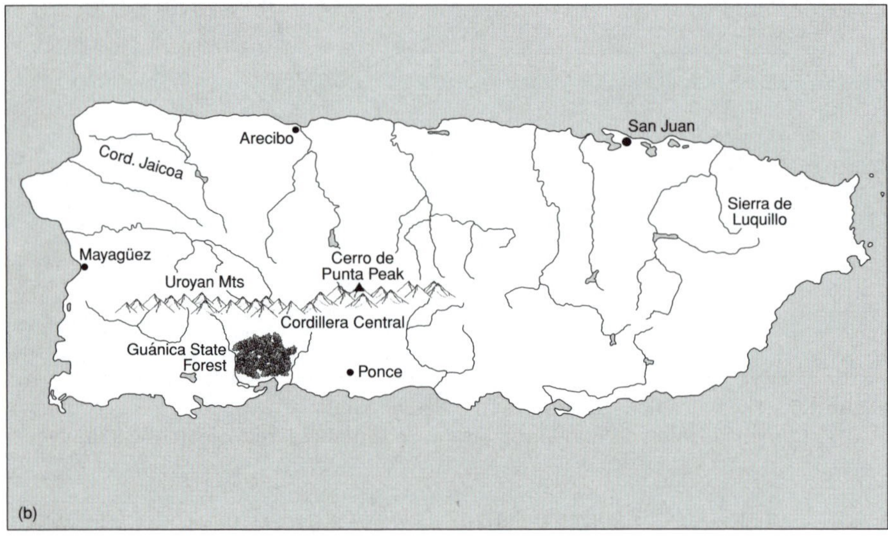

Figure 1.6 Some topographic features of (a) Jamaica, (b) Puerto Rico, (c) Cuba and (d) Hispaniola

The geological origins of the Greater Antilles are quite different to the Lesser Antilles (Draper *et al.* 1994). They are located in the vicinity of a transform plate boundary zone between the Caribbean plate and the North Atlantic plate, separated by the Cayman trough. The plate tectonic history of the Greater Antilles islands is complex (Pindell 1994), and there are significant differences between Jamaica, Hispaniola and Puerto Rico which form part of the Caribbean plate and Cuba, most of which is part of the North American plate.

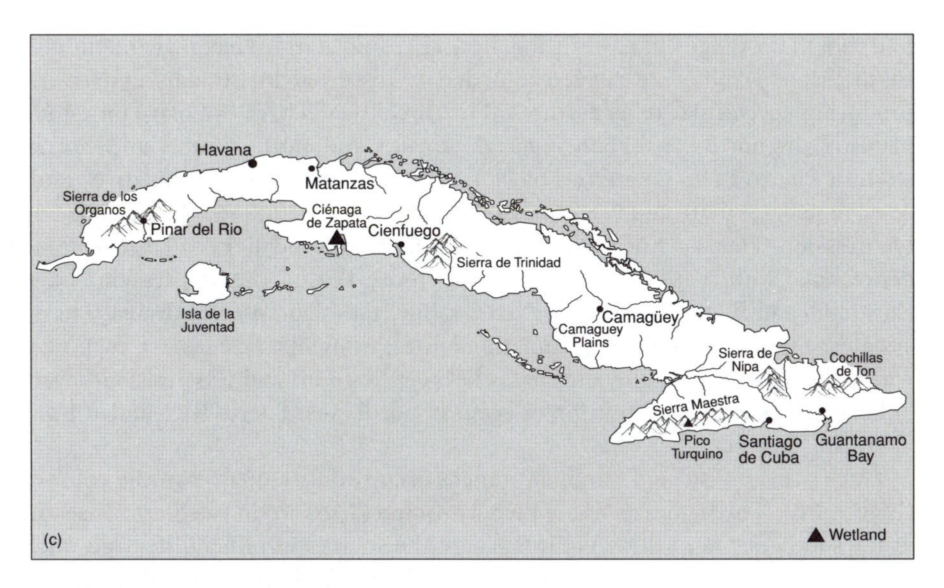

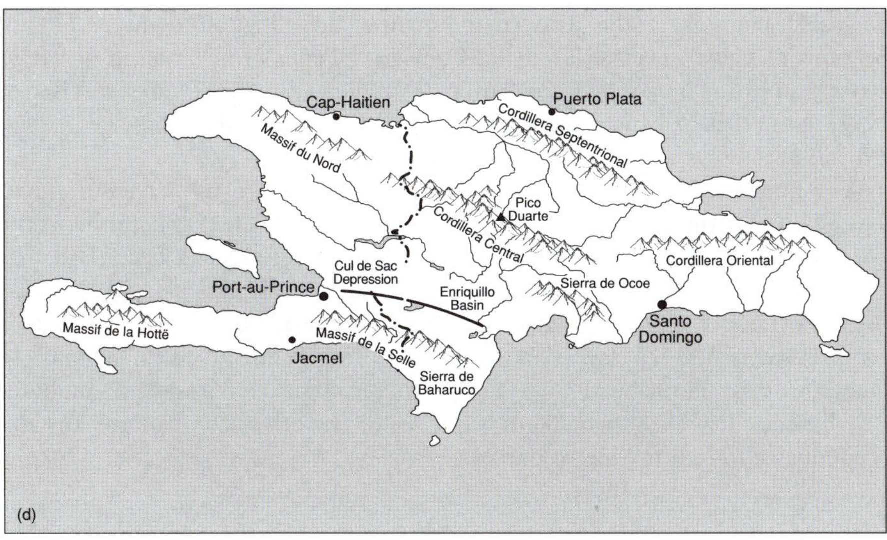

Jamaica, Hispaniola and Puerto Rico are geologically complex islands, composed of sedimentary, igneous and metamorphic rocks that have been folded, faulted and fractured, and subject to recent tectonic uplift. Figure 1.6 shows their main topographical features. Narrow coastal plains rise to high elevations over short distances, and the highest peaks on each island are Pico Duarte (3,175 m) in the Dominican Republic's Cordillera Central, Blue Mountain peak (2,257 m) in Jamaica, and Cerro de Punta (1,353 m) in Puerto Rico. The younger fold mountain ranges in eastern Jamaica, Hispaniola and Puerto Rico (and also south-eastern Cuba) are rugged and deeply dissected by streams and rivers, producing spectacular relief features. The three islands have extensive limestone plateaux, in north-west Haiti, central Jamaica and north-central and north-western Puerto Rico. Numerous block fault mountains are the cumulative result of earthquake activity and vertical displacement along fault lines. The Cul de Sac Depresssion (Haiti) – Enriquillo Basin (Dominican Republic) is a large tectonic depression, with sections 45 m below sea level. Within this depression, Lago Enriquillo and Etang Saumâtre are hypersaline lakes with internal drainage basin systems. Elsewhere, there are dramatic fault escarpments (or cuestas), especially where the local stratigraphy alternates between resistant and less-resistant layers of bedrock (Blume 1974); examples occur in north-west Puerto Rico and at Spur Tree Hill, Jamaica.

Cuba's several isolated mountain ranges occupy only a quarter of the country (Figure 1.6). The highest point is Pico Turquino (2,005 m) in the Sierra Maestro in the south-east, and these mountains are the sources of mineral deposits such as manganese, iron, nickel and copper. The distinctive karst region of the Sierra de los Organos is in the western mountains; their highest point is 728 m. Sections of Cuba's central low-lying Camaguey Plains are composed of flat or gently undulating limestone formed over 100 million years ago (Draper and Barros 1994). Note that Cuba is actually an archipelago of over 4,000 islands and cays, a detail not always apparent on large-scale atlas maps.

The limestone archipelagos of the Bahamas and Turks and Caicos are geologically much younger than the rest of the region (Sealey 1985) and lie to the north-east of the Caribbean Basin on a structural feature known as the Bahamas platform, a carbonate (limestone) platform of similar geological age to the Florida platform (Draper *et al.* 1994). It is stable and tectonically inactive. The Bahamas comprises over 700 separate islands ranging in size from Andros (6,957 km^2) to numerous tiny, uninhabited cays. The offshore physical environment combines a shallow sea floor (less than 200 m below sea level) with warm, tropical seas and is conducive to rich marine resources.

Caribbean karst landscapes and denudation

In the previous section we saw how tectonic activity builds new island landforms through processes like uplift and block faulting and volcanic eruptions. At the same time, denudation processes are lowering, shaping and rearranging physical landforms by weathering, mass movement, erosion, transport and deposition. In the Caribbean region, earth surface materials are denuded by the agents of water, air and waves under the influence of gravity. In mountain regions like the Blue Mountains of Jamaica, recent tectonic activity combined with high rainfall and rapid rates of physical and chemical weathering have produced steep topography and a highly dissected terrain (McGregor 1995).

Table 1.1 Limestone regions in the wider Caribbean

	Karst area (km^2)	Total area (km^2)	Per cent total area
Anguilla	90	91	99
Antigua and Barbuda	120	442	27
Bahamas	11,500	13,878	83
Barbados	370	430	86
Cayman Islands	200	259	77
Cuba	70,000	110,861	63
Dominican Republic	25,000	48,734	51
Guadeloupe	580	1,705	34
Haiti	10,000	27,750	36
Jamaica	7,500	10,990	68
Netherlands Antilles	800	900	89
Puerto Rico	2,500	8,897	28
Trinidad and Tobago	750	5,128	15
Belize	5,000	22,962	22

Sources: Day (1993); Kueny and Day (1998)

Denudation processes are accelerated through forest clearance for plantation agriculture and small farming (Watts 1987). In Chapters 3 and 4 we will explore further the relationships between land degradation, soil erosion and landslides, and human disturbance of the natural landscape, caused by agriculture and human settlement. In this section we will illustrate how earth surface processes influence landform development, focusing specifically on the denudation of karst landscapes, since limestone regions comprise more than 50 per cent of the land area in the wider Caribbean (Table 1.1).

Tropical karst landforms differ from those in temperate latitudes because rainfall is heavier and temperatures are higher. Day (1993) has classified Caribbean karst into three physiographic regions:

- the Greater Antilles and parts of Costa Rica, Nicaragua and Honduras, which have been subject to tectonic stress and block-faulting;
- the Lesser Antilles, with characteristic dry valleys (Fermor 1972);
- the Bahamas, the Yucatan and northern Guatemala, with subdued topography.

The Jurassic formations in Cuba are the oldest limestone rocks in the region (Draper and Barros 1994) and the youngest are the oolitic limestones still being formed on the Bahamas platform as fine-grained sand precipitates out calcium carbonate from the warm, shallow ocean (Sealey 1985). Miller (2004) has reviewed the karst geomorphology of the White Limestone Group in Jamaica, which comprises more than 50 per cent of the surface features of the island, and Monroe (1976) discusses karst landforms in Puerto Rico.

In limestone regions, chemical weathering is more important than physical weathering to landform development, though physical weathering and mass movement should not be ignored. The principal denudation process in tropical karst is corrosion, the wearing away and removal of rock material by chemical and other solvent action. Carbonation occurs when rainwater dissolves carbon dioxide from the atmosphere to form dilute carbonic acid, which reacts with limestone to form calcium bicarbonate. Water readily dissolves the calcium bicarbonate, which runs away in solution in groundwater and run-off. Rapid chemical weathering is accompanied by physical weathering and biochemical solution weathering. The latter is caused by the wedging and the enlargement of cracks and fissures as tree roots penetrate the limestone bedrock, in turn facilitating further solution action. Solution holes are formed by water that percolates and penetrates joints and faults in the limestone bedrock, to form and enlarge underground passageways and cave systems (Fincham 1997).

Day (1993) recognises three limestone terrain types in the region:

- *Doline karst* – areas of subdued topography such as the Yucatán area of Mexico. The landscape is punctuated by numerous enclosed depressions, called dolines, which are circular or oval-shaped, varying in size from a few metres to over one hundred metres in depth and several hundred metres across (Jennings 1985). Dolines are found in both temperate and tropical karst regions and are formed by solution weathering and/or collapse. They have a sink in their lowest part, and sometimes a central shaft that descends into a cave system. Dolines enlarge over time as a result of lateral corrosion and cave collapse. In the Yucatán, forested karst areas harbour Mayan settlement sites, suggesting agriculture was once more widespread.

  The doline karst of the Bahamas is somewhat different (Sealey 1985: 53). Blue holes are typical of Bahamian karst landscapes. They are solution features formed during the Pleistocene, when sea level was much lower, and some have been inundated by recent sea level rise. The island of Andros has at least 178 blue holes on land and another 50 located in the sea (Sealey 1985). Terrestrial blue holes are named for the distinctive colour of the water, usually seawater, indicating they are tidal or that there is a cave connection to the sea.

- *Cone karst (kegel karst)* – occurs in upland limestone regions and is characterised by cone-shaped hills varying in height between 100 and 130 m (Sweeting 1972). The conical hills have cliff-like slopes, with screes, precipices and staircase slopes of ledges and steps. Limestone exposures have pitted surfaces of honeycomb solution weathering. These landforms probably evolved from an earlier doline landscape, in which a central shaft at the base of the depression would be a defining feature, and the dolines would progressively enlarge through a combination of heavy rainfall and high rates of solution weathering. The Cockpit Country of Jamaica (Box 1.1) is a classic world region of cone karst, but similar landscapes occur in Puerto Rico, and elsewhere in Vietnam, Malaya and southern China.

- *Tower karst (turm karst or pinnacle karst)* – isolated, upstanding, steep-sided features sometimes as high as 200–300 m, though they can be much smaller. The general landscape is flatter than cone karst and is formed by lateral solution along the sides of lakes, or in places where rivers come into contact with limestone outcrops (Sweeting 1972). Tower karst areas are thought to be remnant features from a higher land surface (Jennings 1985). The most famous tower karst landscapes are in south China, but another classic area is the Sierra de los Organos, western Cuba, also renowned for its cultivation of tobacco for Cuban cigars.

Another common karst landform is the corrosional plain or polje (Thomas 1994). Since the main directional component of erosion is horizontal (rather than vertical), a polje forms through the progressive enlargement and elongation of a doline. Poljes are located within, or adjacent to, limestone uplands, and a major structural control is the block faulting in the limestones. They have flat valley floors, and underground rivers and springs emerge from along the margins between the valley floor and adjacent uplands. It is along these boundaries that lateral solution is most active in enlarging these depressions. The inflow of water and sediment onto the valley floor has formed fertile alluvial deposits and, historically, poljes have been utilised for sugar cane. These landform features are called hojos in Cuba (Jennings 1985: 124) and interior valleys in Jamaica, examples including the sugar estate areas of the Queen of Spain's valley (Figure 1.6).

Caribbean climate and rainfall patterns

The combined factors of geology, climate, soils and relief contribute significantly to the diversity of Caribbean natural landscapes, especially with respect to types of natural vegetation present. In this section we explain the relationships between climate and rainfall patterns in the region. We will examine the spatial and temporal aspects of precipitation for the entire region and variations within individual islands. Tropical storms and hurricanes, other important elements of Caribbean climate and weather, are explained and discussed as natural hazards in Chapter 4.

The Caribbean islands have a tropical maritime climate and certain features are uniform from Barbados to Belize. The north-east trades blow all year around and are constant and invariant in direction from the Atlantic to Central America. Tropical temperatures are high and at sea level do not vary much geographically or seasonally. Mean sea level temperatures on land reach 28 °C in August

Box 1.1: The Cockpit Country of Jamaica

Early research by Sweeting (1958) brought the Cockpit Country (Figure 1.6) into prominence as a classic area for the study of conical karst. With an estimated 5,000 limestone depressions (Eyre 1989), the area's distinctive landscape resembles an upturned eggbox when seen from an aeroplane or on an aerial photograph. The region is 300–750 m in elevation, dipping northwards due to post-Cretaceous tectonic uplift (Barker and Miller 1995). The white limestone has predominantly north–south fault lines along which some cockpits are enlarged into elongated depressions or glades, as in Barbecue Bottom. The core area lacks surface water and the subterranean stream flow is northwards, emerging along the margins of the white limestone, especially along fault lines. The central shaft in each cockpit is integral to its formation (Miller 1998).

The core area remains a pristine rain forest, with annual rainfall between 2,500 and 3,800 mm, and rainy seasons in May and September/October. The area is highly regarded by local scientists as a 'cradle for biodiversity'. For example, 101 species of plants are endemic to the Cockpit Country itself (Proctor 1986) and some have been collected only a few times or are known only from one locality. Eyre (1995) noted that species diversity among its ferns, bromeliads and other epiphytes, land snails and fireflies is exceptional. The area is refuge for endangered species such as endemic black-billed and yellow-billed parrots, the yellow snake and the giant swallowtail butterfly. The latter has a wingspan of 15 cm and is the largest true swallowtail in the world (Garraway and Bailey 1993). Scientific research continues to report new discoveries (www.cockpitcountry.com); for example, Diesel's (1991) research on the forest canopy has documented the natural history of the bromeliad crab (*Metopaulia monotypica*), a species that lives and breeds in bromeliads, some of which contain 3.5 litres of rainwater. Undoubtedly, more new species are yet to be discovered if the area can be preserved for future generations.

Though Cockpit Country was declared a Forest Reserve over 50 years ago, and was first proposed as a national park in 1970, Eyre (1989) reports a 16 per cent reduction in the forests contiguous to the core area between 1981 and 1987. The southern borders are particularly vulnerable to encroachment by small farmers, although the northern border is probably more stable (Barker 1998; Miller 1998). Yam stick cutting is also a threat (Barker and Beckford 2003), and the forests are sources of charcoal and commercial hardwood timber.

The region's historical and cultural significance lends further weight to its national status as an area worthy of protected area status, and Eyre (1995) has suggested that this unique area be considered a World Heritage Site. Maroons fought the British army to a stalemate in a series of military campaigns during the eighteenth century (Robinson 1969; Eyre 1980) and the 1739 Treaty is still celebrated on 6 January in the village of Accompong. Vestiges of the maroon's communal lands remain part of the local farming system (Barker and Spence 1988). It was British soldiers who coined the term 'Cockpit Country' because the area reminded them of the 'cock pits' used as arenas for cock fighting and gambling in seventeenth-century England.

and about 25 °C in January. Diurnal temperature ranges are greater than seasonal ranges, varying between 34 °C in daytime in the summer months and 20 °C at night (at sea level). Relative humidity is high, typically around 90 per cent at sunrise, and can produce heavy early morning mist and dew in coastal mangroves, valley bottoms, basins or hollows. Coastal locations have a lower relative humidity than inland areas because sea breezes have a mixing effect on coastal air masses.

Caribbean climate and weather are linked to global atmospheric circulation. Figure 1.7 shows that the Caribbean Basin is located between the sub-tropical high-pressure system known as the Azores–Bermuda anticyclone, and the Inter-Tropical Convergence Zone (ITCZ) low-pressure system. The Hadley cell circulation model explains the connections between high- and low-pressure systems. At the ocean's surface, the north-east trades blow from east to west, and are balanced by the return flow of westerly winds aloft, called the anti-trades. The anti-trades stream back towards the Azores–Bermuda anticyclone, where they descend and warm adiabatically. Once they reach the ocean surface as dry air, they diverge and some of the air mass blows back towards the equator again as the north-east trades. An inversion layer is present within the lower portions of north-east trades and is created because the descending anti-trades warm adiabatically. The effect of an increase in temperature with height within the inversion layer is to inhibit condensation and the formation of rain-bearing clouds, conditions that are conducive to drier weather (Barry and Chorley 1992: 230; Barker 1998).

A major influence on Caribbean weather patterns is the seasonal shift of the Azores–Bermuda high and the ITCZ. These shifts help explain the seasonal rhythms of wet and dry seasons and the seasonality of hurricanes and tropical storms. In the winter months, the ITCZ is located over the Amazon basin and the Azores–Bermuda anticyclone strengthens. The accentuated subsidence (as the anti-trades descend to the ocean surface) creates conditions that strengthen the inversion layer and increases the probability of drier weather. During the summer, the Azores–Bermuda high and the ITCZ move further north. The ITCZ is positioned about 12° north in August, and for a time hovers just south of Trinidad. The Azores–Bermuda anticyclone is also located farther north and pressures are not as high. Thus, the trade wind inversion is weaker, its base level is higher, and so condensation and the formation of rain-bearing clouds is more likely (Walsh 1998). Also, tropical waves and tropical storms form in the Atlantic in the summer and temporarily disrupt the temperature inversion.

The regional pattern of climate and weather is subject to temporal oscillations caused by the teleconnections of El Niño and to the longer-term global warming. Tropical cyclone activity is suppressed during El Niño years though rainfall is augmented in the wet season (www.ema.co.tt). Global warming is likely to affect Caribbean weather and climate in several ways, including an increase in the frequency and severity of tropical cyclones. The Intergovernmental Panel on Climate Change (IPCC) (1990) suggested a possible decrease in precipitation of

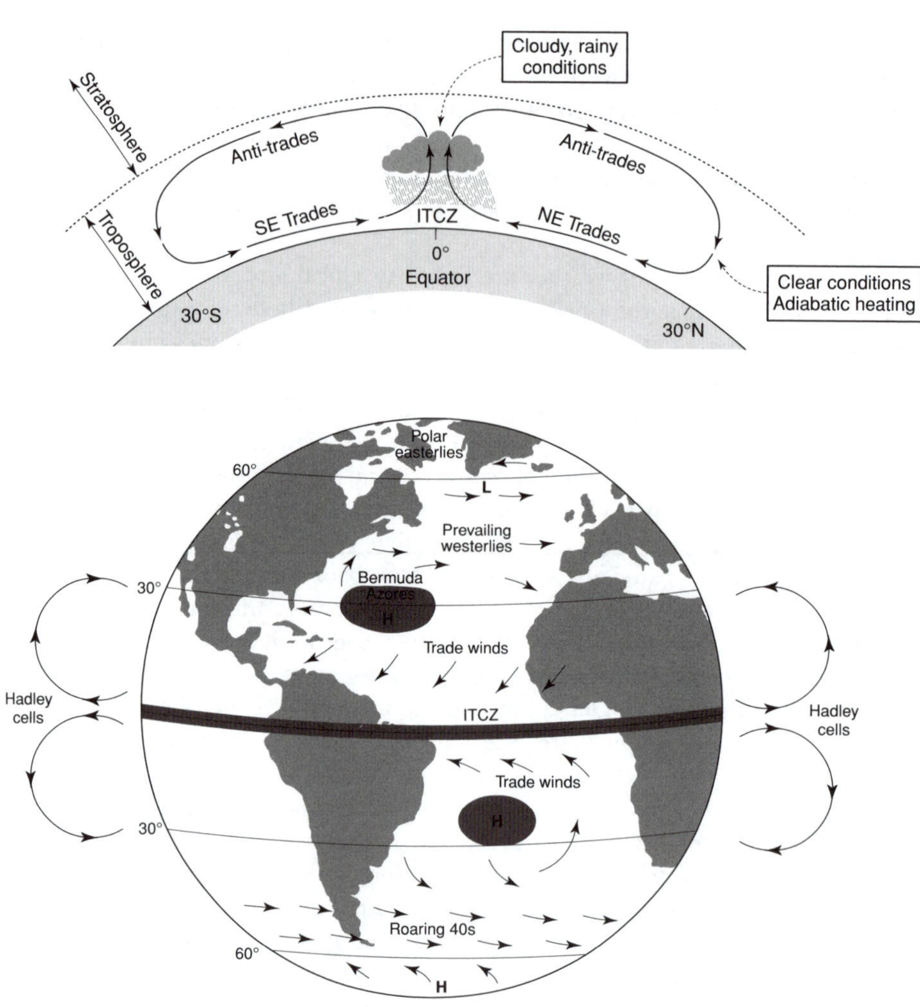

Figure 1.7 Trade winds and Hadley cell circulation
Source: Adapted from a table from *The Science and Wonders of the Atmosphere*, by Gedzelman, S. D., Copyright © 1980 John Wiley & Sons, Inc. This material is reproduced by permission of John Wiley & Sons, Inc.

up to 350 mm per decade in the Caribbean as a result in shifting global weather patterns. Another effect may be a rise in sea level due to greenhouse gas emissions, perhaps by as much as 65 cm by the year 2100 (see www.cep.unep.org). Global warming has numerous potential impacts on natural systems and human populations both direct and indirect (Figure 1.8).

Most of the region has a summer rainfall maximum (Table 1.2) with distinct wet and dry seasons, the timing of which is variable and slightly different from country to country across the region. For example, in Trinidad the rainy season lasts from June to December with peaks in June/July and November. The rainy

Figure 1.8 Potential impacts of global warming

More frequent and more intense tropical cyclone activity
Greater risk to population and settlements from storm hazards
Coral reef bleaching through higher sea temperatures
Drowning of coral reefs if sea level rises not matched by coral growth
Beach erosion
Flooding of low-lying cays and islands
Flooding and destruction of wetland habitats
Reduction in biodiversity
Shifts in natural geographical ranges of flora and fauna
Differential impacts on survival of flora and fauna
Inundation of low-lying inhabited coastal margins
Increased salinity of coastal groundwater aquifers
Loss of coastal infrastructure such as roads, buildings, tourist facilities

season is sometimes broken by 'Petit Careme', a short dry period usually in late September or early October. In Jamaica, September and October are the wettest months, and the island has two rainy seasons, the shorter one usually occurring around May/June. Rainfall in the northern Caribbean is also influenced by the passage of cold fronts or northers moving in from North America. Cuba and the north coast of Jamaica are affected by these events, which can bring torrential rainfall during the winter months and, like tropical depressions and tropical cyclones, can cause severe flooding.

Individual Caribbean islands often have a mosaic of local climates, mainly as a result of the combination of varying rainfall amounts, island size, island shape (geometry) and island topography (since temperature declines with increasing elevation). Two types of precipitation are significant at the local level:

- *Convective activity* – air currents created by daytime heating cause atmospheric cooling, condensation and convective rainfall. The process is more pronounced in summer, when land surfaces are heated directly by the sun, and typically occurs as a heavy afternoon shower.

- *Orographic uplift* – mountain barriers create atmospheric instability conducive to orographic rainfall. As the north-east trades rise over mountains, the height of the inversion layer is elevated, permitting the development of towering, rain-bearing clouds. Orographic precipitation is heavy and prolonged and modifies the dry season effect. It tends to occur on windward slopes of mountain ranges along eastern and northern coasts, and is reinforced by sea breezes (part of the daily tropical, sea breeze–land breeze regime).

Table 1.2 Average rainfall for selected Caribbean stations (mm)

Rainfall (mm)	**Jan**	**Feb**	**Mar**	**April**	**May**	**June**	**July**	**Aug**	**Sept**	**Oct**	**Nov**	**Dec**	**Total**
Nassau, Bahamas	36	43	45	78	117	159	150	135	165	164	85	39	1216
Kingston, Jamaica	20	18	10	37	138	114	51	92	86	168	52	25	811
Belize City, Belize	137	61	38	56	109	198	163	170	244	305	226	186	1893
Bridgetown, Barbados	66	28	33	36	58	112	147	147	170	178	205	96	1276
Piarco, Trinidad	77	61	27	71	129	269	243	213	144	151	212	153	1750
Willemstad, Curaçao	68	31	14	12	18	26	34	48	31	67	98	85	532
Georgetown, Guyana	251	122	113	178	296	346	281	185	88	98	147	313	2420

Source: World Survey of Climatology [most data for period 1951–1960]

Zones of low rainfall occur in several typical locations:

- rainshadow zones on the drier, leeward side of mountain ranges;
- low-lying islands like the Bahamas and Turks and Caicos;
- flatter, penninsula-like coastal areas of islands away from mountain ranges.

Examples of low rainfall zones are shown in Figure 1.9, which compares the geographical distribution of rainfall in St Lucia and Dominica. Note that even small islands with mountainous relief have spatially variable rainfall patterns across short distances. In Dominica, precipitation rises to over 6,250 mm per annum in the mountainous central regions. In St Lucia, elevations and rainfall are lower, increasing to 3,750 mm per annum in the south-central uplands, where the highest point is Mt Girole (950 m). The windward coast of St Lucia is drier than that of Dominica because the coastal plain is lower and wider. Dominica's leeward coast is drier than its windward coast and has a rainshadow zone in the lee of Morne Diablotins. The low-lying south-east and northern tips of St Lucia are the driest parts of both islands. However, St Lucia's drier coastal zone does not extend completely around the island because the famous Pitons area (elevation 196 m) is close to the coastline in the south-west and their elevation produces increased rainfall in their immediate environs.

Caribbean natural vegetation and forest landscapes

Much of the scientific work on the classification of the natural vegetation of the Caribbean is specific to individual islands. Botanists have derived several detailed classifications based on different criteria, but Beard's (1955) terminology is still widely used (Figure 1.10). Recent work by the Nature Conservancy and the EROS Data Center of the USGS is progressing towards the first detailed, standardised Caribbean Vegetation Classification and Atlas (see edcintl.cr.usgs.gov).

Though forests once clothed most of the aboriginal landscape of the islands, some low rainfall–low elevation areas were tropical savannas. As much as one-third of Cuba may have originally been tropical savanna, and this partly contributes to the high level of endemism in Cuban flora. In the Dominican Republic, the ecologically unique Enriquillo plains stretch into Haiti's Cul-de-Sac region and comprise semi-arid and arid vegetation types, sand dune features and the hypersaline Enriquillo Lake, which lies below sea level. Similar but smaller, pocket-sized areas of natural thorn and cactus scrub are scattered around the drier parts of many Caribbean islands. Note that there are large tropical savanna biomes on the South American mainland, such as the Llanos in Venezuela and Colombia, and the Rupununi in Guyana.

Historically, clearance of natural vegetation for agriculture was the principal cause of landscape change, and reduced primary forests to remote uplands and a few drier lowland areas (Watts 1987). Table 1.3 provides data on present forest

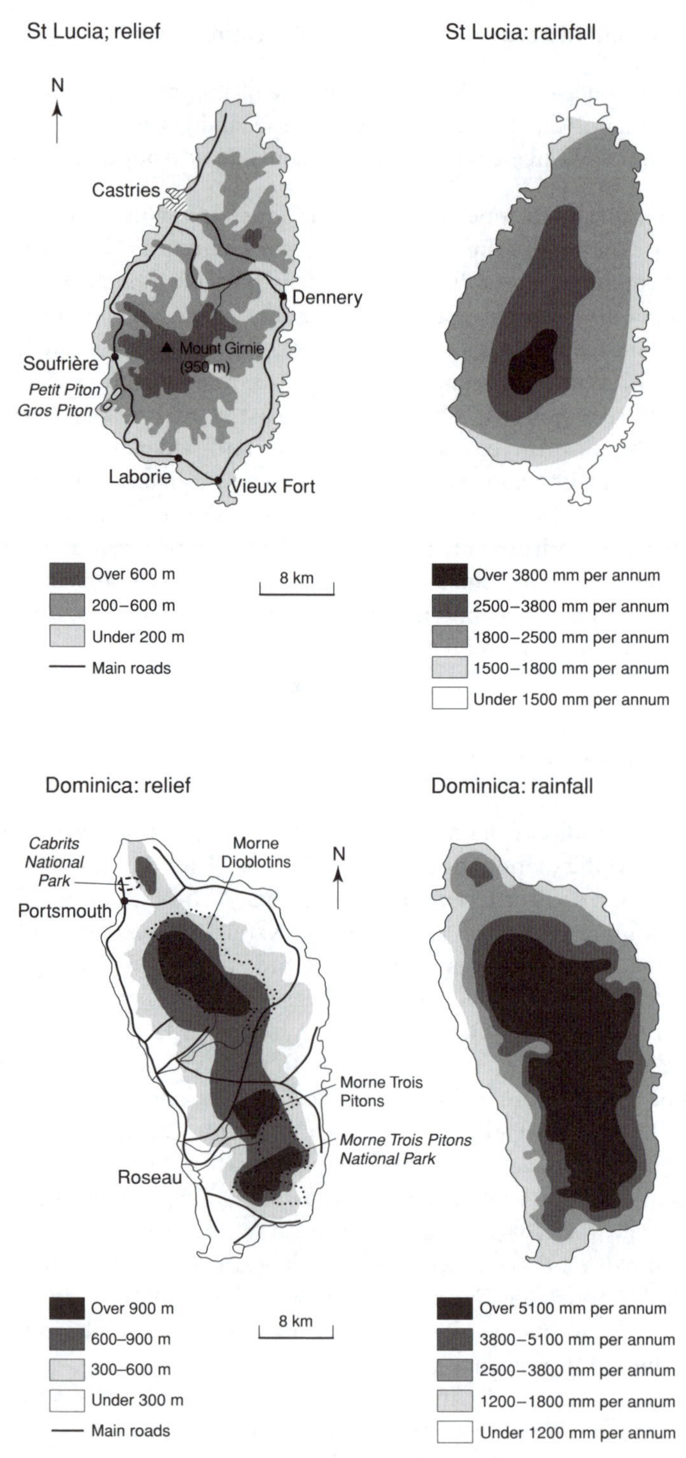

Figure 1.9 Rainfall patterns in St Lucia and Dominica
Source: MacPherson (1980)

Figure 1.10 Beard's vegetation classification for tropical America

I Optimum Formation

1. Rain forest

II Seasonal Formations

2. Evergreen seasonal forest
3. Semi-evergreen seasonal forest
4. Deciduous seasonal forest
5. Thorn woodland
6. Cactus scrub
7. Desert

III Dry Evergreen Seasonal Formations

8. Dry rain forest
9. Dry evergreen forest
10. Dry evergreen woodland and littoral woodland
11. Dry evergreen thicket and littoral thicket
12. Evergreen bushland and littoral hedge
13. Rock pavement vegetation

IV Montane Formations

14. Lower montane rain forest
15. Montane rain forest or cloud forest
16. Montane thicket
17. High mountain forest
18. Elfin woodland or mossy forest
19. Paramo
20. Tundra

V Seasonal Swamp Formations

21. Seasonal swamp forest
22. Seasonal swamp woodland
23. Seasonal swamp thicket
24. Savanna

VI Swamp Formations

25. Swamp forest and mangrove forest
26. Swamp woodland
27. Swamp thicket
28. Herbaceous swamp

Note: not all these categories are found in Caribbean islands
Source: Beard (1955)

cover in the region. In Haiti, less than 1 per cent of the country is still forested compared with more than 20 per cent in the neighbouring Dominican Republic (Wilson *et al.* 2000). Eyre (1998) estimates that only 2,978 km^2 of closed canopy remains in 13 Eastern Caribbean countries. The area of Caribbean forests is tiny compared with the rest of the tropics, but deforestation can still pose a significant local problem. The high figure for Jamaica, for example, is disputed by some sources (Miller 1998; Eyre 1987) and rates of deforestation are technically difficult to calculate. Chalmers (2002) is a comprehensive review of the management problems of the region's forest resources.

Despite past deforestation, the main types of aboriginal forest formation are still present in the contemporary Caribbean and, from the perspective the region's high level of endemism and of species biodiversity (see below), it is important that remaining representative forest types are protected. The main factors determining the type of natural forest occurring in a given location are:

- elevation – inversely related to temperature;
- aspect – whether windward or leeward facing;
- precipitation – volume of rainfall and length of dry season;
- local geology and soils – further modifying factors.

Table 1.3 Forest and marine environments in the wider Caribbean

Country	Land area ('000s ha)	Forests ('000s ha)	Coastline (km)	Mangroves (ha)
Caribbean Islands				
Anguilla	10	nd	56	nd
Antigua and Barbuda	44	10	178	1,500
Aruba	19	nd	76	100
Bahamas	1,393	186	3,542	233,200
Barbados	43		97	12
Cayman Islands	26		160	11,655
Cuba	10,982	1,715	3,735	626,000
Dominica	75	44	153	nd
Dominican Republic	4,838	1,077	1,576	23,500
Grenada	34	6	117	nd
Guadeloupe	178	93	306	5,700
Haiti	2,756	23	1,771	18,000
Jamaica	1,083	239	1,022	20,200
Martinique	110	43	290	2,200
Montserrat	10	nd	49	7
Netherlands Antilles	80	nd	390	2,200
Bonaire				1,000
Curaçao				300
St Maarten and Barthelemy				100
Puerto Rico	890	321	585	6,497
St Kitts and Nevis	26	13	130	20
St Lucia	62	5	140	179
St Vincent and the Grenadines	39	11	135	nd
Trinidad and Tobago	513	155	362	9,000
Turks and Caicos	43		250	nd
Virgin Islands (British)	15		250	nd
Virgin Islands (USA)	35		350	310
Central America				
Belize	2,280	1,996	386	78,317
Costa Rica	5,106	1,428	210	35,000
Guatemala	10,843	4,225	120	16,000
Honduras	11,189	4,605	700	117,000
Nicaragua	11,875	6,013	550	60,000
Panama	7,599	3,117	1,200	297,532
South America				
Guyana	19,685	18,416	459	80,000
Suriname	15,600	14,768	386	115,000
French Guiana	9,000	7,997	378	5,500
Venezuela	88,205	45,690	2,800	673,569
Colombia	103,870	54,064	1,080	501,300

Source: WCMC (1992), CEP/UNEP

Tropical evergreen or broadleaf rain forest is roughly delimited by the 2,000 mm per annum isohyet, which is the lower boundary for wet, humid conditions at least ten months in the year (Watts 1987). There are numerous sub-types in this broad classification. Broadleaf rain forests have a better developed vertical structure than seasonal forests, and at lower elevations there may be three vegetation storeys: the canopy (15–40 m in height), emergents (up to 60 m), and a lower storey (between 3–15 m). A feature of some native Caribbean tree species is enormous buttress roots, extending several metres up the trunk, an example being the silk cotton tree used by Tainos and Caribs to build their huge canoes. Forest trees have root systems that are shallow but extensive, and adapted to low levels of soil nutrients and steep slopes. In their primary condition, these forests contained valuable hardwood tree species such as mahogany. The forest architecture is replete with tree ferns, vines and epiphytes, such as orchids and bromeliads. Many of the latter species harbour specialised animals, such as tree frogs and tree crabs. The wet limestone forests of the Cockpit Country are a particularly rich source of endemism and biodiversity (Box 1.1).

Seasonal rain forest occurs at low elevations (less than 200 m) in areas where rainfall is less than 2,000 mm per annum (Watts 1987). The pronounced dry season can last up to five months. Many tree species are deciduous and shed their leaves in the dry season (not in the winter, as in temperate latitudes). Other physiological responses include leaf curl for protection from the sun. The wetter sub-types of seasonal forests have fewer deciduous species and the canopy can reach 30 m in height. Dry seasonal forest occurs where precipitation falls below 1,250 mm per annum (Watts 1987) and the canopy is lowered to 8–15 m. Seasonal rain forests were formerly much more extensive but their accessibility to human populations has meant that the original forest has been cleared, and secondary acacia thorn scrub has taken over the landscape.

The combination of limestone bedrock and edaphic conditions in low-lying regions give rise to dry limestone forests. The few remaining areas of intact, primary seasonal forest are critical to conservation efforts. There are extensive dry limestone forests in Cuba and the Dominican Republic, and in south-west Puerto Rico the Guánica State Forest National Park affords a measure of protection through the United States Parks Service. However, the pristine dry limestone forests of the Hellshire Hills in Jamaica are under threat from urban development and charcoal burners.

Montane mist or elfin forest is found at higher elevations (Asprey and Robbins 1953). The rain forest canopy is lower (8–15 m) and trees appear bent and misshapen and long trailing mosses drip down from their branches. At some of the highest elevations, in the Dominican Republic and a tiny area on top of Blue Mountain Peak in Jamaica, elfin forest gives way to a more open type of vegetation and an absence of trees, called montane summit savanna (Grossman *et al.* 1991).

Mangrove forests are an important component of coastal wetlands (Table 1.3). Mangroves are halophytic (salt tolerant) trees inhabiting inter-tidal zones of salt-water areas within the tropics and subtropics, although only four of the 80 species of mangrove are native to the New World. They develop only where the physiography of the shoreline provides shelter from wave action (Chapman 1976). Favoured locations are the leeward side of peninsulas and islands, lagoons and tidal bays, and behind spits and tombolos (a tombolo is formed when sediments connect an offshore island or sea stack to the shoreline). Chapman (1976) notes that whilst the mangrove ecosystem may represent a climax community for a coral reef cay, mangroves normally reflect the presence of different stages of vegetation succession. Mangrove species are specialised plants and segregate into geographical zones with respect to tidal height, salinity of the water, and various salinity and aeration characteristics of soil. The stages are:

1 red mangrove (*Rhizophora mangle*)
2 black mangrove (*Avicennia germinans*)
3 white mangrove (*Laguncularia racemosa*)
4 buttonwood mangrove (*Concarpus erecta*)

At the pioneer stage (Figure 1.11), red mangrove colonises mudflats, tidal estuaries and coastal inlets. It builds and consolidates land because its stilt-like prop-roots trap vegetation detritus (sticks, grass and leaves), which decays and mixes with silt sediments to provide nutrients for its growth. Red mangrove seeds start to germinate inside the fruit and when about 20–30 cm long, they drop like darts and grow in the coastal mud or float elsewhere on the coastal tides (Bacon 1978). Gradually they spread out into the tidal waters, extending the land area. The second stage occurs when continued deposition of silt and detritus, the slowing of water circulation and a slight rise in the level of sediment all become less favourable to red mangrove. Black mangroves colonise areas farther from the water's edge. Their horizontal systems of 'breathing roots' or pneumatophores protrude above the low-water mark. At the next stage, white mangrove and buttonwood mangrove take over as sediments and soils consolidate further, and the mangrove forest becomes more similar to a terrestrial forest (Figure 1.11). The gradation from saline conditions to brackish water then freshwater is reflected in an increasing number of botanical species on the landward fringe, where there is greater mingling with terrestrial species.

The climax community in a mangrove succession may be a freshwater swamp (Chapman 1976). Freshwater wetlands can also form behind a blockage like a sandy beach bar to create a lagoon. In a freshwater wetland, sediments will accumulate and there may be a vegetation succession from water plants to lilies, sedges, then taller swamp grasses. Thus a freshwater wetland can develop into a dryland ecosystem through natural processes of sedimentation, then may develop further into a lowland tropical forest. Seasonal flooding in freshwater wetlands creates habitats for fish and shrimp and the larger wetlands are important to traditional fishing grounds (Johnson 1998; Sletto 1998). Depending upon local environmental conditions, the various dynamic ecological and geomorphological coastal processes can produce other temporary or permanent landforms such as salt ponds and,

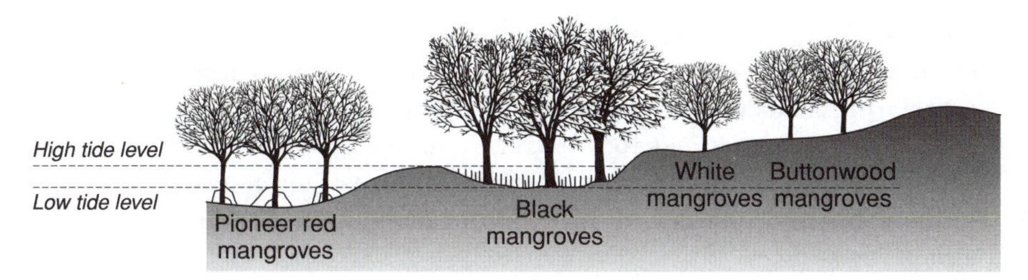

Figure 1.11 Transect showing vegetation succession in mangrove communities

under drier conditions, salt flats or salinas. Larger wetlands are a complex amalgam of different landforms and ecosystems, such as mangroves, lagoons, sawgrass swamp and dryland palm forest. One of the largest in the region is the 600 km^2 Cienaga de Zapata, in Cuba.

Caribbean biodiversity and species conservation

Biodiversity is a concept that has practical application at the level of ecosystems and for individual plant and animal species. The Caribbean region is a major centre or 'hot spot' for biodiversity because a large number of endemic species are to be found within the diverse ecosystems of the islands. The term 'endemic species' refers to plants and animals unique to a particular location, usually a country – although it can apply to a group of countries or islands, or even a smaller locality or site within a country. By contrast, native species are indigenous to a particular country but are also found in the wild in other countries. For example, the American crocodile is native to Jamaica, but it is also found in Cuba and the southern USA. Cuba has 3,233 endemic plant species (WCMC 1992), and approximately 780 of the 3,000 flowering plants in Jamaica are endemic. In Hispaniola, 41 of the 151 native species of butterfly are endemic. Several of the Windward Islands have endemic parrots: the St Lucian parrot or Jacquot (*Amazonia versicolor*), the St Vincent parrot (*Amazonia guildingii*) and two in Dominica, the red-necked parrot (*Amazonia arausiaca*) and imperial parrot (*Amazonia imperialis*). The preservation of rare endemic and native species is important for scientific research and also has potential as an income-earner through the promotion of ecotourism.

There are several theories that try to explain how endemic species originate on oceanic islands (Whittaker 1998). There are two distinctive types of endemic, reflecting their different origins:

- *Palaeo-endemics* – relic species whose mainland counterparts have become extinct. A Caribbean example is a palaeo-endemic family of birds, *Todidae.* The family is found nowhere else in the world, yet there are three separate species of the red-throated todies, found in Hispaniola, Puerto Rico, Cuba and Jamaica.

- *Neo-endemics* – newly evolved species that are often closely related. For example, the rock iguanas (genus *Cyclura*) are found only in the northern Caribbean islands (Vogel *et al.* 1995). A prehistoric prototype evolved into eight separate species and 16 sub-species on different Caribbean islands. Small islands like Anegada in the British Virgin Islands and Acklin Island in the Turks and Caicos have their own endemic iguanas, found nowhere else in the wild.

Oceanic islands tend to be species-poor compared with similar-sized areas of mainland, but have higher levels of endemism (Whittaker 1998). Also, larger islands tend to have more species (including endemics) than smaller islands, and so not surprisingly, Cuba has the most endemic species of the Caribbean islands. Generally speaking, the number of species present on an island depends on at least six factors (WCMC 1992) and increases with:

- increasing geographical size;
- increasing topographic complexity;
- more pronounced climatic variations;
- increasing geographical isolation;
- a longer geological period of isolation as an island;
- presence of particular biological characteristics (e.g., ease of dispersal).

More species are found in mountain environments than in flatter areas because there is greater diversity of habitat. Rain forests tend to have more species than grasslands because forests have several storeys of vegetation, including the little-known forest canopy, where new scientific discoveries continue to be made (Diesel 1991). The geographical isolation of an island is a significant factor in its species composition. For example, Jamaica has been an oceanic island for 12 million years but Trinidad was connected to nearby South America in recent geological times. Trinidad has far greater species diversity compared with many larger Caribbean islands, but fewer endemic species. For example, 400 species of birds have been recorded in Trinidad compared with 250 in Jamaica, even though Trinidad is a smaller island. Yet Jamaica has 30 endemic bird species compared with only one endemic bird in Trinidad (Raffaele *et al.* 1998). Moreover, Trinidad has a hundred native mammal species, including carnivores, marsupials, primates and an armadillo, all found on the nearby South American mainland. By contrast, there are very few native mammals in the other Caribbean islands.

The alarming rate of species extinction has focused attention on the need for conservation of endangered species. Extinction is a natural part of biological evolution. Giant ground sloths (MacPhee and Iturralde-Vinent 2000) and huge flightless owls once roamed the Cuban plains but disappeared long before the earliest humans arrived. However, modern species extinctions are directly caused by human activities, like the destruction and modification of natural habitat and hunting and collecting animals. Even during the last century, a number of endemic Caribbean species apparently became extinct. The Barbados raccoon has not been seen since 1964 (Fraser *et al.* 1990) and the Caribbean monk seal was officially declared extinct in 1996 in the IUCN's Red List of Threatened Animals, the last confirmed sightings being in 1952 on a cay between Honduras and Jamaica (see www.monachus.org). Occasionally, a species thought to be extinct is rediscovered in remote, relatively undisturbed habitats. In a celebrated case in 1990, a small breeding population of the endemic, ground-dwelling Jamaican iguana was found surviving in the Hellshire Hills, one of largest remaining intact dry limestone forests (Vogel *et al.* 1995). A few years later the blue-tailed galliwasp (*Celestus duquesneyi*) (a type of lizard not an insect), which was also thought to be extinct, was rediscovered in the same area (Wilson and Vogel 2001).

One of the few enduring success stories in the region regarding the protection of endangered species is the case of the West Indian flamingo. In 1952, the population of West Indian flamingos was estimated to be around 21,000, with the largest breeding colony of 4,500 individuals at Lake Windsor on the Bahamian island of Inagua. A nature reserve was established on Inagua, and ten years later the Bahamas National Trust turned it into a bird sanctuary, and eventually, as the Inagua National Park, the area was declared a Ramsar wetland site in 1997. The flamingo (called locally the fillymingo) is the national bird of the Bahamas, and the colourful spectacle of more than 60,000 flamingos attracts tourists from all over the world. Flamingos have subsequently returned to the Bahamian islands of Andros, Acklins and Abaco, and flamingo populations are recovering in the Turks and Caicos.

As people migrate around the world they carry plants and animals with them, so non-native species are introduced into new environments. Crosby (1972) discusses the many historical introductions of plant and animal species and their impacts in the West Indies. Farming systems, forests and garden landscapes (Rashford 1994) usually contain introduced species, many of which are productive and useful. However, non-native species are also referred to as invasive or alien species, reflecting the often unintended negative consequences on native species, upsetting the natural balance in ecosystems, and even leading to extinctions. The mongoose, introduced from India around 1872 to combat the rat problem in cane fields, is a notorious example. Ground nesting birds and reptiles are particularly vulnerable to this predator, which has decimated many native reptiles and is probably responsible for the extinction of several species.

Many Caribbean governments are signatories to CITES (Convention on Trade in Endangered Species), which signals a commitment to protect endangered species and halt the trade in such animals and animal products within their borders. Various types of public educational campaign and poster are used to highlight endangered species (Figure 1.12). However, laws protecting endangered species are not always easy to enforce. In Guyana, for example, the problem of protecting wildlife is particularly acute because the wildlife trade is big business. Williams (1998) reports that 153,000 wild animals, including parrots, macaws, poison dart frogs

Animal species of Jamaica

Protected under the Wildlife Protection Act

Sea turtles	 Leatherback (*Dermochelys coriacea*) Loggerhead (*Caretta caretta*) Hawksbill (*Eretmochelys imbricata*) Green turtle (*Chelonia mydas*)
Butterfly	Swallowtail (*Papilio homerus*)
Iguana	Jamaican iguana (*Cyclura collei*)
Coney	Jamaican hutia (*Geocapromys brownii*)
Snake	Jamaican boa (*Epicrates subflavus*)
Parrots	Black-billed (*Amazona agilis*) Yellow-billed (*Amazona collaria*)
Manatee	West Indian manatee (*Trichechus manatus*)
Crocodile	American crocodile (*Crocodylus acutus*)

Figure 1.12 Protected animal species of Jamaica
Source: National Environment and Planning Agency, Jamaica (www.nepa.gov.jm). Photographs from WIDECAST (Wider Caribbean Sea Turtle Conservation Network)

and primates, were exported from Guyana in 1992 alone, with a declared foreign exchange value of US$1.9 million. But illicit smuggling is growing unabated and is difficult to control. Other examples of potential conflicts between conservation and commercial interests include the harvesting of black coral (to make tourist souvenirs), trapping parrots (for house pets) and hunting turtles (for meat and shells).

Caribbean coral reefs and coastal environments

The recreational activities associated with coral reefs are very much a part of the Caribbean tourist product but coral reefs, a unique type of ecosystem, are often called the 'rain forests of the sea' because of their biological productivity and species diversity. They provide food, shelter and nurseries for many species of crustacean and fish, but are very sensitive to environmental change. The Caribbean region has about 14 per cent of the world's coral reefs (WCMC 1992) and the World Resources Institute estimates that two-thirds of the region's reefs are at risk (see www.wri.org/wri/reefsatrisk). Here we look at the characteristics of coral reefs in the context of the mosaic of Caribbean coastal environments, and highlight their environmental benefits.

Corals are communities of animals called polyps that surround a protective skeleton of secreted calcium carbonate (Bacon 1978). Coral reefs can only survive in tropical and sub-tropical marine environments where the sea temperature is between 23 and 25 °C, and need clear, well-illuminated conditions and so are not found in sea depths below 30–40 m. There are dozens of species of coral, including soft coral, anemones and sea fans, but the so-called stony corals are those which build coral reefs. Stony corals have evolved into many beautiful and exotic forms, with common names like brain coral (*Diploria clivosa*), staghorn coral (*Acropora cervicornis*) and elkhorn coral (*Acropora palmata*) reflecting their evocative shapes. Corals are colonial animals and duplicate themselves through asexual division, creating identical copies that add to the limestone mass as the reef grows in size.

Three principal types of coral reef are found in the Caribbean Sea (Figure 1.13):

- fringing reefs, which grow on hard surfaces close to the shore at depths of 3–6 m and shelve into sand;

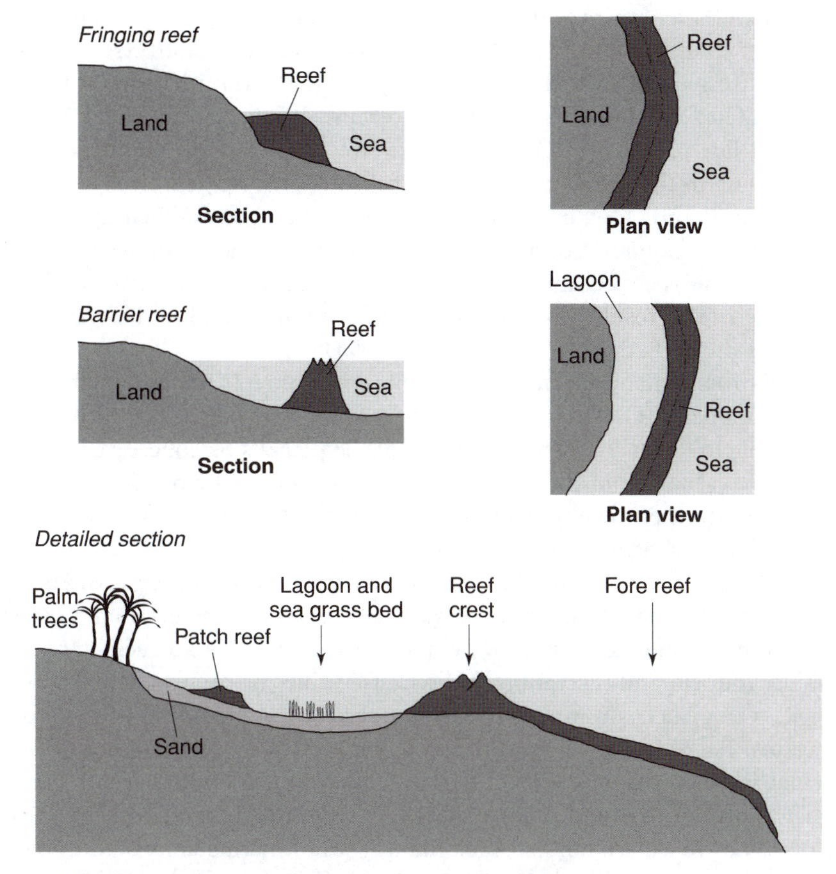

Figure 1.13 The main types of coral reef in the Caribbean

- barrier reefs, which develop along a linear axis and develop in successive stages with climax conditions (Bacon 1978);
- patch reefs, which are isolated boulders of coral that form in shallow water in irregular patches on the sea bed.

The coral reefs on the Bahamas platform are the most extensive area of fringing reefs in the world, whilst the 257 km reef off the coast of Belize is the world's second largest barrier reef and a World Heritage Site. Barrier reefs are a common coastal feature throughout the Caribbean Basin and are generally small and often run parallel to the coastline, located up to a kilometre offshore. When looking out to sea, a line of surf marks the reef crest, beyond which the sea is a deeper colour of blue.

Small rises in sea temperatures caused by global warming have resulted in episodes of coral bleaching. Tiny algal plants (zooxanthellae) live inside the coral polyp tissue in a symbiotic relationship with the corals, and a rise in sea temperature of only 1–2 °C is enough to expel them from their host into the ocean, causing the corals to lose their colour pigmentation. Bleaching has increased in frequency and intensity over the last two decades and although reefs can recover, mortality of corals has long-term negative impacts on reef biodiversity. Another source of stress occurs when overfishing upsets the natural balance within the ecosystem. On a coral reef, the growth of certain algae is kept in check by grazing species like parrot fish. Overfishing severely depletes fish populations, and with a decline in the numbers of grazing fish, the long-spined black sea urchin (*Diadema antillarum*) took over the niche as principal grazer. When, in turn, the sea urchin's population was decimated by a waterborne pathogen in the early 1980s, many Caribbean reefs became smothered with algae since they were no longer being grazed. The result was that species diversity in some reef habitats plummeted, and the reefs degenerated (Hughes 1994).

The environmental importance of Caribbean coral reefs has both a global and a local component. Though their biomass is lower than forests, coral reefs play a similar role to forests in the maintenance of the planet's biosphere because they help remove carbon from the atmosphere and slow down the build-up of greenhouses gases. In the ocean, carbon is stored in the shells of marine animals and in the calcium carbonate of the reef structure.

At the local scale, all ecosystems, including coral reef ecosystems, have a value to society in two contrasting ways. Direct use values are the economic benefits derived from income-generating activities that take place in an ecosystem, such as fishing, tourism and recreation. Indirect use values or environmental services are the less tangible environmental benefits an ecosystem provides, such as protection against flood damage (Figure 1.14). Coral reefs provide a number of environmental services; they are critical to the stability of Caribbean beaches because they act as breakwaters and protect shorelines from rough seas and help prevent beach erosion. Reefs are nurseries for some types of marine animal and possible sources of beach sand material because animals like parrot fish graze coral and

Figure 1.14 Environmental services of Caribbean inshore marine ecosystems

Coral reefs

Protect coastlines against wave action, especially during storms

Stabilise beaches against erosion

Contribute to the formation of sandy beaches

Fish nurseries

Habitats for food sources such as molluscs and crustaceans

Sea grass beds

Act as baffles and cause sediment particles to settle

Root rhizomes help stabilise bottom sediment and hinder re-suspension of sediments

Habitats for halimeda

Food source for marine animals

Nurseries for marine animals

Coastal mangroves

Help stabilise coastal zones with respect to flood control

Help recharge coastal aquifers for water supply

Help filter silt and sediment from terrestrial sources

Nurseries for commercial and subsistence fisheries

Retain sediments and nutrients

Provide protection of coastlines from storms and wave action

Economic evaluation of natural landscapes

Increasingly, efforts are being made to place economic evaluations on natural landscapes. It involves applying techniques to estimate the value of both the natural functions and economic uses of natural landscapes. Conceptually there are three components that need to be assessed:

1. **Direct use values.** These are based on the value of the economic uses of natural landscapes, and include activities such as hunting, fishing, harvesting of wild species for food, shelter, fuel or medicine (WCMC, 1992).
2. **Indirect uses values.** Also called 'environmental services', the idea is that there are economic benefits derived from the various functions provided by a natural landscape for direct economic activities. Thus a wetland might protect an area from flooding, or a coral reef might protect a tourist beach from erosion. Other functions to be valued might include the recycling of carbon, oxygen and nitrogen, buffers on adverse weather conditions like storm surge, and water purification. A number of techniques have been developed to assess the economic values of environmental services.
3. **Non-use or preservation values.** These values are derived from neither current direct nor indirect uses, but relate to the satisfaction derived from knowing that a natural landscape exists for either the present or future generations (Barbier, 1991).

Barbier (1991) provides a methodology appropriate to the economic evaluation of a tropical wetland natural resource system.

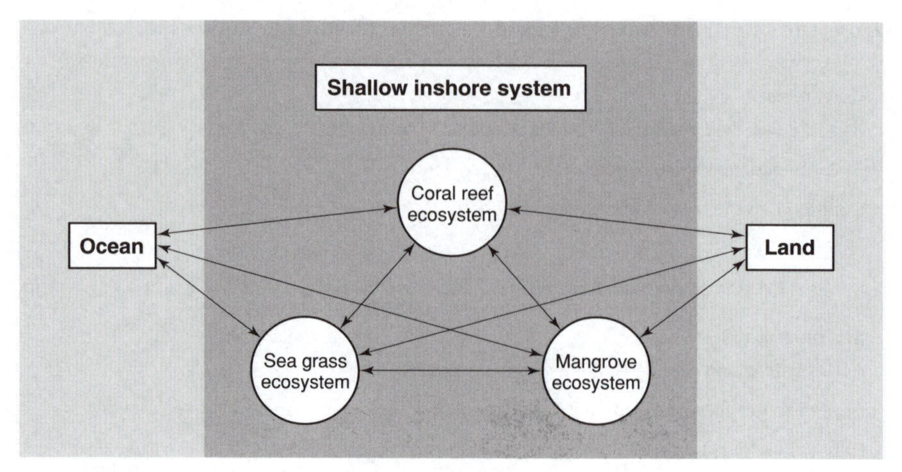

Types of interaction between ecosystems

1. Physical (e.g. inshore circulation of sea water, wave action, rivers)
2. Nutrients
3. Particulate organic matter
4. Dissolved inorganic matter
5. Marine animal migrations
6. Human impacts

Figure 1.15 Interrelationships between coral reef, sea grass and mangrove ecosystems

bite off pieces that are excreted as fine particles. Environmental economists use environmental accounting methods to assess the monetary or economic values of environmental services (Pearce *et al.* 1989).

There are important biological and physical linkages between coral reefs and other coastal environments such as sea grass beds in shallow lagoons and nearby mangrove ecosystems (Figure 1.15). Sea grass beds are important offshore marine environments colonised by various flowering species such as turtle grass (*Thalassia testudinum*) and manatee grass (*Syringodium filiforme*). Swimmers tend to regard sea grass as a nuisance, but sea grass beds also perform critical environmental services. They act as huge filters because the blades of the grasses act as baffles that cause sediment particles to settle. Sea grasses are biologically productive plants whose detritus forms an important food source for marine organisms and their predators (Jupp 1983). These habitats provide nurseries for juvenile commercial fish like snapper, grunt, lobster and conch, and the plants are grazed by turtles and manatees. The plants' rhizome root structures stabilise bottom sediments and prevent re-suspension. Several species of dark green algae called *halimeda* also thrive in this environment. Their calcified platelet discs disintegrate into fine carbonate particles and constitute a major source of the sediment input to white sand beaches, together with foraminifera, which also colonise the sea grass beds. Mangrove ecosystems provide environmental services because they function as buffers between terrestrial and marine ecosystems, helping to mitigate coastal flooding. Mangrove areas act like huge sponges, absorbing water from rivers, filtering silt and sediment, and releasing it slowly into the marine environment, a process that helps prevents silt from smothering an offshore coral reef.

Figure 1.15 depicts how coastal ecosystems are interconnected. There are many ecological links because fish and other marine animals migrate, feed and circulate between diverse marine habitats. Geomorphogical links include the transport of material between these marine environments by offshore currents and terrestrial stream and river flow into coastal waters. The protection of one ecosystem may benefit a nearby ecosystem. But also destructive activity in one ecosystem can have a negative and unintended impact on a contiguous ecosystem. For example, the destruction or dredging of a small wetland to build a hotel could have negative impacts on local fish stocks as the wetland might be a nursery for juvenile fish. Coral reefs may be harmed by silt plumes in rivers carried out to sea during a flood event because silt tends to accumulate on the reef and change the optical clarity of seawater, thereby damaging the reef and possibly harming tourism and fishing. Increased silt load in rivers, in turn, may be the result of accelerated soil erosion in upland farming areas, which underscores the point that coastal ecosystems are also functionally interconnected to terrestrial ecosystems, a factor contributing to the environmental fragility and vulnerability of small tropical islands (McGregor and Barker 1995).

Wetlands are complex and fragile coastal environments, consisting of mosaics of habitats such as mangroves, salt marshes, salt ponds and salinas. Historically, the region's wetlands have disappeared or shrunk in area by conversion to other land uses. Wetlands were once regarded as unhealthy swamps of mainly nuisance value, to be drained, cleared and reclaimed as land for building development, ports and coastal towns, as in the case of Kingston and Montego Bay (Hudson 1983). St Lucia probably has lost 40 per cent of its original wetlands. Wetlands are particularly vulnerable to human impacts and their continued destruction and removal (especially mangrove areas) is a concern today. In Guyana, large areas of wetland have been reclaimed for rice production (Williams 1995) but tourist development poses the biggest threat to wetlands in the Caribbean islands. In Antigua, for example, the Jolly Hill Salt Ponds were converted to a luxury hotel resort (de Albuquerque 1991; Lorah 1995) and McKinnon's pond was once a mangrove and a salt water pond environment, a habitat supporting native flora and fauna and migratory birds. In the early stage of tourism development in the Dickenson Bay area an embankment was built which disrupted local ecology and the drainage between the terrestrial and marine ecosystems, effectively cutting off the pond from the sea (Government of Antigua and Barbuda Northwest Coast Evaluation Project 1998). Amongst other things, the mangrove area has been greatly diminished (Figure 1.16) and a stagnant pond produces offensive odours in the Runaway Bay tourist area.

Attitudes towards wetlands are gradually changing in the Caribbean region as elsewhere in the world. For example, Trinidad and Tobago declared the Nariva Swamp a Ramsar site in 1993 (Box 1.2), and the Bahamas, Belize and Jamaica

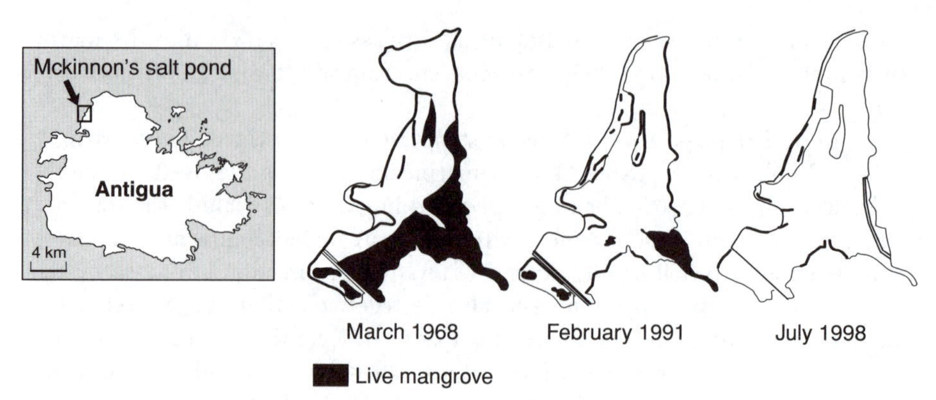

Figure 1.16 The diminution of mangroves in McKinnon's salt pond, Antigua
Source: Government of Antigua and Barbuda Northwest Coast Evaluation Project (1988)

also have signed the Ramsar Convention on Wetlands. We noted earlier the flamingo colony in the Bahamian island of Inagua is a major tourist attraction, whilst other wetland tourist activities include the swamp safaris into the Lower Black River Morass in Jamaica (Johnson 1998), and boat trips into Trinidad's Caroni Swamp to see the quarter of a million scarlet ibis birds that nest there.

Protecting Caribbean landscapes

In this chapter, we have examined the characteristics of the region's diverse natural landscapes and noted that they constitute a world centre for biodiversity, with many endemic and endangered plant and animal species. Natural environments are constantly transformed through human activities, yet the integrity of interconnected tropical island ecosystems is integral to the sustainable development of the countries in which they are located (McGregor and Barker 1995). Figure 1.17 summarises the main contemporary anthropogenic and environmental threats to their continued survival.

Pollution, particularly from farming, is a major threat to the environment. Sources of agricultural pollution include nutrients and pathogens (bacteria, viruses and other micro-organisms) and are derived mainly from the application of agrochemicals used in fertilisers, pesticides, fungicides and herbicides. Pollutants are mobilised by hydrological conditions and may be present in surface water or groundwater flow. The application of excessive amounts of fertiliser leads to nitrogen enrichment of surface and groundwater, causing eutrophication (excessive algal growth that gives stagnant water bodies a greenish hue). Water hyacinth has colonised some rivers and wetland areas as a result of excessive nutrient levels, and needs to be removed continuously, as in the Black River Lower Morass in Jamaica. Monoculture farming (as in cane fields) rapidly depletes soil nutrients and necessitates large amounts of fertiliser, and pure stands of a single crop are

Box 1.2: The Nariva Swamp in Trinidad

The Nariva Swamp was declared a Ramsar site in 1995. Its 6,000 hectares makes it Trinidad's largest, biologically most productive and diverse wetland with distinct vegetation zones of mangrove woodland, herbaceous swamp, palm swamp forest and evergreen rain forest. It is separated from the Atlantic Ocean by a wide sand bar called the Cocal, which was instrumental to the evolution of the wetland because it impounded run-off from the Central Range. The area is low-lying and the water level fluctuates as part of a seasonal regime, with annual flooding. Animal biodiversity is significant with 151 species of bird recorded, 29 species of mammal (including manatees), 16 species of bat, 25 species of reptile (including anaconda) and seven species of amphibian.

The wetland is a natural resource for about 5,000 people who depend upon it for their livelihoods. Plum Mitan and Cocal-Kernahan are villages located on its periphery and their traditional resource management involves farming, fishing and hunting. Small-scale farming activities focus on rice cultivation and vegetable farming, and the aquatic food sources include cascadura (the amoured cat fish), black conch and callaloo crab. Timber resources are utilised for construction and the palm heart (*Roystonea oleracea*) is eaten as a delicacy, especially at Hindu weddings.

During the 1980s, large-scale rice farmers moved into the area. They were not from the local communities and engaged in large-scale, mechanised clearance of wetland tracts. Lagoons disappeared and the complex local hydrology and drainage patterns were disrupted, threatening the environmental services provided by the wetland, such as sediment filtering, protection from erosion and natural flood control for surrounding farming communities. The aerial spraying of pesticides on rice crops caused large fish kills, and habitat disturbance forced wildlife to move into neighbouring farming areas, where they became pests.

Inevitably, bitter social conflicts ensued between the large-scale rice farmers and local communities over the competing demands for land and other natural resources of the wetland. Environmental NGOs and the Forestry Department joined the fray on the side of the traditional communities, in efforts to protect and conserve the ecology of the area whilst maintaining its traditional, sustainable resource management function (Sletto 1998). The groups successfully lobbied to have the area declared a national park in 1993, and a National Wetland Policy was formulated. Declaration of Ramsar status came two years later, and a management plan has been drafted, and the large-scale rice farmers were evicted, pending an environmental impact assessment.

vulnerable to insect, bacteria and fungal pests and so require the application of pesticides. Herbicides are used to keep down the weeds that the use of fertilisers encourage. For example, commercial banana operations try to minimise understorey vegetation by using herbicides to kill weeds because they compete with shallow-rooted bananas for water, nutrients and light.

Figure 1.17 Threat to natural landscapes in the Caribbean region

Marine and coastal ecosystems

Overfishing

Pollution from terrestrial sources (e.g. agriculture pollution, urban areas, tourist resorts)

Global warming

Coastal building developments

Illegal sand mining on beaches

Inappropriate waste management disposal

Charcoal production in mangroves

Natural disasters – storm surge flooding, flash flooding, silting impacts on coral reefs

Environmentally damaging tourist activities (e.g. coral souvenirs, reef damage from ships' anchors, discharge from cruise ships)

Forest ecosystems

Forest clearance for farm land

Forest clearance for bauxite mining

Fires burning out of control

Hunting and collecting wildlife

Commercial timber production, including selective logging of valuable tree species

Uncontrolled charcoal production

Uncontrolled cutting of yam sticks (Jamaica only, see Chapter 3)

Inappropriate garbage disposal by resource users, including tourists

Many Caribbean countries have begun to establish national parks and protected areas to protect their natural landscapes and to conserve native flora and fauna, and currently there are about 640 protected areas in the region, over 100 of which are marine parks. Traditionally, national parks were established in Europe and North America for environmental conservation, domestic recreation and the sustainable use of natural resources (especially forestry). However, in the search for sustainable development, more and more developing countries have established national parks and protected areas, broadening their roles considerably (Hales 1989). We can summarise the most important reasons for setting aside a geographical unit as a national park or protected area:

- protection of individual species;
- protection of ecosystems and biodiversity;
- protection of landforms and geophysical features;
- protection of historical, cultural and religious sites;

- recreation and education;
- protecting watersheds;
- sustainable use of forestry for timber, harvesting wildlife and natural products;
- employment creation and income generation through tourism.

An area may be set aside for more than one of these purposes, and Caribbean territories (like other developing countries) are adopting and adapting the International Union for the Conservation of Nature's (IUCN) classification scheme (see www.iucn.org) for categorising different types of protected area. The IUCN's methodology envisages the creation of a national *system* of protected areas whereby each area may serve a different purpose and/or be designed to protect a specific landscape or ecosystem. In Dominica, for example, the Cabrits National Park is set aside for the protection of the cultural heritage of indigenous Carib people, whereas the Morne Trois Pitons is a mountainous, rain forest national park intended for environmental protection, species conservation and ecotourism. The latter became the first World Heritage Site in the eastern Caribbean in 1998. Cuba probably has the most sophisticated and best-managed system in the region incorporating many different landscapes, recreational uses and cultural and heritage sites (Box 1.3).

The effective management of protected areas in developing countries is a difficult and fraught task for a variety of reasons. For example, there are many actual and potential conflicts over the use of natural resources located in a protected area by different groups and institutions. Resource sustainability and economic development, the protection of indigenous peoples and traditional livelihoods, and the pros and cons of international tourism have all to be balanced against the need for environmental conservation and sound land management. In the Caribbean islands, land and resource management problems in protected areas are particularly acute because land itself is a scarce resource and there are precious few natural landscapes left entirely intact.

One way of trying to mitigate resource use conflicts in protected areas is through participatory planning. In Jamaica, for example, local communities and NGOs were involved in 14 public meetings prior to the publication of the Green Paper on protected areas (Government of Jamaica 1997). Further, governments claim that budgets cannot pay the full costs of managing protected areas, so they often delegate management responsibilities and fund-raising activities to NGOs, as a cost-sharing device (Miller 1998). In this way, Caribbean NGOs like the St Lucia National Trust, the Bahamas National Trust and the Jamaica Conservation and Development Trust are actively involved in protected areas management. Various kinds of community-based management system have been advocated, including co-management schemes. Co-management involves sharing power and responsibility for resource management between government, NGOs and local people. Regional examples include CANARI (Caribbean Natural Resource Institute) working with the St Lucian government and fishing communities in Laborie

Box 1.3: Cuba's protected areas

Cuba is the largest island in the region and has the greatest biodiversity and highest endemism. For example, over 50 per cent of its flora and 32 per cent of its vertebrate fauna are endemic. Furthermore, many animal species are endangered. Cuba's isolated mountain ranges, extensive lowland plains and vast stretches of tropical coastal environments provide a dramatic range of natural vegetation, ecosystems and landscapes, ranging through rain forest and savanna to coastal and karst environments. All these various types are represented in its national system of protected areas. The first national park was designated in 1930, but more direct ownership and control was introduced following the 1959 revolution, the Constitution declaring that the state had a duty to protect national monuments and historic and aesthetic areas of outstanding natural beauty for the benefit of the population (wcmc.org.uk).

There is a complex but coherent system of management for protected areas involving many government ministries and other organisations. Overall responsibility for policy formulation is vested in COMARNA (the National Commission of Environmental Protection and Rational Use of Natural Resources). The National Network of Protected Areas was established in 1981 and encompasses more than 65 protected areas, covering 18 per cent of Cuba's land area (UNEP CEP 1996). The categories of protected area units in this comprehensive system include managed flora reserves, ecological areas, hunting areas, integrated management areas, terrestrial and marine national parks, wildlife refuges, nature reserves and touristic natural areas (www.unep-wcmc.org) .

The Ciénaga de Zapata wetland, one of the largest and best-preserved wetlands in the insular Caribbean, is also a Biosphere Reserve (Figure 1.6). It contains the endangered Cuban crocodile (*crocodylus rhombifer*) and the more numerous American crocodile (*crocodylus acutus*), as well as other endangered animals and 23 of Cuba's 26 endemic species of birds. The area was declared a Biosphere Reserve in the year 2000 and is a multiple-use protected area that helps sustain the livelihoods of the 9,000 people in the area. They make a living from fishing, small-scale farming, charcoal production and, more recently, tourism; about 800,000 domestic and 100,000 international tourists visit the area annually.

Cuba's protected areas are closely aligned to international networks as they include six Biosphere Reserves, nine World Heritage Sites and six Ramsar wetlands (www.ramsar.org). The World Heritage Sites include Old Havana and its fortification and an archaeological landscape area on the site of the first coffee plantation in south-east Cuba. The Alejandro de Humboldt National Park in north-eastern Cuba is also a World Heritage Site and is of interest partly because it was named after the famous nineteenth-century German geographer who travelled in and wrote extensively about the island.

(Figure 1.9), south-west St Lucia (see www.canari.org) and the Caribbean Conservation Association's involvement in similar projects in Belize and Barbados (see www.caribbeanconservation.org).

The IUCN information network incorporates an on-line international database of parks and protected areas which lists only protected areas deemed to be effectively managed. The pejorative term 'paper park' refers to protected areas that appear in glossy national development plans but lack the necessary financial and human resources for effective management. Twenty-five years ago most Caribbean national parks were regarded as paper parks (Brandon *et al.* 1998), but there has been some progress since Agenda 21. The main advances include improved scientific data collection and documentation, the training of a new cadre of young public sector professionals and a growing number of enthusiastic environmentalists and conservationists in the private sector, innovative conceptualisation of national systems of protected areas along the guidelines suggested by the IUCN, and in the greater use of feasibility studies and environmental impact assessments for development projects in environmentally sensitive areas. However, as elsewhere in the developing world, the effective management of national parks and protected areas is proving an elusive goal. Caribbean governments seem to lack the political will to implement comprehensive systems, being preoccupied with other policy issues perceived to have greater priority, citing financial resources as a major constraint (Miller 1999). Thus, to a large extent, the laudable and plausible policy objectives for the region's parks and protected areas are yet to be realised.

References

Asprey, G. F. and Robbins, R. G. (1953) 'The vegetation of Jamaica', *Ecological Monographs*, **23**, 4, 359–411.

Bacon, P. R. (1978) *Flora and Fauna of the Caribbean: An Introduction to the Ecology of the West Indies*, Key Caribbean Publications, Trinidad.

Barbier, E. B. (1991) 'An approach to Economic Evaluation of Tropical Wetlands: with examples from Guatemala and Nicaragua', in Girvan, N. P. and Simmons, D. A. (eds) *Caribbean Ecology and Economics*, Caribbean Conservation Association, Barbados, pp. 207–32.

Barker, D. (1998) 'Yam farmers on the edge of Cockpit Country: aspects of resource use and sustainability', in McGregor, D. F. M., Barker, D. and Lloyd Evans, S. (eds) *Resource Sustainability and Caribbean Development*, UWI Press, Kingston, Jamaica, pp. 357–72.

Barker, D. (1998) 'The north-east trades and temperature inversions: notes for the classroom', *Caribbean Geography*, **9**, 1, 58–65.

Barker, D. and Beckford, C. (2003) 'Yam Production and the Yam Stick Trade in Jamaica: integrated problems for planning and resource management', in Barker, D. and McGregor, D. F. M. (eds) *Resources, Planning and Environmental Management in a Changing Caribbean*, UWI Press, Kingston, Jamaica, pp. 57–74.

Barker, D. and McGregor, D. F. M. (1995) 'A geographical focus for environment and development in the Caribbean', in Barker, D. and McGregor, D. F. M. (1995) (eds) *Environment and Development in the Caribbean: Geographical Perspectives*, UWI Press, Kingston, Jamaica, pp. 3–20.

Barker, D. and Miller, D. J. (1995) 'Farming on the Fringe: small scale agriculture on the edge of Cockpit Country', in Barker, D. and McGregor, D. F. M. (eds) *Environment and Development in the Caribbean: Geographical Perspectives*, UWI Press, Kingston, Jamaica, pp. 271–92.

Barker, D. and Spence, B. (1988) 'Afro-Caribbean Agriculture: A Jamaican maroon community in transition', *Geographical Journal*, **154**, pp. 198–208.

Barry, R. G. and Chorley, R. J. (1992) *Atmosphere, Weather and Climate,* 6th edition, Routledge, London.

Beard, J. S. (1955) 'The classification of tropical American vegetation types', *Ecology*, **36**, 89–100.

Blume, H. (1974) *The Caribbean Islands*, Longman, London.

Brandon, K., Redford, K. H. and Sanderson, S. E. (eds) (1998) *Parks in Peril*, The Nature Conservancy, Washington.

Chapman, V. J. (1976) *Mangrove Vegetation*, J. Cramer, Florida-9490.

Chalmers, W. S. (2002) 'Managing forest resources', in Goodbody, I. and Thomas-Hope, E. (eds) *Natural Resource Management for Sustainable Development in the Caribbean*, UWI Press, Kingston, Jamaica, pp. 219–76.

Crosby, A. W. Jr. (1972) *The Columbian Exchange: Biological and Cultural Consequences of 1492*, Greenwood Press, Westport, Connecticut.

Day, M. J. (1993) 'Human impacts on Caribbean and Central American karst', in Williams, P. W. (ed.) *Karst Terrains, Environmental Changes, Human Impact, Catena Supplement 25*, Germany Cremlingen-Destedt, pp. 109–125.

de Albuquerque, K. (1991) 'Conflicting claims on the Antigua coastal resources: the case of the McKinnon's and Jolly Hill salt ponds', in Girvan, N. P. and Simmons, D. A. (eds) *Caribbean Ecology and Economics*, Caribbean Conservation Association, Barbados, pp. 195–206.

Denevan, W. M. (1992) 'The pristine myth: the landscape of the Americas in 1492', *Annals, Association of American Geographers*, **82**, 369–85.

Diesel, R. (1991) 'A Jamaican speciality: the Bromeliad crab', *Jamaica Naturalist*, **1**, 1, 7–9.

Draper, G., Donovan, S. K. and Jackson, T. A. (1994) 'Geologic provinces of the Caribbean region', in Donovan, S. K. and Jackson, T. A. (eds) *Caribbean Geology: An Introduction*, UWIPA, Kingston, Jamaica.

Draper, G. and Barros, J. A. (1994) 'Cuba', in Donovan, S. K. and Jackson, T. A. (eds) *Caribbean Geology: An Introduction*, UWIPA, Kingston, Jamaica, pp. 65–86.

Eyre, L. A. (1980) 'The maroon wars in Jamaica – a geographical appraisal', *Jamaica Historical Review*, **12**, 80–102.

Eyre, L. A. (1987) 'Jamaica: test case for tropical deforestation?' *Ambio*, **16**, 338–43.

Eyre, L. A. (1995) 'The Cockpit Country: a world heritage site?' in Barker, D. and McGregor, D. F. M. (eds) *Environment and Development in the Caribbean: Geographical Perspectives*, UWI Press, Kingston, pp. 259–70.

Eyre, L. A. (1998) 'The tropical rainforests of the eastern Caribbean: present status and conservation', *Caribbean Geography*, **9**, 2, 101–20.

Fermor, J. (1972) 'The dry valleys of Barbados', *Transactions of Institute British Geographers*, **57**, 153–66.

Fincham, A. G. (1997) *Jamaica Underground: The Caves, Sinkholes and Underground Rivers*, UWI Press, Kingston, Jamaica.

Fraser, H., Carrington, S., Forde, A. and J. Gilmore (1990) *A–Z of Barbadian Heritage*, Heinemann Publishers (Caribbean) Ltd, Kingston, Jamaica.

Garraway, E. and Bailey, A. J. A. (1993) 'The ecology and conservation biology of Jamaica's endangered giant swallowtail butterfly, *Papilio Homerus*', *Jamaica Naturalist*, **3**, pp. 7–12.

Gedzelman, Stanley, D. (1980) *The science and wonders of the atmosphere*, John Wiley, New Jersey, p. 263.

Government of Antigua and Barbuda Northwest Coast Evaluation Project (1998) final report, EMPAL-Jacques Whitford Ltd.

Grossman, D. H., Iremonger, S. and Muchoney, D. M. (1991) *Jamaica: A Rapid Ecological Assessment: Phase 1 – An Islandwide Characterisation and Mapping of Natural Communities and Modified Vegetation Types*, The Nature Conservancy, Arlington, Virginia.

Hales, D. (1989) 'Changing concepts of national parks', in Western, D. and Pearl, M. (eds) *Conservation for the Twenty-first Century*, Oxford University Press, Oxford, pp. 139–44.

Hudson, B. J. (1983) 'Wetland reclamation in Jamaica', *Caribbean Geography*, **1**, 2, 75–88.

Hughes, T. P. (1994) 'Catastrophes, phase shifts and large-scale degradation of a Caribbean coral reef', *Science*, **265**, 1547–51.

Intergovernmental Panel on Climate Change (IPCC) (1990) *Climate Change: The IPCC Scientific Assessment*, Houghton, J. T., Jenkins, G. J. and Ephraums, J. J. (eds), Cambridge University Press, Cambridge.

Jennings, J. N. (1985) *Karst Geomorphology*, Blackwell, Oxford.

Johnson, A. M. (1998) 'The artisanal fishery of the Black River Morass, Jamaica: a traditional system of resource management', in McGregor, D. F. M., Barker, D. and Lloyd Evans, S. (eds) *Resource Sustainability and Caribbean Development*, UWI Press, Jamaica, pp. 390–404.

Jupp, B. (1983) 'The sea grasses of Jamaica', *Jamaica Journal*, **16**, 4, 23–6.

Kueny, J. A. and Day, M. J. (1998) 'An assessment of protected karst landscapes in the Caribbean', *Caribbean Geography*, **9**, 2, 87–101.

Lloyd Evans, S., McGregor, D. F. M. and Barker, D. (1998) 'Sustainable development and the Caribbean; geographical perspectives', in McGregor, D. F. M., Barker, D. and Lloyd Evans, S. (eds) *Resource Sustainability and Caribbean Development*, UWI Press, Jamaica, pp. 3–25.

Lorah, P. (1995) 'An unsustainable path: tourism's vulnerability to environmental decline in Antigua', *Caribbean Geography*, **6**, 1, 28–39.

MacPhee, R. D. E. and Iturralde-Vinent, M. A. (2000) 'A short history of Greater Antilles land mammals: biogeography, paleogeography, radiations and extinctions', *Tropics*, **10**, 1, 145–54.

MacPherson, J. (1980) *Caribbean Lands*, Longman Caribbean.

Mann, P., Schubert, C. and Burke, K. (1990) 'Review of Caribbean neotectonics', in Dengo, G. and Case, J. E. (eds) *The Geology of North America. Volume H, The Caribbean Region,* Geological Society of America, Boulder, Colorado, pp. 307–38.

McGregor, D. F. M. (1995) 'Soil erosion, environmental change and development in the Caribbean: a deepening crisis?' in Barker, D. and McGregor, D. F. M. (eds) (1995) *Environment and Development in the Caribbean: Geographical Perspectives*, UWI Press, Jamaica, pp. 189–208.

McGregor, D. F. M. and Barker, D. (2003) 'Environment, resources, and development: some reflections on the Caribbean research agenda', in Barker, D. and McGregor, D. F. M. (eds) *Resources, Planning and Environmental Management in a Changing Caribbean*, UWI Press, Kingston, Jamaica, pp. 1–10.

Miller, D. J. (2004) 'Karst geomorphology of the white limestone group in Jamaica', *Cainozoic Research*, **3**, in press.

Miller, D. J. (1998) 'Invasion of the Cockpits: patterns of encroachment into the wet limestone forest of Cockpit Country, Jamaica', in McGregor, D., Barker, D. and Lloyd Evans, S. (eds) *Resource Sustainability and Caribbean Development*, UWI Press, Kingston, Jamaica, pp. 373–89.

Miller, L. A. (1999) 'Perspectives on the sustainability of protected areas in Jamaica', *Caribbean Geography*, **10**, 1, 52–62.

Monroe, W. H. (1976) *The Karst Landforms of Puerto Rico*, U. S. Geological Survey, Professional Paper 899.

Pindell, J. L. (1994) 'Evolution of the Gulf of Mexico and the Caribbean', in Donovan, S. K. and Jackson, T. A. (eds) *Caribbean Geology: An Introduction*, UWIPA, Kingston, Jamaica, pp. 13–40.

Pearce, D. W., Markandya, A. and Barbier, E. B. (1989) *Blueprint for a Green Economy*, London: Earthscan Publications.

Potter, R. B., Binns, T., Elliott, J. A. and Smith, D. (2004) *Geographies of Development*, 2nd edition, Pearson/Prentice Hall, New York.

Proctor, G. R. (1986) 'Cockpit Country and its vegetation', in Thompson, D. A., Bretting, P. K. and Humphreys, M. (eds) *Forests of Jamaica*, Jamaican Society of Scientists and Technologists, Kingston, Jamaica, pp. 43–8.

Raffaele, H., Wiley, J., Garrido, O., Keith, A. and Raffaele, J. (1998) *A Guide to the Birds of the West Indies*, Princeton University Press, Princeton, NJ.

Rashford, J. (1994) 'Jamaica's settlement vegetation, agroecology and the origin of agriculture', *Caribbean Geography*, **5**, 1, 32–50.

Robinson, C. (1969) *The Fighting Maroons of Jamaica*, W. Collins and Sangster, Kingston, Jamaica.

Sauer, C. O. (1966) *The Early Spanish Main*, University of California Press, Berkeley and Los Angeles.

Sealey, N. E. (1985) *Bahamian Landscapes*, Collins Caribbean, London.

Sletto, B. (1998) 'Fish, rice and the meaning of place: the political ecology of the Nariva Swamp, Trinidad', *Caribbean Geography*, **9**, 1, 14–29.

Sweeting, M. M. (1972) *Karst Landforms*, Macmillan, London.

Speed, R. C. (1994) 'Barbados and the Lesser Antilles forearc', in Donovan, S. K. and Jackson, T. A. (eds) *Caribbean Geology: An Introduction*, UWIPA, Kingston, Jamaica, pp. 179–92

Thomas, M. (1994) *Geomorphology in the Tropics*, John Wiley, Chichester.

Vogel, P., Nelson, R. and Kerr, R. (1995) 'Conservation strategy for the Jamaican iguana, *Cyclura collei*', in Powell, R. and Henderson, R. W. (eds) *Contributions to West Indian Herpetology: A Tribute to Albert Schwartz*, Society for Study of Amphibians and Reptiles, Ithaca, New York.

United Nations Conference on Environment and Development (1992) AGENDA 21, Chapter 17, 'Protection of the oceans, all kinds of seas, including enclosed and semi-enclosed seas, and coastal areas and the protection, rational use and development of their living resources', United Nations, New York.

Wadge, G. (1994) 'The Lesser Antilles', in Donovan, S. K. and Jackson, T. A. (eds) *Caribbean Geology: An Introduction*, UWIPA, Kingston, Jamaica, pp. 167–78.

Walsh, R. P. D. (1998) 'Climatic changes in the eastern Caribbean over the past 150 years and some implications in planning sustainable development', in McGregor, D. F. M., Barker, D. and Lloyd Evans, S. (eds) *Resource Sustainability and Caribbean Development*, UWI Press, Jamaica, pp. 26–48.

Watts, D. (1987) *The West Indies: Patterns of Development, Culture and Environmental Change Since 1492*, Cambridge University Press, Cambridge.

Whittaker, R. J. (1998) *Island Biogeography: Ecology, Evolution and Conservation*, Oxford University Press, Oxford.

Williams, P. (1995) 'Drainage and irrigation projects in Guyana: environmental considerations', in Barker, D. and McGregor, D. F. M. (eds) *Environment and Development in the Caribbean: Geographical Perspectives*, UWI Press, Kingston, Jamaica.

Williams, P. E. (1998) 'Environmental problems in Guyana's hinterland: some policy considerations', *Caribbean Geography*, **9**, 2, 121–35.

Wilson, B. S. and Vogel, P. (2000) 'A survey of the herpetofauna of the Hellshire Hills, Jamaica, including the rediscovery of the blue-tailed galliwasp (*Celestus duquesneyi* Grant)', *Caribbean Journal of Science*, **36**, **3–4**.

Wilson, J. S., Brothers, T. S. and Marcano, E. J. (2000) 'Land cover contrasts on the Haitian/Dominican border', *Caribbean Geography*, **11**, 1, 244–9.

World Resources Institute (WRI)/**International Union for Conservation of Nature** (IUCN)/**United Nations Environment Program** (UNEP) (1992) *Global Biodiversity Strategy*, Gland, Switzerland.

World Conservation Monitoring Centre (WCMC) (1992) *Global Biodiversity: Status of the Earth's Living Resources*, report compiled by WCMC (ed. Groombridge, B.), Chapman & Hall, London.

Chapter 2

POPULATION AND MIGRATION

Introduction

International migration is one of the Caribbean's most fundamental demographic processes, contributing substantially to the population diversity that characterises the small insular societies of this oceanic region. The Caribbean was settled by immigrants, and the subsequent migrations of its people, in part, grow out of these early mobility patterns. The Caribbean was also incorporated into the world economy by the sixteenth century and, to this day, external influences and connections feature strongly in island lives and livelihoods. Mobility strategies came to be strategic responses of many Caribbean islanders to the realities of island existence, the environment's limits, their small territorial size, and their vulnerabilities. In addition to international movement 'off the island' both within the region and beyond, rural-to-urban migration has also played its part as a survival strategy for Caribbean people, and today's rapidly growing cities (see Chapter 5) are in large part a consequence of this city-ward transfer.

This chapter first documents the history of settlement and illustrates how regional population diversity came to be so culturally rich. Then, a second section documents the socio-historical evolution of migration strategies and how migration becomes a livelihood option for all classes. Thirdly, twentieth-century demographic trends and internal as well as international migration processes are detailed. In addition to models of international migration and circulation, a model of rural-to-urban migration and spontaneous settlement patterns in Caribbean cities is developed in this third section. Lastly, a brief glimpse of the future is projected, where multi-local patterns of transnational movement between the Caribbean, North America and Europe appear well entrenched and self-perpetuating.

Population diversity and changing dynamics in the Caribbean

Settled by waves of immigration since the 'Encounter' in 1492, migration was a fundamental factor in the colonisation, settlement and resettlement of the region, as it was also a fundamental force behind both the development and underdevelopment of Caribbean economies and societies. The purpose of this first section of the chapter is to consider the early patterns of European settlement and the population dynamics of Caribbean societies.

The 'Encounter', and depopulation, patterns of European settlement, plantation riches and population influxes

The Lesser and Greater Antilles chain of islands were originally populated by mobile aboriginal cultures from Middle America and South America: first, ancestors of Guanahatebeys in Cuba, then Tainos and Arawaks moved island to island in apparent stepwise fashion from South America with Tainos finally populating the island of Santo Domingo (present day Haiti and the Dominican Republic) in considerable numbers. Later, small fighting bands of Caribs also moved into the Antilles from South America, and they conquered the Arawak inhabitants of the Lesser Antilles, forcing the latter into assimilation (Kimber 1992). Some islands were densely populated, like Santo Domingo and Cuba; others like Barbados were settled then abandoned. These aboriginal peoples warred among themselves, established coastal settlements and lived tenuous lives in the main. However, in the fifteenth century, the 1492 'Encounter' or 'Discovery' ushered in the first of a series of immigrant waves that was to fundamentally influence the Caribbean's settlement and population history.

Claiming sovereignty over the New World, the Spanish Crown enlisted the services of adventurers such as Christopher Columbus to aid their colonisation. However, for political and religious reasons, only Spanish citizens qualified as immigrant *conquistadores* or as colonists. The first waves of European immigrants were small in number, but large in their power, authority and influence. While their population size continued to increase, albeit modestly, the numerically dominant indigenous Tainos and Arawak populations suffered calamitous declines. The causes were many: European-introduced diseases and epidemics, the destruction of their indigenous systems, and enslavement and mass deportations to the New Spain mainland (Mexico) meted out by the European newcomers. Soon, these declining local populations forced their conquerors to seek new sources of labour from across the oceans. Accordingly, they recruited yeoman farmers from Spain, bought slaves shipped from Africa and recruited indentured servants from other parts of their colonial empires. For the Portuguese in Brazil, Dutch managers first served their masters and developed sugar cane plantations there. After these Dutch aliens were expelled, they were invited to the British West Indian islands to share their agro-industrial prowess, and thereby all parties (except the Portuguese) derived mutual benefits from this highly profitable venture.

The demographic 'turn-around': emergence of an Afro-Caribbean region

By the end of the sixteenth century, sugar plantations relied on an active slave trade. The Spanish operated their system direct from Africa to licensed ports in

Box 2.1: The Afro-Caribbean demographic turnaround

The dramatic turnaround in Cuban demography illustrates this transformation. Between 1780 and 1840, Cuba changed from a Spanish *colonia* of European free men and yeomen settlers to a plantation society that was predominantly black African and enslaved. In 1774, 56 per cent of Cuba's population were white, 20 per cent were free-coloured '*mulatto*', 23 per cent were black slaves, and the island's population totaled 171,420 persons. By 1841, only 42 per cent of Cuba's population were white, 15 per cent were free coloured, 43 per cent were now black slaves and the total exceeded 1 million people (Knight 1970). Elsewhere in British and French colonies, the new demographic profile was even more racially polarised, with 15 per cent a common proportion of white planters, yeomen farmers and merchants, black African slaves constituting between 60 and 70 per cent, and small numbers of free coloureds making up the remainder (Lowenthal 1972). Small islands like Barbados and Antigua soon had African-Caribbean majorities of 90 per cent, and even among the larger West Indian islands like Jamaica, an African-Caribbean majority emerged. Puerto Rico and the Dominican Republic, on the other hand, did not experience such a dramatic racial transition. By the time of the Emancipation (1834 in British colonies, later in others) some of these plantation societies, Barbados, Antigua, St Kitts and Nevis, had some of the highest rural population densities in the world. For example, Barbados' population density averaged 280 slaves per square mile in 1684, but by 1851 this had risen to 817 slaves per square mile.

the New World, such as Santo Domingo, Veracruz and Cartagena. The British, active in both the slave trade and the intensive use of slave labour, shipped slaves to their West Indian plantation colonies from West African ports. The French and Dutch colonists and planters also relied on the large-scale trading of slaves; so that in all, as many as 14 million Africans came across the 'Middle Passage' as a forced transatlantic diaspora of lasting significance. Other parts of the respective colonial empires were continually being scoured for indentured labour, or domestic servants, but demographically this phase of 'forced' importation of Africans brought about changes in regional population concentrations that were to persist to the present: namely, the overwhelming dominance of Afro-Caribbean racial majorities (Box 2.1).

Cultural diversity in the Caribbean insular region

Throughout the fifteenth and sixteenth centuries the slave trade and the sugar plantation economies were so profitable that other European colonial powers successfully challenged Spanish hegemony. Buccaneers and privateers plundered the Caribbean's 'Spanish Main'. French and British navies fought pivotal maritime battles, their armies built strategic fortifications and enlisted the services of indigenous militias, and their politicians traded the dominions, haggling over terms that kept the colonial administrations in a constant state of insecurity. The island archipelago of the Lesser Antilles was one such battleground, the Caribbean coasts of Central America another region of British colonial expansion and mercantile control. Earlier recruitment of European yeomanry to settle and farm these island possessions had relied on land and property enticements, then arranged deportations, and encouragements of persecuted religious minorities. These efforts brought several ethnic minorities – Welsh Royalists, Dutch Jews, Syrians (Levantines) and Madeira Portuguese – to the region. When plantations demonstrated their profitability, however, buying easily replaceable labour from Africa proved to be the Caribbean's answer.

The plantation economies of French, British, Dutch and Spanish colonial possessions in the Caribbean flourished for the next one and a half centuries, producing 'brown gold' (sugar cane) with 'black gold' (slaves). This brought enormous profits to the European colonialists, to slavers, the planters, the Crown governments and administrations, and the merchant classes. Not surprisingly, the environmental consequences of this wholesale transformation of small island landscapes to cultivable acreages were dramatic, and eventually (inevitably) many of the smaller islands' ecological systems were extensively degraded as a consequence (Watts 1987). There was, however, plentiful land for plantation expansion elsewhere in the Caribbean, and newer plantation development in Trinidad, the Guianas, Santo Domingo and Puerto Rico consolidated the need for more supplies of African slaves. Although the British outlawed the slave trade in 1807 and called for the abolition of slavery in 1834, Caribbean plantations still needed labour. With many of the ex-slaves intent on moving 'off the plantation', or 'off the island' if necessary, the newer plantations of the Guianas and Trinidad were able to recruit them, thereby initiating inter-regional migration streams from small islands to these larger territories (Richardson 1983). The plantations, however, needed more labour than could be provided by such inter-regional mechanisms.

Following the abolition of slavery, and after a brief experimentation with 'apprenticeship' arrangements, plantation owners and island administrations turned to the recruitment of indentured labourers to solve their labour shortages. The region therefore experienced another wave of mass immigration, this time from Asia. China was the largest source region, followed by India. Chinese men, and later women, were brought into Cuba and several British territories, Jamaica, British Guiana and Trinidad, to work as plantation labourers. Although mainly recruited for field labour, many of the Chinese who took up their 'indenture' option to remain rather than be repatriated gravitated to urban commerce, such as 'dry goods' businesses, laundries and restaurants. Approximately half a million Indians undertook indentured contracts to work in the plantations of Trinidad, Jamaica and British Guiana. Remaining behind were approximately 150,000 East Indians (so they were called) in Trinidad, 21,500 in Jamaica and 240,000 in British Guiana. Later, but under similar indenture contractual arrangements, 20,000 'Hindustanis'

Plate 2.1 Windmill at Morgan Lewis, Barbados, 1991
Source: *Dennis Conway*

Plate 2.2 Four Square cane factory, Barbados, 1967
Source: *Dennis Conway*

(also from India) and 30,000 Javanese (from that Dutch colony) were brought to the newly opened plantation estates in the coastal lowlands of Dutch Guiana. Unlike the Chinese, most of these Indians stayed in rural surroundings after their indentured contracts expired, some continuing to serve as plantation workers, others forming a small farmer sector.

Indenturing was not the only labour-recruitment strategy, however. There was continued recruitment of immigrant minorities from southern Europe and the European mother countries. In the Spanish colonies of Cuba, the Dominican Republic and Puerto Rico, administrations and planters made efforts to encourage white immigration from Europe, in large part fearing the loss of cultural dominance and the demographic threat of an African-Caribbean racial majority. Of the three, Puerto Rico was never really transformed into a plantation economy, the Dominican Republic also remained white and feudal, and only Cuba encouraged Chinese immigration as a counter to the black presence (Patterson 1978). Elsewhere among British colonies, there were attempts to attract ex-slaves from North America, and even some free-African colonisation was promoted and accomplished (Roberts 1954).

There were, then, successive waves of immigration that diversified the cultural mosaics of many Caribbean island populations. To this day, these minorities, as well as the varied (and numerous) colonial histories – British, French, Dutch, Spanish, Danish – and United States' 'neo-colonialism', provide rich cultural diversity to these Afro-Caribbean societies (Boswell and Conway 1992).

Box 2.2: Haiti's revolutionary beginnings

The turbulent history of Haiti's plantation society and its 'war of national liberation' against French colonialism, which resulted in the formation of an independent state in 1804, was, of course, a rejection of slavery of a different magnitude. Haitian elites, as well as revolutionaries, had long-established ties with the fledgling colonies of the United States. Connections between Haiti and Louisiana were strong, and the bloody conflict precipitated a flood of planters and their slaves, free coloureds and black *affranchis* to North America, some emigrating for good, others leaving for precautionary reasons. Within Haiti, many ex-slaves fled the plantations, preferring the freedom of peasant small holding in remoter regions to plantation labouring (LaGuerre 1983). With the war leaving a destroyed infrastructure and depleted workforce, the ensuing strife, instability and social polarisation within Haiti was in part an internal struggle between black and mulatto races, and in part an external struggle against external forces opposed to emancipation and opposed to the existence of this independent black nation-state with its revolutionary message (Beckles 1993). These problems would persistently influence internal dislocations as well as Haitian refugee flight and emigration to external destinations.

Later, international movement would be further promoted by 20 years of US occupation and administration of Haiti from 1914 onwards. This prompted the next distinctive wave of Haitian international mobility: international circulation across the border to work in the sugar cane fields of neighbouring Dominican Republic, or nearby Cuba (Perusek 1984). Most of this Haitian migration was contracted by US companies in Cuba and the Dominican Republic, many (perhaps two-thirds) returned, so this was the beginning of that people's international circulation tradition as well as a considerable amount of 'emigration leakage' as Haitians stayed behind (see Conway (1988) for a fuller explanation of this strategy alteration), and added to the Afro-Caribbean minorities in these other Caribbean destinations.

Evolution of migration strategies in the Caribbean

The purpose of this second section is to provide a historical geography account of the structural conditions and resultant people's behavioral adjustments that prompted the initiation and evolution of migration as a fundamental adaptive strategy for all classes of Caribbean people. The socio-historical experiences during sequential phases of Caribbean migration and circulation demonstrate that emigration was an induced (forced) survival strategy for the rural under-classes: clearly a legacy of the plantation's coercive authority. For others with more resources (and freedom), international mobility was already an adaptive strategy. Since most came as immigrants, or grew up in immigrant families, resort to emigration or return migration could be expected if hardship and failure in the Caribbean occurred. Members of elite families (the plantocracy) assumed it was their birth-right to 'circulate' between home – meaning their European residence or landed estate – and the colonial residence. For this class, schooling and training, marriage opportunities, careers in the military, clergy or the professions, were to be pursued 'back home' in Europe.

The eighteenth century witnessed the genesis of international connections between the Caribbean and the North American/United States mainland, in terms of both capital investment circulating and people migrating. At the very beginning of the plantation era, the demise of white yeoman small farming in the West Indies had prompted the more adventurous to emigrate to North American colonies to start again. Later, some successful planters expanded their family properties to start mainland plantations. For example, Barbadian planters settled and prospered in South Carolina, and Barbadian merchant families established links with Baltimore. Louisiana–Santo Domingo (Haiti) connections were entrenched from their earliest French colonial beginnings, and migration and circulations between Cuba, Santo Domingo and Louisiana further enmeshed these Caribbean and Gulf coast societies in a common cultural heritage – part Creole, part Hispanic, part Arcadian. The 'triangular trade' of sugar, manufactured goods, foodstuffs, timber, cotton and tobacco, which involved merchant houses with interests in Britain, its North American colonies and the Caribbean, cemented transnational and transatlantic mercantilist linkages. It also served planters' interests and furthered regional interconnections, thereby deepening and entrenching the dependent relations of the Caribbean with external forces and influences. The wealth that was created in the region was always circulating out of it (as Mintz (1985) and Eric Williams (1970) have persuasively argued), and the region's people were to respond likewise, as conditions worsened.

First, wealthy elites and merchant classes led the way, with many cutting their losses as conditions deteriorated, crops failed, banks foreclosed, or hurricanes and floods (and the occasional volcanic eruption) destroyed their island livelihoods. Through time, emigration became a volitional livelihood strategy for the poor, undertaken in response to ecological and social pressures the under-classes suffered in their under-developed Caribbean colonies. For many, overseas journeys became an intrinsic aspect of family histories, with tales of successes, bold adventures and adjustments to strange lands always tempered and upheld by the strength of island ties and tales of successful returns (Sutton and Makiesky 1975; Chamberlain 1995).

The initiation and establishment of migration 'off the island' as a small island tradition

During slavery, escaped slaves had either fled into the interior of larger island territories or had fled the island. For the latter, blacks working on the inter-island schooners had participated in such freedom flights. Ex-slave, '*maroon*' communities remained distinct and separate in the remote interior of Jamaica. The demonstration effect of successful escapes served to strengthen the wish to escape the plantation, or the island, if possible. Emancipation offered freedom to pursue this option.

1834, the year of the abolition of slavery for British colonies, signalled the beginning of a large-scale emigration of slaves, and free coloureds and blacks from the smaller, already ecologically stressed Leeward Islands – St Kitts, Nevis and Antigua – to the relatively undeveloped colonial territories of Trinidad, Guiana and the Windward Islands (Richardson 1983, 1989). Planters in these newer territories also sponsored recruitment campaigns in efforts to replace their own fleeing (ex-slave) workforce. For example, of the over 22,000 slaves in Trinidad plantations, only 10,000 remained after Emancipation. As a result, these recruitment efforts, often facilitated by blacks working on inter-island schooners, meshed with ex-slave hopes and thousands willingly left St Kitts, Antigua and Barbados to get 'away from their plantations'. Between 1835 and 1846, as many as 19,000 persons from these small British plantation colonies migrated to Trinidad and British Guiana. In Trinidad alone, the immigrant labour force of small islanders quadrupled in 40 years, from 11,717 persons in 1886 to 46,900 persons by 1901 (Ramesar 1976).

Labour shortages were not the only problems facing Caribbean plantation economies. By 1850, many of the small islands' ecological and economic systems were in steep decline; soil fertility loss, soil erosion and denudation were widespread. Estates were being abandoned, and droughts in 1844, then again in 1863 and 1865, brought crop failures and ruin. By 1865, Jamaican merchants who had acquired some wealth were leaving for Australia (Thomas-Hope 1978), and some British West Indian colonial administrations began to consider emigration as an appropriate 'safety valve' for their depressed masses (Sutton and Makeisky 1975).

However, while administrations of land-rich colonies such as Trinidad and the British and Dutch Guianas favoured overseas recruitment and labour immigration, the planters of affected small islands like Barbados were definitive in their opposition. By 1844, Barbados was already densely settled, with 740 persons per square mile. Indeed, it had higher rural population densities than the British Crown's South Asian colonies, India and Pakistan. Sensing the potential loss of their seasoned labour, Barbadian planters influenced their legislature to pass laws to restrict their workers' emigration to Trinidad and British Guiana. On the other hand, Guyanese planters convinced their legislature to sponsor an immigration programme to attract seasoned Barbadian field workers. Ships' captains plying trade between Barbados and Guiana were the earliest recruiters, and many hundreds of Barbadians went to the British and Dutch Guianas to work seasonally between 1835 and 1846, after which East Indian labour proved most sufficient and inter-island recruitment unnecessary (Beckles 1990). Later waves of Barbadians would be recruited in 1863–1886 and 1920–1928, both of these latter initiatives prompted by the cessation of Chinese and Portuguese immigration to British Guiana (Rodney 1977).

Caribbean labour recruitment within the region and beyond

Calamitous declines in international commodity prices, and the failure of many merchant banks in London, New York and Boston in the 'depression' of the 1880s, left many planters bankrupt and brought about wrenching structural changes in several small islands: St Vincent and Tobago, for example. In other islands, social unrest brought on by deepening deprivation of the rural masses ripened into open rebellion, one of the most significant being the Morant Bay riots in Jamaica in 1875–1876. Between 1880 and 1924 all the Caribbean's plantation economies became severely depressed and island populations suffered deepening impoverishment. Several colonial administrations advocated or permitted 'safety valve' emigration, though few actually enacted this politically sensitive policy. Voluntary emigration and circulation was an option taken up by the more fortunate. Wealthier traders and businessmen fled the hard times using the improved steamship services to New York, with Jamaicans, Barbadians and Trinidadians emigrating to New York and Boston. Others moved with the help of their ethnic networks to havens in less hard-hit Caribbean territories. Emigration to the colonial mother country for the more privileged elite classes was always an option, and sons and daughters were duly shipped off to be educated in European institutions, as befitting their social station in life.

'International circulation' via short-term labour contracts was the only available opportunity for the impoverished and disenfranchised rural poor in the British West Indies. European and United States' companies recruited British West Indian men, and an appreciable number of women also, to labour in fruit and sugar plantations, build railways and dig canals in Central and South America, in Cuba and the Dominican Republic, and later during the first two decades of the twentieth century, to similar construction and labouring opportunities on the United States mainland. Jamaicans had been departing for Central America since the mid-1850s, when up to 2,000 men went to Panama in connection with the construction of the trans-isthmus railway, and others went to Costa Rica to work on railroad construction for the American Fruit Company (later to become United Fruit) (Wilson 1947) (Box 2.3).

Cuba and the Dominican Republic were favoured by United States duty preferences on their sugar in 1895. Accordingly, the downturn in the British West Indian sugar economies, in addition to the upsurge of US corporate investment, initiated waves of circulating black migrant labour into these Spanish-speaking 'US dependencies'. As many as 121,000 Jamaicans laboured in Cuban sugar fields and factories between 1902 and 1932. They were joined in this massive recruitment campaign of over one and a half million transient workers by other nationals. In addition to the 25 per cent from Jamaica, 60 per cent were from Spain, 6 per cent from Haiti, 3 per cent from the United States and 2 per cent from Puerto Rico (Patterson 1978). In addition, thousands of Leeward Islanders – Kittians, Antiguans, Nevisians and Montserratians – travelled as 'deckers' for

Box 2.3: 'Silver men': West Indians in Panama

Jamaicans, along with Martiniquese and Guadeloupeans, formed the bulk of contract labour in de Lessep's unsuccessful venture to build a French Panama Canal during the period 1881 to 1889, but appreciable numbers returned from Panama on the project's demise (Proudfoot 1970).

Later, when the United States government purchased the canal and railway projects from the defunct French Universal Inter-Oceanic Company, British West Indian men and women were heavily recruited, again. In Barbados there was considerable wariness on the part of Barbadians, who viewed Panama as a 'workers' graveyard'. However, the recruitment skills of William Karner, an American employee of the Isthmian Canal Commission and of S. E. Brewster, a Bridgetown businessman, won them over, with initial reluctance becoming transformed into infectious enthusiasm. Once reinforced by the return of the first recruits with their 'Panama money' and tales of success, many Barbadians overcame their reluctance to travel and international circulation on labour contracts became an island-wide phenomenon (Richardson 1985). It is estimated that between 45,000 and 60,000 Barbadian 'silver men' circulated and laboured in Panama between 1805 and 1914, when the Canal was finally completed, and one estimate of their death toll was 20,000 persons dying between 1906 and 1912 (Thomas-Hope 1978). Jamaicans, Haitians and other British West Indians joined them, many choosing to travel to Panama without contracts in the hope they would find work once they arrived. Even after contract work on the Canal had ceased in 1911, British West Indian men and women still bought, or worked, their passage to Panama. An unknown number among this reciprocal stream opted to remain in Panama after their contracts had expired, taking up labouring and starting small businesses in the Canal Zone cities of Colon and Panama City. Others also chose not to return to the islands but moved on, taking up labouring contracts in railway construction, forest harvesting and fruit plantations elsewhere in Panama, Nicaragua and Costa Rica (Proudfoot 1970).

temporary jobs in the US/Gulf & Western Corporation's sugar plantations in the Dominican Republic (Richardson 1989).

Meanwhile, a specific Cuba-to-Florida migration stream had evolved in the late 1800s, as Havana cigar manufacturers and their workforces relocated to the US mainland to penetrate the high tariff wall the government had imposed on imports. Key West, and later Tampa, became thriving cigar manufacturing centres, and Tampa soon served as the home for a Cuban exile group, who in the 1890s were as preoccupied with the cause of independence and '*Cuba Libre*' as their revolutionary counterparts on the Caribbean island. Gradually, these emigré communities assimilated and they became a Floridian community, rather than representatives of a 'government in exile' (Pérez 1978).

US Virgin Islanders and Bahamians circulated to Miami in the first decade of the twentieth century, joining the construction crews that helped build that new gateway metropolis in Florida. Middle-class Jamaicans, Barbadians, Trinidadians and wealthier small islanders circulated to West Indian communities in New York and Boston, with considerable emigration occurring as the return circuit was postponed indefinitely and the intended repatriation forgone in favour of staying in their new country of adoption. Migration as a livelihood option for all classes of people was becoming embedded in the social fabric of the islands, both among those who stayed behind, those whose family and kin had left, and among those whose were journeying 'away from home'.

The period 1924–1940 was one of continued hard times and limited opportunities for the Caribbean masses. The Caribbean economic state remained grim and there was another crash in world sugar prices in 1921. Coincidently, anti-immigrant, 'nativist' sentiments were heightened in the United States, in large part due to the massive influxes of 'new immigrants' (from Italy, Greece, other Southern and Eastern European regions, and from China) that the country had experienced during the previous two decades. Accordingly, the US National Origins Law (Johnson–Read Act) of 1924 promulgated a discriminatory national quota system for Asians and Caribbean peoples – limiting the number of immigrants from any country to an annual quota based on the number of nationals already in residence in the USA in 1920.

Elsewhere in the Caribbean Basin region, restrictive immigration legislation in Venezuela, Cuba and the Dominican Republic also discriminated against previously welcomed 'black visitors'. Foreigner repatriation was to be encouraged and Haitians, Jamaicans and small islanders returned to swell the ranks of the unemployed back home. The only labouring opportunities available in the region were in oil-related industries in the Netherlands Antilles (Aruba and Curaçao) and Trinidad. Small islanders from places like Grenada, Barbados, Antigua and St Kitts circulated to that employment opportunity. All too soon, the two Netherlands Antilles also tightened their immigration policies and, accordingly, many of these temporary workers were also repatriated in the early 1930s (Marshall 1982). Other small islanders continued patterns of circulation wherever they could find opportunities: inter-island trade and commerce, working passages and joining the merchant marine, often as not relying on networks provided by previous emigrants.

Colonialism, post-colonialism and migration to 'mother countries': emigrés and expatriates, migrant return intentions, refugee flights

The coming of the Second World War (1939–1945) started another international labour recruitment drive in the American hemisphere. The United States' entry into the war in 1941, its burgeoning labour needs in its agricultural sector and service industries and other non-essential industries as its men went to war, made it a magnet for overseas recruitment of contractual labour. '*Braceros*' were recruited from Mexico as farm labour, and Caribbean cane cutters joined this programme as 'B-1' agricultural workers. For many of the unemployed in the Caribbean, circulation via short-term labour contracts to the United States was again possible under the US War Manpower Act, although all too quickly the door would be slammed shut again in 1952! In addition, the Second World War saw many colonial regiments 'doing their duty' in the European theatre, and the demobilisation of these war veterans would soon lead to a restructuring of colonial ties, to the growth of radical social movements and the onset of decolonialisation.

Predictably, the post-Second World War period from 1940 and 1965 was one of mass emigration from Caribbean colonies to a variety of European 'mother countries' – Britain, France and the Netherlands. Denied opportunities in nearby North America by the US McCarran–Walter Act, and driven by their longstanding colonial ties (and 'colonised mentality' as Franz Fanon (1967) would depict it), British West Indians responded to London Transport and British government Health Service recruitment and shipped off to Britain in their thousands. Between 1955 and 1959, West Indian migrations to Britain were in the order of 20,000 to 30,000 per year. Some of the smaller Leeward and Windward Islands actually experienced depopulation due to this exodus. Montserrat lost between 5 and 10 per cent of its inhabitants each year, and Carriacou, a small ward island of Grenada, lost approximately 20 per cent of its residents (Lowenthal 1972).

Thousands of West Indians emigrated, exercising their right to take advantage of employment and educational opportunities in mother Britain. This long-distance emigration of British West Indians to their mother country was a considerable outflow. It consisted of families joining their parent or parents who had preceded them as well as single, young men and women. Many were drawn from the rural and urban lower classes, but an appreciable number of middle-class West Indians emigrated to take advantage of the situation. Many intended to return, retaining a 'return ideology' in which they had every intention to not permanently sever ties with their Caribbean homes (Philpott 1973). Some did return, but significant numbers stayed to become 'black British' (Peach 1986, 1996).

Emigration with an intention to return was also obligatory among colonials from French Départmentes d'Outre Mer (Martinique, Guadeloupe, St Martin) to France and from the Netherlands Antilles (Aruba, Bonaire, Curaçao, St Maarten, St Eustatius) and Dutch Guiana to Holland. Prior to the mid-1960s, these modest flows were selective circulations and emigrations of elites and youth from the small professional and middle classes of these European colonial islands. Privileged positions beckoned many back, but others became acclimatised to European ways and sought opportunities and careers there, sometimes fostering a mentality of metropolitan superiority among these emigrés (Searle 1984).

In North America, a similar 'colony-to-mother country' circuit sprang to life as Puerto Rican hopefuls (and desperates) sought opportunities on the mainland. Although Puerto Ricans had possessed rights of legal access to the USA since the turn of the century, only self-selective streams of the urban working classes with sufficient resources opted for emigration to New York during the first two decades (Sanchez Korrol 1983). Mass emigration to the mainland began after the 1930s, when the collapse of the country's sugar industry and an increasing rate of unemployment that accompanied rapid urbanisation triggered international mobility 'off the island' (Hernandez Alvarez 1967). The reversal of trends was quite dramatic. Between 1880 and 1946, 100,000 Puerto Ricans left for the United States, although many chose to return home after working on the mainland (Senior 1947). Between 1955 and 1970, approximately one-third of the entire Puerto Rican population circulated and/or emigrated to the mainland. Most (approximately 70 per cent) made for New York city, so that by 1970, almost 1.5 million Puerto Ricans were recorded as resident in the United States, with over 800,000 living in New York city (Dominguez 1975; Boswell 1976). By 1980 the resident Puerto Rican population on the mainland was estimated at 1.8 million (Boswell 1985). Temporal fluctuations of emigration volumes were, however, considerable. For example, approximately 470,000 left the island in the 1950s, only 212,000 left during the 1960s, and during the 1970s the net loss dropped to 41,000 (Sandis 1970).

San Juan, the rapidly growing capital of Puerto Rico, was the transit stop for rural *Riquenos* on their way to 'America', and the international circulation between island and mainland was made even easier with the advent of less expensive jet airline travel. Women followed their husbands to seek mainland employment as 'tied movers', but on later visits a growing number sought employment opportunities for themselves (Conway *et al.* 1990). Women also often returned to the island for family reasons, to provide their daughters with an appropriate Catholic upbringing, to care for sick relatives, or to return to their family home (Bailey and Ellis 1993). Families too returned, and many Puerto Rican families came to view their transnational livelihood of split and dispersed families as inevitable, or manageable – maintaining their island home while extending their familial 'space' to incorporate a New York accommodation and a New York life (Ellis *et al.* 1996). Although late-comers, Puerto Ricans embraced this Caribbean tradition of international mobility as a livelihood, which other islanders had held since post-Emancipation times.

In Cuba, political repression and the coup by Batista in 1952 instigated another exodus of would-be revolutionaries to set up an 'opposition in exile' in Miami. Between 1951 and 1959, approximately 10,000 Cubans moved to the mainland and naturalised. A much more important politically motivated exodus occurred, however, after Fidel Castro's takeover in 1959 and with the hardening of his government's ideology from a nationalist-socialist leaning to a communist-centrist institution (Boswell and Curtis 1984). Between 1959 and 1962, those among the wealthier middle classes who opposed Castro's ideological turn to the left fled Cuba. They were labelled as 'golden exiles' because they fled with their wealth, and were over-represented in professional and managerial classes. Perhaps as many as half a million 'escaped', first on regular commercial flights, later on special flights and ships provided by the Cuban-Miami community and also by the US government; 6,000 political prisoners and their families from the failed 'Bay of Pigs' invasion were exchanged between governments, and others left for Mexico, neighbouring Caribbean countries and Puerto Rico.

The October 1962 missile crisis and resulting US blockade interrupted the refugee flow until September 1965, when the Camarioca boat-lift helped another 5,000 flee to political asylum in the United States (Moncarz and Jorge 1982; Bach 1987). For other Caribbean hopefuls in the region, the Immigration and Nationality Act of 1952 (McCarran–Walter Act) reaffirmed the 1924 discriminatory quota system, and it was not until 1965 that Caribbean people, other than Puerto Ricans and Cubans, would be able to seek overseas opportunities in the United States.

Caribbean–North American migration networks: 1965 to the present

Prior to the 1970s, the Caribbean was first a region of significant immigration, then of selective emigration streams and of continued inter-regional immigration. Initially, European mother countries received immigrants from their colonies, but eventually North American hegemony held sway. The global restructuring of the world's economic order, which began in the 1970s, gathered pace through the 1980s and imposed draconian neo-liberal policies in the 1990s and early 2000s, brought about changes in these population trends. Wholesale indebtedness, recessional conditions, increasing polarisation of classes, declines in standards of living – to name a few of the structural realities the region has suffered during the last 20 years – have prompted more transnational circulation and the formation of more multi-local networks. Emigration has again replaced immigration as the region's defining demographic process, although inter-regional immigration is not an unknown problem for a few small islands whose economic fortunes have held up – the Cayman Islands and the Bahamas, for example.

There has been continued economic hardship for many, further exacerbated in some islands by outbreaks of civil unrest and violence, by natural disasters and ecological calamities, which have initiated several streams of emigration and refugee flight. International circulation and more short-term, temporary movements to neighbouring or more distant places has become a common strategy for many regardless of their skill levels, and the well-entrenched intra-regional and extra-regional networks facilitate such sojourning patterns. There has also been a widening of fields of opportunity for more educated and highly skilled Caribbean islanders, who have utilised their opportunities in time-honoured ways, drawing upon their flexibility, their resourcefulness and initiative, even their cultural capital (musical and athletic talents, for example) to seek livelihood options in other parts of the Caribbean, in North America and Europe or farther afield in the globalising, international marketplace for skilled professionals. How this evolution of Caribbean–North American/European transnational networks came about is the final piece of this socio-historical record.

In Britain, the 'invasion' of 'New Commonwealth' immigrants from the West Indies, Pakistan and India in the 1950s eventually aroused racist sentiments and fears. Quite abruptly, 'black immigration' was terminated in 1962 by a restrictive and discriminatory Commonwealth Immigration Act, which all too effectively stemmed the West Indian influx. Revised Canadian immigration legislation in 1962, on the other hand, removed racial discriminatory biases in that country's entry requirements, opening another route to North America for Commonwealth Caribbean immigrants. The 1965 US Immigration and Nationality Act replaced the national quota system with hemispherical ceilings, increased the regional ceiling for the Caribbean, Central and South America to 120,000 persons per year, and provided a set of preference categories and tightened labour certificate requirements. Later in 1976, the US Congress legalised a seven-preference 'family reunion' system of entry for petitioners from Western hemisphere countries, which allocated ceilings for individual countries and enacted favourable conditions for entry of relatives and families of US residents. When the 1978 and 1980 Immigration and Nationality Act Amendments further extended the entry ceilings of this preference system and included newly independent countries in their hemispheric reach, the US 'golden door' was truly thrown open for Caribbean emigrants and circulators.

Well-established Caribbean communities in Miami, New York, Boston, Washington, New Jersey and Hartford provided a wealth of family- and kin-based networks and linkages to support US-bound West Indian migration. Temporary trips (circulations) between island home and these mainland enclaves were open to those with sufficient financial and familial resources. Consequently, legal (up to six months) sojourning and unauthorised overstaying using entry on visitors' visas became common practices. Beginning in the mid-1960s, the steady flow of professionals and youth from the Dominican Republic to New York city (America) rapidly grew to mass circulation and emigration between 'island and city'. By 1980 a net estimate of Dominicans resident on the mainland was 400,000, most in the Washington Heights area of New York.

Prior to the 1970s, professional advancement, alienation with island society, social disgrace and unwillingness to conform had encouraged emigration of a self-selected group of highly educated individuals from the middle and elite classes to either mother countries or Canada and the USA. This propensity translated into mass flights from repressive regimes during the late 1960s and 1970s: Haitians to French Canada; and Cubans, Jamaicans and Guyanese to the USA. These legal flights of highly trained professionals – 'enforced brain drain' – were often swelled in number by flights of middle and lower classes from Haiti, Cuba, the Dominican Republic and Jamaica, many of the latter two nationalities resorting to visiting followed by unauthorised overstaying. Cubans, on the other hand, were encouraged to flee the Castro regime, being guaranteed refugee status and residence in the United States. Poorer Haitians, denied such privileges, resorted to illegal means of passage and entry into the United States, some using the Bahamas, and others chancing their lives in unseaworthy, overburdened vessels.

Illegal entry and sojourns in neighbouring Caribbean countries increased in importance and in volume from the mid-1970s onwards. Many individuals sought opportunities in island economies that were experiencing relative prosperity due to their tourist industry, their colonial support or their oil industry, continuing a long-held tradition that uses migration as a means of escape from impoverishment in declining agricultural sectors at home. Paths were established within and through the Caribbean, and current circulation and migration follows such routes. Nevisians and Kittitians transit to Antigua, St Thomas in the US Virgin Islands or onwards. St Lucians transit to Barbados, Trinidad or onwards. The Grenada–Trinidad circuit is well entrenched, and Vincentians who also go to Trinidad, appear in appreciable concentrations in New York city. Dominicans, on the other hand, have chosen neighbouring Martinique and Guadeloupe as their preferred transit, and some find their way on to Paris and Europe via these Euro-Caribbean islands. Sometimes, routes include ports of call in San Juan, Puerto Rico, Miami, New York or Toronto: metropolitan locales that may be terminals or transits. Retaining the widest set of options possible, Caribbean migrants anticipate or at least do not reject the possibility of return to a Caribbean home. While in the USA – 'that other man's society' – many prefer to retain their Caribbean citizenship. Others take US citizenship to 'sponsor' further migrations of kin and family.

In the United States, passage of the 1986 Immigration Reform and Control Act (IRCA) granted amnesty to 2 million 'unauthorised' alien residents and refugees; particularly Haitians, Dominicans and Cubans. Eventually, through the early years of the 1990s, they would be granted their 'green cards' and permanent residency status (see Table 2.1). Not surprisingly, the 1980s witnessed the largest volumes of visiting and resident entries to the United States from the Caribbean of any decade in the twentieth century. Then, the 1990 Immigration and Naturalisation Act Amendments increased admission quotas for highly skilled young women and men in 'under-supplied' occupations, e.g. nursing, medical technicians, which opened the door to the recruitment of health services technicians and practitioners, scientists and even athletes from the Caribbean.

Still, unauthorised entries troubled US officials, the differential treatment of specific streams of refugees (such as the Haitians and Cubans) troubled civil rights groups, and the downturns of the Californian economy from the mid-1980s onwards aided and abetted anti-immigrant rhetoric. During the first half of the 1990s, anti-immigrant sentiment and political posturing in California, and to a lesser extent in Florida, heightened tensions and fomented restrictive immigration policies and practices. The 1994 congressional elections brought about a dramatic change in congressional leadership, and majority sentiments concerning immigration swung to the right, hardened and became more punitive. In 1996, there was a reversal of US immigration policies towards legal immigrants. Congress passed the Illegal Immigration Reform and Immigrant Responsibility Act (IIRIRA), which enforced the rapid deportation of 'criminal' resident aliens, enacted higher penalties for 'over-stayers', restricted legal immigrants' access to welfare programmes and some

Table 2.1 US immigration volumes from selected Caribbean countries, 1995–2000 (legal admissions to US residency[1])

Origin country	Population in 1991 (1,000s)	1965–1969	1970–1974	1975–1979	1981–1985	1986–1990	1991–1995	1996–2000	1991–2000 per cent of 1991 population	1975–2000 per cent/ (pop) of 1991 population
Antigua	64	–	–	3,394	8,081	4,821	2,929	1,983	7.7%	33.1% (21,208)
Bahamas	252	–	–	2,298	2,660	4,648	3,563	3,180	2.7%	6.5% (16,349)
Barbados	255	5,944	8,034	12,021	9,046	8,076	5,366	4,101	3.7%	15.1% (38,610)
Belize	228	–	–	4,221	7,750	10,320	5,848	3,278	4.0%	13.8% (31,417)
Cuba	10,732	183,499	101,066	176,908	58,978	100,279	68,470	112,391	1.7%	4.8% (517,026)
Dominica[2]	86	–	–	2,827	2,818	3,626	3,572	1,963	6.4%	17.2% (14,806)
Dominican Republic	7,385	57,441	63,792	77,785	104,663	147,140	177,090	122,444	4.1%	8.5% (629,122)
Guyana	750	4,230	12,914	32,964	42,725	52,649	44,138	29,755	9.8%	27.0% (202,231)
Haiti	6,287	24,325	28,917	30,181	43,890	96,273	95,977	85,788	2.9%	5.6% (352,109)
Jamaica	2,489	49,480	65,402	72,654	100,560	113,245	90,731	82,808	7.0%	18.5% (459,998)
St Kitts and Nevis	40	–	–	4,019	7,096	3,146	2,730	2,106	12.1%	54.56% (21,824)
St Lucia	153	–	–	2,727	2,964	3,880	2,906	2,752	3.7%	10.0% (15,229)
Trinidad and Tobago	1,285	15,502	34,646	29,326	17,018	22,515	33,708	29,548	4.9%	10.3% (132,115)
Selected Caribbean	29,749,000									2,452,044 (8.2%)

Sources: *Statistical Yearbooks of the INS*, Annual Series, 1965–2000

[1] The 1986–1990 and 1991–2000 estimates include the 1986 IRCA 'amnesty' petitioners.

[2] The INS records of Dominican and Dominican Republic admissions are most probably inaccurately recorded, because the admission estimates of the extremely small island of Dominica are far above the proportion expected (17%). The deflation of the Dominican Republic estimates is not a serious undercount, but it is troublesome.

of their benefits, and increased surveillance at the Mexican border. Such harshness provoked a political backlash, however, and with the US economy continuing to expand in the late 1990s, the essential contributions of immigrant labour to this expansion became more and more obvious. There was a mobilisation of Latino voters in California that effectively reversed that state's anti-immigrant stance. California's anti-immigrant Proposition 187 was found to be unconstitutional and was summarily dismissed. Congressional support for increases in recruitment of highly skilled immigrants on temporary H-1B visas gained bipartisan support, while at the same time security at the Mexican border was couched more in terms of a 'war on drugs' than it was on the interdiction of illegal entrants.

Anti-immigrant sentiment was still harboured (and promoted) by the movement's diehard proponents (NGOs such as FAIR and CIS, for example), but the political climate that had supported such hardening had, by the end of the century, changed to one of tolerance and practicality. Illegal immigrants were no longer pariahs, but rather hard-working hopefuls. Unfortunately this relaxation of anti-immigrant tension was to be short-lived, however. The tragic events of 11 September 2001, and the ensuing security crisis which gripped the nation, promptly changed the 'welcoming' climate in the United States. The anti-immigrant lobby seized the moment, and ever since stronger enforcement of immigration controls has been part of the 'homeland security' lexicon.

For Caribbean hopefuls, the same (predominantly legal) mechanisms used in the 1970s and 1980s were still in play during the 1990s (see Conway 2002). Haitians were perhaps the only Caribbean people whose journeys and visits to the United States were limited, or curtailed. Dominicans continued to visit New York, and the Washington Heights area became even more firmly identified with that Caribbean society. Similarly, Crown Heights in Brooklyn is the West Indian district, where 'Labor Day in Brooklyn' is celebrated with a carnival procession and the best of Trinidad's carnival traditions, with everyone participating – Vincentians, Guyanese, Barbadians and Haitians as well as Trinidadians. Cuban refugees no longer had the privileged status they had enjoyed with previous administrations, after President Clinton 'rationalised' the State Department's policy in 1998, but the hardships of the 'special period' continued to encourage Cuban defections and refugee flights, with *balseros* (makeshift, inner-tube rafts) being the most dangerous means of attempting flight.

Elsewhere in the Caribbean, visiting patterns (non-immigrant entries) to the United States from all over the region continued to be substantial. At the same time, the numbers of unauthorised 'illegal over-stayers' in the United States from the majority of Caribbean societies was not excessive, e.g. in comparison with Canadians, Salvadorans, Guatemalans and Mexicans. Haitians, however, remained among the 'illegal top 20' in the 1990s (Table 2.2).

New York, Miami, New Jersey, Boston, even Los Angeles, have vibrant Caribbean enclave communities where US-based families welcome their Caribbean-based kith and kin. In Canada, Toronto's multicultural mosaic also has its Caribbean enclaves, and Montreal and Quebec city both sustain a distinctive Haitian

Table 2.2 'Top 20' countries of origin of illegal immigrant populations in the USA: 1988[1] and 1996[2] estimated

1985–1988		**1996**	
Origin (Only over-staying visitors)	**Population**	**Origin (Estimates of all 'illegals')**	**Population**
All countries	374,200	All countries	5,000,000
1. **Mexico**	**55,000**	1. **Mexico**	**2,700,000**
2. *Haiti*	14,300	2. El Salvador	335,000
3. Philippines	14,000	3. Guatemala	165,000
4. Poland	13,900	4. Canada	120,000
5. India	8,400	5. *Haiti*	105,000
6. *Trinidad and Tobago*	7,500	6. Philippines	95,000
7. Canada	6,300	7. Honduras	90,000
8. Ireland	5,400	8. Poland	70,000
9. Colombia	5,100	9. Nicaragua	70,000
10. Pakistan	4,600	10. *The Bahamas*	70,000
11. Nicaragua	4,500	11. Colombia	65,000
12. Ecuador	4,500	12. Ecuador	55,000
13. Italy	4,300	13. *Dominican Republic*	50,000
14. Guatemala	4,000	14. *Trinidad and Tobago*	50,000
15. *Jamaica*	3,600	15. *Jamaica*	50,000
		16. Pakistan	41,000
All others	119,600	17. India	33,000
		18. *Dominica*[3]	32,000
		19. Peru	30,000
		20. Korea	30,000
		All others	744,000

Sources: Warren (1990) and 1996 Statistical Yearbooks of the US Immigration & Naturalization Service

[1] Warren (1990) estimated 'over-stayers' for the period 1985–1988, who arrived with non-immigrant visas and stayed longer than their prescribed six months.

[2] The 1996 INS estimates of the total 'illegal' population combine CPS-derived net estimates of EWI immigrants (entry without inspection) with Census Bureau estimates of those who entered before 1982, and with net over-stayer estimates for 1982 to 1996 of the net number of non-immigrant over-stayers for 99 countries of origin derived from INS databases. The net estimate is then adjusted (reduced) by factoring in a Census Bureau 'emigration' estimate of unauthorised non-immigrants who resided in the USA in October 1988 and October 1992, and who left in the following four-year period.

[3] There is an 'identity' problem with this relatively high estimate of Dominican entrants, which stems from inaccurate INS recording of their island of birth – Dominica being confused with the Dominican Republic.

community. 'Black British' West Indian families are not left out of these transnational networks, so that some if not all Caribbean–North American–European transnational networks are multi-local, and Caribbean people's lives are embedded in these multi-local spaces and locales – at once metropolitan, at another, insular – that circumscribe their identities and circumscribe what 'home', 'away from home' and 'returning home' mean (Fog Olwig 2001). Young professionals, some of them second-generation Caribbean immigrants who have grown up in Britain or North America, are experimenting with returning to the island homes of their parents. How they use their transnational family networks and connections can only be surmised, but certainly their growing presence in islands such as Barbados and St Lucia is evidence of the ever-changing and dynamic character of Caribbean transnational migration in today's globalising world (see Potter 2003; Phillips and Potter 2003).

Twentieth-century demographic and migration processes

Demographic trends

Accompanying urbanisation, modernisation and the widely heralded success of family planning programmes in such notable islands as Barbados (Handwerker 1989), fertility declines were common throughout the region. Only Haiti and Belize maintained relatively high rates of fertility (with total fertility rates of over four births per woman) by the 1990s. The rest of the Caribbean, with the exception of Grenada (with a TFR of 3.5 births per woman) had TFRs of less than three births per woman. By 1994, Antigua and Barbuda, Barbados and Cuba had TFRs below replacement level, at two births per woman (Table 2.3). Positive health transitions also accompanied political independence, modernisation and urbanisation, and most island health care systems grew sufficiently to conquer their pervasive childhood morbidity and mortality problems. These advances brought most countries' infant mortality rates down by the 1990s, with only Haiti (at 86 deaths per 1,000 live births), Guyana (at 45 deaths per 1,000 live births), and the Dominican Republic and Belize (both at 38 deaths per 1,000 live births) lagging behind the others. Some, like Barbados, Cuba, Guadeloupe, Jamaica, Puerto Rico and Trinidad and Tobago, had infant mortality ratios below 15 deaths per 1,000 women – equivalent to, or better than, many advanced countries' death rates. Changes in men's and women's life expectancy also reflected the region-wide improvements of health service provision, and noticeable increases (reaching the low 70s) occurred everywhere between the 1970s and the 1990s. Again, only Haitians continued to experience low life expectancy levels, though there were gains – from 48 years of age in the 1970s to 57 years of age by 1994 (Table 2.3). Like the rest of the world, however, the Caribbean region will experience a dramatic ageing of its population in the next few decades of the twenty-first century, as its child-dominant population profile of the 1950s changes to one in 2010 and beyond

Table 2.3 Demographic trends in the Caribbean, 1970–1994

Country	Infant mortality/ 1,000 live births 1970–1975	Infant mortality/ 1,000 live births 1989–1994	Total fertility rate – births per woman 1970–1975	Total fertility rate – births per woman 1989–1994	Life expectancy 1970–1975	Life expectancy 1989–1994
Antigua and Barbuda	21	19	2.6	1.7	67	75
Bahamas	34	23	3.4	2.1	66	73
Barbados	33	10	2.7	1.8	69	75
Belize	34	38	6.4	4.3	59	69
Cuba	25	10	3.6	1.7	71	76
Dominica	27	17	5.5	2.4	n.a.	73
Dominican Republic	94	38	5.6	2.9	60	70
Grenada	33	n.a.	5.9	3.5	n.a.	n.a.
Guadeloupe	42	11	4.5	2.3	68	75
Guyana	79	45	4.9	2.5	60	66
Haiti	135	86	5.8	4.8	48	57
Jamaica	42	13	5.0	2.5	69	74
Martinique	35	8	4.1	2.0	69	76
St Kitts and Nevis	59	32	3.5	2.5	n.a.	69
St Lucia	60	17	5.5	3.0	62	71
St Vincent and the Grenadines	56	19	5.0	2.4	63	72
Trinidad and Tobago	50	14	3.5	2.6	66	72
Puerto Rico	21	12	3.0	2.1	73	75

Sources: *Social Indicators of Development, 1996, 1987*, World Bank and the Johns Hopkins University Press, Baltimore and London; *Statistical Yearbook of Latin America and the Caribbean, 1996*, UNECLAC, Port of Spain, Trinidad and Tobago

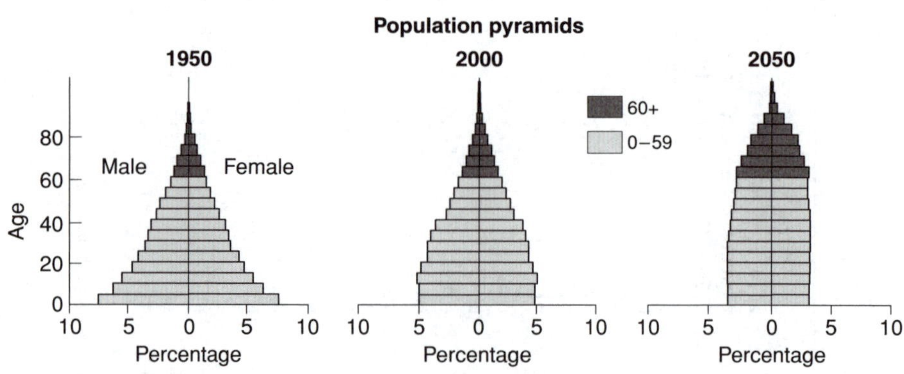

Figure 2.1 Caribbean population pyramids, 1950, 2000, 2050
Source: *World Population Ageing 1950–2050*, Population Division, Department of Economic and Social Affairs, United Nations, New York, p. 91, available at www.un.org/esa/population/publications/worldageing19502050/pdf/022carib.pdf

where the over-50-years-olds predominate (compare the population pyramids in Figure 2.1)

In studies of the effects of long-term emigration on Caribbean fertility decline (Ebanks *et al.* 1975; McElroy and de Albuquerque 1986; McElroy and Radke 1988) it has been demonstrated conclusively that emigration not only reduced population growth directly but also did so indirectly by depressing birth rates. Whether short-term sojourning or longer-term over-staying beyond a year (according to the UN definition of an 'emigrant') are equivalent in terms of their demographic impacts on Caribbean source societies has not been researched. Blake (1955) has suggested that overseas sojourning would theoretically impact fertility levels negatively, according to her version of the 'migration as disruption' hypothesis (see also de Albuquerque *et al.* 1976; Massey and Mullan 1984). Evidence is less definitive on the positive influences Caribbean intra-regional immigration may have had on the fertility regimes of receiving societies in the region (McElroy and Radke 1988).

Male selectivity in the earlier emigrations from many small islands impacted resident sex ratios to the extent that the resultant female surpluses are believed to have contributed to the special family relational structures found in Caribbean societies (Lowenthal 1972; Goossen 1976; Laguerre 1978). Edith Clark's (1957) classic treatise on the matriarchal structure of Caribbean family life, *My Mother Who Fathered Me*, succinctly personified this gendered character of Caribbean populations and family structures. Accompanying modernisation and economic diversification of island economies, declines in infant mortality and fertility have also influenced women's lives and options – positively in many cases. Women's achievements in many Caribbean societies have been considerable, and are worth noting (see Table 2.4 and Chapter 5). The lack of space in this chapter precludes a more comprehensive assessment of women's advancements and achievements in the contemporary Caribbean. Table 2.4 provides summary statistics of some

Table 2.4 Women and capabilities in the Caribbean, 1989–1995

HDI rank – country	Human Development Index 1994	Education – Female net enrolment (%) 1989–1994	Life-course – Female life expectancy at birth 1994	Economic roles – Female administrators and managers 1990	Economic – Female professional and technical workers 1990	Politics – Women in government at ministerial level 1995	Women in government at sub-ministerial level 1995
25. Barbados	0.907	80	78	37%	52%	33%	24%
28. Bahamas	0.894	95	77	26%	57%	20%	38%
29. Antigua and Barbuda	0.892	n.a.	n.a.	n.a.	n.a.	0%	47%
40. Trinidad and Tobago	0.880	78	75	23%	53%	20%	10%
41. Dominica	0.873	n.a.	n.a.	36%	57%	8%	39%
49. St Kitts and Nevis	0.853	n.a.	n.a.	n.a.	n.a.	10%	28%
54. Grenada	0.843	n.a.	n.a.	32%	53%	10%	24%
56. St Lucia	0.838	n.a.	n.a.	n.a.	n.a.	8%	0%
57. St Vincent and the Grenadines	0.836	n.a.	n.a.	n.a.	n.a.	10%	50%
63. Belize	0.806	42	75	37%	39%	0%	14%
83. Jamaica	0.736	70	76	n.a.	60%	6%	16%
87. Dominican Republic	0.718	43	72	21%	50%	3%	16%
104. Guyana	0.649	58	67	13%	48%	11%	21%
156. Haiti	0.338	13	56	33%	39%	17%	11%
All LDCs			52	9%	24%	8%	6%
Industrial countries			78	27%	48%	12%	13%

Source: *Human Development Report, 1997*, United Nations Development Program and Oxford University Press, New York and Oxford, Tables 10 and 11; *Social Indicators of Development, 1996*, World Bank and the Johns Hopkins University Press, Baltimore and London

selective aspects of Caribbean women's progress to date. Further treatment is provided elsewhere (see Chapter 5).

Among the smallest of the region's islands – Montserrat, the Grenadines, St Eustatius and Saba, for example – the dual threat of 'depopulation' and 'ageing' have been cited as consequences of continuous emigration or mass exodus processes (Lowenthal and Comitas 1962). Currently, Caribbean populations are ageing and becoming more urbanised, and their overall rates of growth are no longer excessive; the exception is Haiti, which is still predominantly rural. Island population growth is no longer the problem it was thought to be. There is plentiful evidence that emigration and low rates of natural increase have ameliorated the problems of population pressure, and have stabilised the demographics in most island systems' 'carrying capacity' equation, with the exceptions of Haiti, and possibly the Dominican Republic. That said, the emerging demographic problems of the twenty-first century will be the ageing of island populations and increasing incidences of 'diseases of affluence' – high blood pressure, heart disease, strokes and attacks, obesity, among others (Sinha 1988; Stolnitz and Conway 1991) (see also Chapter 5).

Rural-to-urban migration and spontaneous settlement in Caribbean cities

Today, mobility/migration effects are far more significant influences on internal population growth and transformation patterns than differential rural/urban fertility or mortality rates. Virtually every Caribbean country or territory has experienced migration of its rural population to urban centres, particularly its capital city. In 1960, for example, only 38 per cent of all West Indians lived in cities. By 1987, this proportion had increased to 54 per cent, and by the year 2000 had grown to 65 per cent (Boswell 1989; United Nations 1998).

The evolving patterns of urbanisation throughout the Caribbean are examined elsewhere (see Chapter 7). Here, we explain the processes and patterns of cityward internal movements and subsequent intra-urban movements of the masses who contributed to the rapid growth, area expansion and uncontrolled sprawl common in places like Kingston, Jamaica, San Juan, Puerto Rico, Santo Domingo, the Dominican Republic, and Port of Spain, Trinidad. These primate cities did not have the physical infrastructure and sufficient housing stock to accommodate the rapid urbanisation of the post-1960s period, so suburban and capital city area expansions accompanied the influxes of population. Thus, surrounding hinterland areas accommodated the bulk of in-migrants to the Caribbean's primate cities. This sprawling conurbation pattern also resulted from government policies on island-wide dispersal strategies for low-income housing and road-building programmes (Potter 1989).

Suburban relocation was also underway among the wealthier classes, in part a consequence of their desires to distance themselves from the overcrowded downtowns, and in part a consequence of rising incomes and rising expectations to seek the status provided by such a 'metropolitan' residential move. Such suburban expansion of residential subdivisions was also accelerated by spiralling land and housing market prices fuelled by speculation and excessive inflows of local and external finance capital (Potter 1989). The ensuing patterning of the social geography of several of these major Caribbean cities was found to follow sectoral divisions according to class and socio-economic differentiation (Clarke 1975; Duany 1997), but the persistence of ethnic, racial and immigrant diversity among the low-income masses must also be noted (Clarke 1984; Conway 1989a).

Although the original 'Turner model' was developed as a generalisation for Latin American urban settlement (Turner 1968, 1969), an alternative model of the low-income migrant's intra-urban itinerary for Caribbean cities has been developed by one of the authors (see Conway and Brown 1980; Conway 1985). The model considers both the processes and patterns of intra-urban relocation, the initial and subsequent geographical routes taken into and through their urban environment, and the urban social-spatial structure as it changes and evolves through phases of sustained urbanisation. This alternative Caribbean model builds upon Turner's ideas concerning three basic priorities for the migrant's urban accommodation that conditioned their intra-urban itineraries – accessibility, security of tenure and amenity considerations. *Accessibility* refers to the changing relative locations of residence and workplace. *Security of tenure* refers to the migrant's concern for consolidating their livelihood status in the urban environment by investing in home ownership as a more resistant and flexible mode of survival in a still-vulnerable urban world. *Amenity considerations* initially will be the bare essentials of their day-to-day existence that are necessary to provide shelter and sustenance. A wider set of amenity considerations can surface once economic (job) security is obtained, to become a priority for dwelling improvements and improvements in the residential environment for the migrant's dependent family.

In addition to these dwelling-space priorities, however, the impact of group affiliations, with or without the reinforcement of kinship ties, and the influence of 'chain migration' recruitment of family and kin also affects the relocation decision making of subsequent city-ward migrants. In addition, whether in inner-city slums or informal shanty towns, the low-income in-migrants are living in marginalised urban environments, in which they are likely to respond collectively (and communally) to protect themselves and each other, with their goals and aspirations defined almost entirely in terms of that under-class position (McTaggard 1971). The informal, peripheral low-income settlements of Caribbean cities, therefore, grow and evolve to become distinctive urban social 'spaces', separated from other public and private 'spaces' by race and class cultures of 'informality' (see LaGuerre 1994 for an insightful explanation of the multi-faceted nature of 'informality' in today's spatially unequal cities).

Turner's two-stage process of initial 'bridgeheader' settlement in the inner city and subsequent relocation to a peripheral environment to consolidate the migrant's position is the expected pattern during the early phase of rapid urbanisation

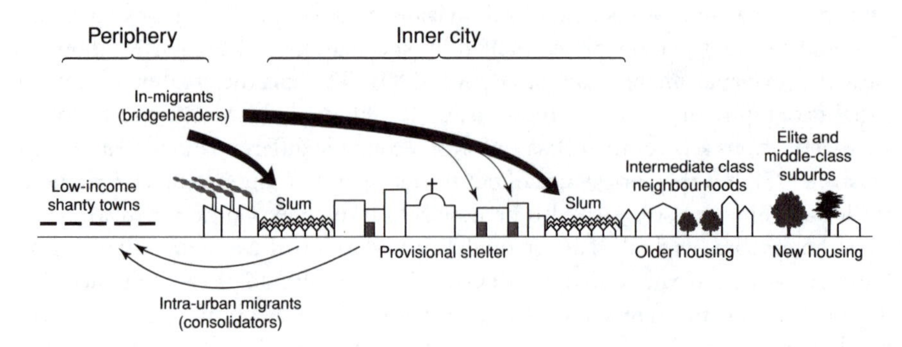

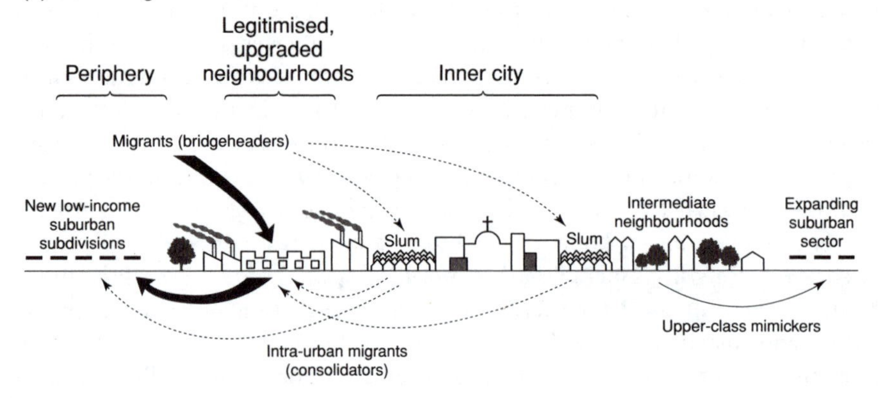

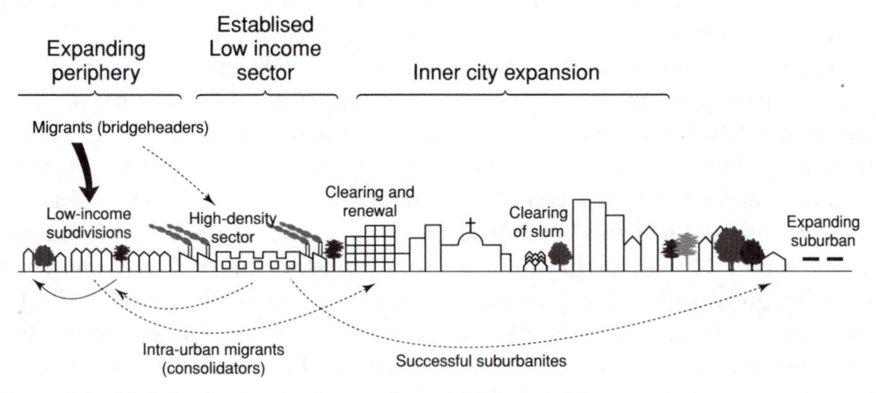

Figure 2.2 Models of urbanisation: early, continuing and late

(Figure 2.2(a)). The inner city is the major reception area for city-ward migrants from the countryside, because this is where temporary employment can be secured. With few relatives or acquaintances to provide information and socio-cultural support, an initial location within walking distance of both jobs and food markets is the priority. Employment and wages acquired provide the main 'security', with the anticipation that opportunities for a steadier income will materialise when the migrant has proved his/her worth. Hence, these 'bridgeheaders' are housed either in self-constructed provisional shelters in unsupervised, or unpoliced, vacant lots, or as renters in inner-city tenements or barracks. If, or when, the income of the bridgeheader becomes fairly steady and there is some accumulation of savings (perhaps, aided by marriage, or family reunion to another wage-earner), the now regularly employed worker may seek to consolidate what has been achieved. The migrant (and family) accordingly trades off the convenience of their inner-city central location for the security of residential stability (and investment flexibility) that home ownership provides in a peripheral low-income settlement. Home building and home improvements may take many years to accomplish, though security of tenure for the land on which the home is being built is a critical factor. In East Port of Spain's Laventille area, illegal occupancy of land plots was not the common pattern, since most of the consolidators who settled the hills rented their land plots (Plate 2.3). There was some illegal

Plate 2.3 East Port of Spain, Laventille Hills low-density settlement, 1967
Source: *Dennis Conway*

'squatting' in John John and Eastern Quarry and Prizgar Lands, but elsewhere small islander and rural in-migrant 'consolidators' legally occupied their small family domains (Conway 1981).

Eventually, however, this two-stage process changes as the evolving urban system becomes a more diversified landscape of employment and housing opportunities. With continuing urbanisation and the continued mass in-migration of new 'bridgeheaders', the inner-city 'zone of transition' is no longer their major reception area (Figure 2.2(b)). The deterioration of rental properties, tenements and barracks, will have occurred, and rent-control legislation is likely to deter repair and maintenance, thereby making the physical conditions of inner-city property worse, as the Trinidad case demonstrates (Conway 1989a). Often encouraged by public and private interests, the commercial core and government administrative domains expand their areas at the expense of the now blighted inner-city tenement slums. The rebuilding and renovating of existing structures to meet more modern needs invariably results in parking lot expansions, government building annexations and mixed-use commercial building projects, and there is a general modernising of the central, 'downtown' area of these Caribbean cityscapes. Stricter police control over vacant land in the central city makes sites for provisional shelters harder to find, or maintain, and vacant land plots may be the first to be built upon, or developed. The inner-city situations in different countries will differ, of course, as some high-density rental properties will continue to exist despite their physical dilapidation; like the 'yards' in Kingston, Jamaica (Clarke and Ward 1980; Brodber 1981), or the corridors of one- and two-room dwellings running along the alleyways of Port-au-Prince, Haiti (Manigat 1997).

At the same time, the earlier peripheral 'shanty-town' settlements gradually become more incorporated into the city, not necessarily as socially integrated communities, but rather existing in uneasy symbiosis – a low-income 'antisystem' (McTaggard 1971), geographically, as marginal, partially incorporated extensions of the Caribbean city. Legitimisation (or regularisation) is achieved through a compromise of interests between the institutional powers and the low-income communities. Conflicts over land tenure rights and legal titles are resolved, occupiers are granted 'squatter's rights', and security of tenure for homes and land plots is achieved. Similarly, after a period of prolonged occupancy, the authorities are likely to accede to community demands and provide these legitimised settlements with essential services. Power, water supplies and infrastructure may be upgraded, roads may be paved and community centres built. Schools, social services and public health clinics may be provided, and eventually these low-income neighborhoods become more fully incorporated residential entities. Markets, small shops, food stands, rum shops and bars and a host of small-scale, informal sector enterprises are established to serve the resident populations (Plate 2.4). The areas function as petty-commodity production centres, where the informality of this complementary economy ensures local people's participation and involvement (LaGuerre 1994; Potter and Lloyd-Evans 1998).

Plate 2.4 Street vendors in Pointe a Pitre, Guadeloupe, 1991
Source: *Dennis Conway*

With sources of both stable and casual employment available in these legitimised/regularised and upgraded neighbourhoods, the newly arrived in-migrants' needs will be met here, rather than in the inner city. With a strengthening of community bonds occurring in the neighbourhoods and the continuation of close kin relationships between these latest arrivals and their now-established new urbanite relatives, socio-cultural and socio-economic 'cushions' are available for newcomers. Thus, these legitimised and upgraded low-income neighbourhoods supersede the inner-city slums or provisional housing arrangements as the primary reception area for these later 'bridgeheaders'. Migrants who have no connections may still seek opportunities in the inner-city transitional areas, but with the dwindling supply of accommodation there, they may be forced to relocate to these other neighbourhoods shortly after arrival. As a result, diversity in housing arrangements is likely. Owner-occupiers become petty landlords, renting spare rooms as well as accommodating in-migrant kin in more flexible contracts, such as 'tenant-at-will' (Conway 1981).

Beyond the city's built-up perimeter, where peripheral land is available, where government authority is weak, or private ownership rights are open to challenge, new low-income, and low-density, suburban shanty towns will continue to mushroom. Although land tenure may often be of dubious legality, these new low-income subdivisions generally exhibit the same social and economic characteristics of the earlier shanty towns: a mix of urban-born and long-term urbanites and successful in-migrants – youthful, consolidating families with securer employment histories seeking a home of their own in these lower-density peripheral environments (Figure 2.2(b)). Since the growing size and increasing local political power of the low-income masses in these settlements can be expected to realise a greater degree of respect from local institutional authority, essential services such as water, paved roads and transportation facilities may be more readily provided to these new subdivisions. Therefore, accessibility from the periphery to other parts of the expanded city is easier for residents of these later settlements. Major arterial roads that might have been upgraded to facilitate movement of the wealthier classes in their automobiles, and the movement of commodities from industrial estates to the sea- and airports, might very well serve as determining locational factors in the siting of these new suburban subdivisions, and in their growth and settlement.

Consolidation by new in-migrants who secure stable employment but remain in blue collar occupations can follow the routes depicted here, but there will be a minority who will actually gain access to the intermediate class, through educational opportunities, or skill acquisition in the higher-skilled professional service sectors. Johnston (1972b) coined the terms 'upper-class mimickers' and 'satisfied suburbanites' to characterise two groups of the intermediate classes who elevated their social ranks sufficiently to rise above their low-income beginnings, but who then made different intra-urban location decisions: the former seeking proximity to the sector(s) of suburbanising elites and upper middle classes, the latter remaining within, or near to, the low-income consolidated settlements. Relative to the numbers who remain working class, however, these upwardly mobile families and individuals scarcely influence the race and class barriers that prevail in Caribbean cities to this day.

In these later stages of continuing urbanisation, the inner city no longer acts as a reception area for the newest waves of in-migrant bridgeheaders (Figure 2.2(c)). Downtown renewal, modernisation of administration buildings and the gentrification of the inner city's parks, city centre and historical buildings provides the new globalising identity to these Caribbean primate cities (Potter 2000). Instead, the in-migrants move directly to the legitimised low-income settlements, with ever increasing proportions moving direct to the more peripheral, lower-density subdivisions (Conway 1982). In this later phase, few among these in-migrants know no-one in the city, or do not have kin networks to find them accommodation. As consolidators move to the periphery, newly arrived kin might very well initially locate with them. Meanwhile, the legitimised settlements are now within the built-up city, and they are likely to experience the diminution of housing supply that the inner-city slum neighbourhoods experienced at an earlier stage of urbanisation. The periphery, therefore, eventually becomes a major reception area for in-migrants, with kinship ties and group affiliations continuing to operate as essential cushions to ease their entry into the urban labour market. Both informal and formal employment opportunities will be accessible for the enterprising (and the well connected) despite the difficulties the Caribbean lower classes face in general, especially the young men (Figure 2.2(c)).

Compared with the limited opportunities in rural areas, the cities still offer more possibilities to the youthful. The in-migrants in these later phases of urbanisation are much less likely to be unfamiliar with urban living and lifestyles. Indeed, the high degree of rural–urban interaction, and the expansion of large conurbations beyond the city limits into the nearby rural districts, has in many respects 'urbanised the countryside'. Commuters from the elite and middle classes have extended their distance from the capital cities, dormitory communities and dispersed residential subdivisions have proliferated, and the modernised transportation infrastructure that has accompanied this expansion of the commuter fields is likely to have affected low-income migrant itineraries. The rural fringe is often more an uncontrolled urbanised mélange of mixed land uses than it is a rural landscape of villages, plantation estates and market centres. Commuting may become a substitute for migration, for some. Others may find employment opportunities in the conurbations, as service-sector and tourism-related jobs outside the city grow at the expense of traditional urban jobs in industry, commerce and service within its boundaries. Indeed, it may be that in the latest phase of urbanisation, which has witnessed the considerable restructuring of national and urban economies under the mandates dictated by neo-liberalism, the development of new large, low-income shanty towns at the urban periphery and beyond the existing low-class sector is less likely to occur. Rather, evidence from such cities as Santo Domingo, the Dominican Republic (Lozano 1997), and Port-au-Prince, Haiti (Manigat 1997), suggest that neo-liberal restructuring has not only contributed to the social demotion of previously middle-class urbanites, it has also contributed to a greater degree of mixing of residential areas (Portes *et al.* 1997) (see Chapter 7).

Migration and circulation in the contemporary Caribbean: two inter-related livelihood behaviours

To explain contemporary Caribbean transnational migration behaviour more thoroughly, a useful theoretical distinction can be made between *circulation*, which refers to reciprocal flows of people undertaking purposeful productive, consumptive or obligatory activities, and *emigration*, which is a relatively permanent displacement of people from one place of residence to another one overseas, or under another state's jurisdiction. In Pacific, Caribbean and African contexts, 'circulation' has been found to be a time-honoured and enduring mode of mobility, deeply rooted in a great variety of cultures and undertaken at all stages of socio-economic change (Chapman and Prothero 1985). More generally, 'circulation' is a reciprocal move beginning and terminating in the same community. Consequently, an unchanged place of residence, or 'home-place', remains as a subjective territorial haven (Chapman and Prothero 1983). On the other hand, 'emigration' as a relatively permanent displacement to establish one's home in another territory is an intended (and realised) departure, with no, or ambivalent, intentions to return. Immigration is, of course, the subsequent successful entry of the Caribbean 'foreigner' or 'newcomer' into the new territory, North America, another Caribbean state, or Europe. Rarely, it seems, is international emigration subjectively assessed as an unequivocal permanent displacement, however. More commonly, return visits are intended, ties with the previous home are not severed, and there is often an expressed intent to return later, when things get better, when fortunes are made, when the move's objectives have been accomplished (Rubenstein 1979; Thomas-Hope 1992).

Treating international emigration/immigration and circulation as two distinct subjective mobility alternatives, both of which may subsequently convert to emigration or circulation due to internal, or external, factors (Desbarats 1983), provides a flexible framework in which to depict contemporary modes of Caribbean transnational mobility. It is the nature of the spatial displacement that categorises these alternative forms of Caribbean migration, and ensuing temporal and spatial changes in mobility behaviour can also be accommodated in this categorical construct (Figures 2.3–2.5). A wider range of mobility alternatives is suggested in Conway (1988). Here, the two major variants are demonstrated.

In today's transnational space, which constitutes the multi-local sphere of Caribbean–North American societal interactions, international circulation between a Caribbean home and a North American accommodation is commonly practised (Conway 2000). It is a preferred strategy for many seeking to take advantage of the 'best of both worlds': culture, identity and a familiar home environment in the former; material wealth, education and technological skill acquisition, and wider opportunities for dependents and circulators alike in the latter. International circulation to more prosperous neighbouring Caribbean territories might also be part of this strategy, as Caribbean hopefuls follow well-worn paths within the region, as well as beyond it to North America and Europe (Conway 1989b).

The resultant networks, which both facilitate this circulation and expand because of it, are likely to evolve from bi-national transnational interchanges to multi-local nets, thus widening the nature of the pan-Caribbean interactions, and incorporating multi-cultural experiences into new community formations. The global and transnational transfers and interchanges that result from this widening 'space' are therefore more than North American host adjustments, or new Caribbean immigrant accommodations, but more two-, three- or multi-way influences of intra-Caribbean, Caribbean, European and North American (Canadian as well as United States) cultural practices, ideas, innovations and traditions.

For international circulators (as shown in Figure 2.3), the Caribbean home environment allows 'familiar' local circulation activity patterns to be maintained

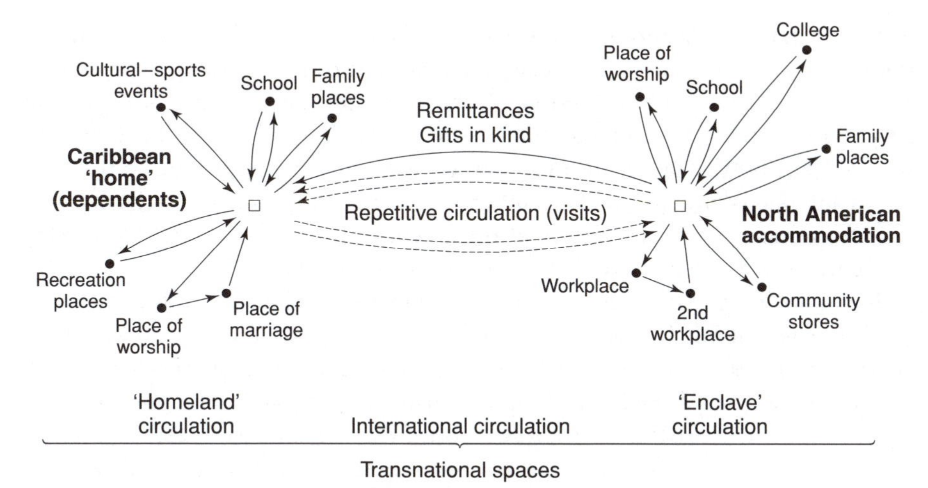

Figure 2.3 Caribbean circulation behaviour: reciprocal movement

and to be continued on return. In North America, at their temporary residence, this rental accommodation is the hub around which the sojourner undertakes limited activity paths: to workplace(s), to relatives' or acquaintances' places, to community stores, and to the neighbourhood church that serves her/his Caribbean congregation. The legal path to the United States for those seeking such a temporary visit is via the B-2 visitor's visa, and since this visa restricts the sojourn to six months in length, this conditions the trip. Canada also has a similar visiting permit. On the other hand, visiting Britain from Commonwealth Caribbean countries is not so straightforward for those without professional standing, and wealth. The embassy's screening of applicants deters many, it seems. Working to save a target amount is a likely objective of much international circulation, and sojourners' behaviour will be very much conditioned by this imperative. A subjective, though important, qualifier to this will be the sojourner's excitement to at least sample, if not wallow in, the recreation and entertainment opportunities the North American cities offer. For the most youthful, a trip to 'the States' is a rite of passage, an adventure, and if subjectively judged a 'success' this first move most probably will initiate subsequent visits or moves later in life. Not surprisingly, given the pervasiveness of North American media penetration and advertising throughout the Caribbean, as well as the demonstration effects of others' successful shopping sprees, consumption desires for modern fashionable goods will likely be present in the sojourner's goals for their visit, especially among the young! Target savings goals may not be met, prompting delays or postponements of returns (Grasmuck and Pessar 1991).

On the other hand, a significant category of Caribbean international circulators are those who seek illegal means of entry into North America, or even other more prosperous Caribbean countries, such as the Bahamas, the Cayman Islands, Trinidad or Barbados. Illegal entry is accomplished via different means, depending upon the relative locations of the source island and the destination and the transportation alternatives, the relative efficiencies of the destination's surveillance methods, and of the clandestine mechanisms. For the most distant Caribbean adventurers attempting to go to North America for 'a visit', acquiring a visitor's visa and having the option to over-stay the six months appears to be the preferred method. An appreciable number of Trinidadians, Guyanese, Barbadians and Vincentians appear to have used this circulation strategy, though how many are repeat visitors is impossible to estimate (Conway 2003).

For more proximate illegal sojourns within the Caribbean Basin, the time-honoured use of small coastal craft and ferry transport across relatively short oceanic distances to remote coastal coves and landing stages is the common mode. Haitians have entered the Bahamas in this manner for decades. Dominicans commonly ship across to Puerto Rico, where they either stay as illegal sojourners or transit on to New York via domestic air carriers. Grenadians and Vincentians have long traditions of sojourning and emigrating to nearby Trinidad, and the larger country's relative prosperity and wider range of employment opportunities continues to attract these 'small islanders'.

Regardless of whether trips are legally sanctioned as tourist or business visits, are within the statutory six months, or are occasionally over the limit but not apprehended, if repetitive returns home and re-entries to North America remain within the means and resources of the individual and her/his family, then this 'optimal' international mobility strategy may continue through the Caribbean person's lifetime as the preferred behaviour. As Commonwealth citizens of the United States, Puerto Ricans commonly practise this strategy. However, 'green card' holders from all over the Caribbean who have acquired resident alien status in the United States also possess the necessary legal rights to undertake repetitive international circulation. For these residents, and their family members, living between two worlds with the Caribbean 'home' as the mythological, or sentimental, haven, and the North American 'second home' or family's place as a transnational extension, is quite possible. In fact, even a reversal of notions about the primacy of one 'homeland' and a switch in territorial allegiances does not necessarily alter international circulation patterns. It may, however, develop over time into an emigration decision as circulators respond to dependents' needs and changing family situations. For example, if second-generation family members eschew their Caribbean ties, see their identities in North American terms and establish their own families in North America, then the first-generation immigrants, now as grandparents, may very well decide not to return to their Caribbean home on retirement.

International circulation might be undertaken as a preferred livelihood strategy by many Caribbean hopefuls. However, as a volitional process subject to many internal and external constraints and obstacles, it is as likely to undergo alterations over time as it is to remain a repetitive behaviour. Obstacles or constraints may foster a permanent emigration intention as the realisation of social, economic and legal restrictions to unfettered international movement becomes apparent. Unanticipated opportunities may also intervene between intentions to return and their realisation, prompting an alteration of the previously rationalised choice. Evaluation of the enlarged transnational field of obstacles and opportunities made after the first, second or successive moves may result in an assessment of mobility options that is at odds with previous circulation cost/benefit calculations. Influencing such alterations in international mobility strategies will be changes in life-course trajectories and changes in personal circumstances, whether planned or capricious.

Accordingly, displacement may come to be perceived as a satisfactory 'new reality' with circulation intentions being retained more as a mythological (apologetic) belief than an intended near-future action. Emigration, as a more permanent international displacement and substitution of a North American 'home' for the Caribbean home-place, thereby becomes a practised reality, despite prior circulation intentions, and even circulations. Establishing a North American residence, with an intention to return to the Caribbean on occasion, thereby comes about as a consequence of changes in context, changes in the decision makers' values, attitudes and behaviour, or as a consequence of interactions between these external and internal factors. Now, with the North American family and its members relying upon their residence's surrounding metropolitan and community enclave resources, a full circulation of localised activities will develop (Figure 2.4). Among the first generation at least, the Caribbean is viewed in nostalgic terms, or critical reminiscences, and occasional returns are intended and even realised. If a return visit is undertaken, the resultant activity pattern in the Caribbean is likely to be a limited familiar circulation, probably selective in purpose and design, to allow the returnee to visit remaining family, attend family funerals, sample the beach and the rum, or attend sporting or festival events, within the limited timeframe of a holiday period. This is likely, sooner or later, because returns to demonstrate 'success' are part of the lexicon of Caribbean migration, and expatriate West

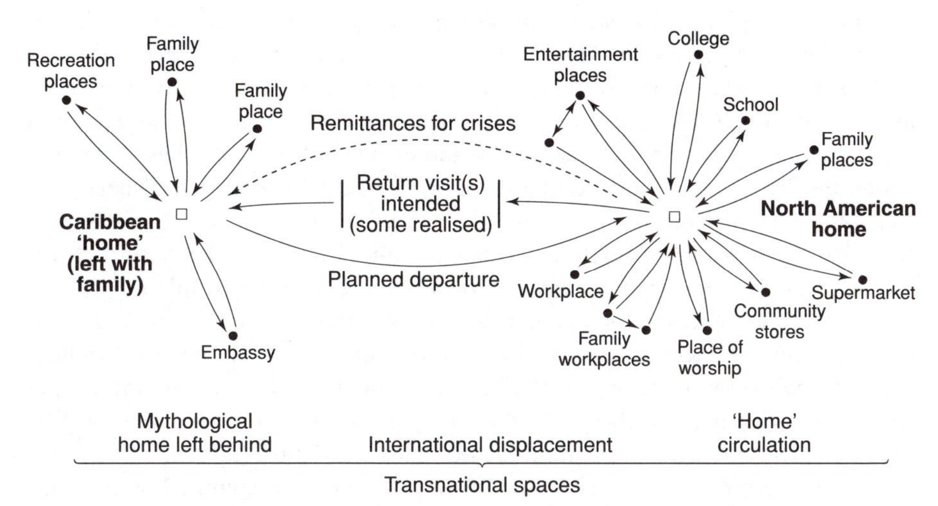

Figure 2.4 Caribbean emigration behaviour: displacement movement

Indians, Dominicans and Haitians in North America, as well as those native sons/daughters visiting Martinique and Guadeloupe from France, Suriname from the Netherlands, and the Commonwealth Caribbean from Britain, hold to this tradition.

Return migration and remittances: consequences for Caribbean societies

Return migration patterns

Return migration, upon retirement, or at other equivalent life-course benchmarks, may also be a common international mobility response after a lifetime sequence of repetitive international circulation, or after years of relatively permanent emigration. Assuming that material wealth, skills and experience have been acquired and achieved during the working life of the migrant, a rationalisation of the situation, on or nearing retirement, may very well induce an actualisation of the perceived mythological intent to live the remainder of one's days 'at home'. The Caribbean retirement home, often a different abode from the one left behind, will likely have been readied by remittances and investments during the time abroad. It will more likely be located in an urban or suburban setting than in the rural community of birth. Although problems of readjustment are to be expected, given the societal changes that have occurred since departure, or since the last visit, returnees will eventually adjust to these contextual alterations and a familial circulation activity pattern will develop centred on this new home and location. Now, occasional visits to the North American mainland are likely for those with sufficient resources to visit their children's residences, attend their weddings, their grandchildren's christenings, high school and college graduation ceremonies, and, perhaps, other places of nostalgic importance to them in North America. Thus, international circulation in transnational space is continued, or reintroduced for such retirees (Figure 2.5). For other retirement returnees, however, the homeward journey effectively severs ties with their mainland existence. This can happen if dependent family ties are not maintained and mainland born, assimilated children dissociate themselves from their parents, or vice versa. If children disperse and become alienated from their parents, then this can prompt the latter's return. Even after a 'lifetime' of relative permanent emigration and/or a long overseas stay with unchanging or declining fortunes, some Caribbean people might subjectively reappraise their situation, mentally unearth their sublimated 'return intentions' and reverse their emigration decision to return home. Among West Indians in Britain who emigrated in the 1950s, a substantial number (4,000 families per year is one estimate (Rosenberg 1983)) are returning home, citing the drab weather and the racist climate as influential factors (Chamberlain 1995).

Return migrants will not always wait for retirement, of course. For example, return migration might be a final intended decision after the first international

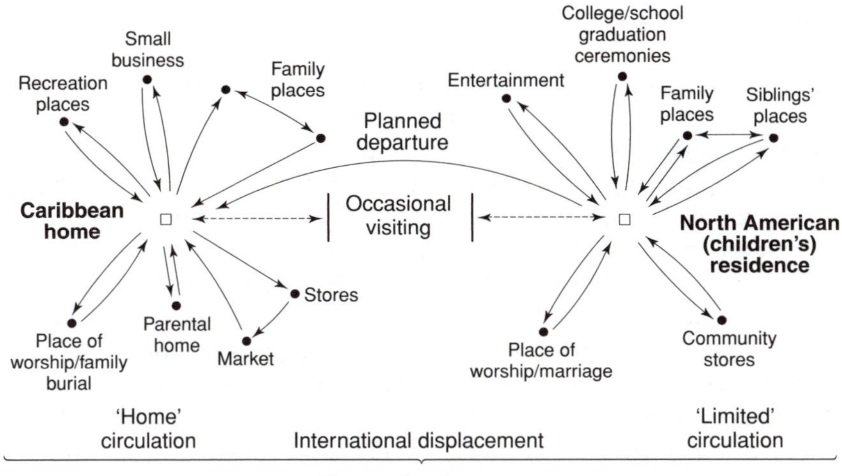

Figure 2.5 Return migration behaviour: relocation movement

circulation proves disastrous, highly disappointing or highly unsuccessful. Indeed, an initial conceptualisation of return migration depicted the flow as a flow of failed adapters, of unsuccessful migrants (see Bovenkerk 1974, 1981; Taylor 1976). Others have not been so critical of returnees' adjustment experiences, and more recent research appears to contradict these earlier selective opinions that returnees are failures and unable to reintegrate themselves into their home societies (see Conway 1985; Gmelch 1987; De Souza 1998). Returnees, even though they have been successful in their sojourns, or their emigration, might at any time during their life-course re-evaluate their North American situation and decide to 'return home for good' to the Caribbean. This can occur when there are significant transitions in their life-course: marriage, divorce, separation, employment lay-off, completion of a tour of duty in the armed services, among others. De Souza (1998) elaborates on the distinctions between return migrants' determinants and suggests there are four patterns of return: (1) seasonal, mobile livelihood circulation driven primarily by economic factors; (2) return visiting, which categorises repetitive visits of Caribbean emigrants; (3) long-term return migrants, who are making their final return decision; and (4) the transnational movement of Caribbean circulators living 'between two worlds'. Segal (1998) has referred to these latter repetitive transnational movers as 'swallows'. Return migration, therefore, is always among the options for international migrants, regardless of their previous intentions and images, their duration of residence abroad, and their previously preferred behaviours of emigration and/or international circulation. In terms of its size and significance, a recent estimate of the size of return migration flows to eastern Caribbean states suggests they are between 7 and 10 per cent of the resident population and have surprisingly wide fields of repatriation (UNECLAC 1998a).

Remittances: their persistence and investment practices

Remittances are the regular transfers of money (hard currency) and in kind transfers of material goods by migrants to their families back home. Migrants and circulators save their earnings and in response to requests and demands from home, or as a prearranged obligation, money or gifts are sent home. Just as migration and circulation are time-honoured strategies of livelihood and survival, so remittance investments have grown to become an essential financial input to Caribbean families of all classes. Commonly they are used to purchase needed goods, food, medicine, clothes and the like for dependents, and the recipient usually spends them to meet such consumption imperatives. In addition, however, remittances are important – and increasingly so – for both investment and savings, and secondly, that in many small island states there are very substantial constraints to alternative 'productive uses' of remittances. Contrary to earlier opinions (see Stinner *et al.* 1982), remittance flows appear to be sustainable over long periods with little indication of remittance decay over time. The strength of the migrant's obligations to continue to remit is often reinforced by social pressures among the migrant community as well as by the family pressures and responsibilities (Philpott 1973). Intentions to return also reinforce remittance persistence. Though remittances are primarily used for consumption there is also substantial investment, in housing, in land purchase and in the well-being of others, some of which enables the perpetuation of more migration. Transnational Caribbean migrants provide their families and kin, at 'home' and 'abroad', with an extended range of options for various forms of familial, personal and community development, starting with some degree of insurance against the uncertainties of life in the island state but also enabling some, albeit limited, investment in local economic development. Most importantly, remittances promote the development of human capital and growth of social and cultural capital stocks in local communities, with investments transferred at the appropriate scale (and scope) of effectiveness, the household level (see Connell and Conway 2000 for more in-depth discussion).

Remittances invested in micro-scale enterprise at the scale of the family or household is an appropriate target for productive investment, and skill and technology transfers via international circulation are likely to be of immediate assistance in the incubation and establishment of new enterprises. While research in this area is generally lacking, a recent research project in the Dominican Republic is illustrative of the potential for remittances, international circulation and return migration to stimulate private micro-economic enterprise on the islands and in North American metropolitan enclave communities. In their investigation, Portes and Guarnizo (1991) found dynamic relationships between the New York-based Dominican community business sector and those thriving in the small business sector in Santo Domingo (see also Chevannes and Ricketts 1997). Such bilateral support included skill acquisitions and transfers, capitalisation initiatives and acquisition of production inputs, among others.

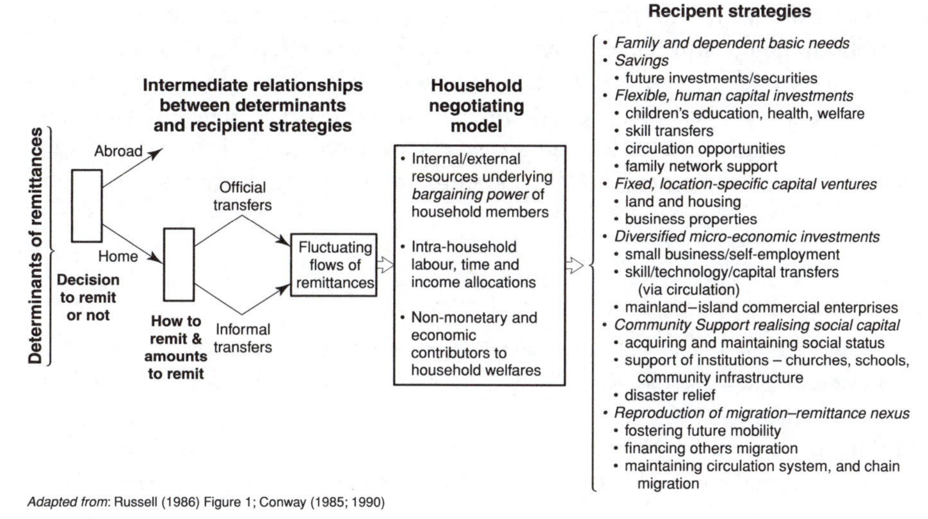

Figure 2.6 Remittance flows, household negotiations and recipient strategies
Source: Adapted from Russell (1986), Figure 1; Conway (1985, 1990)

Conclusion: transnational, multi-local Caribbean networks as livelihood strategies

International circulation, intra-regional migration, return migration, remittance flows and transnational business development are all important migration-related strategies of Caribbean individuals, families and community groups, utilising well-established transnational, multi-local Caribbean–American(–European) networks, which link the individual to family, community and national opportunity fields.

Far from being a simple demographic exodus from limited and vulnerable island territories by able-bodied people seeking a livelihood elsewhere, the complex patterns of short- and long-term international movements of emigration (displacement) and circulation (temporary, reciprocal movement) cement Caribbean societies with North American and European metropolitan societies, in a complex system of transnational interdependence (Goulbourne 2002; Goulbourne and Chamberlain 2001). Short-term visiting appears to be one option to permanent residence overseas, repetitive visits maintain the connections, and living a multi-local life has become a common survival, or sustenance, strategy. Emigration continues to be a long-held tradition for many islanders, too. Migration is a fundamental livelihood strategy in Caribbean societies; the islands' extended social fields are multi-local, and yet distinctive Caribbean identities remain entrenched, and enriched. Moreover, the further incorporation of these small island societies into a globalising new world order, where North American and European values and Caribbean cultures compete, while they are forever undergoing change and retranslation as people interact and reinterpret their lives, has also contributed to the maintenance of migration-related, transnational identities where the Caribbean 'home' still remains the territorial frame of reference (Conway 2003).

Caribbean people continue to demonstrate their flexibility and mobility, their creativeness and their fortitude, despite the perpetual recurrence of social and economic problems they and their societies have faced since their incorporation into the 'other man's world' in the fifteenth and sixteenth centuries.

References

Bach, R. L. (1987) 'The Cuban exodus: political and economic motivations', in Levine, B. (ed.) *The Caribbean Exodus*, Praeger, New York, Westport, Connecticut, and London, pp. 106–30.

Bailey, A. J. and Ellis, M. (1993) 'Going home: the migration of Puerto Rican-born women from the United States to Puerto Rico', *The Professional Geographer*, **45**, 2, 148–58.

Bascom, W. O. (1990) 'Remittance inflows and economic development in selected Anglophone Caribbean countries', Commission for the Study of International Migration and Co-operative Economic Development, Working Paper No. 58. Washington.

Beckles, H. McD. (1990) *A History of Barbados: From Amerindian Settlement to Nation-state*, Cambridge University Press, New York.

Beckles, H. (1993) 'Divided to the vein: the problem of race, colour and class conflict in Haitian nation-building, 1804–1820', in Beckles, H. and Shepherd, V. (eds) *Caribbean Freedom: Society and Economy From Emancipation to the Present*, Ian Randle and James Currey Publishers, Kingston, Jamaica, and London, pp. 494–503.

Boswell, T. D. (1976) 'Residential patterns of Puerto Ricans in New York city', *The Geographical Review*, **66**, 1.

Boswell, T. D. (1985) 'Puerto Ricans living in the United States', in McKee, J. O. (ed.) *Ethnicity in Contemporary America: A Geographical Appraisal*, Kendall Hunt, Dubuque, Iowa, pp. 117–44.

Boswell, T. D. (1989) 'Population and political geography of the present-day West Indies', in West, R. C. and Augelli, J. P. (eds) *Middle America: Its Lands and Peoples,* 3rd edition, Prentice Hall, Englewood Cliffs, NJ, pp. 103–27.

Boswell, T. D. and Curtis, J. R. (1984) *The Cuban-American Experience*, Rowman & Allanheld, Totowa, NJ.

Boswell, T. D. and Conway, D. (1992) *The Caribbean Islands: Endless Geographical Variety*. Rutgers University Press, New Brunswick, NJ.

Bovenkerk, F. (1974) *The Sociology of Return Migration: A Bibliographic Essay*, Martinus Nijhoff, The Hague, Netherlands.

Bovenkerk, F. (1981) 'Why returnees generally do not turn out to be "agents of change": the case of Suriname', *Nieuwe West Indische Gids*, **55**, 3/4, 154–73.

Brodber, E. (1981) *A Study of the Yards in the City of Kingston*, University of the West Indies, Mona, Jamaica, Institute of Social and Economic Research, Working Paper No. 9.

Chamberlain, M. (1995) Family narratives and migration dynamics: Barbadians to Britain', *New West Indies Guide/Nieuwe West-Indische Gids*, **69**, 3/4, 253–75.

Chapman, M. and Prothero, R. M. (1983) 'Themes on circulation in the Third World', *International Migration Review*, **17**, 4, 597–632.

Chapman, M. and Prothero, R. M. (1985) 'Circulation between "home" and other places: some propositions', in Chapman, M. and Prothero, R. M. (eds) *Circulation in Population Movement: Substance and Concepts from the Melanesian Case*, Routledge & Kegan Paul, New York, pp. 1–12.

Chevannes, B. and Ricketts, H. (1997) 'Return migration and small business development in Jamaica', in Pessar, P. R. (ed.) *Caribbean Circuits: New Directions in the Study of Caribbean Migration*, Center for Migration Studies, Staten Island, NY, pp. 161–96.

Clarke, C. G. (1975) *Kingston, Jamaica: Urban Development and Social Change*, University of California Press, Berkeley, Los Angeles and London.

Clarke, C. G. (1984) 'Pluralism and plural societies: Caribbean perspectives', in Clarke, C., Ley, D. and Peach, C. (eds) *Geography & Ethnic Pluralism*, George Allen & Unwin, London, pp. 51–86.

Clarke, C. G. and Ward, P. (1980) 'Stasis in makeshift housing: perspectives from Mexico and the Caribbean', *Comparative Urban Research*, **8**, 117–27.

Clarke, E. (1957) *My Mother who Fathered Me: A Study of the Family in Three Selected Communities in Jamaica*, Allen & Unwin, London.

Connell, J. and Conway, D. (2000) 'Migration and remittances in island microstates: a comparative perspective on the South Pacific and the Caribbean', *International Journal of Urban and Regional Research*, **24**, 1, 52–78.

Conway, D. (1981) 'Fact or opinion on uncontrolled peripheral settlement in Trinidad: or, how different conclusions arise from the same data', *Ekistics*, **286**, 37–43.

Conway, D. (1982) 'Self-help housing, the commodity nature of housing and amelioration of the housing deficit: continuing the Turner–Burgess debate', *Antipode*, **14**, 2, 40–6.

Conway, D. (1985) 'Remittance impacts on development in the eastern Caribbean', *Bulletin of Eastern Caribbean Affairs*, **11**, 4/5, 31–40.

Conway, D. (1988) 'Conceptualizing contemporary patterns of Caribbean international mobility', *Caribbean Geography*, **3**, 2, 145–63.

Conway, D. (1989a) 'Trinidad and Tobago', in Potter, R. B. (ed.) *Urbanization, Planning and Development in the Caribbean*, Mansell, London and New York, pp. 49–76.

Conway, D. (1989b) 'Caribbean international mobility traditions', *Boletin de Estudios Latinoamericanos y del Caribe*, **46**, 2, 17–47.

Conway, D. (2000) 'Notions unbounded: a critical (re)read of transnationalism suggests that US–Caribbean circuits tell the better story', in Agozino, B. (ed.)

Theoretical and Methodological Issues in Migration Research: Interdisciplinary, Intergenerational and International Rerspectives, Ashgate Publishers, Aldershot, UK, and Brookfield, USA, pp. 203–26.

Conway, D. (2002) 'Gettin' there, despite the odds: Caribbean migration to the US in the 1990s', *Journal of Eastern Caribbean Studies*, **27**, 4, 100–34.

Conway, D. (2003) 'Transnational, multi-local migration behaviours: which local context is "home"?' paper presented in the Special Session, 'The Experience of Return Migration: Caribbean Perspectives', at the 99th Annual Meeting of the Association of American Geographers, held in New Orleans, 2003.

Conway, D. and Brown, J. (1980) 'Intraurban relocation and structure: low-income migrants in Latin America and the Caribbean', *Latin American Research Review*, **15**, 3, 95–125.

Conway, D. Ellis, M. and Shiwdhan, N. (1990) 'Caribbean international circulation: are Puerto Rican women tied-circulators', *Geoforum*, **21**, 1, 51–66.

de Albuquerque, K., Mader, K. P. and Stinner, W. (1976) 'Modernization, delayed marriage and fertility in Puerto Rico, 1950 to 1970', *Social and Economic Studies*, **25**, 1, 55–65.

De Souza, R.-M. (1998) 'The spell of the Cascadura: West Indian return migration', in Klak, T. (ed.) *Globalization and Neoliberalism: The Caribbean Context*, Rowman & Littlefield, Lanham, Maryland, pp. 227–53.

Desbarats, J. (1983) 'Spatial choice and constraints on behavior', *Annals, Association of American Geographers*, **73**, 3, 340–57.

Duany, J. (1997) 'From the *Bohío* to the *Caserío*: urban housing conditions in Puerto Rico', in Potter, R. B. and Conway, D. (eds) *Self-Help Housing, the Poor, and the State in the Caribbean*, University of Tennessee Press, Knoxville, pp. 188–216.

Ebanks, G. E., George, P. M. and Nobbe, C. E. (1975) 'Emigration and fertility decline: the case of Barbados', *Demography*, **12**, 3, 431–45.

Ellis, M., Conway, D. and Bailey, A. J. (1996) 'The circular migration of Puerto Rican women: towards a gendered explanation', *International Migration (Geneva)*, **34**, 1, 31–64.

Fanon, F. (1967) *The Wretched of the Earth*, Penguin Books, Harmondsworth.

Fog Olwig, K. (2001) 'New York as a locality in a global family network', in Foner, N. (ed.) *Islands in the City: West Indian Migration to New York*, University of California Press, Berkeley, Los Angeles and London, pp. 142–60.

Gmelch, G. (1987) 'Work, innovation and investment: the impact of return migrants in Barbados', *Human Organization*, **46**, 2, 131–40.

Gmelch, G. (1992) *Double Passage: The Lives of Caribbean Migrants Abroad and Back Home*, University of Michigan Press, Ann Arbor.

Goosen, J. (1976) 'The migration of French West Indian women to metropolitan France', *Anthropological Quarterly*, **49**, 1, 45–52.

Goulbourne, H. (2002) *Caribbean Transnational Experience*, Pluto Press and Arawak Publications, London, Sterling, Virginia, and Kingston, Jamaica.

Goulbourne, H. and Chamberlain, M. (2002) *Caribbean Families in Britain and the Trans-Atlantic World*, Macmillan Caribbean, London.

Grasmuck, S. and Pessar, P. R. (1991) *Between Two Islands: Dominican International Migration*, University of California Press, Berkeley.

Handwerker, W. P. (1989) *Women's Power and Social Revolution: Fertility Transition in the West Indies*, Sage, Newbury Park, London and New Delhi.

Hernandez Alvarez, J. (1967) *Return Migration to Puerto Rico*, University of California, Institute for International Studies, Berkeley.

Johnston, R. J. (1972) *Urban Residential Patterns*, Bell, London.

Kimber, C. (1992) 'Aboriginal and peasant cultures of the Caribbean', in Martinson, T. L. (ed.) *1990 Yearbook of the Conference of LatinAmericanist Geographers*, Vol. 17/18, Ball State University, Muncie, Indiana, pp. 153–63.

Knight, F. (1970) *Slave Society in Cuba during the Nineteenth Century*, University of Wisconsin Press, Madison.

Laguerre, M. (1978) 'The impact of migration on Haitian family and household organization', in Marks, A. F. and Romer, R. A. (eds) *Family and Kinship in Middle America and the Caribbean*, Department of Caribbean Studies, Royal Institute of Linguistics and Anthropology, Leiden, the Netherlands, pp. 446–81.

LaGuerre M. (1994) *The Informal City*, St Martin's Press, New York.

Lowenthal, D. (1972) *West Indian Societies*, Oxford University Press, New York.

Lowenthal, D. and Comitas, L. (1962) 'Emigration and depopulation: some neglected aspects of population geography', *The Geographical Review*, **52**, 2, 195–210.

Lozano, W. (1997) 'Dominican Republic: informal economy, the state, and the urban poor', in Portes, A., Dore-Cabral, C. and Landholt, P. (eds) *The Urban Caribbean: Transition to the New Global Economy*, Johns Hopkins University Press, Baltimore and London, pp. 153–89.

Manigat, S. (1997) 'Haiti: the popular sectors and the crisis in Port-au-Prince', in Portes, A., Dore-Cabral, C. and Landholt, P. (eds) *The Urban Caribbean: Transition to the New Global Economy*, Johns Hopkins University Press, Baltimore and London, pp. 87–123.

Marshall, D. I. (1982) 'The history of Caribbean migrations', *Caribbean Review*, **11**, 1, 6–9/52–3.

Massey, D. S. and Mullan, B. P. (1984) 'A demonstration of the effect of seasonal migration on fertility', *Demography*, **21**, 4, 501–17.

McElroy, J. and de Albuquerque, K. (1986) 'The impact of external migration on the fertility and mortality transitions of insular microstates: an east Caribbean example', in *Human Resource Development in the Caribbean*, Caribbean Studies Association, San Juan, Puerto Rico.

McElroy, J. and Radke, T. (1988) 'Migration and fertility: some evidence', paper presented at the MALAS Meeting, Indiana University, Bloomington, October.

McTaggard, W. A. (1971) 'Squatter's rights, or the context of a problem', *The Professional Geographer*, **23**, 355–59.

Mintz, S. W. (1974) *Caribbean Transformations*, Johns Hopkins University Press, Baltimore and London.

Mintz, S. W. (1985) *Sweetness and Power: The Place of Sugar in Modern History*, Viking Penguin, New York.

Moncarz, R. and Jorge, A. (1982) 'Cuban immigration to the United States', in Cuddy, D. L. (ed.) *Contemporary American Immigration: Interpretive Essays (Non-European)*, Twayne, Boston, pp. 146–75.

Patterson, O. (1978) 'Migration in Caribbean societies: socio-economic and symbolic resource', in McNeill, W. H. and Adams, R. S. (eds) *Human Migration: Patterns and Policies*, Indiana University Press, Bloomington, pp. 106–45.

Peach, C. (1986) 'Patterns of Afro-Caribbean migration and settlement in Great Britain: 1945–1981', in Brock, C. (ed.) *The Caribbean in Europe*, London: Frank Cass, pp. 62–84.

Peach, C. (1996) 'Black-Caribbeans: class, gender and geography', in Peach, C. (ed.) *Ethnicity in the 1991 Census, Volume Two: The Ethnic Minority Populations of Great Britain*, HMSO, London, pp. 25–43.

Pérez, L. A., Jr. (1978) 'Cubans in Tampa: from exiles to immigrants, 1892–1901', *Florida Historical Quarterly*, **57**, 2, 129–40.

Perusek, G. (1984) 'Haitian emigration in the early twentieth century', *International Migration Review*, **18**, 1, 4–18.

Phillips, J. and Potter, R. B. (2003) 'Social dynamics of "foreign-born" and "young" returning nationals to the Caribbean: a review of the literature', Geographical Paper 167, University of Reading.

Philpott, S. (1973) *West Indian Migration: The Montserrat Case*, Athlone Press, London.

Portes, A. and Guarnizo, L. E. (1991) 'Tropical capitalists: US-bound immigration and small-enterprise development in the Dominican Republic', in Diaz-Briquets, S. and Weintraub, S. (eds) *Migration, Remittances, and Small Business Development: Mexico and Caribbean Basin Countries*, Westview, Boulder, Colorado, pp. 101–31.

Portes, A., Dore-Cabral, C. and Landholt, P. (1997) *The Urban Caribbean: Transition to the New Global Economy*, Johns Hopkins University Press, Baltimore and London.

Potter, R. B. (1989) *Urbanization, Planning & Development in the Caribbean*, Mansell, London and New York.

Potter, R. B. (2000) *The Urban Caribbean in an Era of Global Change*, Ashgate, Aldershot, Burlington, USA, Singapore, Sydney.

Potter, R. B. and Lloyd-Evans, S. (1998) *The City in the Developing World*, Addison Wesley Longman, Harlow, Essex.

Potter, R. B. (2003) '"Foreign-born" and "young" returning nationals to Barbados: a pilot study', Geographical Paper 166, University of Reading.

Proudfoot, M. J. (1970) *Population Movements in the Caribbean*, Negro Universities Press, New York.

Ramesar, M. D. (1976) 'Patterns of regional settlement and economic activity by immigrant groups in Trinidad, 1851–1900', *Social and Economic Studies*, **25**, 3.

Richardson, B. C. (1983) *Caribbean Migrants: Environment and Human Survival in St. Kitts and Nevis*, University of Tennessee Press, Knoxville.

Richardson, B. C. (1985) *Panama Money in Barbados, 1900–1920*, University of Tennessee Press, Knoxville.

Richardson, B. C. (1989) 'Human mobility in the Windward Islands, 1884–1902', *Plantation Society*, **2**, 3, 301–19.

Roberts, G. W. (1954) 'Immigration of Africans into the British Caribbean', *Population Studies*, **7**, 3, 235–62.

Rodney, W. (1977) 'Barbadian immigration into British Guiana, 1863–1924', paper presented at the Ninth Annual Conference of Caribbean Historians, the University of the West Indies, Cave Hill, Barbados, April.

Rosenberg, D. (1983) 'Why West Indians are coming home', *Pelican Magazine: The Weekend Nation*, Bridgetown, Barbados, 11 November.

Rubenstein, H. (1979) 'The return ideology in West Indian migration', in Rhoades, R. E. (ed.) *The Anthropology of Return Migration*, Papers in Anthropology, **20**, 330–7.

Sanchez Korrol, V. E. (1983) *From Colonia to Community: The History of Puerto Ricans in New York City, 1817–1948*, Greenwood Press, Westport, Connecticut.

Sandis, E. E. (1970) 'Characteristics of Puerto Rican migrants to, and from, the United States', *International Migration Review*, **4**, 1, 22–42.

Searle, C. (1984) 'Naipaulicity: a form of cultural imperialism', *Race and Class*, **16**, 2, 45–62.

Segal, A. (1998) 'The political economy of contemporary migration', in Klak, T. (ed.) *Globalization and Neoliberalism: The Caribbean Context*, Rowman & Littlefield, Lanham, Maryland, pp. 211–26.

Senior, C. (1947) *Puerto Rican Emigration*, University of Puerto Rico, Social Science Research Center, San Juan.

Sinha, D. P. (1988) *Children of the Caribbean, 1945–1984: Progress in Child Survival, its Determinants and Implications*, Caribbean Food and Nutrition Institute (PAHO/WHO), Kingston, Jamaica, and Bridgetown, Barbados, in collaboration with United Nation's Children's Fund, Caribbean Area Office.

Stinner, W. F., de Albuquerque, K. and Bryce-Laporte, R. S. (1982) *Return Migration and Remittances: Developing a Caribbean Perspective*, Smithsonian Institution, Research Institute on Immigration and Ethnic Studies, Washington, Occasional Papers No. 3.

Stolnitz, G. L. and Conway, D. (1991) *Caribbean Population and Development Trends and Interrelations: a 1990–1991 Assessment*, Indiana University Population Institute for Research and Training, Bloomington, IN/UN-ECLAC, Port of Spain, Trinidad and Tobago.

Sutton, C. and Makiesky, S. R. (1975) 'Migration and West Indian racial and ethnic consciousness', in Safa, H. I. and du Toit, B. (eds) *Migration and Development: Implications for Ethnic Identity and Political Conflict*, Mouton Press, The Hague, Netherlands, pp. 113–44.

Taylor, E. (1976) 'The social adjustment of returned migrants to Jamaica', in Henry, F. (ed.) *Ethnicity in the Americas*, Mouton Press, The Hague, Netherlands, pp. 213–30.

Thomas-Hope, E. M. (1978) 'The establishment of a migration tradition: British West Indian movements to the Hispanic Caribbean in the century after emancipation', in Clarke, C. G. (ed.) *Caribbean Social Relations*, University of Liverpool, Centre for Latin American Studies, Liverpool, Monograph Series, No. 8, pp. 66–81.

Thomas-Hope, E. M. (1983) 'Off the island: population mobility among the Caribbean middle class,' in Marks, A. F. and Vessuri, H. M. C. (eds), *White Collar Migrants in the Americas and the Caribbean*, Leiden, Netherlands: Royal Institute of Linguistics and Anthropology, Department of Caribbean Studies, pp. 39–59.

Thomas-Hope, E. M. (1992) *Explanation in Caribbean Migration*, Macmillan Caribbean, London and Basingstoke.

United Nations (1998) *Population Distribution and Migration*, United Nations, New York, ST/ESA/SER.R/133.

UNECLAC (1998) *A Study of Return Migration to the Organization of Eastern Caribbean States (OECS) Territories and the British Virgin Islands in the Closing Years of the Twentieth Century*, United Nations Economic Commission for Latin America and the Caribbean, Port of Spain, Trinidad and Tobago.

Warren, R. (1990) Annual estimates of non-immigrant overstayers in the United States 1985–1989, in Bean, F. D., Edmonston, B. and Passel, J. S. (eds) *Undocumented Migration to the United States: IRCA and the Experience of the 1980s*, Urban Institute, Washington, pp. 77–100.

Watts, D. (1987) *The West Indies: Patters of Development, Culture and Environmental Change since 1492*, Cambridge University Press, Cambridge.

Williams, E. (1970) *From Columbus to Castro: The History of the Caribbean*, 1492–1969, Andre Deutsch, London.

Wilson, C. (1947) *Empire in Green and Gold*, Holt, New York.

Part II

RURAL AND URBAN BASES OF THE CONTEMPORARY CARIBBEAN

Chapter 3

AGRICULTURE AND AGRARIAN STRUCTURES

St Kitts was one of the first sugar colonies in the West Indies and sugar dominated its economic, social and cultural history (Watts 1987). There is only one sugar factory left, but as recently as 1978 sugar accounted for 77 per cent of export earnings. Crisis in the sugar industry encouraged the government to diversify the economy and by 1987 tourism was the island's main foreign exchange earner. One of St Kitts' main tourist attractions is Brimstone Hill Fortress, a National Park and UNESCO World Heritage Site. From its ramparts, the spectacular panorama of the island's leeward coast portrays a landscape that encapsulates the essence of the region's agricultural legacy. Lush green sugar cane still dominates the coastal plain that surrounds most of the island (Figure 3.1), though the cane fields are not as extensive as in former times. The site of an old windmill hints at the location of a former sugar plantation. Rising from the coast, cane gives way to a steep hillside landscape of mixed cropping and pasture, and at the highest elevations remnants of forest are shrouded in mist and cloud on the volcanic Mt Liamuiga. From colonial times, the slopes above the cane fields have been the domain of small farmers, cultivating land not needed by the precious sugar crop, though the land is susceptible to erosion and marginal to agriculture. Today sugar accounts for only 2 per cent of foreign exchange earnings yet employs 8 per cent of the workforce and occupies 30 per cent of agricultural land, so the legacy of the sugar plantation is still evident in the island's landscape.

Throughout the Caribbean, economies and societies evolved in the context of the plantation and the inequalities created by colonialism (Mintz 1985; Richardson 1992). After Emancipation, new rural land-use patterns and agrarian structures emerged reflecting a different, though still unjust, social order. On most islands a dual agricultural economy emerged, with a large-scale, export-oriented sector juxtaposed against a small-scale farming geared to production for the household and sale on the domestic market. On the more mountainous Caribbean islands like St Kitts, elements of this duality have persisted, giving a distinctive spatial signature to the landscapes (Barker 1989).

Caribbean agriculture has always been articulated within the context of the international economy. Sugar and other primary agricultural products were exported to metropolitan countries during the colonial period both before and after Emancipation. Grenada specialised in nutmeg, Dominica in limes, St Vincent in arrowroot and Jamaica in pimento. On smaller, drier islands like Anguilla and

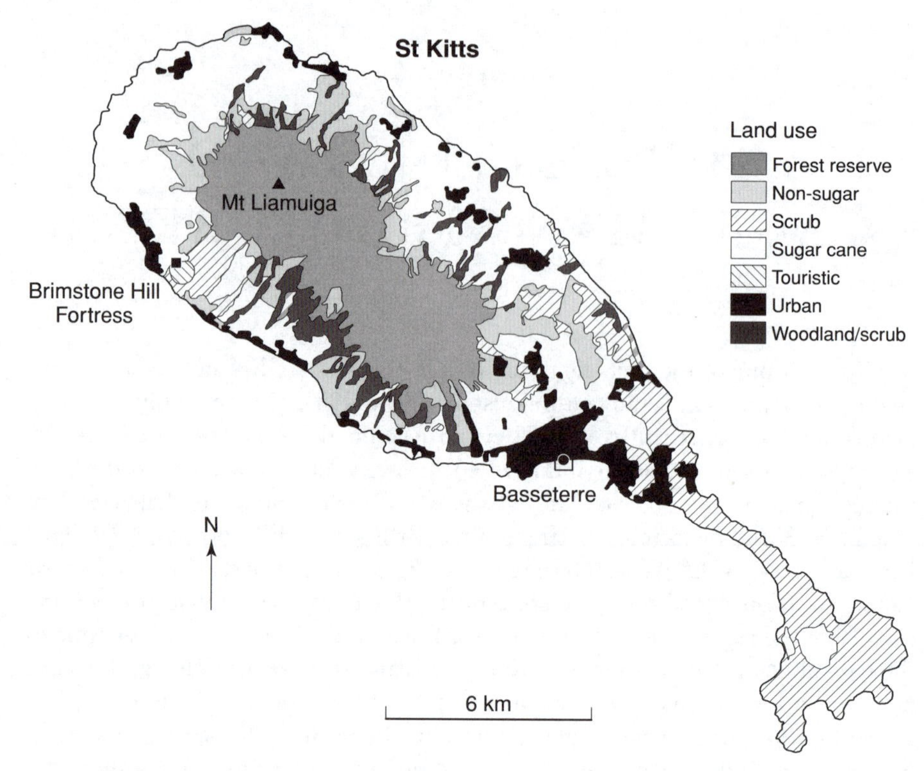

Figure 3.1 Principal land uses in St Kitts
Source: Physical Planning Division, Government of St Kitts and Nevis

the Turks and Caicos, commercial agriculture never prospered and farming was restricted to production for household consumption. The region's agriculture continues to be enmeshed and assailed by forces in the global economy, though there has been a shift away from a dependence on primary export crops such as sugar, and generally speaking, agriculture has stagnated.

Each Caribbean territory has a unique set of agricultural land-use mosaics and rural economies, reflecting economic imperatives, environmental factors and cultural history. Diversity is compounded by the spatial intermixing of small-scale farming and larger-scale operations like sugar plantations, cattle farms and orchard crops. Caribbean agrarian landscapes reflect diverse cultural influences. Farmers with African ancestry grow crops like pumpkins and cassava, which would have been familiar to Tainos and Arawaks (Sauer 1966), yet cultivate yams originally from Africa, carrots, cabbages and onions introduced from Europe, mangoes from India and breadfruit from the Pacific Islands. Some farmers in Guyana and Trinidad are descendants of Indian indentured labourers and rice, water buffalo, Hindu temples and Islamic mosques are part of their agrarian landscapes (Prorok 1991; Prorok and Hemmasi 1993), whilst some farmers in the Hispanic Caribbean have Spanish ancestry. There are smaller groups like the Caribs in St Vincent and Dominica, and Mayan and Garifuna farmers in Belize, each having distinctive rural economies.

Basic themes for Caribbean agriculture

In the past, the Caribbean was classified within the broad framework of world agricultural regions as 'plantation agriculture', 'plantation crops' or 'plantations with small farms' (Grigg 1969, 1974). However, these historically derived categories are unhelpful given the spatial and structural complexity of contemporary Caribbean agriculture. Farming patterns and food production systems are dynamic, and classifying the world into agricultural regions has lost much of its simple utility. The organised application of science and technology to increase food production in order to feed the world's population has transformed farming in many parts of the world, a process that has affected some regions more than others, and prompted the Brundtland Commission (WCED 1987) to suggest that there are now three major types of world agriculture:

- industrial agriculture
- resource-poor agriculture
- Green Revolution agriculture.

Industrial agriculture is technology-dependent and capital-intensive, and is based on the widespread use of gasoline-based, mechanised tractors and other farm equipment. It emerged in North America in the 1920s (Stutz and de Souza 1998). It is normally associated with large-scale farming (like wheat monoculture) but small-scale operations include beef feedlots, dairying, poultry rearing and fish farming. Industrialised agriculture is found mainly in the richer countries, though plantations are examples in developing countries. Figure 3.2 contrasts the characteristics of industrialised agriculture with resource-poor agriculture, better known as 'traditional subsistence farming'. The latter has low levels of technology and food output, and supports two billion people in marginal semi-arid lands, highlands and forests. Green Revolution agriculture is a separate category, since the developing regions affected (especially Asia) have also experienced dramatic transformations in farming and food output. Overall, though, the Green Revolution had differential impacts on regions, socio-economic class and gender. The Green Revolution barely touched the Caribbean because wheat and (to a lesser extent) rice, the main high-yielding crop varieties at the time, are not widely grown as food staples. Richardson's (1972) study of Guyanese rice farming is a rare commentary on the Green Revolution's lack of penetration in the region.

Traditional small farming in the Caribbean is, in many ways, 'resource-poor' though hunger and malnutrition are less prevalent than in crisis-torn Africa. But caution should be exercised in stereotyping Caribbean small farmers as *subsistence* cultivators. Whilst they grow their own food and contribute significantly to domestic food supply, few are purely subsistence producers as defined by

Figure 3.2 Characteristics of industrial agriculture compared with traditional small farming

Industrial agriculture

Increased food output

Increased farm size

Reduced demand for labour

Monoculture

Production for profit

Intensification of capital inputs

High energy use from fossil fuel sources

Demise of family farm

Increasing role of corporate business

Development of industrial substitutes for many agricultural products

Traditional 'resource-poor' small farming

Low levels of food output

Subsistence production and some cash crops

Small farm size

High demand for labour

Mixed cropping and polyculture

Production for home consumption

Low level of capital inputs

Reliance mainly on human, animal and solar sources of energy

Farm based on household unit

Variety of tenure arrangements, some insecure

Associated with poverty and hunger

Ruthenberg (1971). Indeed, there are long traditions of commercial production by small farmers. Mintz (1985) notes that, even before Emancipation, slaves sold crops in native markets and, in the nineteenth century, peasant farmers supplied sugar factories and exported root crops to West Indians labourers constructing the Panama railway (Satchell 1990). Today, small farmers dominate exports in crop sectors such as bananas in the Windward Islands (Grossman 1998a), yams in Jamaica (Barker and Beckford 2003) and rice in Guyana.

Figure 3.3 depicts the main factors that influence the type of farming found in a particular area, identifies processes that transform agricultural systems and highlights concepts that geographers have found useful in analysing agricultural

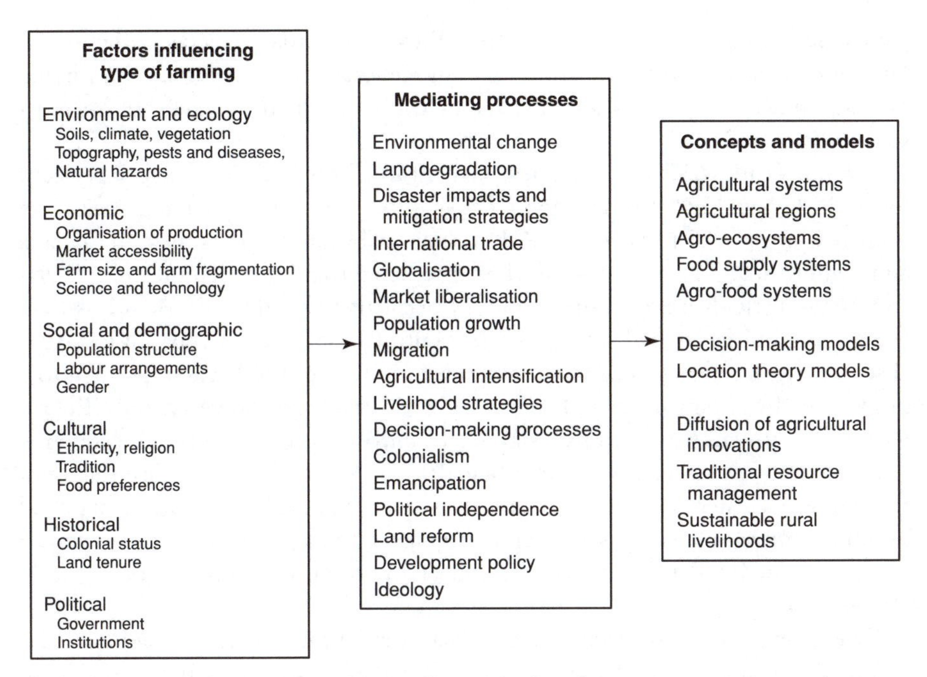

Figure 3.3 Basic themes for understanding agricultural patterns, processes and problems

development issues. In this chapter we select factors, processes and concepts from Figure 3.3 that elucidate important features of Caribbean agriculture and agrarian structure. The most important factor that underpins contemporary farming is the legacy of the colonial plantation system. Plantations had significant impacts on all aspects of economic, social and political life, and elsewhere in the book we examine the plantation legacy in terms of cultural and demographic composition of the population, and social conditions and the colour-class system (Chapter 5). Here we examine how the hegemony of plantations created inequalities in agrarian structure, which in turn prompted innovative tenure responses from Caribbean people. The plantation system was instrumental in dichotomising large-scale commercial export agriculture and small-scale traditional farming. The former became the prime focus for policy and agricultural research, whilst the latter was neglected and under-resourced. This duality has contributed to a pejorative stereotyping of traditional small farmers. Other themes examined include the impact of globalisation on Caribbean farming, and the dynamics of Caribbean small-scale farming systems.

Agriculture's changing role in Caribbean economies

Agriculture benefits Caribbean economies through its contribution to gross domestic product and to export earnings, to domestic food supply and by providing employment. It is a source of industrial raw materials (as in the manufacture of rum), and income from farming generates capital for families living in rural areas. Farming is also the major economic land-use activity on most Caribbean islands.

For most industrialised countries, agriculture contributes less than 5 per cent to GDP (Grigg 1992) but is usually above 40 per cent in Africa and Asia. The average is around 25 per cent of GDP in the Caribbean, being highest in Guyana and Haiti (Table 3.1). There is a declining regional trend; for example, in St Kitts and Nevis agriculture has fallen from 40 per cent of GDP in 1964 to less than 6 per cent, and in Barbados from 38 per cent in 1958 to 4 per cent. In industrialised countries, agriculture's declining contribution to GDP corresponded to a growth in the urban-industrial sector. In the Caribbean, however, only Puerto Rico and Trinidad have acquired significant industrial sectors through economic development, yet agriculture is less than 10 per cent of GDP in more than half the countries listed in Table 3.1. Its demise in the region has been matched by rapid growth of the service sector, which for ten of these countries exceeds 75 per cent of GDP. Only in Guyana and Haiti do services account for less than 50 per cent.

The growth of the service sector reflects structural changes in regional economies. As primary exports like sugar and bananas have stagnated or declined, tourism, light manufacturing and offshore financial and IT services have become more important. McElroy and de Albuquerque (1990) note that in the smaller, drier Leeward Islands, this process proceeded faster than in the more agriculturally favoured Windward Islands. For example, in the 1940s Antigua was a neglected colony entirely dependent on sugar exports, and most people worked for low incomes in the sugar industry. In 1961 agriculture contributed 20 per cent to GDP but soon after the sugar industry collapsed and the last factory closed in the 1970s. Meanwhile, Antigua developed a vibrant tourist industry, though it now imports 80 per cent of its food, including the molasses needed to manufacture Antiguan rum. Lorah (1995) argued that dependence on sugar has been replaced by an equally problematic dependence on tourism.

The growth of tourism was expected to stimulate local farming by generating demand for fresh food in hotels and restaurants, leading to a diversification away from commercial export crops. Belisle (1983, 1984) argued that such linkages had failed to materialise, though de Albuquerque and McElroy (1983) reported an agricultural resurgence in the US Virgin Islands in response to tourism and the creation of regional linkages as in the import of fresh farm produce from the Dominican Republic. Momsen (1998) has analysed linkages between Caribbean tourism and agriculture over the last three decades. She argues the anticipated benefits of diversified agricultural production were largely illusory at first, but notes a turnaround as new linkages between tourism and agriculture are being forged through tourist demand for local foods and hotel chains more actively sourcing local produce rather than relying on imported foods.

Table 3.1 Contribution to GDP by sector, 2000

	Agriculture	Industry	Services
Anguilla	4%	18%	78%
Antigua and Barbuda	4%	12.5%	83.5%
Bahamas	3%	7%	90%
Barbados	4%	16%	80%
Belize	18%	24%	58%
Cayman Islands	1.4%	3.2%	95.4%
Cuba	7%	37%	56%
Dominica	21%	16%	63%
Dominican Republic	11.3%	32.2%	56.5%
Grenada	9.7%	15%	75.3%
Guadeloupe	15%	17%	68%
Guyana	34.7%	32.5%	32.8%
Haiti	32%	20%	48%
Jamaica	7.4%	35.2%	57.4%
Martinique	6%	11%	83%
Montserrat	5.4%	13.6%	81%
Netherlands Antilles	1%	15%	84%
Puerto Rico	1%	45%	54%
St Kitts and Nevis	5.5%	22.5%	72%
St Lucia	10.7%	32.3%	57%
St Vincent and the Grenadines	10.6%	17.5%	71.9%
Trinidad and Tobago	2%	44%	54%
Virgin Islands (British)	1.8%	6.2%	92%

Source: *CIA World Factbook*, 2001
Note: In small island economies, the percentage contribution of agriculture to GDP can vary significantly from year to year, especially as a result of natural hazards such as a hurricane or drought.

The open economies of Caribbean countries have high levels of food imports, for both local populations and the tourism sector. Domestic food crops are produced entirely by traditional small farmers, though their full potential has not been realised, partly due to the historical bias towards commercial export crops. Spence (1999) reported that per capita food production in Jamaica increased between 1985 and 1996, though not as rapidly as per capita food demand, the shortfall being made up through food imports. Semple and Brierley (1998) examine the vagaries of Guyana's domestic food sector between 1960 and 1993. Domestic food production is susceptible to drought and flood rains (Chapter 4), and the drought of 1995/1996 in Jamaica caused domestic food production to decline by 25 per cent (Barker and Beckford 2003).

The legacy of the colonial plantation economy was commercial export crops for an overseas market, and not surprisingly, agriculture still contributes significantly to export earnings, though less so than in former times. In nine of 15 CARICOM states food exports exceed 20 per cent of total exports, a threshold used by the FAO to define those countries it considers too dependent on agricultural exports (see www.fao.org/docrep). The implication is that countries too dependent on agricultural exports need to diversify their economies, or their agricultural sectors. There is an interesting contrast between Belize and Dominica. In Belize, agriculture accounts for 71 per cent of export earnings and in Dominica for 60 per cent, so both are dependent on agricultural exports. However, Belize's agricultural export base is broader and includes sugar, citrus, bananas, cocoa (and marine products such as lobsters from its barrier reef), whereas Dominica's exports are dominated entirely by bananas.

As with agriculture's contribution to GDP, employment as a percentage of the labour force has declined in recent years. Agricultural employment varies across the region, from a low of 2 per cent of the labour force in Puerto Rico and 8 per cent in Trinidad, to between 20–30 per cent in Dominica, Grenada, Jamaica and Belize. Nevertheless, agriculture supports more than a quarter of the Jamaican population, and over half a million people (mostly small farmers) work in agriculture in the Dominican Republic.

Agriculture is a major land use in a region where land generally is a scarce resource. The Caribbean has less than 0.25 per cent of the world's total farmland. Half of all Caribbean farmland is in Cuba, and with the Dominican Republic and Haiti, these three countries account for over 90 per cent of the region's farming area. Of the English-speaking territories, only continental Guyana has over one million hectares (Table 3.2), about three times as much as Jamaica. Farming still occupies 45 per cent of Barbados and Jamaica, even though land is being converted to urban and tourist-related activities. Farming occupies about 30 per cent of the OECS countries, and only 2 per cent of the Bahamas, where much of the arid archipelago is marginal to agriculture. Over the last 50 years the farmland has expanded in some countries and contracted in others (Table 3.2). The largest loss (around 50 per cent) occurred in Puerto Rico (Monk and Alexander 1985), with smaller declines in Grenada and St Kitts and Nevis. In the USVI, the amount of farmland expanded then declined (de Albuquerque and McElroy 1983). The largest increase was in Cuba, where 3 million hectares have been brought into production since 1961. In Guyana, several large-scale rice and irrigation projects have expanded the cultivated area (Williams 1995). In Belize, though agriculture accounts for less than 7 per cent of land area, the area of farmland has doubled since 1961, most notably during the last 20 years.

Table 3.2 Changes in agricultural land, 1961–2000

	2000		1981		1961	
	ha (000s)	**% total**	**ha (000s)**	**% total**	**ha (000s)**	**% total**
Antigua and Barbuda	12	27	11	25	10	23
Aruba	2	10.5	2	10.5	2	10.5
Bahamas	13	1	11	1	10	1
Barbados	19	44	19	44	19	44
Belize	139	6	97	4	79	3.5
Cayman Islands	2	8	2	8	2	8
Cuba	6,665	61	5,938	54	3,550	32
Dominica	17	23	19	25	17	23
Dominican Republic	3,696	76	3,517	73	3,082	64
Grenada	12	35	16	47	22	65
Guadeloupe	49	29	59	35	58	35
Guyana	1,726	9	1,715	9	1,359	7
Haiti	1,400	51	1,403	51	1,255	46
Jamaica	503	46	497	46	533	49
Martinique	33	31	38	36	34	32
Montserrat	3	20	2	20	5	50
Netherlands Antilles	8	10	8	10	6	7.5
Puerto Rico	291	33	467	53	616	69
St Kitts and Nevis	10	28	15	45	20	56
St Lucia	19	31	20	33	17	28
St Vincent and the Grenadines	13	33	12	31	10	26
Trinidad and Tobago	133	26	127	25	102	20
Turks and Caicos	1	2	1	2	1	2
Virgin Islands (British)	9	56	8	53	6	40
Virgin Islands (US)	10	29	16	47	12	35

Source: www.fao.org

The plantation legacy and agrarian structure

Historically, the plantation dominated the agricultural, economic, social, political and colonial organisation of the Caribbean, and indeed plantations were the instrument of political colonisation. Beckford's (1972) classic treatise on plantation economies underscores their crucial role in the region's history. A plantation was more than just a type of agriculture, because it was simultaneously an economic and a social system. As a social system, authoritarian hierarchic organisation vested control in the institution itself. The labour force (slaves then wage earners) were carefully controlled and supervised, to the extent that plantations dominated the lives and livelihoods of the people they embraced. As an economic system, it was geared to the commercial production of export crops for an overseas market, and the bias towards export crops over domestic food crops has persisted among politicians, agricultural planners and technocrats to the present.

Plantations and the dual agricultural economy

The agrarian structure of a country reflects patterns of land ownership, management and control over land resources. Land concentration occurs when most of the land in a country is owned by a relatively small number of wealthy people or institutions. For small farmers, access to, and control over, land is a key issue, especially in the Caribbean, where good agricultural land is limited in supply. Where there is inequality in access to land, the national distribution of landholdings by farm size tends to be positively skewed with two key features:

- many more small farms than large estates;
- large farms and estates occupy the greatest proportion of good farmland.

Table 3.3 illustrates the farm size distribution for Jamaica in 1996. Fully 61 per cent of the country's 187,791 holdings were less than 1 hectare, yet the 77 per cent of holdings under 2 hectares occupy only 20 per cent of the country's farmland. Conversely, the 1,427 large holdings (over 20 hectares) account for 54 per cent of all agricultural land. There are similar disparities in Barbados, where the 1989 Agricultural Census (Ministry of Agriculture, Food and Fisheries 1992) records 99 per cent of landholdings less than 4 hectares, whilst the 117 holdings larger than 20 hectares occupy 81 per cent of agricultural land.

The contemporary inequities in land ownership patterns are a legacy of the colonial period. Plantations and large estates were owned by the wealthy class and geared to commercial export crops. They occupied the best and largest proportion of agricultural land, whilst small farmers usually cultivated poorer quality land under insecure forms of tenancy yet constituted the majority of the rural population. Others were landless labourers working for low wages on large estates. The polarisation of agriculture into large-scale and small-scale sectors is known as a dual agricultural economy, and in Latin America the system is called

Table 3.3 Size distribution and fragmentation of farm holdings in Jamaica, 1996

Size category	Total number of holdings	Total area (acres)	Total number of parcels
Less than 1 acre	71,753	23,229	61,274
1–5 acres	87,273	180,886	114,798
5–10 acres	18,256	115,659	30,987
10–25 acres	7,746	106,247	15,285
25–50 acres	1,339	43,594	2,718
50–100 acres	624	40,691	1,020
100–200 acres	349	45,669	529
200–500 acres	248	75,812	371
Over 500 acres	203	376,557	413
Total	187,791	1,008,344	227,395

Source: *Census of Agriculture 1996*, Vol. 3, STATIN, Kingston, Jamaica, Table 6.4

latifundia (large estates) and minifundia (smallholdings). This system was the basis of agrarian structure in pre-revolutionary Cuba. Large plantations and cattle ranches (haciendas) owned by wealthy families occupied the best agricultural lands and were worked by people exploited for low wages in semi-feudal conditions (Floyd 1978). At the time of the revolution, 70 per cent of Cuban farmers occupied less than 12 per cent of farmland (as tenants or sharecroppers) whereas 1.5 per cent of the total number of landowners controlled 46 per cent of all farmland (Floyd 1978).

In Jamaica, like the rest of the Anglophone Caribbean, the dualist structure of plantations and peasant farming emerged after Emancipation. Sugar plantations monopolised the best agricultural lands on the flat alluvial coastal plains and inland valleys (poljes) in limestone areas. A new rural settlement pattern developed after Emancipation in the 1830s, when a large proportion of the 311,000 ex-slaves began to establish an independent agrarian life on former estate land or in the forested upland regions (Barker 1989). Sugar estates tenaciously retained the best arable land and planters sold estate land only when absolutely necessary (Hall 1959). Church groups, like the Baptists and Methodists, played a prominent role in the purchase of land for the new class of peasant farmers (Marshall 1972), creating a settlement pattern of church-based agricultural communities called 'free villages' (Figure 3.4). Many modern churches in these communities have dates inscribed on their stone walls indicating their foundation in the 1840s.

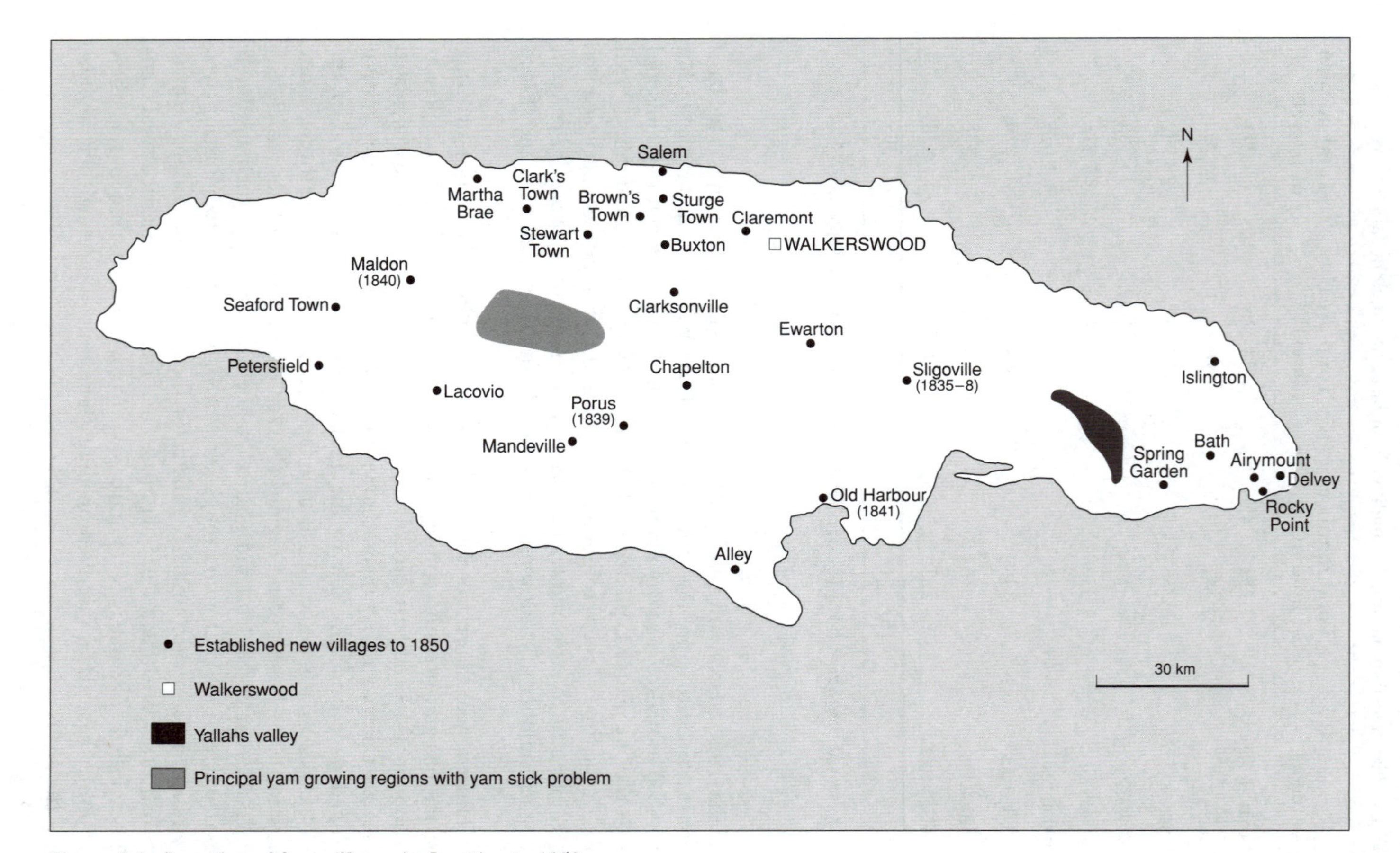

Figure 3.4 Location of free villages in Jamaica to 1950
Source: Adapted from Hall (1959)

Marshall (1972) argues that Caribbean peasant society emerged as a mode of resistance to planter hegemony after Emancipation, so its origin is different to the peasantry in Africa, Asia and Europe. Mintz (1985) eloquently characterises these reconstituted Caribbean peasantries as growing into the historical and ecological crevices of their societies, using land not wanted by the plantations. Thus, dual agrarian economies were created by the unequal competition for scarce land resources between planters and peasant farmers, resulting in distinctive land use patterns in Jamaica, St Kitts and the mountainous Windward Islands, of lowland sugar estates and upland small farming regions, though coffee plantations in Jamaica's Blue Mountains were an exception to this geographical separation. On the lower elevation islands like Barbados and Antigua, dual agrarian economies emerged after Emancipation, but there was a more heterogeneous spatial mixing of plantations and smallholdings. Richardson (1984, 1997) and Berleant-Schiller and Pulsipher (1986) explored contrasts between highland and lowland agricultural adaptations in the eastern Caribbean.

Contemporary patterns of Caribbean land tenure

Land tenure refers to the nature of people's access to land, or land rights. The region's diversity of land tenure reflects the pervasive influence of the colonial plantation system and the adaptation, innovation and experimentation of Caribbean people in response to it. All the basic types of land tenure in the world are present in the region, namely, customary, freehold, landlord–tenant, corporate ownership, socialist and squatting.

Customary tenure (or communal tenure) is where control over land rights is vested in a group (the tribe, clan or village). Farmers have rights to use land (use rights) but do not own individual parcels. Communal tenure was probably the original kind of land tenure, evolving among shifting cultivators before land became a commodity to be bought and sold and where cultivating a particular piece of land was a temporary phenomenon. Communal tenure was egalitarian and provided land for all according to need, and in Africa complex systems reflected different cultures and local ecologies. In many parts of the world there has been a gradual breakdown of communal tenure into more individualised and privatised freehold systems. In the Caribbean, remnant traditional communal systems are associated with Amerindian shifting cultivators in the rain forests of Belize and Guyana. However, the region's most important type of freehold customary tenure is family land and is unique, emerging after Emancipation as an adaptation by newly freed slaves to the hegemony of plantations (Box 3.1).

Another well-documented example of customary tenure is on Barbuda, an island once owned by the Codrington planter family (Berleant-Schiller 1987, 1991). After 1870, the colonial government left the islanders to fend for themselves and a system called 'the commons' evolved. Oral tradition grants all Barbudans equal rights to use common land outside the village for farming or hunting. By custom, common land belongs to everyone socially recognised as Barbudans, even

Box 3.1: Family land: a unique Caribbean customary tenure

Family land is unique to the Caribbean and is otherwise known as generational property in the Bahamas (Craton 1987). In St Lucia, where it is called in dialect 'te family', the census officially defines it as 'land under multiple ownership'. Family land is found amongst Afro-Caribbean rather than East Indian or white Hispanic rural communities, prompting Edith Clarke to suggest West African origins, probably Ashanti (Clarke 1966). However, the pioneering work of Jean Besson (1984, 1987) in Martha Brae in Jamaica has clearly demonstrated that family land originated in the post-Emancipation period and is a genuine Afro-Caribbean institution rather than one inherited from African traditions. People acquired freehold land and sought to maintain it as a valued possession to pass on to their descendants, in the face of the hegemony of the planters. Most forms of communal land tenure evolved in pre-capitalist societies, but family land emerged in the context of a land market, the original purchases made from estate owners. Thus it is an unusual example of conversion from private property to communal land holdings.

Family land is a tenure system in which all the descendants of the original owner inherit the (undivided) land, even if the owner has not made a will and dies intestate. Legal codes recognise that intergenerational transfers of land can be made by word of mouth, according to customary practice rather than through a legal document. Land becomes family land if the owner dies without leaving a will. All the children have a right to use the land, jointly, and in perpetuity. Even illegitimate children of the owner have the same rights to the land provided they are acknowledged and recognised by other family members. The descendants of the original owner have inalienable land rights in perpetuity ('to serve we children's children 'til every generation dead-out') so family land symbolises freedom, security, prestige and pride, and these are more important than its role as a productive asset in farming.

Use rights include the right to farm family land, to build on it, and to harvest food from trees planted by the parent. Children of the wife of the owner (who are not the owner's children) do not have the same use rights except through the generosity of the co-sharers. Strictly speaking, family land is not partitioned (legally) after death, and heirs do not normally acquire tenure to a specific portion of the land. Thus family members will occupy and control different portions as generally agreed by the family. Heirs can use the land for farming/housing and bequeath it to their children, but they cannot sell the land or otherwise dispose of it unless by consent of all the co-sharers. In reality, there are innumerable variations in arrangements with family land, and it is sometimes legally partitioned and parts (or all of it) are sold or willed to individuals. An interesting aspect is that even if the co-sharer has migrated that person retains the use rights to family land and so may return and reclaim a piece of the land and build a house, sometimes for retirement purposes.

if they migrated many years ago. Houses and yards in the island's only village (Codrington) are individually owned, and can be sold or bequeathed. Conflicts have arisen as the national government in Antigua and tourist developers anxious to exploit the island's natural resources have clashed with Barbudan tradition. Other examples of communal lands are found in the Accompong maroon community in Jamaica (Barker and Spence 1988; Spence 1989; Besson 1997) and in the Bahamas (Craton 1987).

Landlord–tenant systems are notoriously repressive sources of social, economic and political injustice and fomented violent political revolutions in China and other developing countries in the twentieth century. Insecurity of tenure is common, and tenant farmers are liable to be evicted at short notice, so this discourages long-term investment in farming. Tenant farmers pay for the right to farm the owner's land and cash payment is normal, as either a rent or a lease. Payments in kind can involve labour arrangements, so a tenant farmer may be required to work for the landlord at certain times of the year for low wages. In sharecropping (metayage) the landowner receives a share of the tenant's produce in lieu of cash rent. Sharecropping was once widespread in the southern United Sates though less common in the Caribbean. After Emancipation in St Lucia, however, some plantation owners leased part of their land to peasant farmers, who had to agree to give up a portion of their cane crop in return for using the land. Originally, as much as a half or one-third of the crop was demanded but, over time, small farmers accumulated capital and became more independent and the share required fell. Today, sharecropping in St Lucia is practised on about 1,000 acres (1.5 per cent of total farmland). In Haiti, inter-generational sharecropping is still a feature of agrarian society (Lundahl 1983).

Rural squatting was historically significant as a means of acquiring farmland. Crown lands and idle estate land were targets for land-hungry farmers (Satchel 1990). Rural squatting exists in the contemporary Caribbean but its significance and geographical extent are largely unknown. Gumbs (1997) suggests there may be 50,000 rural squatters in Trinidad, and in St Lucia, Cole (1994) distinguishes between rural squatters who 'hustle' for a living and those who are genuine farmers on crown lands in forest areas.

Agriculture and globalisation

Industrialised agriculture is said to have originated in North America then spread to Europe (Stutz and de Souza 1998). However, some of its characteristics (Figure 3.2) were present in plantations long before then and in many ways plantations were a prototype for industrial farming. They were capitalist, labour-exploitative and profit-oriented systems geared to export crops for an overseas market. They utilised large amounts of land, and sugar, for example, was cultivated intensively in pure stands and processed into a new product on site. Watts (1987) has described the environmental problems associated with plantation agriculture and the continuous innovation and experimentation needed to maintain yields. Sugar plantations became more 'industrial' and capital-intensive at the end of the nineteenth century, partly as a result of acute labour shortages. Joint-stock companies were formed to raise funds for capital expenditure to modernise production and reap economies of scale, and began to replace family-owned plantations in export agriculture (Courtnay 1965). These companies were early examples of agribusinesses, corporations engaged in food-related enterprises including farming, food processing and food transport, many of which, by the end of the twentieth century, had become transnational corporations (TNCs).

Foreign-owned agribusiness has a long history of engagement in the Caribbean. In sugar, for example, Tate and Lyle acquired the large centralised sugar factories in Frome (western Jamaica) and Caroni (Trinidad) in the mid-1930s (Chalmin 1990). The Boston Fruit Company was involved in the Jamaican banana trade and, at the beginning of the twentieth century, owned hotels in the country's nascent tourist industry. Later, it became the United Fruit Company and then grew into the corporate giant Chiquita Bananas, which continues to play a significant role in bananas in Central America and the world banana trade generally.

In Cuba many large farms became corporate entities in the twentieth century, owned by either Cuban or American businesses at a time when Cuba's relationship with the United States was neo-colonial. Prior to 1959, 13 of the 22 large sugar companies were American-owned (Hall 1989). However, after the revolution, Cuban sugar production became even more industrialised in terms of farming methods. Plantations were nationalised, but as late as 1994 the spatial structure of farming was largely unaltered and 80 per cent of the agricultural land consisted of large state farms roughly corresponding to the expropriated plantations. Agriculture was heavily dependent on imported oil, fertilisers, pesticides and machinery from the Soviet bloc. Sugar prices paid by the Soviet Union were over five times higher than world market prices. Following the collapse of the Soviet Union in 1989, oil imports were immediately reduced by 50 per cent and supplies of fertilisers and pesticides fell by 80 per cent. This disaster reduced sugar production by over 50 per cent, and desperate short-term adjustments included replacing tractors with oxen.

Generally, the increasing dominance of agribusiness has relegated the central role of the farm so that it is just the first stage of a globally integrated economic process that connects traditional farm activities with food consumption in urban societies. These integrated stages involve production, storage, processing, distribution, marketing, and wholesaling and retailing, and restaurants and food outlets. The incorporation of farming into the world capitalist economic system is part of agricultural globalisation, a process of restructuring of global food production and food supply. The main components of agricultural globalisation can be summarised as:

- the dominance of agribusiness;
- the application of science and technology (especially biotechnology) in farming, transport and storage, food processing, and food sales;

- the promotion of trade liberalisation through regulatory bodies like the World Trade Organization (WTO), the World Bank and national government policies.

The relationship between transnational food corporations and developing countries can take several forms. Agribusinesses may own plantations and estates in developing countries, attracted by cheap agricultural land and labour, favourable environmental factors and appropriate infrastructure. They might supply essential farm inputs (agrochemicals, fertilisers, seeds, farm equipment), engage in transport and trade, or purchase farm produce from local producers under contract farming arrangements. A typical procurement strategy is global sourcing, whereby a TNC accesses multiple food production sites (in more than one country) to reduce political and environmental risk (Grossman 1998a).

As plantation economies, Caribbean countries were part of an international economy geared to the export of primary agricultural products from colonies to metropolitan Europe. Globalisation, however, is forging a new international division of labour in agriculture, which is manifested by the trade of high-value, labour-intensive, off-season and non-traditional products from developing countries and a counter-flow of basic grains and oil seed from industrialised countries (Grossman 1998). Momsen (1998) comments on how some Caribbean farmers and local food processors are beginning to break into new overseas markets through souvenir food items such as pickles, sauces and seasonings, as well as aged rums and tropical fruit juices. In Jamaica, the community of Walkerswood is a contemporary example of a highly successful food enterprise that connects local farmers to overseas markets (Box 3.2).

Box 3.2: Walkerswood: linking small farmers, agro-processing and exports

Walkerswood Caribbean Food Ltd is located in the hills of the parish of St Ann, Jamaica (Figure 3.4). The company is perhaps best known for its jerk seasonings, but it also produces more than 20 other spices and sauces, including Coconut Rundown and Escoveitch pickle for fish dishes. Several of its products have won international prizes. The roots of the company lie in the traditions of self-help in the Walkerswood community and efforts to create local employment. It began as a two-person operation engaged in grilling marinated pork in local rum bars (www.forachange.co.uk/index). In 1978 a registered company was established but the spirit of the community cooperative venture has been retained and the company uses local farmers to supply farm produce. Today the company employs over 100 people and is still employee-owned.

The company has expanded its operations and has established a marketing company. The first overseas office was opened in Miami in 1986, and others were added in Toronto and the UK. Products are exported to Europe, North

(Box continued)

Figure 3.5 At the Walkerswood farm and factory 10 km south of Ocho Rios in rural Jamaica, the company's famous jerk seasoning marinates for one year in large plastic drums. Then it is bottled as shown at the top right in the photo. Eighty per cent is consumed overseas
Source: Tom Klak

America and the Caribbean, and are worth US$3 million per annum. In 1997, Walkerswood, in partnership with an enterprising Jamaican fast food chain called Island Grill, opened a restaurant in Brixton, south London, called the Bamboula (www.walkerswood.com). The company has also published a recipe cookbook.

The company also helped the local community to establish a village farm run by the Walkerswood Farm Group, which supplies Jamaica's fiery Scotch bonnet peppers. However, following a series of droughts in the late 1990s, the company experienced supply shortages and high prices, especially for escallion, so it has tried to spread risks by establishing its own farm to supply some of its produce. There is a major expansion underway designed to improve efficiency and move towards a more vertically integrated operation. The old processing factory, located in the grounds of Bromley Great House, is to be replaced by a new factory, and there are plans to build a water catchment to irrigate the company farm, and to build a visitor centre and gift shop.

Like other facets of globalisation (Chapter 10), the odds seem stacked against developing countries, and we can summarise the main issues as they relate to agriculture:

- cheap imported food is attractive to governments in developing countries anxious to feed large, impoverished urban populations;
- cheap imported food is often subsidised by industrialised countries, but depresses commodity markets, with negative impacts on market prices for third world farmers;
- trade liberalisation has a negative impact on high-cost agricultural exports, as in the case of Caribbean bananas;
- trade liberalisation and the globalisation of food consumption habits offers new opportunities in overseas markets.

Caribbean governments recognise the need to diversify the agriculture export sector to take advantage of new markets, but much of the region's export agriculture is still firmly locked in the past and, generally speaking, agricultural globalisation is having negative impacts on export crops like sugar and bananas. The general prospects for Caribbean commercial export crops are discussed in Ahmed and Afroz (1996).

The decline of the sugar plantation

Sugar cane was domesticated in New Guinea and spread to the Caribbean via India, through the Mediterranean to the Atlantic islands and then to the Americas, a diffusion process spanning several millennia (Galloway 1989). Botanically, sugar cane is a grass ideally suited to the Caribbean seasonal climate of wet and dry seasons, and grows within a wide range of precipitation limits. Cane accounts for about 70 per cent of world sugar production, but sugar is also produced from sugar beet, a temperate root crop, and competition comes from sugar substitutes made from corn starch.

In 1961 Caribbean sugar cane accounted for 20 per cent of world production but has since declined to less than 4 per cent (see apps.fao.org). India is the world's largest sugar producer though most of its output is consumed domestically. Brazil, another of the top three producers and the world's largest sugar exporter, has an output larger than the entire Caribbean region. Significant amounts of Brazilian sugar are converted to ethanol and used as an alternative fuel source to power millions of specially converted motor cars. Cuba is by far the region's largest producer, accounting for 75 per cent of Caribbean production in 2001, though its output crashed by over 50 per cent after the collapse of the Soviet Union in 1989. Declines in output and area planted in cane have occurred across the region with the exception of Belize, which now produces three times as much sugar as Barbados. Antigua, St Lucia and the USVI have ceased exporting sugar but the most dramatic decline is in Martinique and Guadeloupe (Table 3.4).

Table 3.4 Sugar production in the Caribbean, 1961–2001

Sugar cane production (Mt)	Year				
	1961	**1971**	**1981**	**1991**	**2001**
Antigua and Barbuda	183,820	134,100	2,801	0	0
Bahamas	0	2,000	52,000	45,000	45,000
Barbados	1,400,117	1,233,500	966,000	587,000	520,000
Belize	247,367	642,783	985,670	1,131,880	1,150,000
Cuba	55,885,920	54,700,000	66,678,496	79,700,000	35,000,000
Dominica	6,400	4,000	4,600	4,400	4,400
Dominican Republic	7,811,195	9,973,725	9,629,000	6,930,457	4,645,332
French Guiana	11,700	5,000	11,250	4,000	5,300
Grenada	10,400	10,535	9,300	6,500	6,750
Guadeloupe	1,914,000	1,733,319	834,045	645,000	798,072
Guyana	3,618,829	4,310,606	4,192,220	2,935,000	3,000,000
Haiti	2,850,000	2,792,400	3,000,000	1,500,000	1,008,100
Honduras	804,300	1,407,112	2,920,360	2,730,136	4,117,000
Jamaica	4,438,093	4,105,846	2,492,370	2,732,000	2,400,000
Martinique	1,140,218	514,375	244,148	189,708	207,000
Puerto Rico	9,755,814	4,156,717	1,848,830	843,061	320,000
St Kitts and Nevis	419,595	276,352	337,500	200,000	188,373
St Lucia	66,700	0	0	0	0
St Vincent and the Grenadines	34,395	0	30,000	23,000	20,000
Suriname	138,782	195,000	146,327	70,000	120,000
Trinidad and Tobago	2,517,954	2,349,221	1,289,521	1,300,900	1,500,000
Virgin Islands (US)	104,000	0	0	0	0

Source: apps1.fao.org

Despite its negative associations with the past, sugar is still an important source of foreign exchange for the remaining producers, generating US$338 million in 1998/99 (Ahmed 2001) and the industry employs some 150,000 unskilled and semi-skilled workers in the English-speaking region. In Guyana, sugar accounts for approximately 16 per cent of GDP and despite declining production in Barbados it still contributes 16 per cent to export earnings and one-third of agricultural production by value.

A modern sugar plantation is a large-scale operation, requiring good, relatively flat, arable land, and a large labour force for cultivation and processing. Its spatial organisation connects a pattern of fields, roads, housing and processing facilities into an integrated farm enterprise. Field layout generally reflects the need to process the harvested cane quickly and a factory will tend to be centrally located within a sugar estate. In research analysing historical plantation maps, Higman (1987) found arcs drawn on some estate maps by land surveyors; these apparently suggested how different types of land use might be located at increasing distances away from the central processing point on the estate. He interprets this as movement minimisation and profit maximisation, similar to the spatial logic of von Thünen's agricultural location theory, whereby land-use intensity diminishes with increasing distance from a central market town.

Sugar cane is also a cash crop for small and medium-sized farmers. Long ago, small farmers ground cane and produced sugar themselves, but later their output was sent to the factory for processing. In the 1930s in Jamaica, independent farmers supplied 75 per cent of the cane used by the island's 34 sugar factories, whilst cane farmers in Trinidad (two-thirds of whom were East Indian), supplied 40 per cent of the cane used by the country's 12 sugar factories (Chalmin 1990). The number of factories and cane farmers has declined considerably since then, but the linkages between the sugar factory and surrounding cane farmers still defines a 'catchment area' whose size is partly determined by transport costs and perishability (since sugar quality deteriorates rapidly 24 hours after reaping). Cane farmers are often located at the economic margins of cultivation with respect to nearby factories, and many farmers have shifted out of sugar into other cash crops.

Auty (1976) examined the pattern of sugar factory closures from the 1930s to 1960s. He found that the threshold size for factory survival had increased, forcing factories unable to increase production to close. The rate of factory closure was inversely related to mean factory size and growth in production. Plantations able to invest in higher-yielding varieties of cane were generally foreign-owned. Some plantations survived by intensifying cane production, whilst others tried to survive by bringing more land into production, even though it was marginal to cane cultivation. At times, strong labour unions and sympathetic governments negotiated better wages for poorly paid sugar workers, but the downside was increased production costs contributing to higher survival thresholds. He depicted the pattern of closure as a four-stage spatial and temporal process. Since his research, the number of sugar factories has declined further and only 23 cane factories remained in the English-speaking region in 1997. Barbados and Jamaica each had 20 working factories in the 1950s, but there are only three factories left in Barbados and eight in Jamaica, several of which face an imminent threat of closure.

Many initiatives and rescue packages have tried to revive the sugar industry, including alternative ownership and management models, such as the failed sugar cooperatives in Jamaica in the 1970s. Transnational corporations withdrew from ownership of sugar estates in the 1980s, and negotiated advantageous management contracts in Jamaica and Guyana, at less risk to themselves and avoiding the responsibility of investing their own money in capital equipment and infrastructure (Ahmed and Afroz 1996). The enduring problem, however, is the high cost of production, partly due to lack of capital investment. Belize has the lowest costs of regional producers (Ahmed 2001) but production costs are higher than the world price of sugar. Globalisation will have an impact on the sugar industry because the Commonwealth Caribbean still enjoys guaranteed prices in the EU, but this situation will change with trade liberalisation. However, the impact of globalisation on Caribbean export agriculture is much more apparent with respect to bananas.

Globalisation and bananas in the Windward Islands

The commercial production and export of bananas in the Caribbean began in the late nineteenth century when Jamaican bananas were exported to the eastern seaboard of the USA. In the Windward Islands commercial banana production started in the 1930s under a system of contract farming (Grossman 1998a) but exports to the UK took off in the 1950s to fill the gap left by the postwar decline in Jamaican banana exports. WINBAN was formed in 1958 and acted on behalf of the four Windward Island Banana Growers Associations (Welch 1996), negotiating preferential treatment with the commercial fruit merchant Geest and the British government, an example of an international trade agreement actually benefiting small farmers.

In the early 1960s Jamaica still produced more bananas than the combined Windward Islands, but by the early 1970s, banana production in the former had collapsed by more than 50 per cent. Figure 3.6 shows the fluctuations in banana production in the Windward Islands from 1961 to the present. After the increases in the 1960s, production fluctuated over the next decade largely in response to the impact of natural disasters such as the volcanic eruption on St Vincent in 1979, which caused a decline from 30,000 to 19,000 metric tonnes. The negative impact of Hurricanes David and Allen in 1980 in St Lucia and Dominica can be clearly seen on these graphs (see Williams 1988). In the late 1980s and early 1990s, there was another boom period for bananas in the Windwards, helped by agricultural innovations like field packing to reduce the bruising of the fruit during transport from the field to the port. The downturn in production as a result of globalisation from the mid-1990s is also evident in Figure 3.6. Grenada, especially, has suffered a catastrophic decline since 1995 because its banana exports failed to meet European quality standards.

In the saga of Caribbean banana exports, the 1975 Lomé Convention was significant because it extended preferential access to the entire European Community. When the EU became a single market in 1993, a new system of tariff-free quotas and licensing arrangements was introduced. In 1996, the USA in conjunction with Mexico, Guatemala, Costa Rica and Ecuador lodged a complaint with the

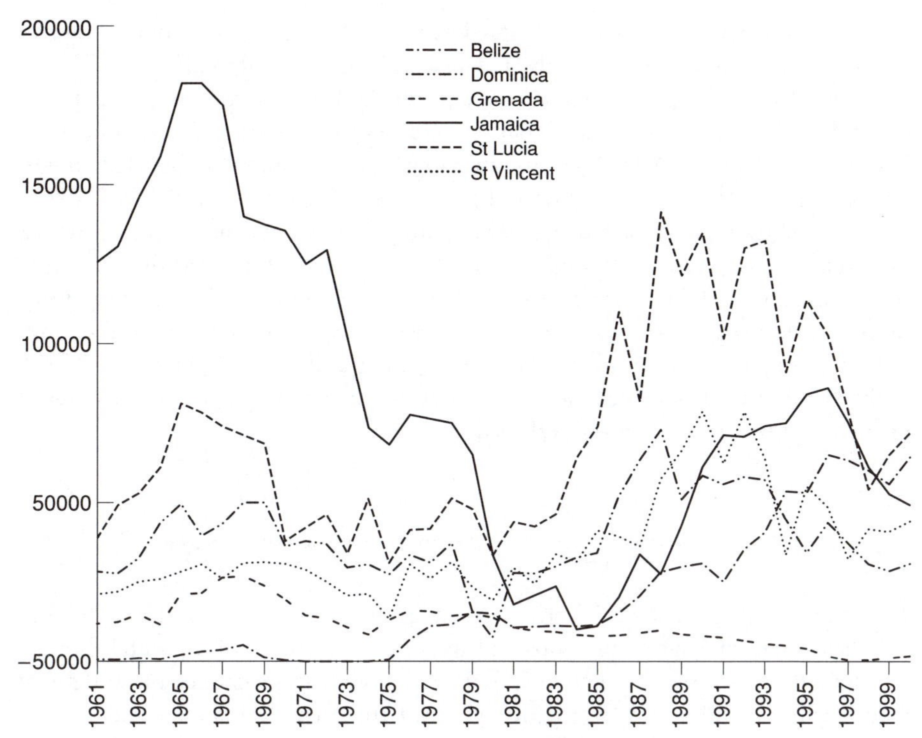

Figure 3.6 Banana production for selected Caribbean countries, 1961–2000
Source: apps.fao.org

WTO about preferential treatment even though Caribbean bananas accounted for only 7–9 per cent of EU banana imports. Thereafter, the preferential regime started to be dismantled. A modified regime introduced in 1997 was also disputed, and another unfavourable WTO ruling followed in 1999 (Ahmed 2001). Another EU agreement in 2000 replaced the Lomé Convention and extended preferential treatment until 2008, to provide a 'breathing space' for countries to reassess their economic options. Thereafter, the full impact of competition and globalisation on Caribbean bananas is likely to be felt.

Trade liberalisation is a handmaiden of globalisation and is based on the theory of comparative advantage, which argues that, in a system of free trade, regions should specialise in those agricultural commodities best suited to local environmental conditions. In practice, trade liberalisation favours low-cost banana producers such as Costa Rica and Nicaragua. In Central America, banana production is undertaken by large corporations on plantations as big as 12,000 acres. This industrial scale of production requires expensive heavy capital investment on roads, irrigation and drainage, cableways and field packing sheds, though there are negative environmental impacts and wages are very low. In the year 2000,

Costa Rican banana exports were valued at over US$500 million, more than five times greater than the Windwards, Jamaica and Belize combined.

In the Windward Islands, bananas are produced by small farmers; 82 per cent of the 30,000 eastern Caribbean banana producers have less than 5 acres of mainly hillside land (Ahmed 2001). They use traditional mixed cropping rather than monoculture, and yields are much lower than in Costa Rica. During the late 1990s, the world market price for bananas was around US$450 per metric tonne, whilst preferential EU prices were higher than US$800. Alarmingly, production costs in the Eastern Caribbean are over U$500 per metric tonne, higher than the price on the world market (Ahmed 2001). Though bananas currently account for about half of all exports, one-third of total employment and over 10 per cent of GDP in the Windward Islands, the long-term prospects are dismal unless costs are reduced or niche 'green' markets are developed.

The dynamics of small-scale farming systems

An agricultural system is a functionally integrated system for producing food in which farmers manage their crops and livestock and the environmental, human and financial resources at their disposal. An agricultural system can be defined at different spatial scales, from the field and the farm, watershed and region, to nation-state and the world. The term agro-ecosystem is also used, since agricultural systems have reciprocal relationships with the physical environment and irrevocably modify natural landscapes.

Small-scale farming systems in developing countries are being transformed by population growth and technological innovations, and Figure 3.7 shows how they adapt to population increase. To increase food output, small farmers either bring more land into agricultural production or intensify farm inputs, both of which can lead to land degradation if farming methods are not environmentally sustainable. Agricultural intensification is defined as an increase in labour inputs and/or an increase in capital inputs per unit of land. Indigenous intensification is spontaneous and internal to the farming community (Adams and Mortimore 1997), although intensification is usually induced externally by the application of imported capital inputs (agrochemicals, farm machinery), or by the transfer of technology packages as in the Green Revolution.

Reduction of the fallow period

A typical response of small farming systems to population growth is a reduction in the length of the fallow period (Figure 3.7). Traditional small farming usually involves a system in which land is rotated between a cultivation period and a fallow period. The cultivation period is normally shorter than the fallow period. In

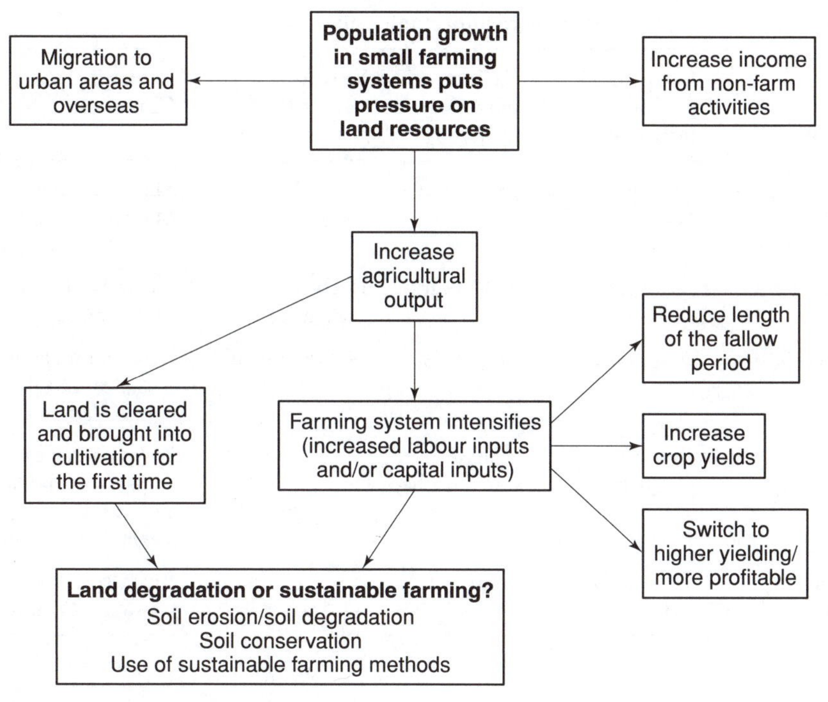

Figure 3.7 Responses of small farming systems to population pressure

its primal form, shifting cultivation incorporates a long forest fallow (Ruthenberg 1971; Grigg 1974). A farmer clears a small patch in the forest, cultivates it for two or three years, then the land is abandoned because soil fertility is depleted or weed problems are too difficult to manage (Kellman and Tackaberry 1997). He then shifts his farming operations to another part of the forest, where land is cleared and cultivated again until that too is abandoned. Simple technology (axes and machetes) are combined with fire to clear the land (hence the term 'slash and burn'). The fallow phase improves soil in a variety of ways. Leaf litter and root decomposition helps recover soil nutrients, recycles nutrients leached into the subsoil, increases soil biological activity and improves the soil's physical structure. In time, farmers return to the same piece of land, clear and cultivate it for a few years before abandonment again to fallow, for another cycle. Thus land rotates between cultivation and fallow, and the agricultural system comprises both cultivated and fallow land.

Boserup's (1965) seminal model suggests that small-scale farming systems pass through several stages of intensification as a result of population growth. These stages are associated with reductions in the length of fallow and the intensification of labour inputs (Table 3.5). A forest fallow requires a large tract of land to support a relatively small population. Today, these systems are confined to tribal

Table 3.5 Boserup's model of agricultural intensification

Type	Cropping period (approx)	Fallow period (approx)	Fallow vegetation	Examples in Caribbean
Forest fallow (shifting cultivation)	1–3 years	15–25 years	Secondary forest	Mayan farmers in Mayan Mts, Belize; Arawaks in Guyana
Bush fallow	3–5 years	6–10 years	Bushes Immature trees	Uplands of Haiti and Jamaica
Grass fallow	2–5 years	1–4 years	Grasses, shrubs	Drier, rain shadow areas, St Elizabeth parish, Jamaica
Annual cropping	One crop per year	none or few months		Yams, sugar cane, bananas, in Jamaica and Windwards; Vegetables, Trinidad
Multiple cropping	Two or more crops per year	none		Rice production Guyana, Trinidad

peoples in the world's remaining rain forests, and are found in Guyana and the Mayan Mountains in Belize. In Belize shifting cultivation is called *milpa* agriculture and a variant is 'slash and mulch' (Spurney and Cavender 2000). In a bush fallow, land is cultivated for three or five years then abandoned for about six to ten years. Since the forest has not fully regenerated, bushes and immature trees are cleared at the start of the next cultivation period. In a grass fallow, land is left uncultivated for a few years to be recolonised by tall grass and small shrubs before being cleared again for farming. Eventually intensification proceeds to the point at which fallow is no longer practised and the land is permanently cultivated (Table 3.5).

Caribbean farmers refer to fallowing as 'resting' the land, which is not to be confused with idle land or 'ruinate', which is often in a similar condition of abandonment and revegetation but withdrawn permanently from production. Bush and grass fallowing are common techniques for restoring soil fertility among Caribbean farmers, but fallow periods are shorter than a generation ago and have disappeared in the commercial yam farming region of central Jamaica (Barker 1998; Barker and Beckford 2003). Intensive kitchen gardens and tree crop farming are examples of permanent agriculture discussed below. Multiple cropping (Table 3.5) is when a crop is grown more than once on the same farm plot in a calendar year. It is common in Guyanese rice farming and among vegetable farmers growing short-term cash crops for urban areas or tourist hotels.

Land degradation and soil erosion

Land degradation and soil erosion are natural processes, and the Caribbean region has high rates of geomorphic activity as a result of the combination of geology and climate (McGregor 1995). Tectonic activity, rapid physical and chemical weathering and soil erosion have resulted in landscapes with steep slopes and a highly dissected terrain (Chapter 1). Land degradation is the loss or reduction in the quality or utility of land, usually with respect to agricultural potential, in terms of the physical, chemical and biological properties of land. Soil degradation refers to the deterioration in soil quantity and/or soil quality. Deterioration in quality can occur through prolonged production, in a single crop, as with sugar cane (Richardson 1992). Land degradation is most commonly associated with soil erosion and slope failure resulting in mass movement (Chapter 4).

Severe land degradation is caused by human disturbance of the landscape, especially for farming. Deforestation is a key component and a trigger for accelerated soil erosion because forest ecosystems act as a natural buffer against soil loss. For example, the tree canopy intercepts and absorbs the energy of raindrop impact, and tree roots bind the soil together and open up passageways, allowing rainwater to penetrate into groundwater aquifers. Vegetation clearance removes the protection provided by leaves and roots, exposing topsoil to the elements especially during torrential tropical rainfall. Erosion is initiated by the impact of rain splash, which dislodges soil particles. There is a progression through splash erosion, overland flow and sheetwash, to rill erosion and gully erosion.

In the contemporary Caribbean, land degradation is widespread, especially on the mountainous islands, and Haiti is widely regarded as the worse case (Paskett and Philoctete 1990) with only 1 per cent of its forest intact. In Jamaica, 17 of its 26 watershed units are considered critically degraded and the Scotland district in Barbados is an area with a long history of soil erosion problems (Patel 1995). Land degradation began with the colonial plantation system, when lowland forests were cleared to establish sugar plantations. Even on Barbados' subdued topography, soil erosion was rampant 250 years ago, and 'holing' was an innovation devised to impede overland flow and sheetwash erosion in cane fields. Watts (1987: 403) explains how cane-holes (about 1.5 m square and 15 cm deep) were excavated by hand and cane was planted in them, and the system of two-directional ridges between the cane holes hindered downslope soil wash. In Jamaica's Blue Mountains in the eighteenth century, coffee plants were sometimes washed out of the ground by torrential rainfall before newly planted coffee bushes were harvested (Barker and McGregor 1988; McGregor and Barker 1991). After Emancipation, the clearance of forests in upland areas by land-hungry small farmers also triggered soil erosion and land degradation in hitherto unaffected areas, such as central Jamaica and Haiti (Barker 1989; Edwards 1995; Lindskog 1998).

Upland watersheds and steep hillsides are particularly susceptible to land degradation and soil erosion, and small farming systems are associated with land degradation whenever land is not cultivated using environmentally sound methods. In Jamaica, slopes over 20° cover half the island, and as much as a quarter of the surface area has slopes over 30°, yet these same hillsides are the principal areas of small farming in which reside about a quarter of the island's population. For the Eastern Caribbean, Gumbs (1997: 6) notes that slopes exceeding 20° occupy 70 per cent of St Vincent's and Grenada's surface areas, again often areas of small-scale hillside farming.

A number of factors influence rates of soil erosion, including slope angle and length of slope. Soil erosion rates are difficult to assess and there is an absence of reliable scientific data in the Caribbean. Many published rates of soil erosion are no more than intelligent guesswork (McGregor *et al.* 1998). In an experimental study, Gumbs and Lindsay (1982) measured erosion rates in Trinidad's Northern Range, comparing maize and cowpeas with bare plots, on slopes of 11 per cent, 22 per cent and 52 per cent. During the field trials, two major rainfall events accounted for 56 per cent of the total rainfall and 70–92 per cent of the soil loss from bare plots, depending on slope angle. Cropped plots had a full canopy during the two storm events so there was little soil loss. Bare slopes lost an estimated 28 to 55 tons/ha on the steepest slope, whereas maize and cowpea losses were estimated at 19 and 11 tons/ha, respectively. In a critical review of soil erosion data for Jamaica, McGregor (1995) explains the problems of extrapolating from controlled experiments at the scale of the farm plot on an agricultural research station to general assessments for erosion for an entire watershed.

Figure 3.8 shows the three basic approaches to soil conservation: technical and engineering methods, soil management methods and agronomic methods, many of which have been utilised in the Caribbean (Gumbs 1997; Sheng 1972). Floyd (1970) and Baxter (1975) examined the diffusion of soil conservation methods in the Yallahs valley in eastern Jamaica (Figure 3.8). In the past, technical engineering

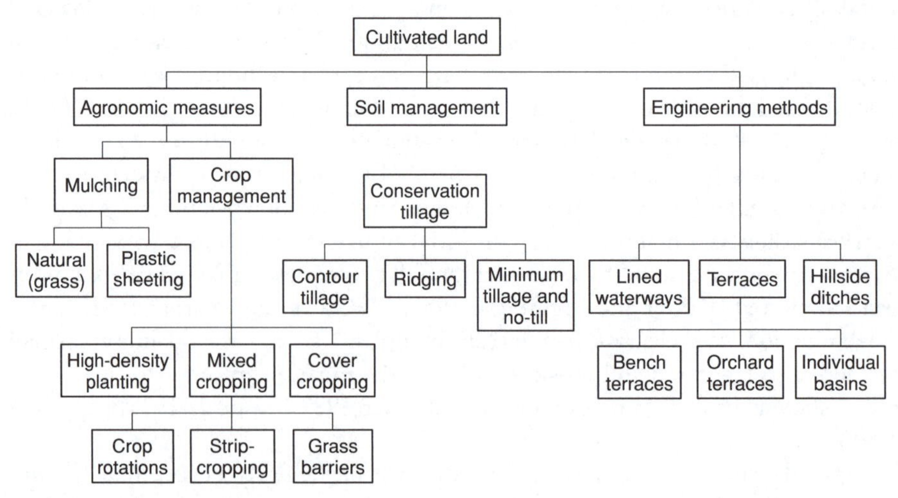

Figure 3.8 Classification of soil conservation methods
Source: Adapted from Morgan (1986)

methods were favoured among planners, for example, bench terraces, which are effective on steep slopes up to 25°. However, they are expensive to construct and experience shows that maintenance costs are too high for small farmers (Edwards 1995). Contemporary interest in sustainable farming has focused attention on agronomic methods of soil conservation because they are cheaper and often involve traditional methods familiar to small farmers.

Small farming traditions and sustainable agriculture

In the past, traditional small farmers in developing countries were stereotyped in negative and pejorative ways (Hills and Iton 1983; Richards 1985). They were stigmatised as inefficient, subsistence-oriented, conservative and reluctant to adopt agricultural innovations. Traditional small farmers are generally 'resource-poor' with low levels of technology and formal education, but negative stereotypes are unhelpful in understanding the dynamics of small farming systems. For example, small farmers are often associated with, and contribute to, soil erosion and land degradation through poor farming techniques, but a wider perspective is needed to fully appreciate the complex linkages between land degradation and social, economic and political processes (Blaikie 1985; Blaikie and Brookfield 1987). In the Caribbean, the historical dominance of the plantation system and its appropriation of the best farm land, and the ramifications of agricultural globalisation, have placed severe constraints on the options available to small farming communities and their survival strategies. Yet despite this, there are many examples of small farming systems that do not degrade the environment, generate farm income and are productive in terms of food output.

Other stereotypes are also misleading. Grossman (2000) criticises conventional assumptions that equate women farmers with subsistence cultivation, among commercial banana producers in St Vincent. Bliss (1992) shows that Garifuna women in Belize responded positively to economic change by becoming involved in the citrus industry whilst maintaining their traditional role in subsistence production, especially in the cultivation and processing of cassava. Momsen (1988) has analysed the key role of women in Caribbean agriculture through the plantation period to the present. Similarly, older farmers are stereotyped as conservative and, supposedly, are less able to cope with the rigours of farming than younger people (an assertion that might be disputed in the Caribbean). Spence (1999) reports complex relationships between farm size, age, crop types, local ecology and age of entry into farming in Jamaica and concludes that, irrespective of ecological setting, small farmers are rational decision makers in their use of human, agro-physical and infrastructural resources, and adjust their land-use decisions accordingly.

Decision making in traditional farming systems

The theme of the inherent rationality of farmers' decision making is the cornerstone to understanding the complexities of the small farming systems. Several factors

Figure 3.9 Some factors that influence the decision making of small farmers

Socio-personal
- Age
- Gender
- Personality factors
- Formal education and agricultural training
- Life experiences
- Knowledge and reasoning ability
- Family size and responsibilities
- General level of health
- Community cultural/religious beliefs
- Sources and channels of information

Economic
- Alternative income sources
- Market demand
- Market location
- Profit levels
- Availability of land
- Size and number of farm plots
- Location of residence in relation to farm plots
- Availability of labour
- Availability of capital
- Transport costs
- Marketing methods
- Government policies
- Cost of conservation techniques

Environmental
- Soil type and drainage
- Slope angle and aspect
- Predictability/uncertainty of weather
- Seasonability of climate/weather
- Rainfall amount, distribution, intensity
- Strong winds/wind direction
- Temperature (diurnal/seasonal range)
- Biological pests
- Natural hazards

influence decision making (Figure 3.9), and these are often sources of risk and uncertainty, including changes in personal circumstances. Small farming systems may be well adapted to local soils, climatic or topographical conditions, but vulnerable to a natural hazard or a disaster caused by a hurricane or a prolonged drought (Chapter 4). Production may be profitable but vulnerable to a collapse in market price resulting from cheap food imports. Game theory illustrates how small farmers in developing countries often seek to spread or minimise risk rather than maximise profits (Ilbery 1985).

Three basic assumptions underpin small farmers' decision making:

- Farmers are generally rational even though decisions like the refusal to adopt a new technological innovation may not appear to be rational to an outsider.
- Farmers make choices based on information and knowledge, but the complete 'perfect' knowledge assumed in models of perfect competition is unrealistic.
- Information used in decision making is partly derived from external sources but is critically based on their attitudes, perception, cognition and past experience.

Small farming is a system of traditional resource management and farmers manage their local environmental and human resources in adaptive, innovative ways in order to survive. For example, labour costs and shortages are endemic problems partly because of the stigma attached to agricultural work and partly because young people perceive more glamorous opportunities in tourism or in working abroad. Thus, small farmers use family labour wherever possible. Farming communities also share labour resources cooperatively to minimise cash payments and to spread the risk of labour being unavailable for certain physically demanding or time-consuming activities. Though less common than in former times, in Jamaica, 'morning sport' and 'day-for-day' are reciprocal, communal labour arrangements whereby groups of farmers work together to perform seasonal tasks. No money exchanges hands though food and libation is usually provided for those contributing. Lundahl (1983) describes similar cooperative labour arrangements in Haiti, known as *coumbites* and *escouades*.

Another example of farmers' rationality is seen in the debate over farm fragmentation, when a farmer's land is divided into several geographically separate plots or parcels. The conventional view is that small, scattered plots are spatially inefficient and 'uneconomic' since a farmer 'wastes' time travelling between home and distant fields (Chisholm 1979). In Caribbean hilly terrain, a farmer may spend more than an hour walking or riding a donkey to a remote yam or banana plot. However, fragmented farming occurs in the developed and developing world, and in small- and large-scale farming (King and Burton 1982; Ilbery 1984). In the Caribbean, fragmentation is the norm (Barbados is an exception) and in Jamaica, the average number of plots per farmer is 1.21. Larger farms are more fragmented than smaller plots (Table 3.3) and the average number of plots for farms over 25 acres is 1.64 compared with an average of 1.11 for farms less than 5 acres.

An alternative view is that fragmented plots are an environmental adaptation and reflect rational decisions to utilise parcels of land in different geographical locations to take advantage of an ecologically differentiated environment (Igborzurike 1970). Rainfall, temperature, soil quality and slope angles are spatially variable over relatively short distances in the tropics. In mountainous areas, higher slopes are cooler and wetter than lower elevations, and forests and woodland margins offer different farming conditions compared with more open terrain. Farmers use their knowledge and past experience about local weather patterns, topography and soil types to decide what to plant on different farm plots. They can spread risk by growing the same crops in more than one location, to reduce the potential loss due a localised hazard, such as a pest outbreak, plant disease, flash flood or landslide. Brierley (1987) comments on the social value that Grenadian farmers place on fragmented 'mountain land'. Farmers say they cultivate these plots because visits to them represent a kind of 'pilgrimage', allowing them to stay overnight in a cooler place away from the social pressures of village life. This outweighs the marginal economic returns they derive from cultivating distant mountain land.

Careful and detailed research in the Caribbean and elsewhere has forced a reappraisal of the value of traditional skills and knowledge in small-scale farming (Hills and Iton 1983; Chambers 1983, 1997; Richards 1985). Cropping systems are ecologically complex and techniques like intercropping require skilful use of several crops at the same time and in a limited space. Crop rotation cycles incorporate legumes to help replenish depleted soils nutrients, whilst mulching conserves soil moisture. Animals are tethered on fallow land (Berleant-Schiller and Pulsipher 1986), a practice called 'ramming' or 'fly penning' in Jamaica (Omrod 1979). It has several advantages including rotational manuring (the animals are moved from time to time) and trampling by hooves compacts soil and reduces soil loss from wind erosion. Farmers have a wealth of local knowledge about soil types and their spatial variability, even though they categorise soils differently from scientific classification (Davis-Morrisson and Barker 1997).

Problems are often solved in improvised and innovative ways. A decision not to adopt an agricultural innovation may be based on sound reasoning by the farmer even if it is difficult for an outsider to comprehend. In Jamaica, yam farmers have been resourceful in trying to solve the 'yam stick problem' whilst steadfastly refusing to adopt minisett, an innovation promoted by agricultural scientists (Box 3.3). Similarly, high costs of fertilisers and pesticides may prompt local solutions, such as the use of animal manure and bat guano from limestone caves, and use of wood ash to deter insect pests.

Indigenous technical knowledge and sustainable farming

Recognition of the validity of farmers' traditional knowledge has opened new avenues of research (Chambers *et al.* 1989). Indigenous technical knowledge (ITK) refers to the unique, traditional knowledge of people who have lived in a community over a long period. It is based on inherited oral traditions and has enabled people to develop successful rural livelihood survival strategies in the face of environmental uncertainty and economic hardship. It is a dynamic body of knowledge that adapts to, and copes with, new problems as they arise (Richards 1985). In farming systems, ITK involves the management of soil fertility and water resources, traditional methods of pest control, crop selection and crop combination, land and crop rotation. ITK is not restricted to farming; it encompasses knowledge and skills in the use of all local natural resources including traditional medicines and multiple practical uses of plant and animal species.

The importance and credibility of indigenous technical knowledge and traditional farming skills has emerged strongly since the Rio Summit in 1992 with its focus on sustainable agriculture. Three basic criteria must be present to meet requirements that agriculture is sustainable:

- economically viable
- environmentally sound
- socially acceptable.

Box 3.3: Yam farming and the yam stick problem

Despite a faltering agricultural sector, yam production in Jamaica increased from 130,000 tonnes to 212,000 tonnes between 1983 and 1997 (Barker and Beckford 2003). Yams are grown entirely by small farmers for the domestic and export markets in the UK and North America. Exports are currently worth over US$10 million. Agricultural intensification has occurred in the geographically favoured central parishes. Production specialised in yellow yam, cultivated often in pure stands, fertiliser use increased and fallow has virtually disappeared in these areas (Barker 1998).

In Jamaica, yams are cultivated in traditional ways by planting setts in a yam hill and supporting the aerial biomass of the yam vine with a stick. The yam stick can be four to six metres in length and the sturdy hardwood species used in the past were known to last several decades. Increased yam production has increased the demand for yam sticks. In former times, farmers cut their own sticks but today purchase them from informal part-time commercial traders (Beckford 2000).

The need to purchase yam sticks has added to farmers' production costs. They are faced with a 'yam stick problem' with three components: *scarcity* of sticks, *high price* of stick and *inferior quality* of sticks; they have had to improvise to survive. Inferior quality sticks supplied by traders often last one season, or are easily broken and so farmers use several sticks to stake one yam hill. Furthermore, poor quality sticks means they need to be replaced more frequently. Barker and Beckford (2003) estimate an annual stick replacement rate of 63 per cent. The price of sticks has climbed so high that yam farming may become unprofitable. With a current annual national demand for sticks estimated between 41 million and 63 million, the yam stick problem has environmental implications for the sustainability of Jamaica's forest resources (Barker and Miller 1995), especially as a principal source area is Cockpit Country, an area slated to become a National Park (see Chapter 1).

A number of alternatives to traditional yam sticks are feasible (Barker and Beckford 2003). Minisett is a scientific innovation that uses smaller yam setts and does not require staking since the yam vines trail on the ground. It was developed at the International Institute for Tropical Agriculture in Nigeria and has been promoted by agricultural planners as a high-yielding package suitable for the export trade because the smaller tubers are easier to package. Farmers have entirely rejected this innovation for a variety of sound reasons: for example, it involves the higher costs of using plastic sheets for mulching, and they reason that environmental factors like soil drainage and slope aspect are not conducive to minisett. They also perceive smaller tubers to be associated with lower yields, and fear that commercial exporters will not buy small yams from them since they are different to the traditional varieties that are in demand overseas.

Economic viability implies food is produced and income generated at the farm level whilst, at the national level, agriculture contributes to GDP and to exports. To be environmentally sound, farming practices should not degrade land resources and local ecologies, and productive capacity and the physical environment should be maintained for the benefit of future generations. Social acceptability means that farming systems are appropriate to the human and financial resources at the disposal of local people, and that the farming practices are in harmony with cultural values and community needs.

What constitutes sustainable farming practices varies from location to location depending on many environmental and societal factors. More than one type of system may be appropriate at the same place at a given time. Agricultural practices considered sustainable include reduced or minimum tillage, crop rotations, soil and nutrient management, the efficient use of water resources, integrated pest, disease and weed management, slope management, and organic farming. Furthermore many of the techniques and farm practices advocated as sustainable are, in fact, part of the traditional repertoire of techniques familiar to small farmers – in other words, indigenous technical knowledge. Thus, many of the traditional farming techniques are not only potentially environmentally sound and capable of generating food and income but also satisfy the third criterion, of social acceptability. Figure 3.10 lists traditional practices used by Caribbean small farmers that can be considered sustainable. Notable examples include kitchen gardens and food forests, both types of traditional agroforestry.

Figure 3.10 Sustainable traditional agricultural practices used in the Caribbean

Intercropping and polyculture – symbiotic relations between plants (shade, rooting systems), plant diversity encourages natural biological control of insect pests, provides year round food supply

Crop rotation with legumes – helps retain soil fertility and year-round food supply, (rotations involve red peas, gungo peas, cowpeas, string beans, etc)

Spatial organisation of crops in fields – strip cropping, grass barriers, contour planting – all contribute to soil conservation, planting trees to act as wind breaks

Fallowing – helps restore soil fertility if sufficient period elapses, helps maintain vegetative cover to reduce erosion

Mulching – helps reduce evapotranspiration and soil loss from wind erosion, adds nutrients to soil, minimises the impact of splash erosion

Ramming, fly penning – integrates crops and livestock into household production, reduces potential erosion by trampling, animal faeces manure the land

Kitchen gardens and food forests – traditional types of agroforestry

Silvo-pasture – combining food trees with pasture, e.g. coconuts and cattle

Kitchen gardens and food forests

A kitchen garden is the area around a house cultivated intensively, mainly for food for the household (Brierley 1976). It is widespread in the tropics and is found in Fiji and the Pacific islands, Sri Lanka, Indonesia, Indonesia, Africa and Central America. Kitchen gardens are also called multi-storey gardens and tropical home gardens (Landauer and Brazil 1990). In a comparative study, Berleant-Schiller and Pulsipher (1986) referred to them as 'Antillean gardens', whilst other regional terms include backyard gardens, dooryard gardens (Grenada), *jardins nou* (Dominica) and *jardins de case* (Martinique). Kitchen gardens were evident on plantations even before Emancipation, because slaves farmed the tiny areas around their huts. Because the areas they were allowed to cultivate by the plantation owners were so tiny, dozens of crops, trees and other plants were packed together and used for food, beverages, medicine and numerous other purposes.

A similar type of Caribbean farming is the food forest, which Hills (1988) suggests reflects ecological artistry on the part of the small farmer rather than random chaos, as thoughtful, clever adaptations to limited land resources. A definition is provided in Hills and Iton (1983):

> A food forest is ideally a sophisticated agronomic device, varying according to physical background, social and economic conditions and comprising a variety of crops, in number from 6 to 65 different species of different height, physiognomy and root characteristics, which are spatially, temporally and numerically organised in such a way that they will ecologically benefit each other and the whole . . .

A kitchen garden is a food forest whenever it displays a multi-storeyed configuration of food trees. However, not all kitchen gardens are food forests, and likewise not all food forests are kitchen gardens. A food forest may be located some distance from the house and be non-contiguous with the house plot. Also, food forests can be several acres in size and, in Jamaica, are sufficiently distinctive and important that the official agricultural census records them as a separate land-use category. In fully developed food forests and kitchen gardens, the cultivated area is packed with useful plants in a horizontal and vertical dimension used for dozens of non-culinary purposes (Figures 3.11, 3.12 and 3.13).

The different biological cycles of food plants ensure a continuous food supply for the household. Figure 3.13 summarises the numerous ecological, environmental and other benefits of this type of food production system. Food forests and kitchen gardens play an important social role in strengthening reciprocal relationships within extended families and rural communities. Produce is given to relatives visiting a rural area from town, and to friends, neighbours, the poor, elderly and sick members of the local community, sometimes as a kind of informal community food aid. Seeds and planting material are exchanged to reduce purchasing costs and inadvertently help redistribute crop genetic material. Children are socialised into farming via kitchen gardens, perhaps by helping mother or grandmother in simple tasks like weeding or harvesting (Brierley 1991). In effect, the kitchen

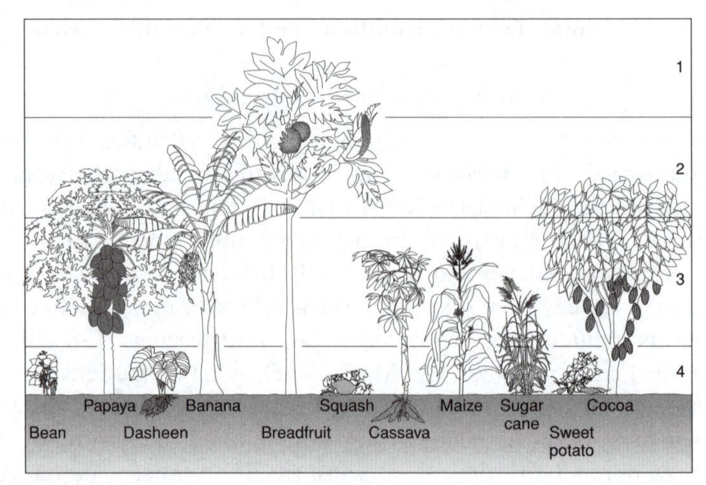

Figure 3.11 Vertical section through a food forest
Source: Adapted from Niñez (1985)

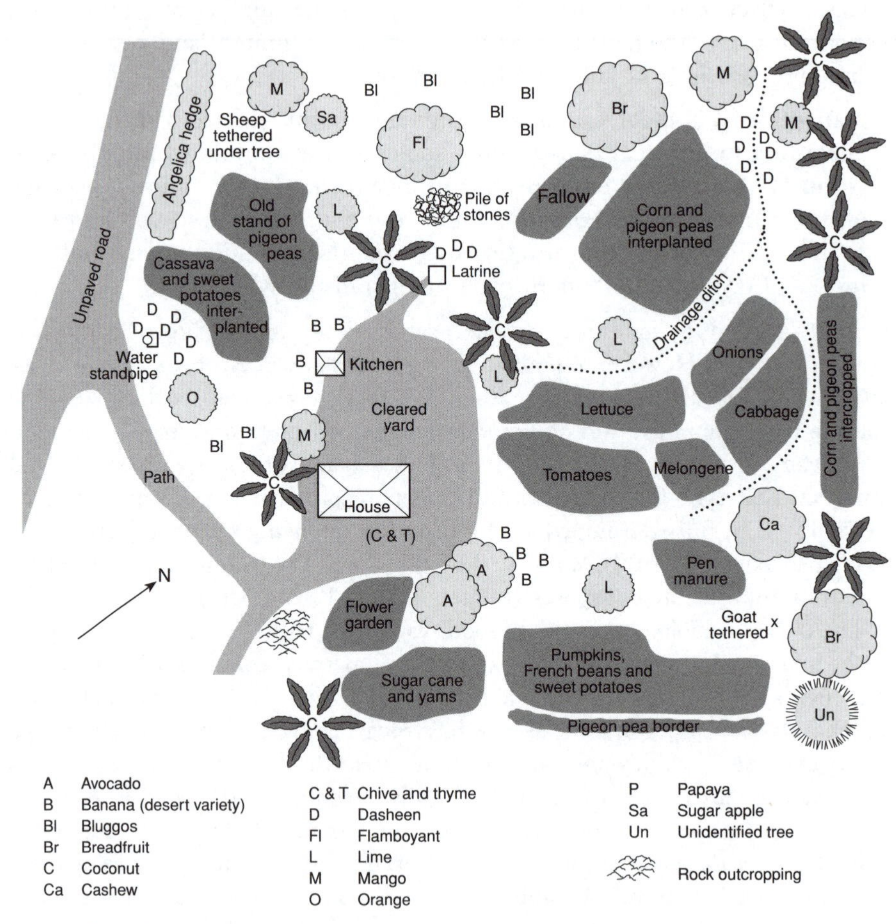

Figure 3.12 Plan of a kitchen garden
Source: Adapted from Brierley (1985, *Food and Nutrition Bulletin*)

Figure 3.13 The functions, structure and products of Caribbean food forests

FUNCTIONS

Subsistence	Boundary markers	Decorative	Beverages	Construction	Dyes
Commercial	Protective	Religious	Medicinal	Furniture	

STRUCTURE AND PHYSIOGNOMY (Examples)

Tall trees (50–100 ft)

Breadfruit	Cedar	Hog plum	Lignum vitae	Mahogany	Palms
Breadnut	Cinnamon	Jackfruit	Mahoe	Mango	Silky oak

Medium trees and bushes (30–50 ft)

Ackee	Guinep	Sapodilla
Coconut	Pimento	Star apple

Small trees and bushes (15–30 ft)

Avocado	Papaya	Soursop
Banana	Plantain	Sugar apple

Small bushes and shrubs (6–15 ft)

Cocoa	Citrus	Pigeon pea
Cashew	Coffee	Tree tomato
Castor bean	Okra	

Tall grasses (3–6 ft)

Corn	Guinea grass
Elephant grass	Napier grass
Guinea corn	Sugar cane

Legumes – bushy, creepers

Broad bean	French bean
Butter bean	Haricot bean
Cowpea	Peanut

Rosette type

Pineapple
Sisal

Spreading plants

Christophene (cho cho)	Pumpkin
Melon	Squash

Roots and tubers

Dasheen	Sweet potatoes
Eddoe	Yams

Vegetables

Beetroot	Carrot	Celery	Lettuce	Shallot	Radish	Calalloo
Cauliflower		Aubergine	Onion	Spinach	Tomato	

Vines	**Beverages**	**Medicinal**	**Herbs and spices**		
Granadilla	Mauby	Aloe	Basil	Dill	Thyme
Black pepper	Sorrel	Physic nut	Bayleaf	Garlic	
		Sarsaparilla		Tamarind	

garden becomes a botanical classroom and demonstration farm plot. Thomasson's (1994) study in pre-disaster Montserrat documents their important role in the aftermath of a hurricane, as emergency food supply, community self-reliance and even spiritual and psychological regeneration.

Traditional agroforestry techniques that incorporate food trees (like food forests and kitchen gardens) are one of several cropping systems that offer the prospects

Figure 3.13 (*continued*)

Human foods

Cereals
Fruit
Nuts
Oil seeds
Pulses
Roots
Spices
Tubers
Vegetables

Animal foods

Corn
Cowpea
Guinea corn
Jackfruit

Beverages

a) Non-alcoholic
  Coconut
  Ginger
  Fruits
  Sorrel
  Vines

b) Alcoholic
  Banana
  Sugar
  Cassava

Medicinal

75 per cent of plants serve some medicinal purpose

Thatching

Banana
Coconut
Plantain

Fibre

Coconut
 – coir fibre

Sisal

Timber

Bearings
Construction
Flooring
Furniture
Mallets
Ornaments
Propeller shafts
Pulleys
Walking sticks

Containers

Calabash
Coconut shell

Miscellaneous

Boat caulking
 – breadfruit bark
Cosmetics
 – bixa
 – Ganja
Dyes
 – jackfruit
 – guinep
Fish poison
 – pulverised ackee
Flowers
Gums
Indelible ink
 – cashew
Mulch
Musical instruments
 – calabash
Novelties
Coconut
Ornamentals
Preservatives
 – papaya
 – cassava
Resinous latex
 – jackfruit
Soap
 – ackee skin
Starch
 – cassava
Waterproofing
 – cashew

Source: Adapted from Hills (1988)

of sustainable solutions to hillside farming in the Caribbean (Sheng 1972). The techniques are familiar to small farmers, offer income-earning opportunities and sources of food for household consumption, and provide protection for hillsides from erosion and degradation. Like other facets of indigenous technical knowledge, small farming in the Caribbean has the potential for developing sustainable agricultural systems. Their positive attributes need to be identified and sensitively explored in conjunction with those aspects of modern science and technology that are appropriate to local farming systems. Further, as Momsen (1998) notes, there are encouraging signs that small farmers and enterprising local agro-processors are beginning to take advantage of opportunities afforded by globalisation, in ethnic and 'green' niche markets abroad, and by new demands for local food in the tourism sector at home. Perhaps, in these ways, local farming will become an economically viable, environmentally friendly and socially acceptable alternative to the faltering legacy of commercial export agriculture.

References

Adams, W. M. and Mortimore, M. J. (1997) 'Agricultural intensification and flexibility in the Nigerian Sahel', *Geographical Journal*, **163**, 150–60.

Ahmed, B. (2001) 'The impact of globalization on the Caribbean sugar and banana industries', The Society for Caribbean Studies, Annual Conference papers, Vol. 2, available at csonline.freeserve.co.uk/olvol2.html.

Ahmed, B. and Afroz, S. (1996) *The Political Economy of Food and Agriculture in the Caribbean*, Ian Randle, Kingston, Jamaica.

Auty, R. M. (1976) 'Caribbean sugar factory size and survival', *Annals of the Association of American Geographers*, **66**, 1, 76–88.

Barker, D. (1989) 'A periphery in genesis and exodus: reflections on rural–urban relations in Jamaica', in Potter, R. B. and Unwin, T. (eds) *The Geography of Urban–Rural Interaction in Developing Countries*, Routledge, London, pp. 294–322.

Barker, D. (1998) 'Yam farmers on the edge of Cockpit Country, Jamaica', in McGregor, D. F. M., Barker, D. and Lloyd Evans, S. (eds) *Resource Sustainability and Caribbean Development*, UWI Press, Kingston, Jamaica, pp. 357–72.

Barker, D. and Beckford, C. (2003) 'Yam production and the yam stick trade in Jamaica: integrated problems for resource management', in Barker, D. and McGregor, D. F. M. (eds) *Resources, Planning and Environmental Management in a Changing Caribbean*, UWI Press, Kingston, Jamaica, pp. 57–74.

Barker, D. and McGregor, D. F. M. (1988) 'Land degradation in the Yallahs basin, Jamaica: historical notes and contemporary observations', *Geography*, **73**, 116–24.

Barker, D. and Miller, D. J. (1995) 'Farming on the fringe: small scale agriculture on the edge of Cockpit Country', in Barker, D. and McGregor, D. F. M. (eds) *Environment and Development in the Caribbean: Geographical Perspectives*, UWI Press, Kingston, Jamaica, pp. 271–92.

Barker, D. and Spence, B. A. B. (1988) 'Afro-Caribbean agriculture: a Jamaican maroon community in transition', *Geographical Journal*, **154**, 198–208.

Baxter, A. E. (1975) 'The diffusion of innovations: soil conservation techniques, the Yallahs Valley, Jamaica', *Jamaica Journal*, **9**, 4, 51–6.

Beckford, C. (2000) 'Yam cultivation, the yam stick trade and resource depletion in the yam growing region of central Jamaica: integrated problems for planning and resource management', Ph.D. thesis, Department of Geography and Geology, University of the West Indies, Mona campus, Jamaica.

Beckford, G. L. (1972) *Persistent Poverty: Underdevelopment in Plantation Economies of the Third World*, Oxford University Press, Oxford.

Belisle, F. J. (1983) 'Tourism and food production in the Caribbean', *Annals of Tourism Research*, **10**, 497–513.

Belisle, F. J. (1984) 'The significance and structure of hotel food supply in Jamaica', *Caribbean Geography*, **1**, 4, 219–33.

Berleant-Schiller, R. (1987) 'Ecology and politics in Barbudan land tenure', in Besson, J. and Momsen, J. (eds) *Land and Development in the Caribbean*, Macmillan, London, pp. 116–31.

Berleant-Schiller, R. (1991) 'Statehood, the commons and the landscape in Barbuda', *Caribbean Geography*, **3**, 1, 43–52.

Berleant-Schiller, R. and Pulsipher, L. D. (1986) 'Subsistence cultivation in the Caribbean', *New West Indian Guide*, **60**, 1–2, 1–40.

Besson, J. (1984) 'Family land and Caribbean society: towards an ethnography of Afro-Caribbean peasantries', in Thomas-Hope, E. M. (ed.) *Perspectives on Caribbean Regional Identity*, Liverpool University Press, Liverpool, pp. 57–83.

Besson, J. (1987) 'A parodox in Caribbean attitudes to land', in Besson, J. and Momsen, J. (eds) *Land and Development in the Caribbean*, Macmillan, London, pp. 13–45.

Besson, J. (1997) 'Caribbean common tenures and capitalism: the Accompong maroons of Jamaica', *Plantation Society in the Americas*, Vol. IV, 201–32.

Blaikie, P. (1985) *The Political Economy of Soil Erosion in Developing Countries*, Longman, Harlow, UK.

Blaikie, P. and Brookfield, H. (1987) *Land Degradation and Society*, Methuen, London.

Bliss, E. (1992) 'Adaptation to agricultural change among Garifuna women in Hopkins, Belize', *Caribbean Geography*, **3**, 2, 143–59.

Boserup, E. (1965) *The Conditions of Agricultural Growth*, Allen & Unwin, London.

Brierley, J. S. (1985) 'Kitchen gardens in the West Indies, with a contemporary study from Grenada', *Journal of Tropical Geography*, **43**, 30–40.

Brierley, J. S. (1985) 'West Indian kitchen gardens: A historical perspective with current insights from Grenada, *Food and Nutrition Bulletin*, **7**, (3).

Brierley, J. S. (1987) 'Land fragmentation and land use patterns in Grenada', in Besson, J. and Momsen, J. (eds) *Land and Development in the Caribbean*, Macmillan, London, pp. 194–209.

Brierley, J. S. (1991) 'Kitchen gardens in the Caribbean, past and present: their role in small farm development', *Caribbean Geography*, **3**, 1, 15–28.

Chalmin, P. (1990) *The Making of a Sugar Giant: Tate & Lyle 1859–1989*, Harwood Academic Publishers, Chur, Switzerland.

Chambers, R. (1983) *Rural Development: Putting the Last First*, Longman, Harlow, UK.

Chambers, R. (1997) *Whose Reality Counts? Putting the First Last*, Intermediate Technology Publications, London.

Chambers, R., Pacey, A. and Thrupp, L. A. (eds) (1989) *Farmer First: Farmer Innovation and Agricultural Research*, Intermediate Technology Publications, London.

Chisholm, M. (1979) *Rural Settlement and Land Use: An Essay in Location*, 3rd edition, Hutchinson, London.

Clarke, E. (1966) *My Mother Who Fathered Me: A Study of the Family in Three Selected Communities in Jamaica*, 2nd edition, Allen & Unwin, London.

Cole, J. (1994) 'Socio-political problems of the tenurial system in St Lucia', in Alleyne, F. W. (ed.) *Land Tenure and Development in the Eastern Caribbean*, Carib Research & Publications Inc., Bridgetown, pp. 32–44.

Courtenay, P. P. (1965) *Plantation Agriculture*, Bell & Sons, London.

Craton, M. (1987) 'White law and black custom – the evolution of Bahamian land tenures', in Besson, J. and Momsen, J. (eds) *Land and Development in the Caribbean*, Macmillan, London, pp. 88–115.

Davis-Morrisson, V. and Barker, D. (1997) 'Resource management, environmental knowledge and decision-making in the Rio Grande valley, Jamaica', *Caribbean Geography*, **8**, 2, 96–106.

de Albuquerque, K. and McElroy, J. (1983) 'Agricultural resurgence in the United States' Virgin Islands, *Caribbean Geography*, **1**, 2, 121–32.

Edwards, D. T. (1995) 'Small farmers and the protection of the watersheds: the experience of Jamaica since the 1950s', Occasional Paper Series, No. 1, UWI Press and Centre for Environment & Development (UWICED), Kingston.

Floyd, B. (1970) 'Agricultural innovation in Jamaica: the Yallahs Valley Land Authority', *Economic Geography*, **46**, 64–77.

Floyd, B. (1978) 'Cuba si? – Socialist transformation of agriculture in Cuba', paper presented to Annual Conference for Caribbean Studies, UK.

Galloway, J. H. (1989) *The Sugar Cane Industry*, Cambridge University Press, Cambridge.

Grigg, D. B. (1969) 'The agricultural regions of the world: review and reflections', *Economic Geography*, **45**, 2, 95–132.

Grigg, D. B. (1974) *The Agricultural Systems of the World: An Evolutionary Approach*, Cambridge University Press, Cambridge.

Grigg, D. B. (1992) 'Agriculture in the world economy: an historical geography of decline', *Geography*, **77**, 210–22.

Grossman, L. S. (1998) *The Political Ecology of Bananas: Contract Farming, Peasants and Agrarian Change in the Eastern Caribbean*, University of North Carolina Press, Chapel Hill.

Grossman, L. S. (2000) 'Women and export agriculture: the case of banana production on St Vincent in the eastern Caribbean', in Spring, A. (ed.) *Women Farmers and Commercial Ventures*, Lynne Rienner, London.

Gumbs, F. A. (1997) 'Farmers and soil conservation in the Caribbean', Occasional Paper Series No. 3, UWI Press, UWICED, University of the West Indies, Kingston.

Gumbs, F. A. and Lindsay, J. I. (1982) 'Runoff and soil loss in Trinidad under different crops and soil management', *Soil Science Society of America, Journal*, **46**, 1, 264–66.

Hall, D. (1959) *Free Jamaica 1838–1865: An Economic History*, Caribbean Universities Press, Ginn & Co., Aylesbury, UK.

Hall, D. R. (1989) 'Cuba', in Potter, R. B. (ed.) *Urbanization, Planning and Development in the Caribbean*, Mansell, London.

Higman, B. W. (1987) 'The spatial economy of Jamaican sugar plantations: cartographic evidence from the eighteenth and nineteenth centuries', *Journal of Historical Geography*, **13**, 17–39.

Hills, T. L. (1988) 'The Caribbean food forest: ecological artistry or random chaos?' in Brierley, J. S. and Rubenstein, H. (eds) *Small Farming and Peasant Resources in the Caribbean*, Manitoba Geographical Studies 10, University of Manitoba, Winnipeg, pp. 1–28.

Hills, T. L. and Iton, S. (1983) 'A reassessment of the "traditional" in Caribbean small agriculture', *Caribbean Geography*, **1**, 1, 24–35.

Igbozurike, M. U. (1970) 'Fragmentation in tropical Africa: an overrated phenomenon', *Professional Geographer*, **22**, 321–5.

Ilbery, B. W. (1984) 'Farm fragmentation in the Vale of Evesham', *Area*, **16**, 159–65.

Ilbery, B. W. (1985) *Agricultural Geography: A Social and Economic Analysis*, Oxford University Press, Oxford.

Kellman, M. and Tackaberry, R. (1997) *Tropical Environments: The Functioning and Management of Tropical Ecosystems*, Routledge, New York.

King, R. L. and Burton, S. P. (1982) 'Land fragmentation: fundamental spatial problem', *Progress in Human Geography*, **6**, 475–94.

Landauer, K. and Brazil, M. (eds) (1990) *Tropical Home Gardens*, United Nations University Press, Tokyo.

Lindskog, P. D. (1998) 'From Saint Domingue to Haiti: some consequences of European colonisation on the physical environment of Hispaniola', *Caribbean Geography*, **9**, 2, 71–86.

Lorah, P. (1995) 'An unsustainable path: tourism's vulnerability to environmental decline in Antigua', *Caribbean Geography*, **6**, 1, 28–39.

Lundahl, M. (1983) *The Haitian Economy: Man, Land and Markets*, Croom Helm, London.

Marshall, W. K. (1972) 'Peasant movements and agrarian problems in the West Indies, Part One: aspects of the development of the peasantry', *Caribbean Quarterly*, 18, 30–46.

McElroy, J. L. and de Albuquerque, K. (1990) 'Sustainable small-scale agriculture in small Caribbean islands', *Society and Natural Resources*, **3**, 109–29.

McGregor, D. F. M. (1995) 'Soil erosion, environmental change, and development in the Caribbean: a deepening crisis? in Barker, D. and McGregor, D. F. M. (eds) *Environment and Development in the Caribbean: Geographical Perspectives*, UWI Press, Kingston, pp. 189–208.

McGregor, D. F. M. and Barker, D. (1991) 'Land degradation and hillside farming in the Fall River basin, Jamaica', *Applied Geography*, **11**, 143–56.

McGregor, D. F. M., McCoubrey, A. and Stidwell, R. (1998) 'Developing an index of land degradation, *Caribbean Geography*, **9**, 2, 121–35.

Ministry of Agriculture, Food & Fisheries (1992) *Barbados 1989 Agricultural Census*, jointly published by Ministry of Agriculture, Food & Fisheries and Barbados Statistic Department, Bridgetown, Barbados.

Mintz, S. W. (1985) 'From plantations to peasantries in the Caribbean', in Mintz, S. W. and Price, S. (eds) *Caribbean Contours*, Johns Hopkins University Press, Baltimore, pp. 127–54.

Momsen, J. H. (1988) 'Changing gender roles in Caribbean peasant agriculture', in Brierley, J. S. and Rubenstein, H. (eds) *Small Farming and Peasant Resources in the Caribbean*, pp. 83–100, Manitoba Geographical Studies 10, University of Manitoba, Winnipeg.

Momsen, J. H. (1998) 'Caribbean tourism and agriculture: new linkages in the global era?' in Klak, T. (ed.) *Globalization and Liberalization: The Caribbean Context*, Rowman & Littlefield, Lanham, Maryland, pp. 115–34.

Monk, J. and Alexander, C. (1985) 'Land abandonment in western Puerto Rico', *Caribbean Geography*, **2**, 1, 2–15.

Morgan, R. P. C. (1986) *Soil erosion and conservation*, Longman, p. 169.

Niñez, V. (1985) 'Introduction: Household gardens and small-scale food production', *Food and Nutrition Bulletin*, **7**, (3).

Omrod, R. K. (1979) 'The evolution of soil management practices in early Jamaican sugar planting', *Journal of Historical Geography*, **5**, 157–70.

Patel, F. (1995) 'Coastal development and geomorphological processes: Scotland District, Barbados', in Barker, D. and McGregor, D. F. M. (eds) *Environment and Development in the Caribbean: Geographical Perspectives*, UWI Press, Kingston, pp. 209–32.

Paskett, C. J. and Philoctete, C.-E. (1990) 'Soil conservation in Haiti', *Journal of Soil and Water Conservation*, **45**, 457–9.

Prorok, C. V. (1991) 'Evolution of the Hindu temple in Trinidad', *Caribbean Geography*', **3**, 2, 73–94.

Prorok, C. V. and Hemmasi, M. (1993) 'East Indian Muslims and their mosques in Trinidad: a geography of religious structures and the politics of ethnic identity', *Caribbean Geography*, **4**, 1, 28–48.

Richards, P. (1985) *Indigenous Agricultural Revolution*, Hutchinson, London.

Richardson, B. C. (1972) 'Guyana's "green revolution": social and ecological problems in an agricultural development programme', *Caribbean Quarterly*, **18**, 14–23.

Richardson, B. C. (1984) 'Slavery to freedom in the British Caribbean: ecological considerations', *Caribbean Geography*, **1**, 3, 164–75.

Richardson, B. C. (1992) *The Caribbean in the Wider World, 1492–1992: A Regional Geography*, Cambridge University Press, Cambridge.

Richardson, B. C. (1997) *Economy and environment in the Caribbean: Barbados and the Windward Islands in the Late 1980s*, UWI Press, Kingston and University Press of Florida, Gainesville.

Ruthenberg, H. (1971) *Farming Systems in the Tropics*, Clarendon Press, Oxford.

Satchell, V. M. (1990) *From Plots to Plantations: Land Transactions in Jamaica 1866–1900*, ISER publications, University of the West Indies, Mona campus.

Sauer, C. O. (1966) *The Early Spanish Main*, University of California Press, Berkeley.

Semple, H. M. and Brierley, J. S. (1998) 'An overview of domestic food production in Guyana, 1960–1993', *Caribbean Geography*, **9**, 1, 30–43.

Sheng, T. C. (1972) 'A treatment-oriented land capability classification scheme for hilly marginal lands in the humid tropics', *Journal of the Science Research Council of Jamaica*, **3**, 93–112.

Spence, B. A. B. (1989) 'Predicting traditional farmers' responses to modernisation: case of a Jamaican maroon village', *Caribbean Geography*, **2**, 4, 217–28.

Spence, B. (1999) 'Spatio-evolutionary model of Jamaican small farming', *Geographical Journal*, **165**, 3, 296–305.

Spurney, J. A. and Cavender, J. C. (2000) 'Effects of milpa and conventional agriculture on soil organic matter, structure and mycorrhizal activity in Belize', *Caribbean Geography*, **11**, 1, 21–33.

Stutz, F. P and de Souza, A. R. (1998) *The World Economy: Resources, Location, Trade and Development*, 3rd edition, Prentice Hall, New Jersey.

Thomasson, D. A. (1994) 'Montserrat kitchen gardens: social functions and development potential', *Caribbean Geography*, **5**, 1, 20–31.

Watts, D. (1987) *The West Indies: Patterns of Development, Culture and Environmental Change Since 1492*, Cambridge University Press, Cambridge.

Welch, B. M. (1996) *Survival by Association: Supply Management Landscapes of the Eastern Caribbean*, UWI Press, Jamaica and McGill University Press, Montreal.

Williams, M. C. (1988) 'The impact of Hurricane Allen on the St Lucia banana industry', *Caribbean Geography*, **2**, 3, 164–73.

Williams, P. E. (1995) 'Drainage and irrigation projects in Guyana: environmental considerations', in Barker, D. and McGregor, D. F. M. (eds) *Environment and Development in the Caribbean: Geographical Perspectives*, Macmillan, London, 189–208.

WCED (World Commission on Environment and Development) (1987) *Our Common Future*, Oxford University Press, Oxford.

Chapter 4

NATURAL HAZARDS AND DISASTER MANAGEMENT

Introduction: basic concepts

The beginning and end of the twentieth century were marked by major volcanic eruptions that devastated several Caribbean islands. In 1902, both Mt Pelée (Martinique) and Soufrière (St Vincent) erupted within 18 hours of each other. A hundred years later, the eruptions of the Soufrière Hills volcano in Montserrat changed the lives of the island's population for ever. During Mt Pelée's eruption, a cloud of lethal superheated gas and ash (a pyroclastic flow) raced down the volcano's flank at speeds over 100 mph (Bullard 1976). The prosperous town of St Pierre, then Martinique's principal commercial centre, was annihilated in less than two minutes. Nearly 30,000 people were incinerated, asphyxiated or killed by the blast. Every building was destroyed and the heat was so intense it melted glass, fresh market produce was instantly carbonised and the wooden decks of ships in the harbour spontaneously burst into flames. Everyone in the town on that day was killed except for two lucky survivors, one of whom was Auguste Ciparis, who was incarcerated in jail at the time of the eruption, and the thick-walled, windowless cell saved his life although he was badly burned.

In 1902 the science of vulcanology was in its infancy, but the obvious need to protect the vulnerable town was blatantly ignored by people in authority. For at least a month before the eruption, signs of impending disaster accumulated. Frequent earthquakes occurred and minor eruptions and blasts covered St Pierre in ash. But an election was impending and votes were needed, so local politicians tried to dispel fears that an eruption was imminent. Nature provided other bizarre warnings. According to historical reports, hundreds of deadly fer-de-lances (a type of pit viper) suddenly appeared in the town and apparently killed 50 people (mainly children) and 200 animals before soldiers shot scores of them and restored some degree of calm (Time-Life Books 1982). Next day, the soldiers were ordered by the political authorities to the outskirts of St Pierre to turn back people who were trying to evacuate. Two days later, the town and its entire population were engulfed by the pyroclastic flow.

Mt Pelée's eruption was a catalyst in promoting the scientific study of volcanoes, especially the explosive andesitic volcanoes typical of the Caribbean. Scientific knowledge and society's ability to cope with natural disasters has improved significantly over the last 100 years. So, when Montserrat's eruptions began in 1995, eventually devastating two-thirds of the island and destroying the capital, Plymouth, the destructive potential of Caribbean volcanoes was better understood and there was little loss of life. The authorities had safely evacuated towns and villages in the danger zones based on timely warnings provided by local scientists.

Other types of natural hazard, like earthquakes, hurricanes, floods, landslides, even droughts and tsunamis, also affect the region. A study of 153 Caribbean disasters recorded between 1900 and 1988 found that two-thirds were the result of hurricanes and tropical storms (Office of Disaster Assistance 1988). Haiti (25 events) and Jamaica (23) experienced most disasters. Tomblin (1981) reported that volcanoes and earthquakes have been responsible, historically, for the most loss of life. Natural disasters threaten a country's economic and social fabric by destroying infrastructure, interrupting production and other economic activities, and creating irreversible changes to natural resources (Barker 1993). For small Caribbean countries, disasters can have an overwhelming impact on economic and social development and national economies. Hurricane Gilbert cost Jamaica US$956 million (Barker and Miller 1990). Inflation increased by 30 per cent and the public sector deficit grew from 2.8 per cent of GDP to 10.6 per cent of GDP (Vermeiren 1989). In Antigua and Barbuda, a double strike by hurricanes Luis and Marilyn in 1995 incurred damage equivalent to 65 per cent of the country's GDP, and between 15 and 25 per cent of the labour force lost their jobs.

This chapter introduces natural hazards and their impacts in the contemporary Caribbean. Figure 4.1 is a typology of hazardous events that affect the contemporary Caribbean. Here, we restrict our discussion to 'natural' rather than 'technological' disasters such as oil spills and the disposal of hazardous waste, although this distinction is fairly arbitrary because human activities contribute significantly to the incidence and severity of many natural disasters. Also, we focus on geological and meteorological hazards rather than biological hazards.

Basic concepts in disaster management and mitigation

Natural hazards involve extreme atmospheric, hydrological or geological events such as those listed in Figure 4.1. A hazard is a naturally occurring or human-induced event or process with potential to create loss of life or damage to property and disruption to normal life (Smith 1996). Natural disasters are not caused simply by an extreme natural event impacting negatively on society but are partly (and sometimes entirely) produced by the economic, social and political environments within which an extreme event takes place (Blaikie *et al.* 1994). Thus the impact of human populations and their economic activities can:

- increase the frequency and severity of natural hazards;
- create natural hazards where none existed before;
- reduce the mitigating effect of natural ecosystems.

For example, most landslides in the Caribbean are caused by human disturbance of the natural landscape. Natural events like torrential rains or an earthquake will

Seismic

Earthquakes – Ground shaking, liquefaction and ground failure, tsunamis, terrestrial and submarine landslides
Volcanic – Lava flows, pyroclastic flows and surges, lahars, ash clouds, tsunamis

Meteorological and hydrological

Tropical storms, hurricanes, tropical depressions, northers, high winds, flash floods, storm surges, riverine flooding, coastal flooding, drought, waterlogging

Geomorphological

Mass movement, landslides, slumping, mudflows, debris flows, rock falls, soil erosion, silting of rivers, silting of harbours, silting of coral reefs, beach erosion, coastal erosion

Biological

Epidemics, human diseases, insect pest outbreaks, plant and animal diseases in farming, forest, bush and grass fires, plant and animal invasions

Technological

Oil and other toxic spills, ground and atmospheric pollution, waterborne pollution, industrial explosions, fires in urban and rural areas, collapse of buildings and other infrastructure, pipeline leakages, poor public health systems

Figure 4.1 A typology of hazardous events affecting the Caribbean region

trigger a landslide, but the underlying cause may be unsustainable farming practices, residential development or road construction on steep hillsides.

Extreme natural events become natural disasters only when they affect people, whether by loss of life, damage to property or disruption of normal social and economic activities. Therefore, specific operational definitions are needed for a hazardous event to be classified as a disaster. The OFDA/Centre for Research on the Epidemiology of Disasters (CRED) international disaster database at the Université Catholique De Louvain in Belgium (www.cred.be), for example, records a disaster when one or more of the following conditions apply:

- when 10 or more people are killed;
- when 100 people are reported affected;
- when a call is made for international assistance;
- when a state of emergency is declared.

To understand how natural hazards become natural disasters, we need to understand the interrelationships between natural systems and socio-economic systems. Two aspects of the interrelationship are critical:

- physical exposure to the event
- human vulnerability.

Physical exposure to a hazard has a temporal and a spatial component. The recurrence interval of a hazard at a particular location (the return period) is a function of its magnitude, duration and frequency. For a given hazard, the return period is the average time that elapses between successive events. It can be calculated from historical data, time series or a catalogue of events over many years. Thus, a city located in an earthquake zone may have a return period of 1 in 100 years for a major earthquake, or 1 in 10 years for a direct hurricane strike.

Physical exposure has a geographical dimension too, because some locations are more hazardous places than others. Whilst the entire Caribbean Basin is seismically active, there is no volcano hazard in the Greater Antilles. Upland areas under certain geological conditions may be prone to landslides, and most coastal zones are prone to storm surges. Large urban areas tend to be more hazardous locations than sparsely populated rural areas because of their population size and the potential scale of damage. Cities like Kingston and San Juan in Puerto Rico are particularly hazardous locations because they are prone to multiple hazards (Box 4.1).

A global trend is for a dramatic increase in the number and scale of disasters in the contemporary world, whether measured by loss of life and/or by economic costs of damage to property. To highlight these alarming trends the UN declared the 1990s the International Decade for Natural Disaster Reduction (Housner 1989). Smith (1996: 39) notes that the number of people affected by disasters increased from 50 million to 250 million between the 1960s and the 1990s and lists a number of explanatory factors for their increased incidence:

- rapid increase in the population of developing countries, many of which are located in the world's major hazard zones;
- increased levels of urbanisation, especially in developing countries, including large numbers of people residing in squatter settlements often located in hazardous places such as steep hillsides and floodplains of rivers;
- increasing population pressure on land resources in rural areas, resulting in unsustainable land-use practices that contribute to soil erosion hazard, landslides, mudslides and flash flooding;
- increased levels of inequality and poverty in developing countries whereby large numbers of people exist below the poverty line and lack access to resources to cope with hazardous events;
- climate change (notably global warming) contributing to increased severity of tropical cyclones and modifications to geographical patterns of drought and seasonal flooding.

Of these factors, only climate change is directly related to the increasing incidence of natural extreme events or processes, and even that has anthropogenic causes. The other factors relate to prevailing conditions in developing countries, where burgeoning populations seem increasingly vulnerable to hazards and disasters.

Hazard and risk are different concepts. A hazard is a pre-disaster situation and refers to a potentially dangerous process or event. Risk refers to a situation where there is potential for loss or damage. Risk assessment involves consideration of the hazard itself and human vulnerability to the hazard, and may be conceptualised

Box 4.1: The multiple hazards and vulnerability of Kingston

Kingston (pop. 750,000) is a classic example of a multiple hazard location, being vulnerable to seismic, atmospheric and hydrological hazards. Urban expansion has filled most of the available land space, and the new residential dormitory settlement of Portmore has been constructed on drained and in-filled coastal wetlands. To the north and east of the plain, middle and high-class residences have expanded into surrounding hills to take advantage of lower temperatures and magnificent panoramic views. Critical infrastructure facilities, including the island's main international airport, electric power generation plants, port facilities, cement factory, flour mills and commercial office blocks, are located around the perimeter of Kingston's large natural harbour. A major urban redevelopment project in the 1960s extended the downtown waterfront into the harbour on reclaimed land, and the airport and container port are also built on landfill.

Kingston has experienced two major earthquakes in recorded history. In 1692 an earthquake destroyed the buccaneer town of Port Royal and the disaster led directly to the foundation of Kingston, based on a classic grid town plan. The combined effects of the great Kingston earthquake of 1907 (6.5 on the Richter scale) and a subsequent major fire destroyed 85 per cent of the city's buildings and killed 1,000 people, about 2 per cent of Kingston's population. There are historical reports of liquefaction and ground failure during both earthquakes, and the tilted tourist attraction of Giddy House, an old artillery store in Port Royal, is clear evidence of such phenomena. Commercial high-rise buildings and other infrastructure resting on landfill or alluvium deposits and the airport road located on the Palisadoes tombolo are also susceptible to liquefaction. Shepherd and Aspinall (1980) calculated the return period for an modified Mercalli VII earthquake striking Kingston as 38 years (and 87 years for a modified Mercalli VIII event), so there is a high probability that the city will be struck by another major earthquake sometime in the near future.

Residential areas in the hills on the urban fringe are also susceptible to landslide hazard. These hills consist of limestone, clastic sedimentary and volcanic rocks, through which run geologically active fault lines. Detailed landslide susceptibility maps produced by the Unit for Disaster Studies at the University of the West Indies highlight the geographical extent and severity of the hazard. Some houses and roads are built on ancient landslide features that have been reactivated, for example in the Jack's Hill area in 1963 (Hurricane Flora), in 1973 (Tropical Storm Gilda), and again in 1988 following a combination of an earthquake and heavy rains. An earthquake in 1993 (5.4 on the Richter scale) triggered over 40 landslides in the Kingston metropolitan area.

Kingston has had many brushes with passing hurricanes, including direct hits in 1951 (Hurricane Charlie) and in 1988 (Hurricane Gilbert). On both

(Box continued)

occasions wind and rain were the primary causes of damage, but fortunately damage from storm surges was minimal. Nevertheless, the hazard posed by hurricane storm surges in the low-lying coastal areas of Kingston is real, if yet unrealised, in contemporary times. A hurricane travelling close to the south coast, or a direct hit from a storm tracking south to north across the island and west of Kingston, could produce a significant storm surge. Yet, no dykes or levees have been constructed to protect the 150,000 people living in the low-lying dormitory communities of Portmore. Localised flooding from streams and gullies also occurs periodically, especially in squatter communities and low-lying residences, often caused by careless disposal of garbage and solid waste blocking drains and gullies.

as the costs and benefits of inhabiting a location that is at risk from a disaster. Thus vulnerability involves risk assessment and the social and economic ability to cope with a hazardous event and can be applied at different levels, from individuals and households, to larger organisations and communities, towns and cities, or to an entire country or island.

The study of disasters is multidisciplinary with six main sub-fields: geographical, sociological, anthropological, development studies, disaster medicine and epidemiology, and technical. The latter is natural science-based and focuses on trigger events and engineering solutions (Alexander 1993). Smith (1996) argues that two different paradigms have emerged reflecting different interpretations of vulnerability. The behavioural paradigm originated in the 1960s through the formative work of geographers (see for example, Burton *et al.* 1978), who elevated the field from its hitherto narrow focus on engineering approaches to mitigating flood hazard in the American Midwest. People's attitudes and perception of hazards are major explanatory variables in understanding why disasters happen. Thus, people may continue to live in hazardous locations because of their poor understanding (perception) of the nature of the risks involved, or their reluctance to do anything about them.

A more radical structural paradigm emerged during the 1970s, and linked environmental disasters more specifically to development theory and dependency (Blaikie *et al.* 1994). These ideas interpret disasters largely as the product of the machinations of the global economy and the marginalisation of poor people. Less emphasis is placed on the causal role of extreme natural events or processes in a disaster. Rather, emphasis is placed on how increasing poverty contributes to people's greater vulnerability to disasters, especially in the developing world. For example, poor people are constrained in their ability to mitigate responses to hazards through their own actions by more powerful and often oppressive institutional forces. Solutions lie in increasing people's access to resources rather spending huge sums of money on science and technology to mitigate specific hazards. Smith (1996) argues that the structuralist paradigm has greatly improved understanding of how specific disasters have unfolded in developing countries. Its main contribution, perhaps, is that it puts extreme natural events and processes into a better perspective, as trigger mechanisms or events, rather than sole causes of disaster. However, the approach appears to be conspicuously absent from most of the disaster-related research that has been undertaken in the insular Caribbean.

The hazard reduction process

The field of hazard management and disaster planning has become a specialist technical field in its own right, though disaster management professionals still broadly reflect the tenets and methods of the behavioural rather than the structuralist paradigm, or what Alexander (1993) terms 'the technocratic approach' to mitigation solutions. As an applied science, disaster management seeks to mitigate hazard impacts through data collection, data analysis and hazard prediction. Structural and non-structural hazard mitigation strategies involve the application of modern engineering and building design, better land-use planning, improving information flows to modify the behaviour of ordinary people towards hazards, and to better inform institutions responsible for mitigating hazards. Modern science and technology methods provide the tools for advance warning and prediction.

Risk assessment is an important component of the hazard reduction process. Three considerations are critical in quantitatively assessing the risks from a given hazard at a specific place:

- What particular hazardous events of certain magnitudes are likely to occur?
- How can the risks of each event occurring be estimated?
- What is the extent of the damage or loss likely to be incurred for each event?

The techniques of risk assessment utilise mathematical estimates of statistical probabilities, and establish scientific connections between a hazardous event and its disastrous impact on, say, buildings and infrastructure. The methods employed are specialised and technical (Smith 1996).

Two other aspects of hazard assessment are important:

- *Hazard mapping* – the production of maps that reflect the spatial distribution of risk and the magnitude and frequency of events likely to occur.
- *Vulnerability assessment* – detailed inventories of buildings and infrastructure, especially network structures like water pipelines, electricity and telephone lines, are needed to assess the degree of loss or damage due to an event.

At one time, coping with disasters in developing countries simply meant providing post-disaster relief. Then, elements of pre-disaster preparedness were added to the process. The contemporary approach is even more holistic (Blaikie *et al.*

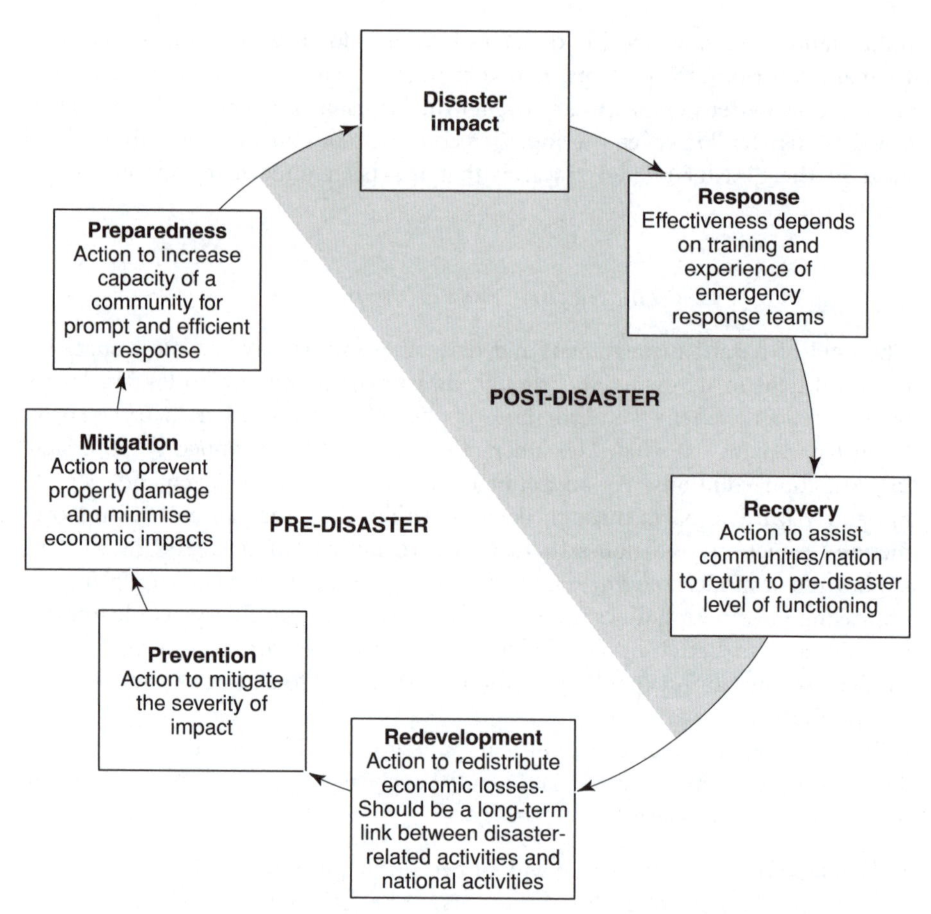

Figure 4.2 The disaster management cycle

1994; Smith 1996; Collymore 1998). The disaster management cycle (Figure 4.2) encompasses hazard assessment, risk reduction and rehabilitation, through the connected activities of prevention, mitigation, preparedness, response, recovery and redevelopment. Further, Figure 4.3 characterises various general ways in which disasters can be minimised or avoided by physical and social adjustments.

In the contemporary Caribbean, all territories have embraced the idea of disaster reduction through hazard mitigation and disaster management, and established national disaster agencies and developed various types of strategy and plan to cope with emergencies. Several important regional initiatives have proactively supported and contributed to this trend. The Pan Caribbean Disaster Prevention and Preparedness Project (PCDPPP) pioneered modern approaches to disaster management in the region. It was established in Antigua in 1981 after severe hurricanes and floods occurred in the region in 1979/80. The project initiated hazard mitigation and loss reduction programmes via technical assistance and training in emergency health and water supply provision (Toulmin 1987; Collymore

Figure 4.3 Possible physical and social adjustments to hazards

Physical adjustments include:
Building and construction techniques to withstand a hazard
Identifying (mapping) and avoiding sites where hazards are likely to occur
Predicting the occurrence of a hazard
Preventing or altering the characteristics of a hazard

Social adjustments include:
Land-use zoning and restrictions for hazardous sites
Establishing minimum building standards for hazardous sites
Public awareness through public education
Issuing early warnings of imminent threats
Evacuation plans, preparations for emergency food and shelter
Emergency preparedness programmes to protect life and property
Spreading economic loss more equitably through insurance, taxation and grants
Reconstructing a community so it is less vulnerable

Source: Adapted from Housner (1989)

1995) and disseminated information through newsletters and regional workshops (Barker 1989). Its successor, the Caribbean Disaster Emergency Response Agency (CDERA), was created in 1991 and is an inter-governmental disaster management organisation based in Barbados. Its wide-ranging activities include collating and channelling information to governmental agencies and regional NGOs, mobilising and coordinating disaster relief, disaster mitigation efforts and education, training and institutional strengthening among its 16 participating states (www.cdera.org).

The University of the West Indies has also engaged in hazard and disaster management activities through workshops (Barker 1989; Ahmed 1992; 1997) and teaching programmes, including collaboration with the Caribbean Disaster Management Programme (CDMP) based in Kingston from 1992 to 1999. CDMP activities were themselves significant and included retrofitting and strengthening low-income housing in vulnerable communities, community participation and disaster mitigation in the Dominican Republic, workshops for the insurance industry, and strengthening regional resources for hazard mapping capabilities (Worrell-Campbell 1997).

In the rest of this chapter we examine the region's main natural hazards: earthquakes, volcanoes, hurricanes, flooding and landslides. We introduce and outline the basic scientific concepts that help explain these events, discuss the nature and impacts of each hazard, and comment on aspects of disaster reduction and mitigation.

Earthquakes and seismic hazards

Earthquakes, like volcanoes, are a type of seismic hazard. There is a close correlation between plate boundaries and the distribution of earthquakes and volcanoes (Chapter 1). Central America is more seriously affected by seismic hazards than the insular Caribbean. For example, the 1972 Managuan earthquake in Nicaragua killed 20,000 people of the capital city's population of 400,000 and rendered 250,000 people homeless, whilst in 2001 an earthquake off the coast of El Salvador killed over 700 people, injured 4,000 and destroyed 69,000 homes and damaged another 500,000.

Tectonic earthquakes originate inside the earth's crust and are triggered by the movement of crustal plates relative to each other, or by rocks at a given location slipping or sliding along a geological fault. Stresses build up along active fault lines until friction is overcome, and the sudden displacement of rocks on either side of the fault plane causes rocks to lurch into a new position. Displacement in a single event may be a few centimetres or several metres. Volcanic earthquakes are caused specifically by volcanic activity, generated by the movement of magma in the lithosphere.

An earthquake releases an enormous amount of seismic energy into the surrounding crust. The energy radiates outwards as a kind of ripple or seismic wave. The point of origin, or focus, of an earthquake within the crust is called a hypocentre and its equivalent location on the earth's surface is the epicentre. At least three seismic stations are needed to triangulate and plot the position of the focus. The energy released radiates outwards from the hypocentre, gradually diminishing with increasing distance, but can be strong enough to register on seismographs hundreds of kilometres away. The epicentre is normally the location where the strongest shock is felt. Foreshocks sometimes occur before the main earthquake, and aftershocks can last weeks or months after the main shock.

Earthquakes are classified by depth of focus. Deep earthquakes (at depths greater than 300 km) tend to be associated with subduction zones (Chapter 1) and are thus more common in the eastern Caribbean. Intermediate earthquakes occur at depths between 70 and 300 km. Shallow earthquakes (less than 70 km depth) occur along the slip zone of a transform plate boundary as in the northern Caribbean, or near a marine trench like the Puerto Rican trench (Chapter 1). Worldwide, shallow-focus earthquakes are potentially more dangerous than deeper-focus events (Smith 1996). Shallow earthquakes occur throughout the Greater Antilles, but Puerto Rico is extremely vulnerable (Molinelli 1989) and four of the major historical earthquake disasters in the region occurred there in 1670, 1787, 1867 and 1918. Similarly, in Jamaica, the devastating Port Royal earthquake of 1692 and the Great Kingston earthquake in 1907 were probably shallow-focus earthquakes with epicentres north of the island.

The magnitude of an earthquake is measured on the Richter scale. Calculation of an earthquake's magnitude is scientifically complex but for simplicity is reduced to and expressed as a single value on a logarithmic scale. Thus, each whole number on the Richter scale represents a tenfold increase in magnitude. A severe earthquake of 7.6 is ten times more powerful in terms of energy release than one measuring 6.6. Events measuring 7.0 or higher cause major damage but those around 3.5–5.4 can be felt and cause relatively little damage.

The modified Mercalli (MM) scale measures earthquake intensity in terms of effects on people, buildings and the land at a given location. Observations made during an earthquake and assessments of physical damage are used to classify an earthquake along a ranked, ordinal measurement scale (Table 4.1). The intensity with which an earthquake is felt at a particular place will partly depend on its magnitude, but will also depend on other factors like distance from the epicentre and local geological site conditions. Note that each earthquake has only a single value on the Richter scale but will have different values on the modified Mercalli scale, depending on the location of the observer.

Earthquake hazard and mitigation

Figure 4.4 shows an earthquake hazard map for the Caribbean and Central America. It was produced as part of the Global Seismic Hazard Assessment Program (GSHAP) by scientists at the Instituto Panamericano de Geografia y Historia (IPGH) in collaboration with other regional scientific organisations (www.geohazard.cr.usga.gov/paigh). Detailed seismicity and earthquake hazard maps of the eastern Caribbean have also been published by the Seismic Research Unit at UWI's St Augustine campus (www.seismic.com).

Figure 4.5 summarises the main hazards and impacts associated with Caribbean earthquakes. The main earthquake-induced processes that constitute seismic

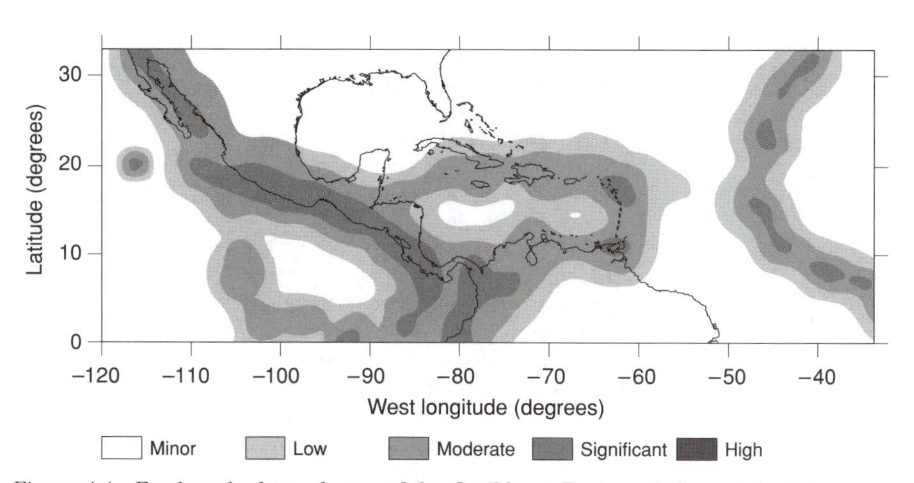

Figure 4.4 Earthquake hazard map of the Caribbean Basin and Central America
Source: geohazards.cr.usgs.gov/paigh/project.gif

Table 4.1 Modified Mercalli scale of earthquake intensity

Intensity value		Description	Corresponding Richter scale (magnitude)
I		Not felt. Marginal and long-period effects of large earthquakes.	
II	Feeble	Felt by persons at rest, on upper floors, or favourably placed.	
III	Slight	Felt indoors. Hanging objects swing. Vibration like passing of light trucks. Duration estimated. May not be recognised as an earthquake.	< 4.2
IV	Moderate	Hanging objects swing. Vibration like passing of heavy trucks, or sensation of a jolt like a heavy shell striking the walls. Standing motor cars rock. Windows, dishes, doors rattle. Glasses clink. Crockery clashes. In the upper range of IV, wooden walls and frames creak.	
V	Slightly strong	Felt outdoors; direction estimated. Sleepers wakened. Liquids disturbed, some spilled. Small unstable objects displaced or upset. Doors swing, close, open. Shutters, pictures move. Pendulum clocks stop, start, change rate.	< 4.8
VI	Strong	Felt by all. Many frightened and run outdoors. Persons walk unsteadily. Windows, dishes, glassware broken. Knickknacks, books, etc., off shelves. Pictures off walls. Furniture moved or overturned. Weak plaster and masonry cracked. Small bells ring (church, school). Trees, bushes shaken visibly, or heard to rustle.	< 5.4
VII	Very strong	Difficult to stand. Noticed by drivers of motor cars. Hanging objects quiver. Furniture broken. Weak chimneys broken at roof line. Fall of plaster, loose bricks, stones, tiles, cornices, also unbraced parapets and architectural ornaments. Some cracks in masonry. Waves on ponds; water turbid with mud. Small slides and caving in along sand or gravel banks. Large bells ring. Concrete irrigation ditches damaged.	< 6.1

Table 4.1 (*continued*)

Intensity value		**Description**	**Corresponding Richter scale (magnitude)**
VIII	Destructive	Steering of motor cars affected. Fall of stucco and some masonry walls. Twisting, fall of chimneys, factory stacks, monuments, towers, elevated tanks. Frame houses moved on foundations if not bolted down; loose panel walls thrown out. Decayed pilings broken off. Branches broken from trees. Changes in flow or temperature of springs and wells. Cracks in wet ground and on steep slopes.	
IX	Ruinous	General panic. General damage to foundations. Frame structures, if not bolted, shifted off foundations. Frames cracked. Serious damage to reservoirs. Underground pipes broken. Conspicuous cracks in ground. In alluviated areas, sand and mud ejected, earthquake fountains, sand craters.	< 6.9
X	Disastrous	Most masonry and frame structures destroyed with their foundations. Some well-built wooden structures and bridges destroyed. Serious damage to dams, dikes, embankments. Large landslides. Water thrown on banks of canals, rivers, lakes, etc. Sand and mud shifted horizontally on beaches and flat land. Rails bent slightly.	< 7.3
XI	Very disastrous	Rails bent greatly. Underground pipelines completely out of service.	< 8.1
XII	Catastrophic	Damage nearly total. Large rock masses displaced. Lines of sight and level distorted. Objects thrown into the air.	> 8.1

hazard are ground shaking, ground failure, surface fault rupture and tsunamis. Ground failure takes several forms, including liquefaction, landslides, cracks and fissures in the ground, subsidence and submarine landslides.

Earthquakes vary in terms of their duration, that is, the length of time the ground shakes violently. The hazard from ground shaking has a number of components because an earthquake generates several different types of wave that

Figure 4.5 Caribbean earthquake hazards and impacts

Primary hazard

Ground shaking

Surface faulting

Secondary hazard

Ground failure and soil liquefaction

Landslides and rockfalls

Debris flows and mudflows

Tsunamis

Impacts

Total or partial destruction of building structures

Interruption of water supply

Breakage of sewage disposal systems

Loss of public utilities such as electricity or gas

Floods from collapsed dams

Release of hazardous material

Fires

Spread of chronic illnesses

travel through the earth. From the point of view of a person or building standing on the ground, each type of wave shakes the earth in a different way. P and S waves are called body waves and travel through the interior of the earth. In addition, there are several types of L or longitudinal waves that travel along the earth's surface and cause most of the damage in an earthquake because they shake the ground horizontally at right angles to the wave or in an elliptical motion (Smith 1996).

Unsafe buildings constitute a major hazard and over 95 per cent of deaths in earthquakes are due to building failures (Alexander 1993). Most residential and commercial buildings in developing countries are not constructed to exacting standards and become highly dangerous structures in the event of an earthquake. Earthquake disasters in Turkey (1999) and India (2001) tragically illustrate extreme examples of the problem. Secondary hazards associated with the aftermath of earthquakes include fire (through the rupture of gas lines or electricity cables). In the famous San Francisco earthquake of 1906, 80 per cent of the property damage was due to fire. Similarly, fire helped destroy buildings in the Great Kingston earthquake of 1907 (Box 4.1).

The response and reaction of a particular site to ground shaking also depends upon the physical properties (or site characteristics) of the soil and rock. A number of factors are important, including the depth of the soil layer, its moisture content and whether the underlying rock is unconsolidated or hard. Significant amplification of a seismic wave can occur under certain site conditions that increase the severity of the earthquake. Examples of environments prone to amplification of seismic waves are steep topography, especially the crest of ridges, and unconsolidated materials like alluvial fans and deltas.

Liquefaction occurs when water-saturated sediments temporarily lose strength because of strong ground shaking and behave like a fluid. It is a serious hazard in loose sedimentary deposits, especially silt and sand deposits. Old deltas and alluvial fans less than 10,000 years old are environments prone to liquefaction, and several major cities in the region, including Kingston (Box 4.1), San Juan and Port-au Prince, are located on such sites. Several types of liquefaction have been identified (Tinsley *et al.* 1985, cited in Smith 1996: 130), including lateral spread in the subsoil, which causes damage to pipelines and the foundations of buildings, loss of bearing strength, which causes buildings to tilt, and the most catastrophic of all, flow failure (Smith 1996). The latter can occur simultaneously on the surface and in subsurface material, and even under the sea. Indeed it was probably a submarine landslide that led to part of the buccaneer town of Port Royal sliding beneath the waves in the 1692 earthquake.

The jerky movements that cause earthquakes along fault planes and transform boundaries are not entirely predictable. There is often regularity between successive large events (i.e. severe earthquakes), however, which allows for the calculation of return probabilities for the occurrence of an earthquake of a particular severity at a particular geographical location: in effect, the 'average' time between successive earthquakes over a long period. These figures can be converted into an earthquake hazard map, either at the regional level (Figure 4.4) or for individual countries or city regions. Another source of data for earthquake hazard assessment for towns and cities is from urban surveys of unsafe structures, as undertaken for Santiago de Cuba (Alonso 1989).

The impact of earthquakes can be reduced by various means, including:

- land-use zoning to avoid high-risk sites;
- better engineered buildings;
- stabilising unstable ground;
- redevelopment of vulnerable sites;
- establishment of scientific warning systems.

Building codes are intended to set engineering standards for new or retrofitted buildings, to mitigate damage from earthquakes (and wind damage from hurricanes). Most countries in the region have national building codes, but they are difficult to enforce. Efforts to introduce the Caribbean Uniform Building Code (CUBiC) date back to 1983 but these too have faltered, and although higher costs are a factor, the reasons for the lack of progress are not clear (Chin 1997).

Tsunamis are another type of seismic hazard, which can be triggered by both earthquakes (seismic tsunamis) and volcanoes (volcanigenic tsunamis). They are commonly known as tidal waves. Fortunately, tsunamis are rare in the Caribbean, although Lander (1997) catalogued 50 events in the historical record since 1530. For example, in 1867, a tsunami affected the Virgin Islands. The initial wave was 2.4 m high, then the sea receded about 100 m and returned as a 6 m wave, swamping boats in the harbour at Charlotte Amalie, penetrating about 76 m inland and killing a dozen people. Another in Puerto Rico in 1918, triggered by an earthquake measuring 7.5 on the Richter scale, caused 40 fatalities and destroyed 300 homes, whilst a third in the Dominican Republic in 1946 killed 100 people. There is a potential threat of tsunamis from Kick 'em Jenny, a submarine volcano north of Grenada (Chapter 1), and modern urban and tourism coastal developments are contributing to higher potential risk from tsunamis than in the past.

Volcanoes

Volcanic activity in the islands is restricted to the eastern Caribbean, where there are 25 potentially active volcanoes (Figure 4.6). They tend to be quiet for long periods, sometimes hundreds of years, and then become active for several years, when increased activity may be accompanied by high-magnitude eruptions. There have been 17 eruptions on Caribbean islands in recorded history, with the Mt Pelée eruption accounting for the highest death toll. In modern times population evacuations have occurred on St Vincent in 1971 and 1979, Guadeloupe in 1976, and Montserrat from 1995.

Volcanic activity is found at three types of location. Volcanoes at divergent plate boundaries are called rift volcanoes on land and are associated with sea-floor spreading on the ocean floor. Hot spot volcanoes occur within crustal plates; the most famous are those in the Hawaiian Islands. The third group are island arc volcanoes, associated with subduction at convergent plate boundaries (Chapter 1), and they represent 80 per cent of the world's 500 active volcanoes (Smith 1996). All the volcanoes of the eastern Caribbean and Central America belong to this group. The region also has a submarine volcano, Kick 'em Jenny (Chapter 1).

The cone of a volcano is actually one end of a vent that descends through the earth's crust into the mantle and asthenosphere. The molten material extruded is called magma or lava and contains a mixture of minerals (especially silica, magnesium and iron), gases and crystallised minerals. Volcanoes erupt in different ways depending on the chemical constituents of the magma and its viscosity. Volcanic eruptions of free-flowing but slow-moving lava are called effusive eruptions (Christopherson 1997). The lava contains little gas, is low in silica content, and high in magnesium and iron. Unfortunately, Caribbean volcanoes erupt more violently, in ways more similar to the Mount St Helens eruption in the USA in 1980. These are explosive eruptions. Phreatic eruptions occur early in an eruption cycle where standing water or a crater lake is present and cold groundwater

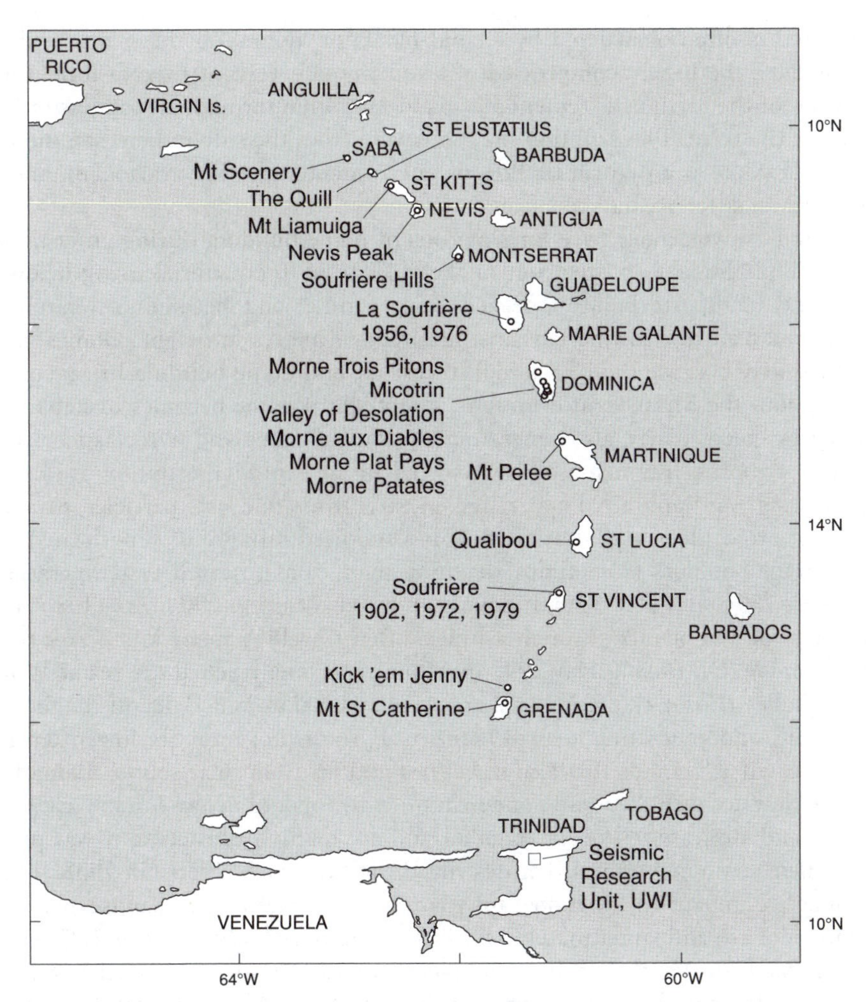

Figure 4.6 Active and potentially active volcanic centres in the eastern Caribbean
Source: Adapted from Robertson *et al.* (1997)

or surface water comes into contact with hot magma. Recent examples include La Soufrière in Guadeloupe (1975–1976), and the eruptions in the early phase of the Montserrat emergency. Pelean eruptions (named after Mt Pelée) are more common than phreatic eruptions, and Smith (1996) notes that they are the most dangerous of all volcanic eruptions.

In the case of island arc volcanoes, the crustal plate is subducted and descends into the mantle, to be metamorphosed at high temperature and pressure. The lava is thick and highly viscous. Silica minerals in the descending plate are the first to melt and become concentrated in the magma, before it rises to the surface. The subduction of ocean floor sediments is an additional source of silica. As the viscous magma rises, gas is trapped and combines with minerals in the lava, and is often obstructed by a solid block, or 'dome', in the main conduit. Eventually, the highly compressed gases in the effervescent magma force their way out of the earth in a tremendous explosion, often through a weakness on the flank of the cone. The eruption may thus vent from the side rather than the top of the volcano, as a pyroclastic flow or nuée ardente (glowing avalanche), one of nature's most terrifying spectacles.

Island arc volcanoes have long periods of dome building during an eruption cycle. The 'dome' is pushed out of the volcano by the material rising below it (imagine toothpaste being squeezed from a tube) and becomes a temporary landscape feature, growing perhaps hundreds of metres in height. Domes grow fairly slowly because the lava is highly viscous, and dome building has occurred throughout the Montserrat eruptions. Eventually a dome becomes unstable and collapses spectacularly, also generating a pyroclastic flow and pyroclastic surges.

The airborne fragments of lava ejected in an explosive eruption are called 'pyroclasts' or 'tephra'. They range in size from fine ash particles to solid ballistic projectiles. A pyroclastic flow is a fluidised mixture of superheated ash and larger boulders (sometimes weighing many tons), heated to temperatures between 750 and 1,000 °C and moving at speeds of up to 200 m/s. They 'flow' down a volcano or race through a valley, often travelling many kilometres from their source. On Boxing Day 1997 in Montserrat, one reached the sea at White River Valley (Box 4.2), creating a 2 km-wide and 600 m-long delta off the former shoreline, and generating a small tsunami. Pyroclastic surges are lower density, superheated ash clouds that also travel fast and far from their source. Being less dense, they are turbulent and not constrained by topography, so can move uphill. Flows and surges often occur together and are highly destructive; it was probably their combined effects that destroyed the town of St Pierre in 1902. Their cumulative impact on the topography is to create a bizzare new landscape; deep deposits of ash and larger material fill valleys and bury buildings (Box 4.2). Many of the ghauts on Montserrat have completely vanished under the immense volume of deposited material and, in time, geomorphological processes will fashion an entirely different landscape.

Box 4.2: Montserrat's volcano emergency

Montserrat is only 16 km long, 10 km wide and about 103 km^2 in area; in tourist brochures it was once called the 'Emerald Isle' because of its historical connection with Ireland. The Soufrière Hills volcano eruptions began on 17 July 1995, rendering 64 per cent of the land surface in the southern and central parts uninhabitable.

Politically, Montserrat is an Overseas British Dependent Territory. Its population peaked in 1946 at 14,333 and has been declining ever since. In the 1970s

(Box continued)

and 1980s the country enjoyed high rates of economic growth and was a small but a thriving tourist destination. Hurricane Hugo devastated the island in 1989, precipitating an exodus of more than 1,000 people, but the most significant migration 'push' in contemporary times has been the volcano emergency. By 1997, fewer than 3,500 people remained on the island, the rest reluctantly forced to migrate to neighbouring islands like Antigua and St Kitts, or further afield to the United Kingdom (through a somewhat controversial voluntary evacuation scheme). The population has since recovered to its current level of about 4,500.

The most intensive period of volcanic activity was during the second half of 1997. Successive pyroclastic flows eventually reached and destroyed the abandoned capital and port of Plymouth, igniting buildings and covering the landscape in tens of metres of ash. On the other side of the island, Bramble international airport was destroyed, connections to the outside world being maintained by the introduction of a ferry service to Antigua from Little Bay, in a northern corner of the island.

Two key organisations were the Emergency Operations Centre (EOC) and the Montserrat Volcano Observatory (MVO), the latter staffed by scientists from the Seismic Research Unit of the University of the West Indies and the British Geological Survey (BGS). Scientists needed to monitor and revise their evaluations and risk assessments constantly as volcanic conditions changed almost on a daily basis. Hazard and risk assessment maps played an important role in defining exclusion zones, safe zones and daylight entry zones (Figure 4.7(a) and (b)). Plymouth was evacuated three times without loss of life, in August and December 1995 and again in 1996. For a while there was daytime access to Plymouth for certain individuals engaged in essential services, including offloading cargo from the port. But eventually the island's only port was also destroyed. Based on scientific assessments and the general good sense of the population there were only 19 fatalities during the emergency, and all those who perished were farmers tending their land in the 'exclusion zone' at the time of a major eruption in 1997.

Another hazard arising from the eruptions was volcanic dust, potentially a cause of silicosis and respiratory problems for asthma sufferers, due to the presence of the mineral cristobalite (silica crystals) in the volcanic ash. The dust was spread with devastating effect during frequent pyroclastic surges. Gigantic, towering ash clouds blocked out the sun and reduced visibility in the northern sections of the island to almost zero on several occasions and airborne dust reached Antigua and even St Vincent. Face masks became an essential part of life for the population, especially during clean-up operations.

The task of reconstruction for Montserrat and its people is enormous and the prospects of achieving economic recovery daunting. The capital has been laid waste and together with it most of the island's infrastructure, commerce and housing. There is little land space in the tiny northern area, into which population, industry, essential services and government agencies are all

(Box continued)

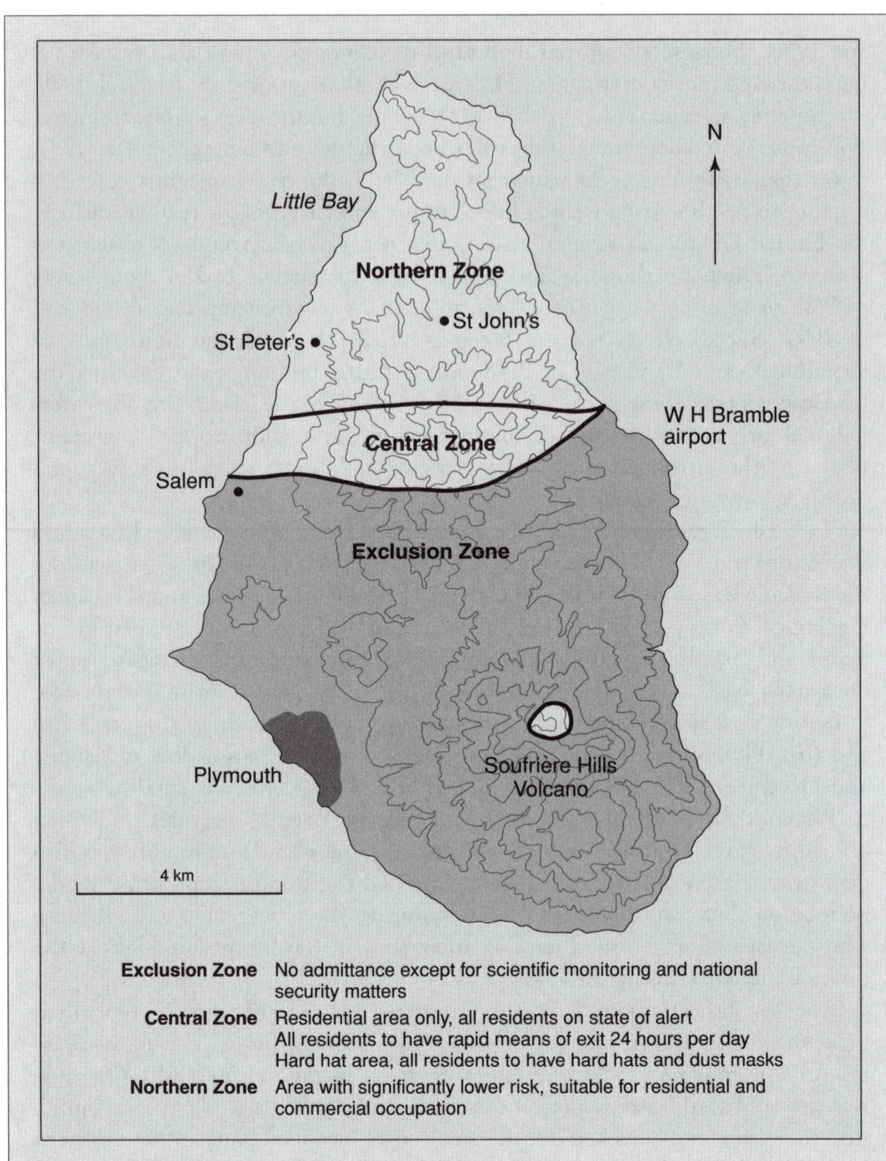

Figure 4.7(a) Volcanic risk map, September 1997
Source: Montserrat Volcano Observatory

crowded. Yet the inhabited area is a hive of construction activity, one of the many projects being the symbolically important new houses of CARICOM village at Davy Hill, built with the assistance of the Caribbean community and the Cuban and UK governments.

(Box continued)

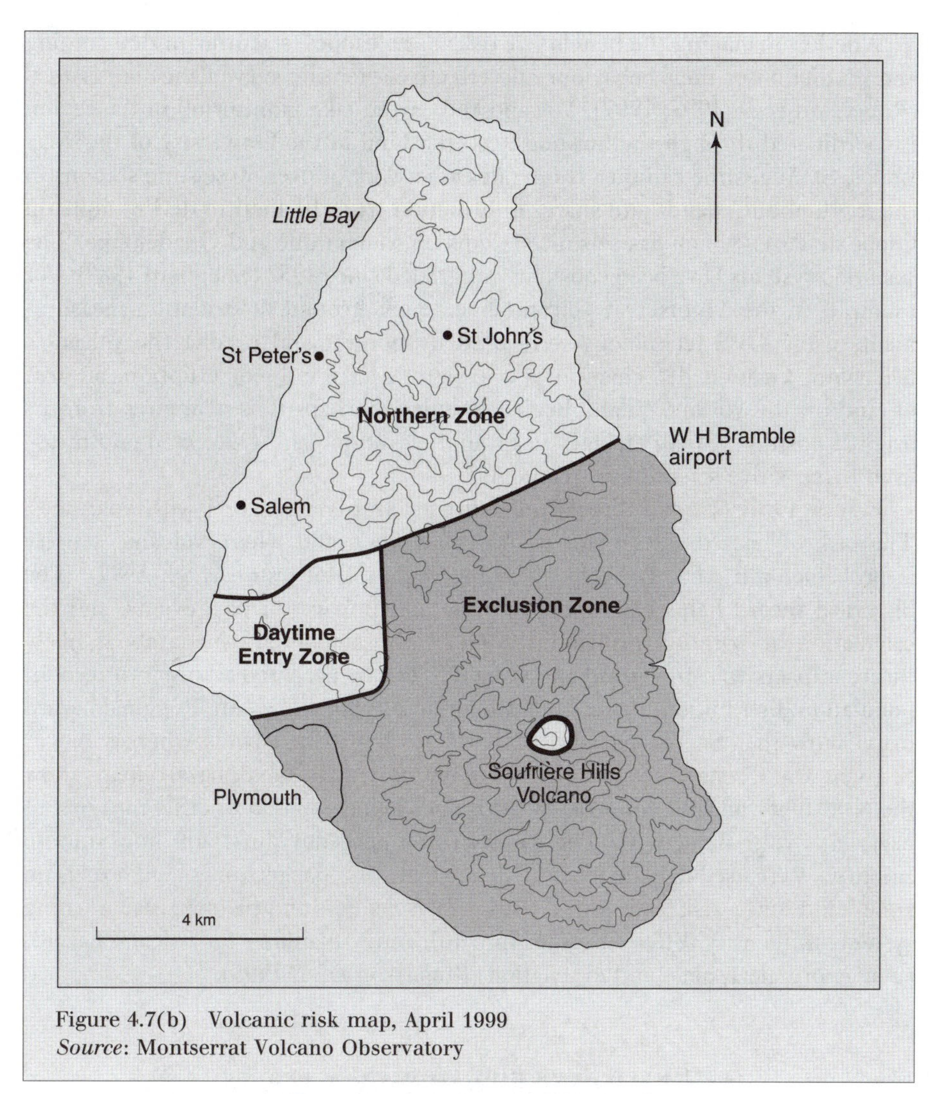

Figure 4.7(b) Volcanic risk map, April 1999
Source: Montserrat Volcano Observatory

Volcano hazard and emergency management

For some people, the risks associated with volcanoes are discounted against their benefits, and they also view them as assets. Volcanic soils are extremely fertile and so not surprisingly farming communities populate the slopes of dormant volcanoes. In the Caribbean islands, some 25,000 people probably live on the slopes of volcanoes (Robertson *et al.* 1997), though the numbers are much higher in Central America. Hawaii and the Philippines advertise and market their volcanoes as scenic attractions in tourist brochures, in effect commodifying risk as a tourist product. Montserrat may move in a similar direction in its post-disaster phase in trying to rebuild tourism, and advertise itself as a volcano isle and adventure tourism destination.

Whether managing the benefits or risks of volcanoes, scientific understanding and prediction of their behaviour and effective scientific surveillance are critical (Robertson *et al.* 1995, 1997). Volcano and earthquake monitoring in the region is coordinated through the Seismic Research Unit at the University of the West Indies, St Augustine campus. It operates a network of over 30 seismic stations in the Anglophone islands and shares information with the Institut de Physique du Globe de Paris, which has seismic stations on Martinique and Guadeloupe. The scientific systems have been constantly upgraded since 1952 (Shepherd 1989). For example, in the Montserrat volcano emergency, ground deformation measurements using GPS technology were used to monitor and predict the volcano's behaviour. Ground deformation is one sign of an imminent eruption; an area of land on the volcano's flank begins to bulge, perhaps tens of metres from its original position. Small changes in ground elevation can be detected and measured by sensitive scientific instruments.

Figure 4.8 lists some direct and indirect hazards associated with volcanoes. The task of hazard assessment is to decide when and where volcanic activity is probable, and what is likely to be the impact (Roberston *et al.* 1997). The historical record helps clarify the location and impacts of past events, and the calculation of eruption probabilities. Once assessments are made about likely future impacts volcano hazard and risk maps can be prepared taking into account population distribution, patterns of economic activity and so on. Regional hazard maps show that St Vincent and St Kitts are high-risk islands whereas Nevis, St Lucia and Grenada are lower risk. For an individual island, hazard maps show the probability of different areas of the islands being exposed to different types of hazardous volcanic activity. The Montserrat emergency illustrates how volcano risk maps were used to good effect to define high-risk 'danger zones' and 'exclusion zones' and lower-risk 'safe zones' (Box 4.2). Scientists and planners use warning systems to identify different stages in contingency planning for volcano hazard, such as pre-alert, alert and evacuation (Robertson *et al.* 1995).

Hurricanes and tropical storms

Hurricanes are the most violent natural hazards to occur in the Caribbean region with regularity. Indeed, the term has a Caribbean etymology, derived from the Taino (Arawak) name 'hurakan' meaning devil wind. A hurricane is an intense low-pressure weather system with bands of strong winds that spiral around a central area, called the eye. Winds of awesome destructive power are capable of delivering enormous amounts of rain. Damage by high winds and floods are aspects of the hurricane hazard, and damage can be caused also by storm surges in coastal areas.

Figure 4.9 shows areas of the world affected by hurricanes, called cyclones in the Indian Ocean, typhoons in the western Pacific and China Sea, and willy-willies

Figure 4.8 Some hazards associated with volcanic activity in the eastern Caribbean

Direct

Ballistic projectiles and tephra falls

Pyroclastic flow and surges

Lahars (mudflows and debris flows)

Lava flows

Volcanic gases

Earthquakes

Laterally directed blasts and dome collapse

Phreatic explosions

Indirect

Atmospheric ash falls

Landslides and debris flows

Tsunamis

Acid rainfall

Socio-economic

Destruction of towns and villages

Loss of life

Loss of farmland, timber resources

Destruction of infrastructure – roads, airstrips, port facilities

Disruption of communications

Source: Adapted from Robertson *et al.* (1997)

in Australia. Note that hurricanes form only between latitudes 5° and 20° north or south of the equator. Virtually all the islands in the Caribbean Basin are susceptible to hurricane hazard, as well as the Bahamas and Turks and Caicos archipelagos, Central America, the Gulf of Mexico, Florida and the Carolinas, and Bermuda. Trinidad is rarely impacted though Tobago is at risk. Hurricanes do not form within 5° of the equator because the Coriolis force is too weak to deflect air sucked into a weather system into a spiral motion, so Guyana and Suriname are not affected by this hazard.

The occurrence of tropical cyclones is seasonal in nature. The official Atlantic hurricane season runs from 1 June to 30 November, although occasionally storms form before or after those dates. Around 79 per cent of all hurricanes occur between August and October, with a peak in mid-September. The deadly November storms

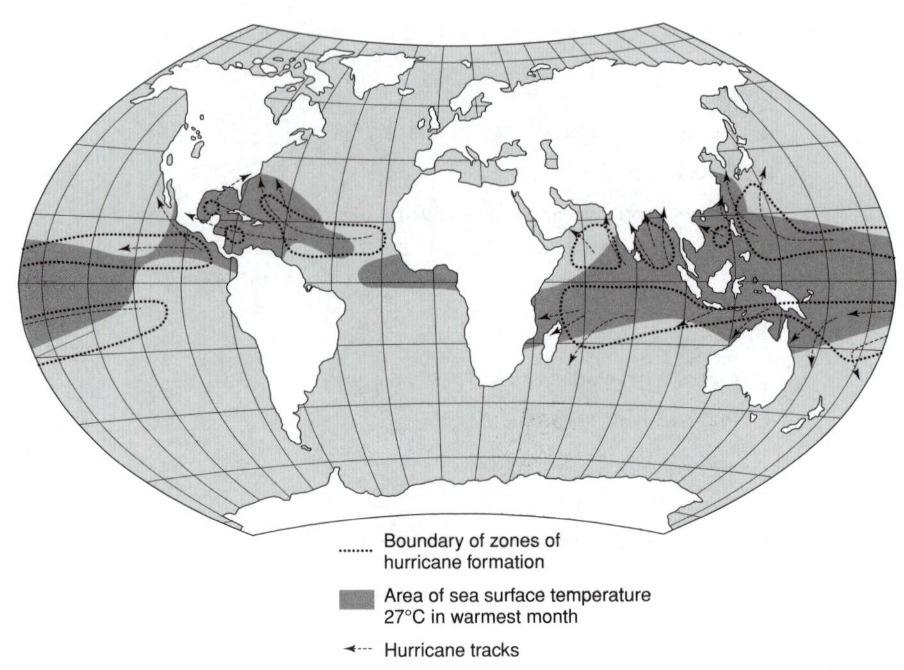

Figure 4.9 Areas of tropical cyclone formation
Source: After Briggs, D. and Smithson, P. (1985)

Michelle (2001) and Lenny (1999) are reminders to avoid the complacency of the Caribbean folk rhyme about hurricanes:

June, too soon
July, stand by
August, come you must
September, remember
October, all over . . .

The seasonality of tropical cyclones is partly due to the movement of the Inter-Tropical Convergence Zone (ITCZ) north of the equator in the summer (see Chapter 1). Analysis of time series data for Atlantic hurricanes shows that there are secular trends in hurricane activity over long time periods. For example, low hurricane frequencies occurred in the region in the 1870s and 1910s, with higher frequencies in the 1890s and again for the 1930s–1950s (Reading 1990; Reading and Walsh 1995). Frequencies declined after 1960, but the apparent increase in the intensity of hurricanes over the last decade prompted the IPCC to suggest that cyclonic activity may be on the increase as a result of global warming and higher sea surface temperatures (SSTs) (McGregor and Potter 1997), although their recent pronouncements (IPCC 2001) are more cautious regarding the possible effect of global warming on cyclonic activity. El Niño also has an effect on the pattern of hurricane activity but the scientific relationships and ramifications are not fully understood at this time.

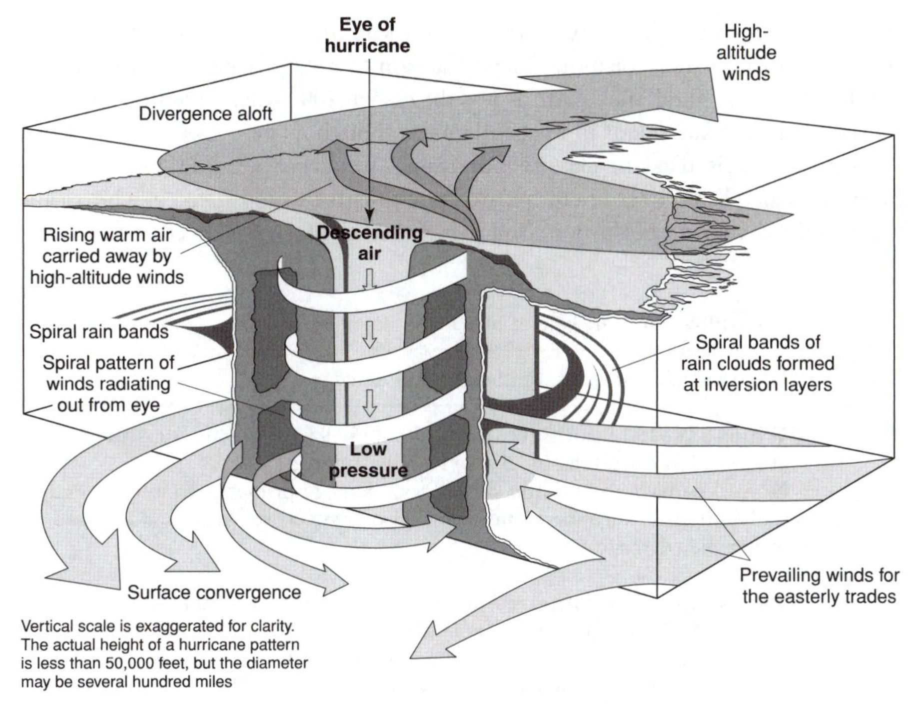

Figure 4.10 The structure of a hurricane

The structure of a hurricane

The main source of a hurricane's energy is the latent heat that becomes available when water vapour condenses in an immense updraft of convective activity. Hurricanes are atmospheric heat engines. In a single day, they release energy equivalent to several hundred hydrogen bombs, enough to supply the electrical needs of the USA for six months (Coch 1995). Despite their destructive capacity, hurricanes are highly efficient natural systems for transferring the huge surplus energy that builds up in the tropics towards the poles.

Hurricanes have a complex structure of horizontal and vertical air circulation (Figure 4.10). The eye is a central area of calm descending air with little wind and clear skies and has the lowest pressure in the system. The eye may be 30–50 km in diameter over the ocean. Immediately surrounding and defining a boundary zone is the eye wall, where there is an enormous pressure gradient producing the strongest winds, and where huge cumulonimbus clouds may reach heights of 10–12 km. Inside the eye, the descending air is warmed adiabatically, a process that accentuates the high temperature in the core. The outflow of air in the upper troposphere at heights of 10–15 km is in an opposite clockwise rotation compared with inflowing air at the ocean surface.

The spiralling inflows of air at the ocean surface extend to atmospheric heights of 2 km and arrange themselves into spiral rain-bearing clouds. Two trailing, outer spiral rain bands can be quite distinctive on satellite photographs. They rotate slowly with the system, bringing torrential rain to a very large geographical area not directly in the hurricane's path, as was the case in 2001 when Hurricane Michelle caused landslides and floods in eastern Jamaica though its path took it across Cuba.

Wind speed is used to classify hurricane strength according to the Saffir–Simpson scale (Table 4.2). There are two components to wind speed in a hurricane: the rotational velocity of the spiralling winds and the forward speed of the storm.

Table 4.2 The Saffir–Simpson scale of hurricane intensity

Category	Wind velocity, storm surge height and damage
1	**Winds 119–153 kilometres per hour (74–95 miles/hr), or storm surge 1.2–1.5 metres (4–5 feet) above normal** No real damage to building structures. Damage primarily to unanchored mobile homes, shrubbery and trees. Also some coastal road flooding and minor pier damage.
2	**Winds 154–177 kilometres per hour (96–110 miles/hr), or storm surge 1.8–2.4 metres (6–8 feet) above normal** Some damage to roofing material, and door and window damage to buildings. Considerable damage to vegetation, mobile homes and piers. Coastal and low-lying escape routes flood 2 to 4 hours before arrival of hurricane eye. Small craft in unprotected anchorages break moorings.
3	**Winds 178–209 kilometres per hour (111–130 miles/hr), or storm surge 2.7–3.6 metres (9–12 feet) above normal** Some structural damage to small residences and utility buildings with a minor amount of curtain wall failures. Mobile homes are destroyed. Flooding near the coast destroys smaller structures, with larger structures damaged by floating debris. Terrain continuously lower than 1.5 metres (5 feet) above sea level may be flooded inland as far as 9.6 kilometres (6 miles).
4	**Winds 210–249 kilometres per hour (131–155 miles/hr), or storm surge 3.9–5.5 metres (13–18 feet) above normal** More extensive curtain wall failures with erosion of beach areas. Major damage to lower floors of structures near the shore. Terrain continuously below 3 metres (10 feet) above sea level may be flooded, requiring massive evacuation of residential areas inland as far as 9.6 kilometres (6 miles).
5	**Winds greater than 249 kilometres (155 miles/hr), or storm surge greater than 5.5 metres (18 feet) above normal** Complete roof failure on many residences and industrial buildings. Some complete building failures, with small utility buildings blown over or away. Major damage to lower floors of all structures located less than 4.5 metres (15 feet) above sea level and within 457 metres (500 yards) of the shoreline. Massive evacuation of low areas on low ground within 8–16 kilometres (5–10 miles) of the shoreline may be required.

Source: www.noaa.gov, www.hnc.noaa.gov

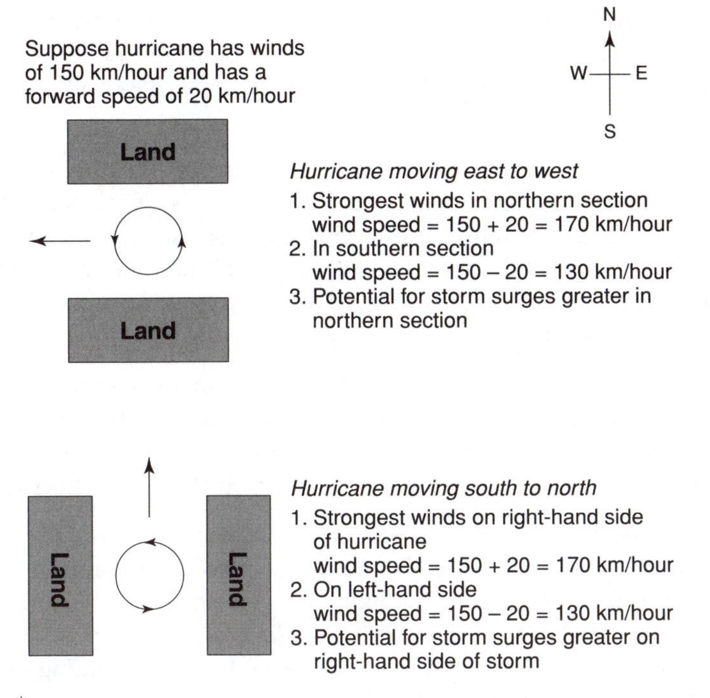

Figure 4.11 Wind speed and storm surge potential in a hurricane

Speed of forward movement can range from 8 to 32 km/hr (5–20 mph) and needs to be taken into account in assessing wind damage potential (Figure 4.11). Also, wind speed declines sharply with distance from the eye of the storm, so that an island may only experience tropical storm-strength winds rather than hurricane-strength winds if the eye passes some distance away. Generally, each hurricane has a unique spatial distribution of wind speed velocities and these patterns will change as the system moves its position.

The life cycle of a hurricane

Hurricanes form over the ocean not over land, where sea temperatures are at least 26 °C to a depth of 60 m, under conditions of high and constant relative humidity of about 75–80 per cent. Whilst all these conditions are necessary for a hurricane to form they are not sufficient. In fact, it is more useful to conceptualise the development of a full-strength hurricane as the final stage in a process of transition from a tropical disturbance.

1. *An initial tropical disturbance* – a tropical disturbance may originate within a tropical wave. Tropical disturbances are associated with heavy and sustained thunderstorm activity and emerge every three or four days

when low-pressure areas move off the West African coast into the Atlantic, or when a low-pressure system from higher latitudes meanders into the trade wind belt. Not every tropical disturbance that is formed will intensify into stage two.

2. *Formation of a tropical depression* – the weather disturbance becomes more organised and intensifies. The inflow of air at the ocean's surface starts to develop rotary circulation, initiated by wind deflection due to the Coriolis force. In the northern hemisphere the deflection is to the right, hence the rotation motion is counterclockwise. Again, not all tropical depressions will intensify further into stage three.

3. *Formation of a tropical storm* – further intensification of a tropical depression can produce a tropical storm. A tropical storm is formed when wind speeds exceed 62 km/hr (39 mph). On average ten tropical storms form in the region each year and six will intensify further into a hurricane.

4. *Formation of a hurricane* – a hurricane forms when wind speeds of the weather system exceed 119 km/hr (74 mph), due to a further drop in pressure. At this stage, the eye of the hurricane is normally visible on satellite imagery. Further intensification of the hurricane is possible, so several categories of hurricane intensity have been defined, according to the Saffir–Simpson scale (Table 4.2).

During these stages of intensification of weather systems, there must be high pressure (an anticyclone) in the upper troposphere, at elevations of 9–15 km, before a hurricane can develop. This enables the inflow of air at the ocean's surface to be balanced by an outflow of air in the upper troposphere. Also, there must be light upper-level winds with a weak shear. Absence of a high-pressure system aloft and strong wind shear will inhibit further intensification of the weather system.

Once a tropical storm has formed, the weather system begins to move along a particular geographical path, so that all tropical storms and hurricanes have a distinctive storm track. The entire period of intensification, movement and eventual demise can be thought of as a life cycle and may last a few days or a couple of weeks, and the life cycle can be plotted on a map using storm track data. At any time during the weather system's life cycle, it can regress from hurricane to tropical storm, or re-intensify again into a hurricane, depending on meteorological conditions, as was the case with Hurricane Iris in 1995. The life cycle has several stages:

1. *Formative stage* – all the atmospheric and oceanographic conditions described above must be present. The process can take several days or less than 12 hours. As the counterclockwise wind system becomes more organised, atmospheric pressure at the centre falls to around 1,000 mb.

2. *Immature stage* – should the system continue to intensify, then the central pressure continues to fall. Hurricane-force winds are packed around the eye, and cloud and rain patterns become organised into bands that spiral

inwards. Total geographical size will be fairly small, perhaps a diameter of 60–100 km. Many small cyclones do not intensify beyond this stage. These hurricanes are minimal category 1 or 2 on the Saffir–Simpson scale, and an example was Hurricane Iris in 1995.

3. *Mature stage* – the mature stage is the longest stage and may last for two weeks. The geographical size of the entire system is variable, though a typical diameter may be 650 km. The largest Caribbean system ever was Hurricane Gilbert, which at one stage recorded the lowest central pressure ever, at 888 mb (Eyre 1989).

4. *Terminal stage* – there are several ways a hurricane can lose power and be downgraded. A prolonged period over land negates the basic energy source, the supply of warm, humid air from the sea. A strike on the Mexican or USA mainland is likely to degrade the system (but not the Yucatán and Florida). Similarly, when a hurricane moves into the North Atlantic, where sea surface temperatures are below 27 °C, it will weaken because its energy source is less potent. Demise of a hurricane can also occur when conditions in the upper atmosphere change and wind shear (caused by a change in direction of the upper-level winds) in effect 'decapitates' the critical outflow region and literally tears the hurricane apart. Alternatively, if a hurricane moves into an area with an upper-level trough, there can be a similar stifling effect.

At the beginning of the season, the main geographical area where hurricanes form is the western Caribbean and Gulf of Mexico. By the mid-season the principal area of cyclogenesis has shifted into the Atlantic, sometimes in the vicinity of the Cape Verde Islands. Through October and November the western Caribbean and the Gulf of Mexico again become areas where hurricanes are likely to originate, perhaps because the ITCZ often encroaches into the western Caribbean in the early and late season, increasing the likelihood of tropical disturbances forming in those areas.

Each tropical storm is named (Box 4.3). The paths of named tropical storms and hurricanes are plotted easily on a tracking chart because island meteorological offices issue regular advisories giving precise latitudinal and longitudinal coordinates. The National Hurricane Center in Miami produces accurate short-term forecasts of tracks using computer modelling and data from sophisticated instrumentation on board hurricane hunter aircraft that fly into the eye. The penetration of cable and satellite TV into the region has greatly increased the amount of information available to Caribbean people, many of whom routinely follow the Weather Channel's hourly 'Storm Reports' during the hurricane season.

A typical mid-season hurricane such as Hurricane Floyd (Figure 4.13) follows a southerly track parallel to the isobars of the Azores–Bermuda subtropical anticyclone then the system recurves northwards and enters the mid-latitude westerlies. In the more northerly latitudes the system is normally downgraded to a vigorous depression, and then travels across the Atlantic to Western Europe.

Box 4.3: What's in a hurricane name?

In historical times, hurricanes were often named after the particular saint's day on which they struck an island, an example being Hurricane San Narcisco, which devastated Puerto Rico in 1867. An Australian meteorologist, Clement Wragge, apparently named hurricanes after politicians he did not like, whilst during the Second World War, American pilots flying in the Pacific gave tropical cyclones the names of their wives and girlfriends. In the early 1951, the US Weather Bureau introduced the military phonetic alphabet (Able, Baker, Charlie, etc.) to identify tropical storms, so it was Hurricane Charlie that devastated eastern Jamaica in August 1951. This naming system was abandoned in 1953 and female names were used until 1978, when gender bias was abandoned, and lists of male and female appellations alternated alphabetically.

The method of using two separate naming systems for Atlantic and Pacific storms has evolved into a more complex nomenclature used today by the World Meteorological Organization. Different sub-regions are allocated separate name lists: currently, one for the Atlantic region, three zones in the Pacific, two for Australia, and one each for Fiji, Papua New Guinea and India (www.hnc.noaa.gov or www.wmo.org). For Atlantic storms lists of names are used in a six-year cycle, so the list used in 2003 will be used again in 2009. A name is 'retired' from the list if a hurricane is particularly costly or deadly; thus, for example, after 1995 'Luis' was replaced by the name 'Lorenzo'.

The use of formal lists of personal names to identify particular hurricanes has an interesting cultural adaptation in the region. West Indians often use humour to make light of hardship, so severe hurricanes are sometimes imbued with a personality as a psychological coping mechanism. In 1999, Hurricane Lenny's unprecedented track westwards across the Caribbean prompted Leeward Islanders to give it the nickname 'Left Hand Lenny' (Beckford, 2000). In 1988, Hurricane Gilbert, then the largest storm ever recorded in the western

Left handed Lenny

Islands of the Caribbean are counting the cost in human life and damage after the most unusual hurricane for nearly 100 years visited and busied itself for nearly a week in the region.

At least 10 deaths including that of a Dominican have been reported from as far away as Columbia in South America to St Maarten as a result of Hurricane Lenny. Lenny called 'left handed' and more derogatory terms by wits in the islands retraced the path normally taken by the Atlantic-born storms. The last time a storm took that course was about 1903.

Formed in the waters of the Gulf of Mexico, Lenny decided to travel backwards, easterly towards the open Atlantic wher it dissipated last Sunday, but not before generating massive swells which battered the islands through Puerto Rico to Grenada.

Martinique and Guadeloupe through to the Windward Islands are counting the cost caused by sea and wind to lives, coastline housing, roads, hotels, restaurants, sea defences, utility poles and service, fisheries,

Figure 4.12 Newspaper extract with 'Left handed Lenny' headline
Source: *Tropical Star*, 24th November 1999, vol. VII, no. 12

(Box continued)

hemisphere, was given a variety of aliases by Jamaicans: 'Kilbert' and 'Rambo' suggested a rampaging badman in the manner of a notorious Kingston gunman, and 'Rufus' destroyed tens of thousand of homes of all social classes, including their roofs. Perhaps the most memorable alias was coined by Jamaican singer/songwriter Lloyd Lovindeer in his socially witty rendition of 'Wild Gilbert', with its references to satellite dishes and bully beef, with classic lyrics like 'mi roof migrated without a visa'. Jamaicans anthropomorphised the disaster by giving it a human personality. Gilbert personified nature, the omnipotent social leveller, no respecter of persons and a humbler of the mighty, so people were able to internalise the event by creating a wry social context for introspection and discussion of their own experiences (Barker and Miller 1990).

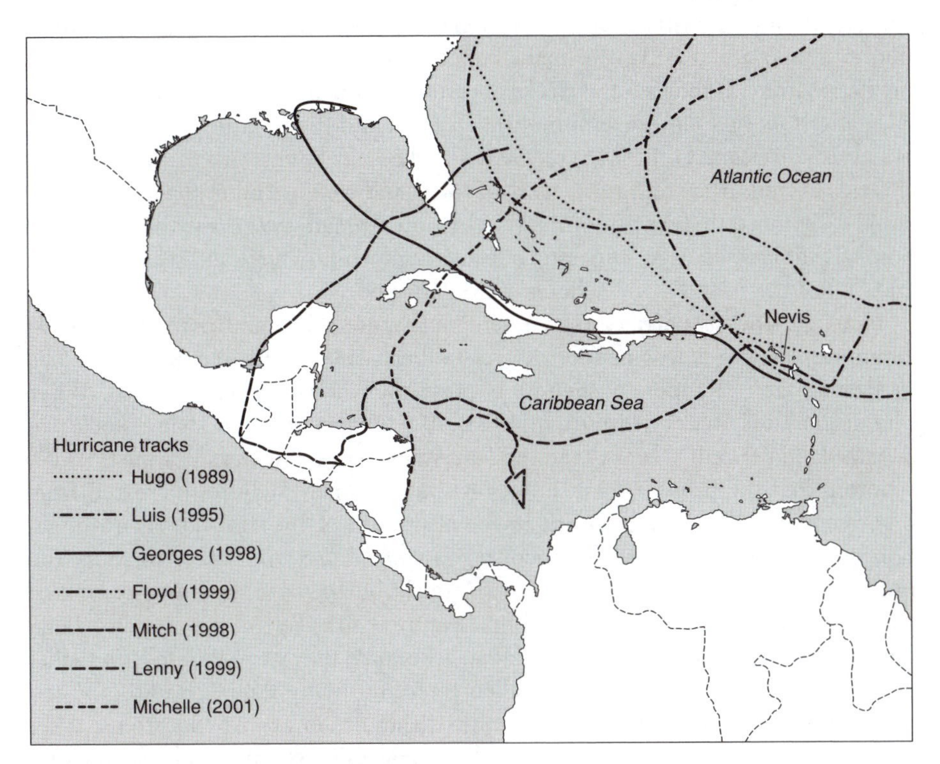

Figure 4.13 Storm tracks of selected hurricanes

Late-season hurricanes forming in the western Caribbean tend to have different tracks to their mid-season counterparts. Hurricane Mitch (1998) caused considerable damage in recent years, and Hurricane Lenny (1999) had an unprecedented track (Figure 4.13 and Box 4.3). Since tropical cyclones are also steered by meteorological conditions in the upper atmosphere, the net result of all these factors is that each storm track is unique.

Hurricane hazard, preparedness and mitigation

The winds and rains associated with hurricanes can have a damaging impact on an island even when the eye passes at distances of 100 km or more. However, the worst damage usually occurs as the result of a direct hit, when the eye passes directly over land, in which case impact is likely to be catastrophic. The Saffir–Simpson Scale (Table 4.2) indicates possible damage for each category. The comprehensive nature of losses may prompt the international mobilisation of aid to assist relief efforts, and Caribbean people living abroad are quick to respond and increase the flow of remittances and 'barrels' to relatives and friends.

Hurricanes can cause loss of life and injury, and pose health risks through contaminated water supplies. Hurricane Georges in 1998 killed an estimated 465 people, mainly in the Dominican Republic and Haiti. However, damage to, and restoration of, critical facilities (public utilities, health services, communication systems) are usually the main concern in the immediate aftermath of a hurricane strike. It took several weeks to restore power and water services in Puerto Rico after Georges, and several months in Haiti and the Dominican Republic. In Jamaica, Hurricane Gilbert (Eyre 1989) knocked out Kingston's main airport for three days, and urban and rural areas of the island were without power and water supply for more than two months (Carby and Ahmad 1995). Furthermore, ten hospitals suffered damage, impairing effective treatment of hurricane-related injuries, and 500 of the island's 530 schools were damaged.

Housing stock is particularly vulnerable (Figure 4.15), and perhaps the most common problem is loss of roofs. After Gilbert, 800,000 people in Jamaica had to spend time in temporary shelters because their houses were damaged. When Hurricane Luis struck Antigua in 1995, 90 per cent of the housing stock was damaged (40 per cent severely) and the Pan American Health Organization (PAHO) reported that the proportion of homes without sanitary facilities increased from 8 per cent in 1993 to 12 per cent in the aftermath of the disaster. As a mitigation measure, homeowners are encouraged to fix hurricane straps to roofs and put storm shutters over windows, all relatively inexpensive measures compared with the costs of house reconstruction (Figure 4.14). Telephone and electricity power lines are also vulnerable to damage because they are mounted on utility poles susceptible to strong winds, fallen trees and flying debris. Radio, TV and microwave transmitters, and even satellite dishes are in exposed locations. After Jamaica's traumatic experience with Hurricane Gilbert, many of the region's utility companies introduced their own company disaster plans and mitigation measures, erecting more wind-resistant poles and placing cables underground.

We noted in Chapter 3 that agricultural losses due to hurricanes and tropical storms can be significant. Hurricane Georges destroyed an estimated 90 per cent of the agricultural sector in the Dominican Republic, and in Haiti, three-quarters of the rice crop and 80 per cent of the banana crop was lost. In Puerto Rico, 95 per cent of the banana crop was lost and 75 per cent of the coffee, whilst St Kitts lost 50 per cent of its sugar crop.

Figure 4.14 Caribbean disaster management agencies disseminate information to the general public to help people prepare for natural hazards
Source: The Office of Disaster Preparedness and Emergency Management, Jamaica

A storm surge is a large dome of water in the ocean in the vicinity of the hurricane system and may be 80–160 km wide. It is caused mainly by winds pushing the ocean surface ahead of the storm (Figure 4.11). The low barometric pressure of the eye also contributes to a storm surge, although a barometric pressure of 900 mb will raise sea level only about one metre. Storm surges are a major

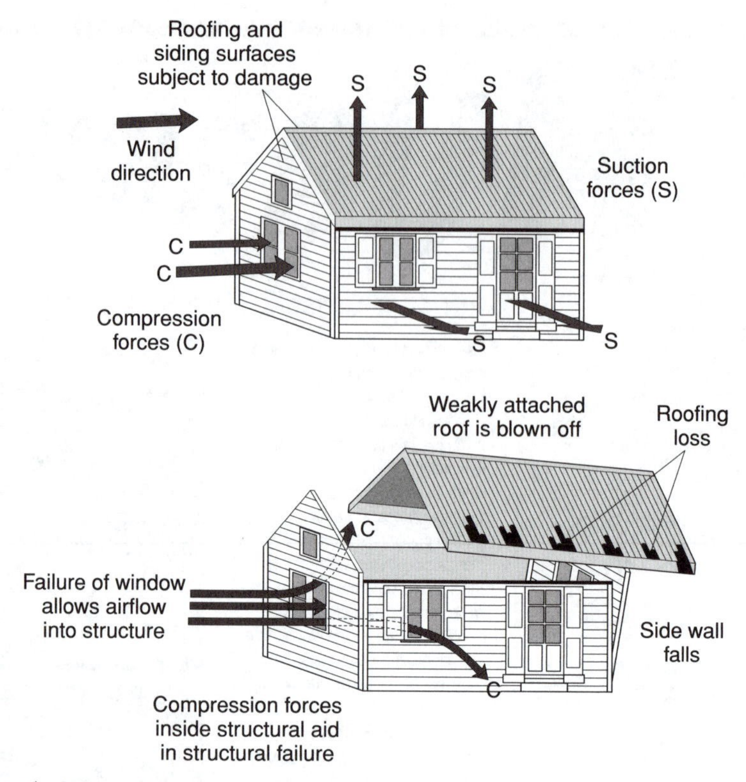

Figure 4.15 Potential damage to a house by a hurricane

component of hurricane hazard, causing coastal flooding and impacting the shoreline near where a hurricane makes landfall. It is defined as the increase of shoreline water level above the normally seasonally adjusted average high-tide level (Alexander 1993) and was reportedly 3–4 m during Hurricane Georges in Puerto Rico. Hurricane Lenny produced surges as high as 5–7 m at Fredericksted in St Croix, in the US Virgin Islands. The severity of a storm surge on a particular coast depends on a number of factors:

- offshore submarine topography – shallow continental shelves 'pile up' water;
- distance from the eye of the storm;
- rate of forward movement of storm;
- position and angle of approach of storm;
- geometry of the coastline – worse in narrow/concave bays;
- tidal stage – worse at times of high tides;
- barometric pressure – storms of greater intensity (lower pressure) are worse.

The National Oceanic and Atmospheric Administration (www.noaa.gov) has developed a computer model known as SLOSH (Sea, Lake and Overland Surges from Hurricanes) to simulate the effects of storm surges to help the planning efforts of communities affected by storm surges. It has been tried for Puerto Rico (Oxman 1989), but generally speaking measures to limit storm surge hazard in the region are often notably absent, as Kingston's predicament illustrates (Box 4.1). Storm surges are one of several ways in which the region is affected by flood hazard.

Floods and flood hazard

Compared with South-East Asia, where floods affect 4 million hectares of farmland and 17 million people annually (Smith 1976), flooding in the insular Caribbean is small-scale though undoubtedly the most frequently occurring and persistent hazard in the region. Flooding tends to affect poor communities in rural and urban environments, who usually do not have insurance for homes, crops or livestock. There is little published research on flood hazard in the Caribbean literature, but a few examples of flood impact studies include Pelling (1996, 2003) for Guyana, Collymore and Griffith (1987) for Barbados, and Bertrand and Romano (1986) for Trinidad and Tobago.

The Caribbean region is exposed to three types of flooding: coastal, riverine and flash flooding. Flooding always occurs during hurricanes and tropical storms, and a typical hurricane will dump between 200 and 400 mm of rain over an area, and the slower the passage of a storm, the more rain is likely. After Hurricane Georges, FEMA paid out more than US$25.4 million in flood claims in Puerto Rico. Hurricane storm surges can be devastating for coastal towns and tourism facilities (see above) and tsunamis also pose a similar potential threat. Other weather systems bringing torrential rains are tropical waves and tropical depressions, and cold fronts (or northers), which affect the northern Caribbean in winter (Chapter 1). Tropical depressions are slow-moving and unpredictable, and when stationary for a few days can dump more rain on an island than faster-moving hurricanes. Jamaica, for example, is prone to such weather systems around May/June and experienced severe flood rains related to tropical depressions in 1979, 1986 and 1991.

Many types of human activity contribute to flood hazard. In rural areas, deforested drainage basins may be flood-prone in the lower parts of the valley or on small coastal deltas, features called flood plains. During heavy rainfall, the erosion and deposition of sediment in river channels reduces channel capacity and adds to the risk of flooding. The destruction of coastal mangroves for tourism may contribute to flood hazard and cause coastal flooding, especially during the passage of a tropical depression (Chapter 1).

Urban settlements may exacerbate flood hazard because they create extensive, hard impermeable surfaces such as roads and roofs. Rivers or gullies that run through towns have modified channels and are fixed in location and width to facilitate bridge construction. Thus a river channel is denied its natural ability to modify itself to accommodate large volumes of water during periods of heavy rainfall. Also, surface systems of storm drains, lined gullies and urban spillways

Figure 4.16 Small-scale construction techniques used in flood control in the Caribbean

Retaining walls

Stone and concrete structures to protect on steep slopes against erosion of the base of the slope.

Check dams

Small gravity dams of variable height built across the width of small gullies. Usually rock and wire mesh structures, but wood and old tyres can be used effectively.

Bunding or gabion baskets

Galvanised wire mesh, stone and riverbed material used to protect streams and riverbanks from bank erosion and landslips. Various types of groyne shape may be used.

Paved drains and culverts

U-shaped concrete structures designed to move water quickly from roadsides, under bridges and steep slopes susceptible to erosion.

Levee

Concrete or earth embankment on side of river to prevent overflow.

deliver water more quickly to the main channel and reduce the lag time between storm event and peak flow. Blocked drains and clogged gully courses in urban areas are other frequent causes of localised flooding in residential areas.

Flood hazard mapping is a standard procedure in the disaster reduction process in many parts of the world. It can depict areas likely to be affected by storm surges, or flood plains likely to be inundated in a flood with a recurrence interval of, say, 50 years or 100 years. In general, flood hazard mapping in the region is not very advanced, except in Belize, where the problem is more widespread (see www.oas.org/en/cdmp/bulletin/bzfld.htm).

With regard to available structural mitigation measures to control flooding, the Mississippi valley and the Netherlands are good illustrations of what can be achieved given substantial financial and technological resources. However, even small-scale structural mitigation measures against flooding have a cost, and in the Caribbean tend to involve low-technology/labour-intensive construction techniques. Common construction techniques in the region are listed in Figure 4.16.

Land-use planning as a technique for mitigating flood hazard is used and enforced in industrialised countries. In the developing world, planners could prohibit new residential developments from flood-prone areas but the information and political will is generally lacking. Most Caribbean governments adopt a non-mitigation strategy that basically accepts flood losses as inevitable, and absorb costs using the budgetary and technical resources at their disposal (perhaps topped up by foreign assistance). In this way, it is common for governments to help rehouse families or communities, provide food aid and material compensation, or compensate farmers who lose crops or livestock.

The management of flood hazard provides a good example of agencies' use of early warning systems to inform the general public of a potential hazard threat. The region's metereological offices issue flash flood watches and flash flood warning to indicate communities at risk during periods of heavy rainfall. A flash flood watch indicates the possibility of a flood, whilst a warning indicates flooding is imminent or already in progress. There is considerable scope for organised local community action with respect to flood hazard, based on local knowledge and observations of local conditions. Community-based early warning systems can be effective in preventing loss of life. In Jamaica, community-level responses are organised through the Office of Disaster Preparedness and Emergency Management (ODPEM). A system of parish disaster committees and zonal committees encourages networks of people and institutions like service clubs to engage in community service during flood emergencies Thus, from time to time, community task forces help clean drains, gullies, culverts and sinkholes.

Flood rains are probably the most widely reported small-scale disasters in the region. Carby and Ahmad (1995) have discussed the problem in relation to roads and water systems and provide some estimates of the costs involved in three separate flood events. Their analysis also considers landslide damage, and in practice flood hazard and landslide hazard together are both a consequence of periods of heavy rains.

Landslides

Landslides are natural events and are ubiquitous in mountainous regions, but human modification of landscape has multiplied the frequency of landslides many times, and created landslide hazards where previously they did not exist. The landslide hazard is especially acute in developing countries, because population increase in certain types of location has resulted in more homes being built on potentially unstable hillsides. For example, in many South American cities, shanties and squatter areas have clawed their way up flanking mountains and steep hillsides, and landslide disasters in such settings can be catastrophic and tragic. Many of the 30,000 people killed in the flood rains in Venezuela in December 1999 died as a result of landslides and mudslides in peripheral urban shanties. In the Caribbean, landslides are recognised as a significant recurrent and persistent problem (Ahmad 1995).

The landslide hazard involves loss of life, damage to property and infrastructure, and disruption of transport and communications, and it causes localised losses of land and soil resources. In Japan, average annual losses are as high as US$4 billion (Smith 1996). Accurate quantitative estimates of damage are less common for developing countries though DeGraff (1989) estimated annual repair costs for roads throughout the Caribbean to be around US$15 million, whilst Naughton (1984) assessed costs based on historical records in Jamaica.

Table 4.3 Varnes' classification of mass wasting

Type of movement			Type of material: Bedrock	Soils: Coarse	Soils: Fine
Falls			Rockfall	Debris fall	Earth fall
Topples			Rock topple	Debris topple	Earth topple
Slides	Rotational	Few units	Rock slump	Debris slump	Earth slump
	Translational	Few units	Rock block glide	Debris block glide	Earth block glide
		Many units	Rock slide	Debris slide	Earth slide
Lateral spread			Rock spread	Debris spread	Earth spread
Flows			Rock flow (Deep creep)	Debris flow (Soil creep)	Earth flow (Soil creep)
Complex			Combination of two or more types		

Source: Adapted from Selby (1985)

A landslide is one of several types of mass movement that displaces material down a hillside. The material may be soil, regolith (weathered bedrock) or rocks. Slope failure occurs when there is a change in the balance between the forces of gravity and the forces of resistance (or strength) of the rock and soil on a hillside. Movement takes place along a plane surface and is called shear, whereas the forces that promote the movement are called stresses. Thus a landslide (i.e. slope failure) occurs when the threshold angle of stability of a slope is exceeded, and the balance tips in favour of gravity as shear stresses overcome shear strength.

Modern landslide research investigates all types of mass movement that lead to slope failure, including falls, slides, slumps, creeps, spreads and flows, and Varnes' (1978) classification of landslides, based on the type of movement and nature of material, is widely used (Table 4.3).

A common basic type is a rotational slide, where a curved plane is created as the mass rotates backwards around a common point with an axis parallel to the slope (Figure 4.17). On the other hand, a debris flow does not have a discrete failure plane and occurs when surface materials become saturated and move downslope as a viscous mass. Larger debris flows may follow pre-existing channels such as gullies and stream courses. Small debris flows are very common, but several measuring 2–3 km were recorded in Jamaica as a result of the landslides associated with Hurricane Michelle in 2001. A mudflow is composed of saturated

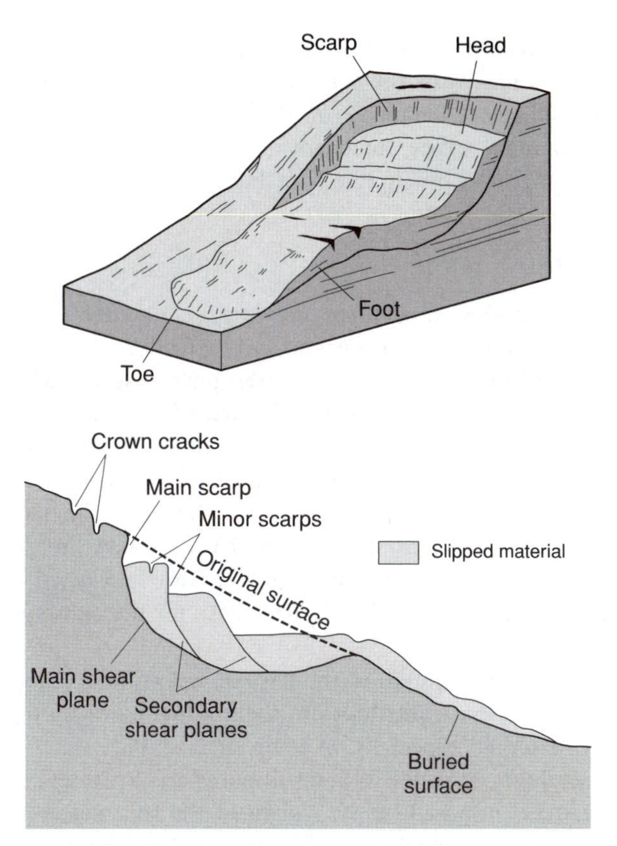

Figure 4.17 A rotational slump landslide
Source: After Briggs, D. and Smithson, P. (1985)

finer particles, mainly clay and silt. All types of flow failure tend to move faster and farther than landslides, and their sudden onset can be a major cause of loss of life. Of the 10,000 people who lost their lives in Hurricane Mitch in Central America in 1998, 1,500 people perished on the slopes of the Casita volcano in Nicaragua when a mud avalanche some 20 km in length and 2–3 km wide engulfed the town of Posoltega and its surrounding villages.

Site geology, rock structure and other geomorphological factors such as slope angle and slope aspect are critical elements of landslide-prone areas (Ahmad 1995). In Jamaica, 80 per cent of the slopes are over 20° and the Blue Mountains are prone to landslides because they are geologically young and heavily fractured, and the bedrock is deeply weathered (McGregor and Barker 1991). The rugged terrain of ridges and valleys is decorated with deep landslide scars (Ahmad 1999a). Lithology is a critical factor. For example, in conglomerate rocks wedge failures (a type of translational slide) are widespread, especially in moderately weathered volcanic rocks (like andesites), and occur because water percolates along fractures, joints and cracks and bedding planes, increasing the probability of slope failure.

Detailed landslide mapping has been undertaken by geologists at the University of the West Indies and reveals the complex nature of present and past landslide activity even in small areas (Manning *et al.* 1992; Maharaj 1993, 1995; Ahmad 1995).

There are two main trigger mechanisms that change the balance between gravity and resistance and precipitate landslides:

- seismic activity
- torrential rainfall.

An earthquake occurring either during heavy rains or after a prolonged period of rainfall (called antecedent rain) is particularly effective in triggering landslides. The region's largest historic landslide is an example of a seismically induced landslide. Judgment Cliff, in eastern Jamaica (Figure 4.9), is the site of a 1692 landslide (Zans 1959). Rupture scars and displaced material are still evident in the field today. Judgment Cliff was so named because folklore held it was God's judgment on a wicked Dutch planter, who perished when an estimated 80 million m^3 of material destroyed his plantation. It seems that the heavy limestone sediments overlying less consolidated clayey–shaly beds became lubricated during rains in the preceding weeks and provided a slippery base for the subsequent slope failure (Zans 1959).

The role of antecedent rainfall in the days or weeks before slope failure is an important factor because it contributes to the gradual build-up of pore pressure within hillslope materials, especially on a slip surface. Pore water pressure refers to the incompressibility of water. When soil pores are saturated, soil water takes up some of the stress imposed by the weight of the soil, destabilising it, so the soil readjusts to a more stable position by moving downslope (Alexander 1993).

Torrential rainfall, of short duration and high intensity, tends to produce shallow soil slips and debris flows in surface materials. Longer rainfall events can produce slope failure in bedrock, and deeper, larger slides. For example, more than 400 landslides were reported in the Luquillo Experimental Forest Area in Puerto Rico after the fast-moving Hurricane Hugo in 1989. Of these 91 per cent were debris flows (Ahmad *et al.* 1993a). Larsen and Torres Sanchez (1992) reported the largest mass movement as a debris avalanche that moved 30,000 m^3 of material more than 600 m. The latter was a reactivation of an old landslide, another feature common after heavy rainfall (Maharaj 1995).

Landslides can create temporary dams across rivers. A recent example occurred in Dominica in 1997. Landslides at Carholm-Huxley dammed the Layou and Mathieu rivers (James and Serrant 1997). The dams burst after a few days, causing flooding downstream and on the coast, and several hundred people were evacuated. Rogers (1997) suggests an important site factor in this slope failure was that less consolidated pyroclastic deposits underlie the landslide zone.

Human disturbance on hillsides also creates conditions that are conducive to landslides by upsetting the balance between the forces of gravity and the forces of resistance, though a trigger event is usually the proxy cause. Human activities that potentially contribute to slope failure include:

- *An increase in slope angle* – steepness and length of slope are important parameters affecting slope stability. In the same way that a stream can erode the base of a slope and remove lateral support, causing a riverbank to collapse, so too a road cut into a hillside can destabilise a slope.
- *Additional weight placed on a slope* – rainfall naturally adds weight to hillslope material as it soaks into and saturates soil and raises the water table. Burst and leaking pipes and sewers have a similar effect. Any type of building, dumped waste or infrastructure such as a road or a utility pole also adds weight to the physical mass resting on a slope.
- *Removal of vegetation* – vegetation roots bind soils together so deforestation, logging, vegetation clearance and land preparation for cultivation, and overgrazing all loosen surface materials, making them more susceptible to erosion and slope failure (see Chapter 3).
- *Exposure of joints and bedding planes* – this may occur as a result of the removal of surface material or vegetation and is problematic when the bedrock joints are dipping at an angle out of the slope.

Landslide hazard mapping and mitigation measures

Considerable progress has been made in landslide hazard mapping in the region, notably in Jamaica, the eastern Caribbean and Puerto Rico. For example, DeGraff *et al.* (1989, 1991) produced a series of general landslide hazards maps for the eastern Caribbean. Rogers (1997) compiled a specific debris flow map for St Lucia (Figure 4.18). Detailed landslide susceptibility maps for deep-seated and shallow landslides have been compiled for Kingston (Box 4.1) using GIS methods, techniques likely to become much more important in this field in the near future (Ahmad 1999a; 1999b).

Mitigation strategies to address the landslide hazard include a wide range of structural geotechnical and engineering techniques to improve slope stability or to construct hazard-resistant buildings. As with flood-control works, small-scale construction activity to mitigate landslide hazard is popular in the region because it is labour-intensive and provides temporary local employment. Some of the methods below are also used in flood hazard control:

- Physical restraining structures such as retaining walls, piles and buttresses, bunding and gabion structures.
- Excavation and filling to 'grade' steep slopes into gentler ones, or to break up slopes into stairs or steps. Reshaping hillsides is difficult and expensive.
- Drainage techniques to relieve the build-up of pore pressure. These include interceptor drains or trenches at the top of a slope, the construction of subsurface drainage, porous pipes and removal of surface water.

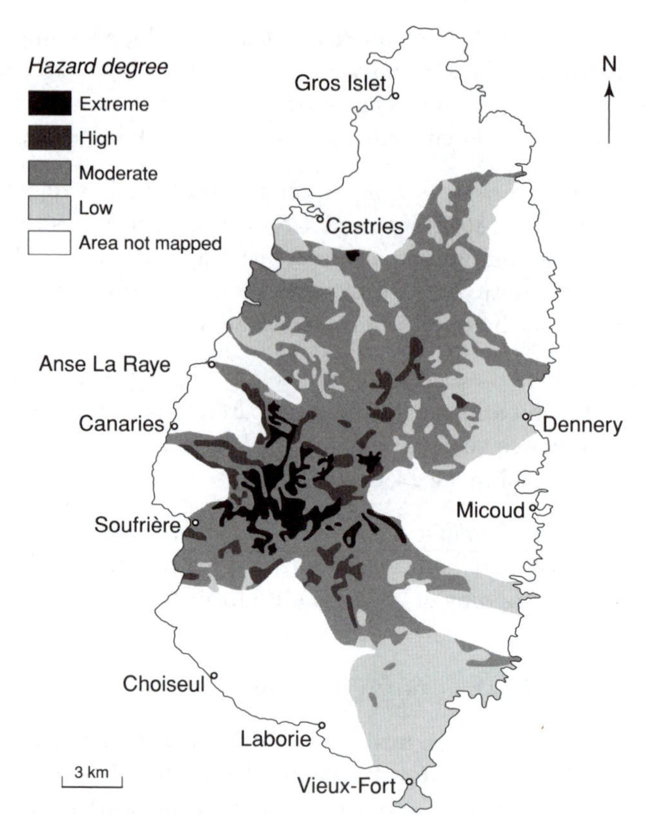

Figure 4.18 Landslide hazard map of St Lucia
Source: Adapted from Rogers (1997)

- A range of techniques are available for watershed management, including re-afforestation and the promotion of agro-forestry. Hillslopes can be re-vegetated by planting grasses whose roots bind surface soil, and trees with deep taproots. Agro-forestry generates income for farmers and offers protective cover on farmed hillsides (see Chapter 3).

For both landslides and flooding, physical planners can use non-structural mitigation measures to mitigate the hazard by trying to prevent or deter the location of residential communities in threatened areas. Hazard avoidance can be achieved by zoning, to steer residential and commercial development away from landslide-prone (or flood-prone) areas. Landslide zoning ordinances are very common, for example, in the Los Angeles area of California. However, widespread non-compliance with planning legislation and building codes in developing countries is a prime factor contributing to the increased incidence and magnitude of landslide and other disasters. Many squatter settlements are located on marginal urban lands precisely because the rest of society does not want the land, perhaps because it is hazard-prone. Even middle- and upper-class residential areas are sometimes built in landslide-prone areas (Box 4.1). Generally speaking, once homes are built, it is unrealistic to expect people to relocate unless alternative housing is provided, especially unplanned spontaneous settlements. In an extreme situation, an entire community may be relocated after a disaster. At the time of the Preston Lands landslide episode in 1986 in Jamaica people in the farming community affected were relocated (Ahmad *et al.* 1993b), and following a landslide disaster at Mameyes, near Ponce in Puerto Rico, the site was cleared and dedicated as a park and memorial to the victims (OAS 1991).

To conclude, in this chapter we have examined the principal natural hazards experienced in the region. The Caribbean is, in many ways, a microcosm of the multiple hazards facing populations, governments and disaster managers and planners in developing countries. Note that each of the phenomena discussed (volcanic eruptions, earthquakes, hurricanes, etc.) can trigger more than one type of hazard, and note too that most places in the region may be affected by more than one of these events in any given year. The insular nature of the Caribbean region, whereby most of the population is located in the coastal zone (see Chapter 7), is another factor compounding vulnerability. Nevertheless, there has been considerable progress over the last two decades with regard to disaster management and disaster mitigation, both at the regional level and in individual countries, though more financial and human resources need to be accessed and mobilised in order to strengthen programmes already in place.

References

Ahmad, R. (ed.) (1992) *Natural Hazards in the Caribbean, Journal of the Geological Society of Jamaica*, Special Issue No. 12, 34–53.

Ahmad, R. (1995) 'Landslides in Jamaica: extent, significance and geological zonation', in Barker, D. and McGregor, D. F. M. (eds) *Environment and Development in the Caribbean: Geographical Perspectives*, UWI Press, Kingston, Jamaica, pp. 147–69.

Ahmad, R. (ed.) (1997) *Natural Hazards and Hazards Management in the Greater Caribbean and Latin America*, Unit for Disaster Studies, University of the West Indies, Mona campus.

Ahmad, R. (1999a) 'Landslide loss reduction: a guide for the Kingston metropolitan area, Jamaica', Unit of Disaster Studies, **6**, OAS/USAID/UWI.

Ahmad, R. (1999b) 'Landslide susceptibility maps for the Kingston metropolitan area, Jamaica, with notes on their use', Unit of Disaster Studies, **5**, OAS/USAID/UWI.

Ahmad, R., Scatena, F. N. and Gupta, A. (1993a) 'Morphology and sedimentation in Caribbean montane streams: examples from Jamaica and Puerto Rico', *Sedimentary Geology*, **85**, 157–69.

Ahmad, R. Carby, B. E. and Saunders, P. H. (1993b) 'The impact of slope movements on a rural community: lesson from Jamaica', in Merriman, P. A. and

Browitt, C. W. A. (eds) *Natural Disasters: Protecting Vulnerable Communities*, Thomas Telford, London, pp. 447–60.

Alexander, D. (1993) *Natural Disasters*, University College London Press, London.

Alonso, D. (1989) 'Preliminary analysis of the cartographic representation of vulnerability of buildings in Santiago de Cuba', in Barker, D. (ed.) *Proceedings of a Meeting of Experts on Hazard Mapping in the Caribbean 1987*, PCDPPP/ODP/Dept of Geography, UWI, pp. 87–8.

Barker, D. (ed.) (1989) *Proceedings of a Meeting of Experts on Hazard Mapping in the Caribbean 1987*, PCDPPP/ODP/Dept of Geography, UWI.

Barker, D. (1993) 'Dualism and disasters on a tropical island: constraints on agricultural development in Jamaica', *TESG*, **85**, 5, 332–40.

Barker, D. and Miller, D. J. (1990) 'Hurricane Gilbert: anthropomorphizing a natural disaster', *Area*, **22**, 107–16.

Barry, R. G. and Chorley, R. J. (1982) *Atmosphere, Weather and Climate*, 4th edition, Methuen, London.

Beckford, Y. H. M. (2000) 'The effects of storm surges on Caribbean coastal areas: the case of Hurricane Lenny', *Caribbean Geography*, **11**(2), pp. 100–7.

Bertrand, D. and Romano, H. (1986) 'Landslide and flood distribution in the west coastal area of Trinidad', Technical Report, Institute of Marine Affairs, Trinidad.

Blaikie, P., Cannon, T., Davis, I. and Wisner, B. (1994) *At Risk: Natural Hazards, People's Vulnerability, and Disasters*, Routledge, London.

Briggs, D. and Smithson, P. (1985) *Fundamentals of Physical Geography*, Routledge.

Bullard, F. M. (1976) *Volcanoes of the Earth*, University of Texas, Austin.

Burton, I., Kates, R. W. and White, G. (1978) *The Environment as Hazard*, Oxford University Press, New York.

Carby, B. E. and Ahmad, R. (1995) 'Vulnerability of roads and water systems to hydro-geological hazards in Jamaica', *Built Environment*, **21**, 145–53.

Chin, M. W. (1997) 'Possible mitigation strategies for hurricanes and earthquakes in the Caribbean', in Ahmad, R. (ed.) *Natural Hazards and Hazards Management in the Greater Caribbean and Latin America*, Unit for Disaster Studies, University of the West Indies, Mona campus, pp. 88–95.

Christopherson, R. W. (1997) *Geosystems: An Introduction to Physical Geography*, 4th edition, Prentice Hall, New Jersey.

Coch, N. K. (1995) *Geohazards: Natural and Human*, Prentice Hall, Englewood Cliffs, NJ.

Collymore, J. McA. (1995) 'Disaster mitigation and cost–benefit analysis: conceptual perspectives', in Barker, D. and McGregor, D. F. M. (eds) *Environment and Development in the Caribbean: Geographical Perspectives*, UWI Press, Kingston, Jamaica, pp. 111–23.

Collymore, J. McA. (1998) 'Emergency assistance as a critical catalyst in future loss reduction planning', *Caribbean Geography*, **9**, 2, 148–55.

Collymore, J. McA. and Griffith, M. D. (1987) 'Flooding in Speightstown; towards a flood management strategy, in Barker, D. (ed.) *Proceedings of a Meeting of Experts* *on Hazard Mapping in the Caribbean 1987*, PCDPPP/ODP/Dept of Geography, UWI, pp. 117–25.

DeGraff, J. V. (1989) 'Assessing landslide hazard for regional development planning in the eastern Caribbean', in Barker, D. (ed.) *Proceedings of a Meeting of Experts on Hazard Mapping in the Caribbean 1987*, PCDPPP/ODP/Dept of Geography, UWI, pp. 40–5.

DeGraff, J. V. (1991) 'Determining the significance of landslide activity: examples from the eastern Caribbean, *Caribbean Geography*, **3**, 1, 29–42.

Eyre, L. A. (1989) 'Hurricane Gilbert: Caribbean record breaker', *Weather*, **44**, 160–4.

Housner, G. W. (1989) 'An international decade of natural disaster reduction: 1999–2000', *Natural Hazards*, **2**, 45–75.

James, A. and Serrant, T. (1997) 'The Carholm–Huxley landslides: an assessment of their impacts on the Lower Layou River valley, Dominica', *Caribbean Geography*, **8**, 2, 120–7.

Lander, J. F. (1997) 'Caribbean tsunamis: an initial history', in Ahmad, R. (ed.) *Natural Hazards and Hazard Management in the Greater Caribbean and Latin America*, Unit for Disaster Studies, Department of Geography and Geology, UWI, Monacampus, Jamaica.

Larsen, M. C. and Torres Sanchez (1992) 'Landslides triggered by hurricane Hugo in eastern Puerto Rico, September 1989', *Caribbean Journal of Science*, **28**, 113–25.

Maharaj, R. J. (1993) 'Landslide processes and landslide susceptibility analysis from an upland watershed: a case study from St Andrew, Jamaica, West Indies', *Engineering Geology*, **34**, 1–2, 53–79.

Maharaj, R. J. (1995) 'Evaluating landslide hazard for land use planning: upper St Andrew, Jamaica', in Barker, D. and McGregor, D. F. M. (eds) *Environment and Development in the Caribbean: Geographical Perspectives*, UWI Press, Kingston, Jamaica, pp. 170–86.

Manning, P. A. S., McCain, T. and Ahmad, R. (1992) 'Landslides triggered by 1988 Hurricane Gilbert along roads in the Above Rocks area, Jamaica', in Ahmad, R. (ed.) *Natural Hazards in the Caribbean, Journal of the Geological Society of Jamaica*, Special Issue No. 12, pp. 34–53.

McGregor, D. F. M. and Barker, D. (1991) 'Land degradation and hillside farming in the Fall River Basin, Jamaica, *Applied Geography*, **11**, 143–56.

McGregor, D. F. M. and Potter, R. B. (1997) 'Environmental change and sustainability in the Caribbean: terrestrial perspectives', in Ratter, B. M. W. and Sahr, W.-D. (eds) *Land, Sea and Human Effort in the Caribbean*, Institut für Geographie der Universität Hamburg, Hamburg, pp. 1–15.

Molinelli, J. (1989) 'Earthquake vulnerability study for the metropolitan area of San Juan, Puerto Rico', in Barker, D. (ed.) *Proceedings of a Meeting of Experts on Hazard Mapping in the Caribbean 1987*, PCDPPP/ODP/Dept of Geography, UWI, pp. 71–86.

Naughton, P. W. (1984) 'Flood landslide damage repair cost correlations for Kingston, Jamaica', *Caribbean Geography*, **1**, 198–202.

Office of Foreign Disaster Assistance (OFDA) (1988) *Disaster History: Significant Data on Major Disasters Worldwide, 1900 – Present*, USAID, Washington.

Organisation of American States (1991) *Primer on Natural Hazard Management in Integrated Regional Development Planning*, OAS/Department of Regional Development and Environment, Washington.

Oxman, B. L. (1989) 'The vulnerability of Puerto Rico to natural hazards', in Barker, D. (ed.) *Proceedings of a Meeting of Experts on Hazard Mapping in the Caribbean 1987*, PCDPPP/ODP/Dept of Geography, UWI, pp. 3–9.

Pelling, M. (1996) 'Coastal flood hazard in Guyana: environmental and economic causes', *Caribbean Geography*, 7, 1, 3–22.

Pelling, M. (2003) 'Vulnerability, urbanization and environmental hazards in coastal Guyana', in Barker, D. and McGregor, D. F. M. (eds) *Resources, Planning and Environmental Management in a Changing Caribbean*, UWI Press, Kingston, Jamaica, pp. 133–52.

Reading, A. J. (1990) 'Caribbean tropical storm activity over the past four centuries', *International Journal of Climatology*, **10**, 365–76.

Reading, A. J. and Walsh, R. P. D. (1995) 'Tropical cyclone activity within the Caribbean Basin since 1500', in Barker, D. and McGregor, D. F. M. (eds) *Environment and Development in the Caribbean: Geographical Perspectives*, UWI Press, Kingston, Jamaica, pp. 124–46.

Robertson, R. A., Ambeh, W. B. and Lynch, L. (1995) 'Strategic planning for volcanic emergencies in the Commonwealth eastern Caribbean', *Caribbean Geography*, **6**, 2, 77–96.

Robertson, R. E. A., Lynch, L. and Latchman, J. (1997) 'Volcano surveillance and hazard mitigation in the eastern Caribbean', *Caribbean Geography*, **8**, 1, 1–17.

Rogers, C. T. (1997) 'Landslide hazard data for watershed management and development planning, St Lucia, West Indies', in Ahmad, R. (ed.) *Natural Hazards and Hazards Management in the Greater Caribbean and Latin America*, Unit for Disaster Studies, University of the West Indies, Mona campus, pp. 150–64.

Selby, M. J. (1985) *Earth's Changing Surface*, Oxford University Press, Oxford.

Shepherd, J. B. (1989) 'Earthquake and volcanic hazard assessment and monitoring in the Commonwealth Caribbean: current status and needs for the future', in Barker, D. (ed.) *Proceedings of a Meeting of Experts on Hazard Mapping in the Caribbean 1987*, PCDPPP/ODP/Dept of Geography, UWI, pp. 50–61.

Shepherd, J. B. and Aspinall, W. P. (1980) 'Seismicity and seismic intersities in Jamaica, West Indies: a problem in risk assessment', *Earthquake Engineering and Structural Dynamics*, **8**, 315–35.

Smith, K. (1996) *Environmental Hazards: Assessing Risk and Reducing Disaster*, 2nd edition, Routledge, London.

Time-Life Books (1982) *Volcano*, Planet Earth Series, Time-Life Books Inc, Alexandria, Virginia.

Tinsley, J. C. ***et al.*** (1985) 'Evaluation of liquefaction potential', in Ziony, J. I. (ed.) *Evaluating Earthquake Hazards in the Los Angeles Region: An Earth-Science*

Perspective, USGS Professional Paper No. 1360, Dept of Interior, Washington, pp. 263–315.

Tomblin, J. (1981) 'Earthquakes, volcanoes, and hurrricanes: a review of natural hazards and vulnerability in the West Indies', *Ambio*, **10**, 6, 340–5.

Toulmin, L.-M. (1987) 'Disaster preparedness and regional training on nine Caribbean islands: a long term evaluation', *Disasters*, **11**, 3, 221–34.

Varnes, D. J. (1978) 'Slope movements and types and processes', *Landslide Analysis and Control*, Transportation Research Board, Special Report 176, National Academy of Sciences, Washington, pp. 11–13.

Vermeiren (1989) 'Natural disasters: linking economics and the environment with a vengeance', paper presented at Conference on Economics and the Environment, 1989, Caribbean Conservation Association, Barbados, available at www.oas.org/en/cdmp/document/papers/vengeance.html

Worell-Cambell, J. (1997) 'Caribbean disaster mitigation project: making inroads into the development process', in Ahmad, R. (ed.) *Natural Hazards and Hazards Management in the Greater Caribbean and Latin America*, Unit for Disaster Studies, University of the West Indies, Mona campus, pp. 96–106.

Zans, V. A. (1959) 'Judgment Cliff landslide in the Yallahs valley', *Geonotes, Quarterly Journal of the Jamaica Group of the Geologists' Association*, **2**, 7, 43–8.

Chapter 5

SOCIAL CONDITIONS

Introduction

Taken as a whole, the contemporary Caribbean region is by no means as poor as many parts of what is frequently referred to as the 'developing world'. Indeed, the latest statistics show that incomes have risen quite sharply in the Caribbean along with those in South America. Between 1960 and 1998, average incomes in the Latin America and Caribbean region doubled, from around US$2,200 to US$5,000. During the same period, the region's income kept in line with the increase in the income of the wealthy industrial nations, standing at around one-quarter of these at both the beginning and end dates (United Nations 2001). If we turn to data summarising contemporary global levels of poverty for Latin America and the Caribbean, the relatively enhanced standing of the region is discernible. Thus, in 1998, some 16 per cent of the total population of the region was living on less than US$1 a day. This compares with 40 per cent in South Asia and 46 per cent in sub-Saharan Africa. Further comparative data are shown in Table 5.1.

Thankfully, with relatively few exceptions, the development problem in the Caribbean is not expressed in terms of starvation, mass poverty and overt under-nutrition. This is largely due to the fact that Caribbean peoples and states as a whole have found innovative solutions to their economic problems, whether by means of migrating to other nations and regions (see Chapter 2), changing and enhancing agricultural practices (see Chapter 3), or by means of the development of new economic activities, for example, in the manufacturing, tourism and off-shore banking sectors, in an era of increasing globalisation (see Chapters 8 to 11). The positive solutions found to common problems by Caribbean peoples is a recurrent theme of this chapter.

Like all regions of the developing world, however, the Caribbean shows very marked inequalities, and this represents the central theme of this chapter. Within the Caribbean, inequalities or marked differences exist, for example, between the territories that make up the region. For instance, at one extreme there is Haiti, which is often referred to as one of the poorest nations in the western hemisphere, with a gross domestic product per capita of US$1,464. Life expectancy in Haiti still hovers at just over 50 years (see Table 5.2), while adult literacy is below half of the population at 48.8 per cent (United Nations 2001). In contrast, as shown

Table 5.1 Proportion of the population living on less than US$1 per day for Latin America, the Caribbean and other developing regions, 1990–1998

	Percentage living on less than US$1 per day	
Region	**1990**	**1998**
Latin American and the Caribbean	17	16
East Asia and the Pacific	28	15
Eastern Europe and Central Asia	2	5
Middle East and North Africa	2	2
South Asia	44	40
Sub-Saharan Africa	48	46
All developing nations	29	24

Source: Department for International Development (DFID), 2000

by the statistics contained in Table 5.2, Barbados and the Bahamas have incomes over US$14,000, akin to those recorded by European nations such as Spain, Portugal, Hungary and Malta. Thus, whilst according to the United Nations Haiti is a nation at a low level of human development, Barbados and the Bahamas are classified as being characterised by high levels of human development (see Table 5.2).

Secondly, marked and persistent inequalities exist from place to place in most territories. Frequently, the coastal urban areas are more densely populated and have far higher concentrations of economically active and well-off populations. Since the age of mercantilism, these narrow coastal zones have been the locus of rapid development and change. The implications of this process for the ways in which urban development has occurred in the Caribbean region are dealt with in Chapter 7 (see also Potter 2000). Chapter 7 also illustrates how marked social inequalities occur between different areas within Caribbean towns and cities. Differences between wealth and poverty are also displayed in the Caribbean with real force if we consider the houses people live in. Houses vary from makeshift dwellings constructed of any materials to hand and built on land which presents problems of ownership or safety, through to luxury housing, resorts and complexes (Potter and Conway 1997). The ways in which inequalities are played out in the housing market are exemplified in Chapter 6.

In other words, whatever geographical or spatial scale we consider, pronounced inequalities and differences are the norm. The aim of this chapter is to discuss the nature and foundations of these differences in social conditions. These not only relate to income, occupations and social class. In the Caribbean, the historical forces of slavery and colonialism have given rise to strong conditions

Table 5.2 Human development-related variables for a selection of Caribbean nations, *c.* 2000

Country	Life expectancy (yrs) 2000	Combined primary, secondary and tertiary gross emolument	Gross domestic product per capita, US$	Human Development Index
High human development				
Barbados	76.6	77	14,353	0.864
Bahamas	69.2	74	15,258	0.820
Medium human development				
Trinidad and Tobago	74.1	65	8,176	0.798
Belize	73.8	73	4,959	0.776
Jamaica	75.1	62	3,561	0.738
Dominican Republic	67.2	72	5,507	0.722
Low human development				
Haiti	52.4	52	1,464	0.467
Latin America and the Caribbean	69.6	74	6,880	0.760

Source: United Nations (2001)

involving differences in race and colour. In turn these differences are linked directly to the concept of colour-class. However, some commentators might be tempted to argue that in the post-independence era, colour is becoming less important in its own right, and that class is becoming more important. Other very salient dimensions are expressed at the level of families and households, as well as gender/sexuality. In the past there has been an observable tendency to interpret Caribbean societies from a pathological point of view. The account presented here argues strongly against this, seeing such social conditions as localised responses to conditions in the region. The chapter then turns to consider contemporary issues such as the incidence of poverty and the survival strategies that are employed by the poor. The account involves the role of the informal sector, although this is also covered in detail in the next chapter on housing. The influence of structural adjustment programmes (SAPs) and the role of social capital are also considered. Finally, the account turns to health and welfare issues in the contemporary Caribbean.

The bases of stratification

Introduction

As already stressed, reflecting their status as former colonial and now post-colonial societies, Caribbean territories display marked and enduring differences between the various groups that make up the population.

As a point of departure, we might look at so-called industrialised, or Western, nations. Statistical analyses of census data for such countries suggest that the main thing that differs over space and between peoples is their social standing with regard to one another, that is their social class. This is to say that members of the population have become sifted and sorted by inferred social standing, as measured by the prestige of their occupations and their overall levels of income and wealth. In wealthy societies, this aspect of difference is followed by family life-cycle status, with populations being grouped according to whether they are young and single or older and living in a family set-up with children, for example. Only as a third statistical trend, it is argued, are people systematically separated on the basis of their racial and ethnic characteristics. This is an interesting dimension of comparison, for a number of authors have argued in the past that the single biggest differentiator between people within Caribbean societies is the interplay of the colour of their skin and their inferred social standing. This is frequently referred to as the colour-class system, the origins of which go back to the realities of plantation slavery. For this reason, we first look at the history of race and colour in the Caribbean context before turning to consider contemporary views on colour-class and social class in the region.

Historical perspectives

Slavery came with sugar and capitalists and planters turned to Africa as a source of labour (Lowenthal 1972). Slaves were cheaper than indentured servants and both they and their offspring were bound for life. Lowenthal states that a sugar plantation of 500–1,000 acres might require 250 hands in field and factory.

Slave conditions varied with the nationality and religion of the owner, but barbarities occurred in every West Indian territory. Lowenthal describes West Indian slave conditions as appalling, probably being the worst in the New World. Jamaican overseers are quoted as explicitly aiming to work slaves to death and depending on replenishment from West Africa. It is frequently suggested that high rates of owner absenteeism and a high ratio of slaves to freemen were directly related to the harshness of treatment. These two conditions were both met in the Caribbean region. Lowenthal notes how plantation 'great houses' were only great in contrast with slave huts. Most affluent owners lost little or no time in returning to Europe, or never set foot in the Caribbean in the first place. Equally, colonial planters sent their children back across the Atlantic for schooling and marriage. In this sense, to speak of a 'colonial elite' is almost a contradiction in terms, and one can all too readily observe the outward look to Europe as part of the social pattern in the contemporary Caribbean. The associated high incidence of slave rebellion reflects the harshness of treatment, as does the number of maroons, or runaway slaves:

> A comparison with American rates of survival underscores the harshness of West Indian conditions. When the slaves were emancipated, the Caribbean contained scarcely one-third the number imported; the United States had eleven times the number brought in (Lowenthal 1972: 43).

It is argued that these conditions had a series of implications for race and colour in the colonial West Indies. Thus, Lowenthal (*ibid.*: 47) maintains that the high proportion of non-white people induced West Indian whites to make colour distinctions that were meaningless in the context of America. Furthermore, the high levels of absenteeism meant that there were relatively few European women in residence in the Caribbean, and Lowenthal (*ibid.*: 48) observes that 'in the West Indies, interracial sexual liaisons were openly countenanced, especially where white women were few. Whites customarily had coloured mistresses, and white fathers regularly placed coloured daughters as concubines'. Given this, the same author notes that 'passing for white was important in the West Indies' (*ibid.*: 48). At the same time, Lowenthal states how 'whites believed that colour distinctions among the free were essential to maintain controls over slaves' (*ibid.*: 50).

Between 1791 and 1863, all West Indian slaves were emancipated. Although made free in the legal sense, Lowenthal notes how West Indian non-whites were kept from any form of economic or political power, and also experienced a meagre share of social goods and benefits. Thus, health, education and welfare allocations remained rudimentary after Emancipation, and 'few West Indians received either schooling or medical care' (*ibid.*: 66). The same author notes how as late as the 1940s, one West Indian in every three could not read or write; one child in four never went to school, while half the remainder only went on an irregular basis, so that most received less than four years of schooling in total.

As we shall see shortly, the argument runs that, right up to the present, such gross inequalities between members of the black, mixed and white populations are still clearly apparent in the contemporary Caribbean. Indeed, in respect of race, the argument is presented that rather than reducing distinctions and prejudice, the period following Emancipation served to further codify such distinctions:

> Indeed, emancipation increased racial prejudice; with slavery gone, colour criteria took on greater importance in West Indian society, not less. Failure to create an instant utopia and the supposed decline of the West Indian economy were cited as evidence of Negro inferiority, unfitness to self-rule, and hereditary ineducability (*ibid.*: 67).

> Countless observers have expressed amazement at how things stayed the same. In other countries travellers look assiduously for traces of the past; in the Caribbean the past is a living presence (*ibid.*: 68).

> Since all slaves were black or coloured and all whites were free, racial distinctions were taken for granted and required that free black and coloured people be set apart from both free white and slave black. Emancipation destroyed the legal but strengthened the pragmatic justification for colour distinctions (*ibid.*: 71).

It is in such an argument that the origins of what is commonly referred to as the colour-class system as a primary basis of social differentiation in the Caribbean are cogently expressed. However, some authors now argue that the colour-class system has been strongly modified in the post-independence era, with the emergence of black political elites, and must therefore be set aside as a monolithic explanation of social structure in the Caribbean region. However, others maintain that the contemporary Caribbean is today characterised by a combination of the two principles of stratification. But we will return to this argument after looking at the evolution and operation of the colour-class system in the post-Emancipation era.

Colour-class as the basis of social differentiation

The colonial plantation system, therefore, created grossly unequal and inegalitarian social hierarchies that were primarily premised on race and skin colour. Up to the seventeenth century, essentially two principal social strata existed, defined by race: the free whites and the black slaves (Clarke 1984, 1986). By the mid-eighteenth century, miscegenation had become more common, notably in the French West Indies, and a mixed or 'coloured' element within the population was formed, which also occupied an intermediate social position.

Thus, as Cross (1979) observes, it comes as something of a surprise to realise that there are about as many people in the Caribbean who would describe themselves as 'white' to a census enumerator as there are those who would identify themselves as 'black' or African. Cross notes there were about 10 million falling into each category within the Caribbean at the time of writing, comprising together just a little less than 80 per cent of the population.

The essential correspondence between colour and class is shown diagrammatically in Figure 5.1. This simple depiction of the colour-class system during the period of slavery is derived from West and Augelli (1976) and Potter and Binns (1988). Below the 'legal line', all slaves were non-white. The next two levels of the social class spectrum, consisting of servants and labourers, and then artisans, were drawn predominantly from the non-white populations, plus a few whites. The small planter, merchant and other professional category is shown as almost exclusively white, topped by learned professionals, planters and government officials, who were totally drawn from the white population (Figure 5.1).

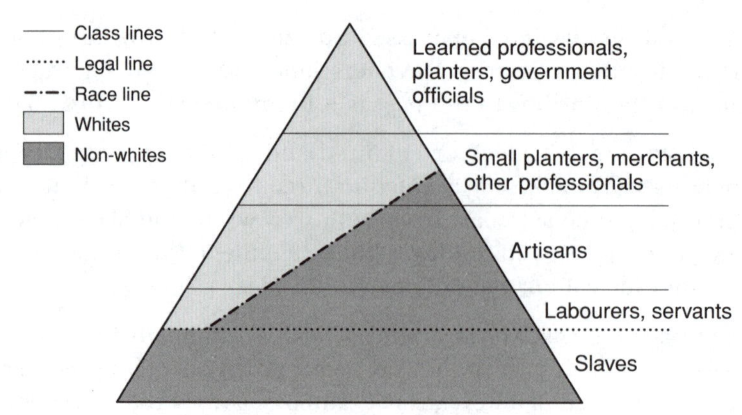

Figure 5.1 Race and social class in Caribbean slave societies
Source: Potter and Binns (1988)

Although the diagram depicted in Figure 5.1 is very simple, writers such as Clarke (1984) have shown that up to the period of political independence in the 1960s and 1970s, it was still highly appropriate in mapping the correspondence between race and class. Thus, Clarke shows how race and social status in Jamaica in 1800 was just like that shown in the simple figure. At the top of the social pyramid shown in Figure 5.2(a) was a small white group who were free with full civil rights. Below this was a larger 'coloured' population with a very small black contingent who were free, but with limited civil rights. At the bottom of the social class pyramid was the large unfree black population of slaves. After Emancipation, Clarke shows the applicability of exactly the same colour-class pyramid for Jamaica in 1860, the only difference being the disappearance of the symbols depicting slavery at the bottom of the pyramid (Figure 5.2(a)). And even for Jamaica in 1970, Clarke's diagram shows that relatively little had changed, other than the addition of various minority groups such as Syrians, Chinese and members of the Jewish population. The same author shows identical social strata in 1800 and 1860 in respect of Trinidad (Figure 5.2(b), upper part). However, a fundamental augmentation of the social stratum by the addition of Hindu, Muslim and Christian members of the East Indian indentured population after Emancipation had occurred by 1970 (Figure 5.2(b), lower part). But in the figure, this population has been grafted onto the pre-existing colour-class hierarchy.

The importance of colour-class stratification is strongly exemplified by Lowenthal (1972) in his book *West Indian Societies*, which still represents one of the most comprehensive accounts concerning social conditions in the Caribbean region (see Lowenthal 1972, chapter 3; see also Lowenthal and Comitas 1973). Lowenthal comments that 'racial distinctions have mattered longer in the West Indies than anywhere else in America' (p. 27). Accordingly, in the early 1970s, Lowenthal subdivided West Indian societies into five broad categories, proceeding from what he refers to as the least to the most complex:

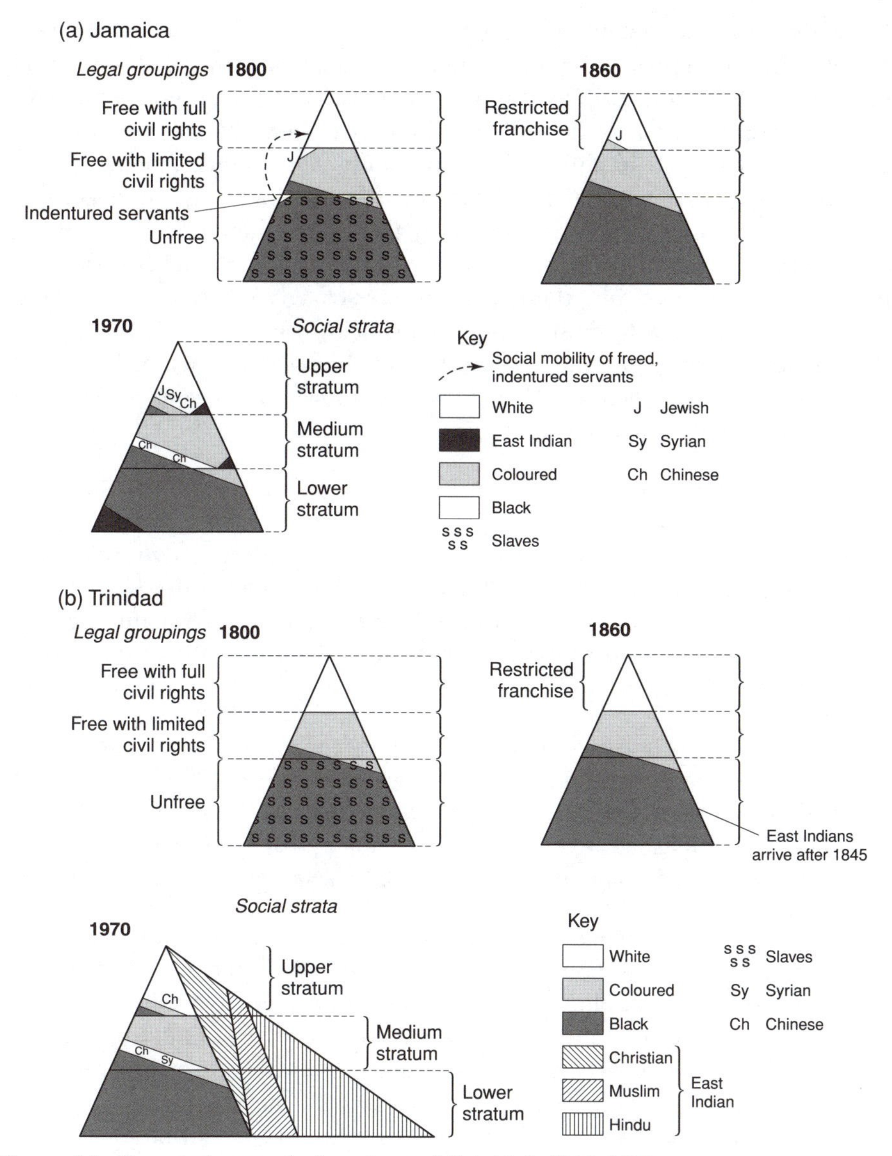

Figure 5.2 Race and status in Jamaica and Trinidad, 1800–1970
Source: Clarke (1984)

1. homogeneous societies without distinctions of colour or class;
2. societies differentiated by colour, but not stratified by class;
3. societies stratified by both colour and class;
4. societies stratified by both colour and class, but lacking white creole elites;
5. societies stratified by both colour and class, and also containing sizeable ethnic groups in large measure outside the colour-class hierarchy, with distinctive cultures, attitudes and values.

Lowenthal's classification of Caribbean territories into these five categories is shown in Table 5.3. A number of territories were described by Lowenthal as relatively homogeneous, including Carriacou and Barbuda. And the author noted a few that were regarded as showing differentiation by colour, but little by class, such as Anguilla, Bequai, Saba and Desirade. But most conspicuously, the table serves to demonstrate that according to Lowenthal, by far and away the majority of Caribbean territories were characterised by the juxtaposition of colour and class stratification. In the colour-class system, there is a clear hierarchy of social gradings, and for the most part, these divisions are strongly reflected in skin colour and social standing, with those of high status being white, and those with black skins generally occupying the lowest levels of the social spectrum. In order to give a common or garden example of the operation of the colour-class system, Lowenthal cites the following scenario:

> In any expensive hotel or restaurant all the guests are white, the manager is brown, the waiters and the dish-washers are black. Nothing to do with colour, you are told, it is simply an economic fact that black people cannot afford expensive restaurants. The real reasons, of course go beyond this, but the extent of economic inequality emerges sharply in various statistics of landholding, income and education (Lowenthal 1972: 81).

It is perhaps worthwhile each of us interrogating the contemporary appropriateness of this type of description of the relations existing between race, colour and occupational status in Caribbean societies with which we are personally familiar. Although now published over 30 years ago, it is perhaps still worth pondering Lowenthal's cautionary comments concerning race and colour in the Caribbean:

> But there are few aspects of West Indian life that race and colour do not significantly touch. As a form of identification, racial classification is perhaps unavoidable. . . . To be aware of difference is not automatically to discriminate; whereas to be (or profess to be) colour-blind may signal grave anxiety or conceal latent hostility. But racial distinctions are usually meant to be invidious: they carry an implication that one group is not just different from, but better than, another. And because individual effort can never wholly eliminate racial barriers, stratification by race and colour impedes economic and cultural development, depresses enterprise, and restricts personal opportunities (Lowenthal 1972: 1).

In citing this polarised pattern in the early 1970s, Lowenthal presented a veritable battery of statistics exemplifying the essential correspondence between colour and income/wealth in the Caribbean, citing the same type of social conditions as those pondered in relation to the period immediately following independence. In the case of Barbados, he noted, 186 estates of over 100 acres were owned by members of the white racial group, while blacks owned only eight and 25 were

Table 5.3 Classification of social types in the Caribbean according to Lowenthal

Territory	Social Type[1]	Poor & metropolitan white pop.[2]	Folk speech
Anguilla	C/XC		English
Antigua	C/C	M	English
Aruba	C/C	M	Papiamento
Bahamas	C/C	M	English
Barbuda	HS		English
Barbados	C/C	MP	English
Bay Islands	C/C		English
Bequia	C/XC	M	English
British Honduras	C/C/+E		English, Spanish
Carriacou	HS		English
Cayman Islands	C/XC	P	English
Curaçao	C/C	M	Papiamento
Désirade	C/XC	P	French
Dominica	C/C/-CE		French
French Guiana	C/C/-CE		French
Gonâve, Île de la	HS		French
Grenada	C/C/-CE		French, English
Grenadines	HS		English
Guadeloupe	C/C	MP	French
Guyana	C/C/+E		English
Haiti	C/C/-CE		French
Jamaica	C/C	M	English
Marie Galante	HS		French
Martinique	C/C	M	French
Montserrat	C/C/-CE	M	English
Nevis	C/C/-CE		English
Providencia	C/XC	P	English
Saba	C/XC		English
Les Saintes	C/XC	P	French
St Barthélemy	HS	P	French
St Kitts	C/C		English
St Lucia	C/C/-CE		French
St Martin/St Maarten	C/C/-CE	P	English
St Vincent	C/C		English
San Andrés	C/C/-CE	M	English
St Eustatius	HS		English
Surinam	C/C/+E		Sranan
Tobago	C/C	M	English
Tortue, Île de la	HS		French
Trinidad	C/C/+E		English
Turks and Caicos	HS		English
Virgin Islands (British)	HS		English
Virgin Islands (US)	C/C/-CE	MP	English

Source: Adapted from Lowenthal (1972)

[1] HS Homogeneous societies
C/XC Differentiated by colour but not by class
C/C Stratified by both colour and class
C/C-CE Stratified by colour and class but with white crude elites absent or insignificant
C/C/+E Stratified by colour and class and containing sizeable ethnic groups generally outside the colour-class hierarchy

[2] P Poor whites significant
M Metropolitan whites significant

described as being owned by those of mixed race (described as 'coloured'). These figures were inversely proportional to population size. Lowenthal also cited an impressive array of income figures showing that the darker the skin type, the lower the average wage. Thus, in the case of Trinidad and Tobago, figures derived from Harewood's (1971) study show that the median income for 'European' males was TT$500 per month at the time of writing. This compared with a figure of $113 for mixed (coloured), $104 for blacks/Africans and $77 for East Indians. Lowenthal noted that educational differences were even more pronounced. Thus, while 48 per cent of white Jamaicans attended secondary school back in 1943, this was only 10 per cent for the mixed and 1 per cent for the black populations. Such statistics give an impression of the degree to which income, education and wealth are differentiated along lines of racial cleavage in the Caribbean region.

Although as we shall observe shortly, colour and class are not matched exactly in all Caribbean societies, in many there is a persisting 'white bias' with status and power being inversely related to group size:

> But colour distinctions continue to correlate with class differences and dominate most personal associations. If prejudice is less blatant it is still visible. . . . Yet small as it is, the West Indian white minority is omnipotent, and the West Indian has learnt, by sheer habit, to take that white presence for granted. . . . A local panel in Jamaica concludes that 'colour and class prejudice are rampant. . . . The West Indian social hierarchy is generally arranged by colour from white through mixed to black' (Lowenthal 1972: 19–20).
>
> . . . The West Indian obsession with differences in shade sustains an atmosphere that, if less polarizing, perpetuates other serious problems of identity and action, problems for white, coloured, and minority groups as well as for black West Indians (*ibid.*: 24).

As Clarke (1986) notes, the persistence of economic and political elitism, bolstered by colonialism, served to shore up the framework of the culturally pluralistic contemporary Caribbean. The classic plural society is one where members of different groups, be they racial or religious, live in relative cultural isolation from one another (Furnivall 1948; Smith 1965a, 1965b; Lowenthal 1972). In recent research dealing with second-generation West Indians born in the United Kingdom and the United States, it has been shown that the recent arrivals from the industrialised world soon become acutely aware of colour differences and the societal significance of colour gradations (Potter 2003a, 2003b). This example is further elaborated in Box 5.1.

Even in the early twentieth century property restrictions in the Anglophone Caribbean meant that less than 10 per cent of the total population had the right to vote, and politics remained an elite pastime. The reins of government remained firmly in expatriate hands (Clarke 1986: 23). In Barbados, although the white population has been displaced from political power, this group still owns virtually

Box 5.1: The importance of issues of colour-class to foreign-born and young returning nationals to St Lucia

A relatively new migration path is now linking the Caribbean to the United Kingdom, Canada and the United States. This is the migration of relatively young second-generation West Indians who were born in the lands to which their parents had migrated in the 1950s and 1960s. These migrants have recently been studied in the context of St Lucia and Barbados (see Potter 2003a, 2003b; Potter and Phillips 2002). In the case of St Lucia, the interviews were carried out in early 2000 (Potter 2003b).

In these studies, the accent has been placed on the reasons why such young returning nationals have chosen to move, along with details of their employment trajectories. In this sense, such returnees can be considered to be the narrators of 'tales of two societies' – firstly, where they were born, and secondly, the part of the Caribbean from which their parents originally hailed and to which they have 'returned'.

In the case of the young returnees to St Lucia, when asked about their experience of living in the lands from which their parents had hailed, the most frequently cited adjustment that had to be faced among the young returnees was the operation of the colour-class system (Potter 2003b). The interviewees indicated that the operation of some facets of such racial differentiation in St Lucia had come as something of a surprise. For example, one respondent who was living on the island, but who had been moving backwards and forwards, noted with some surprise how shocked he had been at the operation of racial typing in St Lucia, and maintained that 'it's this that keeps me away from moving here permanently'. Another interviewee talked about the operational expression of race typing at work, whereby, when visiting his office, some indigenous St Lucians would discount his presence, due to his dark skin: 'Some St Lucians ignore me at work. They ask for my partners who are white, and complain about *nobody* being around!' (Potter 2003b).

Interestingly, two of the informants directly mentioned the operation of the colour-class system by name. One spoke of differences between black and white being judged by the colour-class system. The other commented that there is an in-built class system in the form of the colour-class system (Potter 2003b). The fact that these individuals, who had been brought up in England of West Indian parents, were apparently so aware of the existence of colour-class principles of social stratification is salient in itself (Potter 2003b). In neither case had these individuals studied for a social science degree, which might have entailed them reading about colour-class social grading systems.

all the sugar estates, rum distilleries, major business firms, import and export agencies and large retail stores (Potter and Dann 1990; Karch 1981, 1985). This has come about due to the processes of legislation, corporatisation and intermarriage, interlocking directorships and a virtual colour bar (see Barrow 1983).

Thus, commentators on the contemporary social scene in Barbados still stress the importance of elitist cleavages between whites and blacks (Layne 1979; Karch 1985). Lowenthal (1972: 82–83) stresses that 'coastal tourist resorts divide Barbados physically between the white elite (and tourists) and black folk'. However, it has to be noted that the family ownership of economic enterprises is less extensive in most other Caribbean countries, rather than the norm as in Barbados. Lowenthal also noted in the early 1970s how social clubs on the island 'still exclude dark people' (p. 20). Whilst an editorial in a newspaper claimed that the effective colour bar had all but disappeared over the preceding 15 years, it was followed almost immediately by a counter charge that prejudice was rife in almost every sector of Barbadian life (Lowenthal 1972: 20). It was also noted that social reform was very difficult because the nature of the problem is not openly admitted in an effort to maintain 'the myth of racial harmony' (p. 24). It is by such mechanisms that the colour-class system has been maintained over the recent past.

Two examples can be provided in the context of Barbados to illustrate the persistence of racial lines of cleavage within society. The first involves what was hailed as the first ever indigenous 'Yellow Pages' (or commercial telephone directory). By means of a detailed content analysis, Dann and Potter (1990) pointed to both racial and gender stereotyping in the space fillers that accompanied the commercial listings. They concluded by referring to this as the continued existence of 'Yellow Man [i.e. white] in the Yellow Pages'. The second illustration is provided by the continued existence of poor whites in Barbados, the so-called 'Redlegs of Barbados' (Keagy 1975). These descendants of white indentured labourers to Barbados in the seventeenth century live mainly on the windward eastern side of the island, in the rural parish of St John. As Keagy stresses, although they are poorer than most of the black majority, they do not intermarry with poor blacks: 'they are differentiated from the black peasantry not by culture but by colour, and by their attitude to miscegenation' (Keagy 1975: 14).

But we must, of course, be aware of all important variations within the pan-Caribbean region. Hoetink (1967) emphasised the contrast between what is seen as the more open and mobile racial system of the Spanish-speaking Caribbean and what was regarded as the harsher 'North European variant' encountered in the British, French and Dutch West Indies. But Cross (1979) has argued that work on the Hispanic Caribbean does not support the thesis that racial evaluations are irrelevant in respect of status placement. In presenting this argument, Cross cites Lewis (1974) in maintaining that Puerto Rico is a class-stratified dependent society, and in being unequivocal that to 'be black in Puerto Rico . . . is to suffer the stigmata of a negative image' (Lewis 1974: 143). Equally, the thesis that socialist Cuba with its anti-American capitalist ideology rejects racism is put to one side by Booth (1976). Cross (1979: 120) concludes that 'to be black is still to be placed lower in a hierarchy of prestige than any other group'.

The ascendancy of class in the post-colonial/post-independence period?

But in the period since the early 1970s, is it possible to discern real elements of change in social stratification in the Caribbean region? As noted, although no understanding of Caribbean class relations and structures would be complete without recourse to colour-class, this framework has to be set to one side as a comprehensive explanation of present-day social stratification in the region. The emergence of a black ruling elite in the political sphere has been a major change in the post-independence period in the Caribbean. The parallel rise of a new cohort of educated black professionals and business people has also occurred in many societies (Lewis 1994, 2001; Lowenthal 1973). Writing even in the 1970s, and emphasising throughout as he did the salience of colour-class stratification, Lowenthal clearly denoted this mobility at the very beginnings of the post-colonial period:

> Although hierarchical in structure and European in focus, West Indian society is locally ameliorative and encourages some social mobility.... coloured and later black West Indians gradually moved into positions of power and prestige, in the process emulating white outlooks and attitudes (Lowenthal 1972: 70).

> Today non-whites predominate in all the governments and occupy most places of public eminence.... Although this transformation leaves Caribbean social structures essentially intact, it has induced great changes in public attitudes (*ibid.*: 75).

Among the changes in attitudes, anti-colonial, anti-white, anti-establishment and pan-African sentiments, such as those enshrined in Negritude and Rastafarianism, can be mentioned. In this connection, the black power demonstrations that disrupted Trinidad in 1970 have been given specific attention by Lowenthal (1972). But the main change has been the expansion of the black upper and middle classes since the 1970s. This is reflected in Carl Stone's work on Jamaica's social class structure at the start of the 1980s, as shown in Table 5.4 (Stone 1980). Although some 76 per cent of the population of Jamaica was, at that time, classified as 'lower' class, just under a quarter fell into what was described as the lower middle class, consisting of independent property owners and middle-level capitalists and a so-called 'labour aristocracy'. At the top of the class structure, an upper- and upper middle-class elite comprising 1 per cent of the population was identified by Stone. Not that the emergence of the black middle class has been without critical comment. Thus, some have argued that a true bourgeoisie should be revolutionary in the sense of creating a productive capacity for wealth creation and for wider social change. However, as exemplified in the quote from Lowenthal above, it has been argued that the emerging middle classes have stressed high levels of consumption and imitative patterns of consumerism (James 1973; Cross 1979).

Table 5.4 Stone's categorisation of Jamaican class structure

Category/ sub-category	Social class	Percentage of total population
A.	**Upper and upper middle class**	
1	Owners/managers of Large/medium businesses	0.5
2	Administrative	0.5
B.	**Lower middle class**	
3	Independent property owners and middle-level capitalists	5.0
4	Labour aristocracy	18.0
C.	**Lower class**	
5	Own-account workers/petty capitalists	28.0
6	Working class	23.0
7	Long-term or indefinitely unemployed	25.0

Source: Stone (1980); Thomas (1988)

In a paper published at the end of 1999, Klaus de Albuquerque and Jerry McElroy sought to reappraise race, ethnicity and social stratification by means of the examination of the census data for three majority Afro-Caribbean societies, Dominica, St Lucia and St Vincent. The authors argue at the outset that traditional models of social stratification based on race/colour and colonial privilege are outdated. After two decades of political independence and economic modernisation, the top tier of the social hierarchy is composed of an educated elite of black professionals, politicians and businessmen. It is argued that typologies based on colour and creole elites fall short of contemporary realities, ignoring black control of the political sphere and their penetration of the economy. Specifically, they observe that:

> The political elite since the 1950s have been primarily of African origin, and since 1970 there have been fewer 'mixed' persons drawn from the brown/mulatto elite holding office (de Albuquerque and McElroy 1999: 3).

They also argue that the new black elite is highly influenced by North American consumption patterns, building large homes, owning fancy (sports utility) vehicles, and being characterised by fashionable clothes and frequent trips to North America. In all three small island states examined, it is observed that with over 94 per cent of the population black, an elite black minority dominates upper middle-class, middle-class, working-class and poor groups. In this sense, it is argued that the social sphere is class-based and not race-based alone.

Table 5.5 Median income for groups defined by race/ethnicity for Dominica, St Lucia and St Vincent

	Median income ($EC)		
Race/ethnic group	**Dominica**	**St Lucia**	**St Vincent***
African/negro/black	8,332	7,498	5,796
Amerindian/Carib	3,642	7,840	3,924
East Indian	–	10,028	9,408
Chinese	–	8,001	–
Portuguese	–	17,501	8,904
Syrian–Lebanese	–	17,858	16,200
White	21,334	30,662	13,386
Mixed	7,800	7,710	7,176
Other	9,801	10,626	–

* Authors stated they were not confident with the income figures for St Vincent. They advised only using them to examine relative differences
Source: Adapted from de Albuquerque and McElroy (1999)

This sort of modification to a single monolithic colour-class system is clearly fitting, for all the reasons cited in this section. However, it should be noted that this type of argument fits well those small island societies where minority groups are negligible, and the black population is the majority, recording 96.5 per cent, 96.2 per cent and 94.0 per cent of the total populations of Dominica, St Lucia and St Vincent respectively. It may be less fitting in larger and more heterogeneous societies.

However, when data for income were examined by the same authors, they found that race and ethnicity are still the main predictors of affluence. The incomes of white groups living in these three Windward Islands were shown to be generally double to quadruple those of the majority black population. This is illustrated by the data shown in Table 5.5. Thus, taken over the three countries, the median income (in East Caribbean dollars) ranges between $13,386 and $21,334. For members of the black population the equivalent range is $5,796 and $8,332, and for members of the 'mixed' population, $7,176 to $7,800. This direct association between race/ethnicity and income is mainly explained by the high incomes commanded by white expatriate professionals and white retirees. The authors conclude however that in overall terms, whiteness is still associated with prosperity and blackness with relative penury, but it is argued that this is now explained by differences in education rather than race *per se*.

Thomas (1988) in his survey of *The Poor and the Powerless* in the Caribbean adds some points that are worth considering in the contemporary context. Thomas argues that, by the 1970s, the class structure of the region remained weaker and more complex than those of the capitalist economies of Europe and North America. He also argues that another feature of the region's social structure is that its class groups are more fluid, especially in terms of the rapid development of new groups – for example, urban-based groups such as higglers and unemployed youths from the countryside. Another characteristic is seen as the dominant role that is played by foreign capital, whether resident in the region or not – an echo of the income data reviewed above.

This idea of relative fluidity in the contemporary social context, together with the fact that income is still positively skewed to white expatriate sub-sets of the population, can be linked to an argument presented by Cross (1979). He suggested that the structure of both urban and rural class systems in the Caribbean is increasingly dominated by multinational corporations. The basic thesis is depicted in Figure 5.3. The upper stratum, consisting of urban manufacturing, commercial and service management, and rural plantation management, are seen as directly linked to multinational corporations. An expanding middle class is pinpointed in the urban context, followed by a proletariat and lumpenproletariat. The two systems are directly linked, principally via the displaced peasantry and estate workers, who are forced to migrate to the city in order to seek employment. This argument was elaborated by Kowalewski (1982), who emphasised the increasing involvement of multinationals in the region. It was concluded that intranational inequalities are, therefore, inseparable from international ones. Thus, for example, multinational expatriates form an essential elite element in the Caribbean class system. At the same time, millions of first world tourists, although coming and going individually, remain as a permanent elite group in the Caribbean region (see also Dann and Potter 1994). This globalisation of the Caribbean class

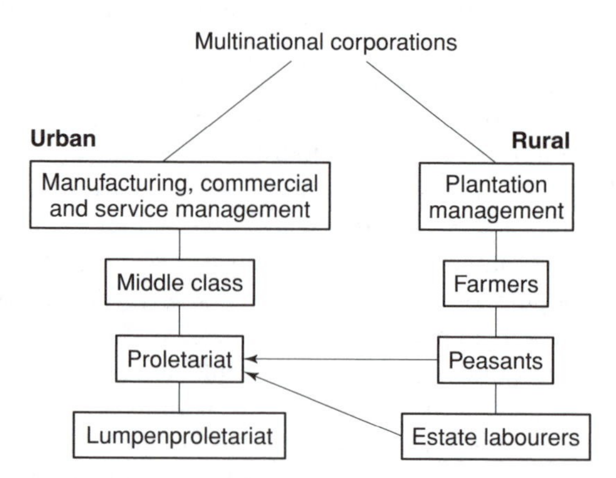

Figure 5.3 Urban and rural class systems in the Caribbean in relation to multinational corporations
Source: Cross (1979)

system is an important aspect of contemporary change, linked to global patterns of consumption in the region among those who can afford them (Potter 1993).

Thus, while class is becoming a more important dimension within Caribbean societies, whether it be via multinationals, the emergence of a black political and economic elite, expatriates, retirees or tourists, as Cross (1979: 117) comments, the evidence suggests the persistence of a strong positive association between wealth and income on the one hand and white or coloured phenotype on the other. The same author provides what remains a strong concluding statement to this discussion concerning social class and stratification in the Caribbean region:

> The development of capitalist class relations in dependent economies alters the caste-like proscriptions of the early colonial period, but nowhere does it destroy them. . . . deference is firmly based on 'race' and the degree of approximation to white ethnicity. It is a system which has endured and justified discriminatory action from Carriacou to Cuba (Cross 1979: 118).
>
> However, colour never did *determine* class boundaries but once reflected them, and to a considerable extent still does (*ibid.*: 119).

Households, family structure and gender issues

There is a rich and extensive literature on Caribbean household groupings, family forms and gender issues (see, for example, Barrow 1996, 1998; Momsen 1993). However, as noted by Cross (1979), studies have tended to concentrate on black families of relatively low status and income. As noted at the start of this chapter, some early approaches seemed to stress what they saw as the 'pathological nature' of many facets of family life – an interpretation that is roundly rejected in the present account.

In this connection, it is frequently suggested that one of the most distinctive features of Caribbean family structure has been that a wide variety of union types prevail (e.g. Cross 1979). Thus, co-residence may occur without registered marriage, and sexual union may occur without co-residence. Further, Cross (1979) argues that the region is also distinguished by a developmental sequence, which moves from sexual union without co-residence to one of the other forms of cohabitation, sometimes with different partners. Thus, apart from legal marriage, there are frequent 'common law' unions. These are defined by co-residence without marriage. The term 'visiting union' is frequently used where a sexual relationship is present without cohabitation.

Lowenthal (1972: 105) maintains that while monogamy is the ideal type for all social classes in the Caribbean, 'each approaches it by a different route and to a different degree'. The argument goes that formal, legal marriage among rural folk is commonly deferred until the male can afford to buy or build a substantial house, organise a wedding ceremony and support a wife who need not work

Table 5.6 Distribution of women by union type and age group in Jamaica

Type of union	Age group 15–19	20–24	25–29	30–34	35–39	40–44
Married	1.3	11.5	26.1	39.4	47.5	51.2
Common law	10.3	28.8	36.4	32.0	27.3	21.2
Visiting	86.8	54.8	30.2	19.7	14.9	13.3
No longer living with partner	1.6	5.0	7.4	8.8	10.3	14.3
Total	100.0	100.0	100.0	100.0	100.0	100.0

Source: Cross (1979: 81), based on Commonwealth Caribbean Census

outside of the home. In these circumstances, the legal marriage ceremony often serves to mark the culmination of a relationship rather than its commencement.

Cross (1972: 81) uses data from the 1970 Census for Jamaica to illustrate how union status changes with age. Less than half of all Jamaican women between the ages of 15 and 39 were married (Table 5.6). Only for those in their forties did formal marriage prevail. Common law unions predominated for women in the 25–29 age range. On the other hand, 'visiting' unions characterised Jamaican women up to the age of 24 years. Thus, the visiting pattern is superseded by either common law or legal marriage, but rarely both (Cross 1979; Roberts and Sinclair 1977). The latter authors show almost exactly the same patterns for Trinidad.

The inevitable consequence of the typical Caribbean union pattern is that a minority of births are 'legitimate'. In the 1970s, Lowenthal (1972) argued that members of the working class attached little or no shame or condemnation to illegitimacy, but that the middle class professed to regard it as shameful, promiscuous and immoral. However, it would seem that attitudes to both formal marriage and childbirth outside of legal marriage have undergone radical reappraisals in many parts of the capitalist industrial world in the last quarter of a century, and are now much more in line with those characteristic of the Caribbean in the past.

Some research has suggested that 'visiting' and 'common law' relationships frequently occur where there are low or uncertain economic returns. Cross (1979) cites Schlesinger's (1962) work in Jamaica, where it was argued that casual and uncertain wages make for short-term relationships. Commonly, men spend a good deal of time away from the home in folk societies, frequently following a pattern of extra-residential mating (Cross 1979; Lowenthal 1972; Dann 1987). In a study of Barbadian males, Dann (1987) reported that 51.4 per cent of respondents indicated that they knew their fathers had one or more relationships outside the home. Such patterns of extra-household union have given rise to the argument that men's 'reputations' are built on their sexual conquests and the children they father (Cross 1979). On the other hand, women are the keepers of 'respectability', based on traditional values of family life (see Wilson 1969, 1973). It is these features of sex role differentiation and the acceptance of extra-familial unions for men that give rise to the frequently expressed view that West Indian family life is Victorian or even pre-Victorian in its pattern and form (Lowenthal 1972; Cross 1979).

Such sexual separation is linked to two other major characteristics of Caribbean family life. The first is its matrifocal nature, meaning that mature women and mothers emerge as the central focus of domestic groupings. But this does not mean that West Indian societies are matriarchies; indeed, far from it, the dominance of men is seldom seriously questioned. Thus, Lowenthal (1972) notes how men and women outside of the middle classes traditionally spend little leisure time together. As a corollary of all this, children are often raised solely by women – mothers, grandmothers or aunties – a situation that gave rise to Edith Clarke's (1966) classic study, based on three Jamaican communities, under the memorable title, *My Mother Who Fathered Me*.

The second outcome is the acknowledged high level of female-headed households in the Caribbean, either due to the maintenance of visiting relationships, or the permanent migration/departure of males. It is estimated that 35 per cent of rural east Caribbean households are headed by women (Momsen, 1993). In addition, women frequently show a high level of economic autonomy (Momsen 1987, 1993).

But illegitimacy, the mother-centred home and male absenteeism must not be regarded as overt symptoms of social disorganisation. As Lowenthal (1972: 114) cogently argues, they have to be seen as accepted features of folk life/society in the Caribbean. It is argued that unlike in the United States, not having a father living in the home on a permanent basis is no predictor of social deviance or mental ill-health, so that 'whether or not the father is present, the child's closest ties are with the mother, and her kin'. Similarly, Cross (1979) argues that male absence cannot be regarded as irresponsible and that men usually make at least some provision for their children, regardless of where they live.

Indeed, Caribbean households are frequently highly religious in orientation. Cross (1979) avers that attendance at church seldom falls below 70 per cent of the population and that religious ritual and belief often provide the backbone of community organisation in the Caribbean. However, the denominations subscribed to are many and varied. While former British and Dutch territories are mainly Protestant, and the Spanish and French Roman Catholic, there are variations. Likewise, alternative religious denominations such as Baptism, Methodism, Seventh Day Adventism and Pentecostalism have all proved to be very popular. A further strand in alternative religion, one that links to the African past, is also emphasised by Cross (1979). This includes the Pocomania cult in Jamaica, Voodoo in Haiti, Shango in Trinidad, and Rastafarianism. The latter, of course, started in the poor tenements and yards of West Kingston in the 1930s.

Contemporary social conditions

As already discussed, in absolute terms, relatively few people in the Caribbean live in abject poverty. However, over and above Haiti being classified as one of the poorest nations in the western hemisphere, considerable pockets of relative poverty exist elsewhere in the contemporary Caribbean, within both rural and urban areas.

As we have seen, such inequalities are a direct outcome of the colonial period, during which social stratification and difference over space came to be firmly etched out, especially in response to the plantation system, slavery, indentureship, absentee landowners and the flight of both profits and capital to Europe. As Potter and McAslan (2000) argue in the case of urban Barbados, such pockets of poverty and hardship are frequently hidden away well out of sight, in what the authors describe as an essentially colonial pattern.

The historical bases of poverty and inequality in the Commonwealth Caribbean were picked up in the landmark *Report of the Moyne Commission* (West Indian Royal Commission 1945). The widespread labour disturbances that occurred in the British West Indies during the 1930s prompted the British government to appoint a Royal Committee of Inquiry in 1938. Set in the context of the hardships of the Great Depression of the 1930s, between 1934 and 1939 there was widespread revolt and rebellion in the then British West Indies, with specific outbreaks affecting Trinidad, St Kitts, Jamaica, St Vincent, Barbados and Guyana (Thomas 1988).

In fact, the report of the Commission did not appear until after the Second World War in 1945, for fear that earlier publication might provide grounds for criticism of Britain during the war years. The report has been described by Thomas (1988: 50) as a 'truly remarkable document for its time, and unparalleled in its detailed description of the poverty and powerlessness of the West Indian masses'. It provided a detailed description of labour conditions in the region, stressing the problems posed by piece rates of pay as a legacy of slavery, rather than time rates, along with under-unionisation, and generally low rates of pay, especially in rural areas. The overall poor nature of housing in the Caribbean, its origins in the era of slavery and the lack of coherent public policy (see Chapter 6) was also pinpointed by the Moyne Commisson.

The report was particularly scathing about education and health in the region. In the field of education, the poor state of secondary education was specifically bemoaned, in particular, the failure to provide for local needs, with virtually no Caribbean-based materials being employed in schools. High rates of infant and maternal mortality were seen as grave causes for concern, especially in Guyana. But the Moyne Commission report put a special emphasis on the need to curb population growth, while in the economic sphere it recommended that the accent should remain squarely on agriculture and not on incipient industrialisation.

Most commentators tend to agree that the Moyne Commission report was strong on analysis, but weak on policy recommendations. The suggestion that the region should not attempt to industrialise is specifically cited in this connection (Thomas 1988). It is argued that the report failed to address the root causes of people's poverty and distress. For example, it did not suggest that the root causes of inequality in uneven land ownership should be dealt with. Rather, the Commission argued that labour needed to be organised under 'responsible leadership', in a manner that clearly connoted the extant British system. It recommended that social services should be improved and aid increased, calls that led directly to the appointment of the highly paternalistic Colonial Development and Welfare Committee by the British government in the immediate postwar period.

The fact that, in the period since 1945, Caribbean societies have remained highly unequal and uneven in their patterns of development has been exemplified in several recent studies. For example, at the urban scale, Laguerre (1990) has provided a social anthropological study of two poor urban neighbourhoods in Fort-de-France, the capital of Martinique, one an inner city slum, the other a squatter settlement. Laguerre sees the family-household unit as a multi-product firm involved in coping with the structural phenomenon of poverty. Potter (1995) has looked at poor housing communities in the urban areas of the Windward Islands (see also Chapter 6).

At the national level, Potter and Jacyno (1996) have provided a detailed analysis of social conditions in contemporary St Lucia, using data for 34 socio-economic and demographic variables drawn from the 1991 Census at the quarter or administrative level. The data were individually mapped for each variable, and patterns of association between sets of linked variables were then searched for using multivariate statistical methods.

Consistent island-wide patterns were repeated for many of the variables, as shown in Figure 5.4, which reproduces the quarter-level maps showing the proportion of the population attending university, average earnings, the proportion of the labour force in employment and the ownership of televisions. Each of these variables reflecting relative affluence show the same basic spatial or geographical patterning. High levels of relative affluence are more typical of the north-western and northern quarters of Gros Islet, Castries metropolitian, Castries suburban and Castries rural.

The fact that there is strong overlap between these socio-economic variables was attested by means of the application of factor analysis. This revealed a first component reflecting high levels of affluence and amenity. The first general factor was associated with the ownership of televisions, the use of electrical lighting, telephones, gas for cooking, video recorders, radios and relatively high gross earnings, as illustrated in Table 5.7. When mapped, factor 1 pinpoints the relative affluence of the northern administrative quarters (Figure 5.5). Factor 2 seems to represent modern and youthful areas, being directly associated with recent housing and low levels of old housing and older people (Table 5.7). Again the north of the island stands out, but so does the east coast as a whole (Figure 5.5). Low scores on factor 1 were revealed to be typical of the three south-western administrative quarters.

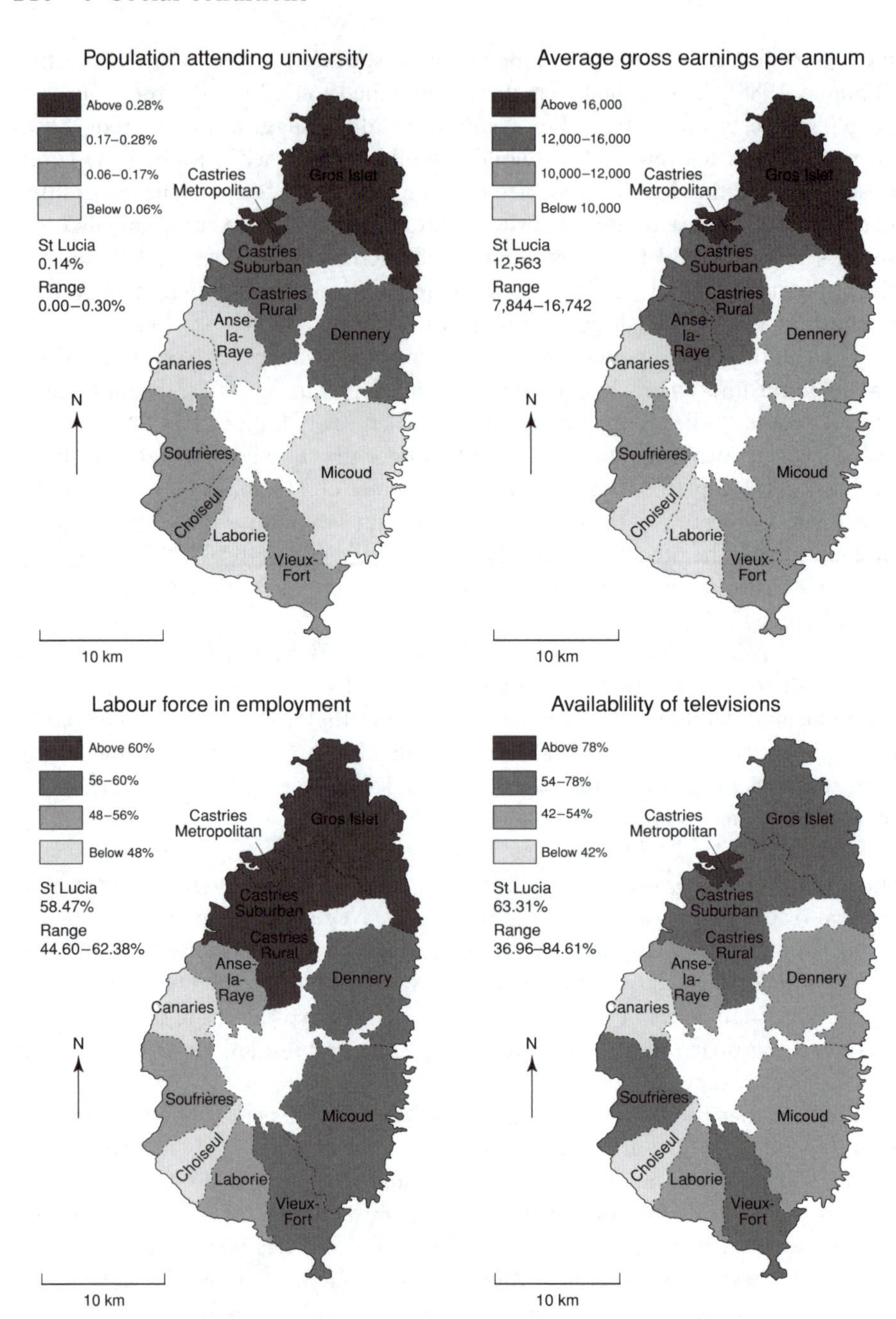

Figure 5.4 Selected social conditions for the quarters of St Lucia
Source: Potter and Jacyno (1996)

Table 5.7 Conformity of the original socio-economic and demographic variables with the first two factors

Factor	Original variable	Loading	Label
1	Television	+0.98	Affluence/amenity
	Electric light	+0.98	
	Telephone	+0.93	
	Outside kitchen	−0.93	
	Gas for cooking	+0.89	
	Charcoal for cooking	−0.89	
	Video recorder	+0.85	
	Gross earnings	+0.76	
	Radio		
2	Pre-1970 housing	−0.92	Modern/youthful
	Post-1980 housing	+0.90	
	Population over 60	−0.90	
	Population over 85	−0.89	

Source: Potter (1999)

When the quarters were cross-classified on the two factors, a clear typology of social areas was revealed for St Lucia (Figure 5.5). The association of affluence with the primate urban area is the most clearly discernible feature, and the north-western urban corridor stands out from the remainder of the island. Essentially, the analysis attests to the sharp spatial polarisation that characterises contemporary Caribbean societies. A similar analysis at the elemental enumeration district level for Barbados is presented as Box 5.2. The outcome is again a highly polarised map of relative wealth and poverty at the national level.

Many analysts have argued that social and spatial differences in wealth and affluence have increased in most Caribbean territories since the 1980s, since then more and more emphasis has been placed on the neo-liberal model of economic growth promoted by Western political leaders, notably Margaret Thatcher and Ronald Reagan (Potter *et al.* 2004a). The New Right argued that the market should be left as much as possible to its own devices and that the state should withdraw from the marketplace as far as practicable. Under this formulation, the accent is placed squarely on trade liberalisation and export-oriented economic growth (Potter *et al.* 2004b; Desai and Potter 2002). During the decade of the 1980s, these ideas became the new orthodoxy. When poorer nations experienced economic and financial crises and approached the World Bank and International Monetary Fund for assistance, these organisations placed strong conditions for economic reform on such developing nations before providing such loans.

Collectively, the package of conditionalities placed on such nations were referred to as structural adjustment programmes (SAPs). The basic measures involved in SAPs included liberalising trade, cutting public sector employment,

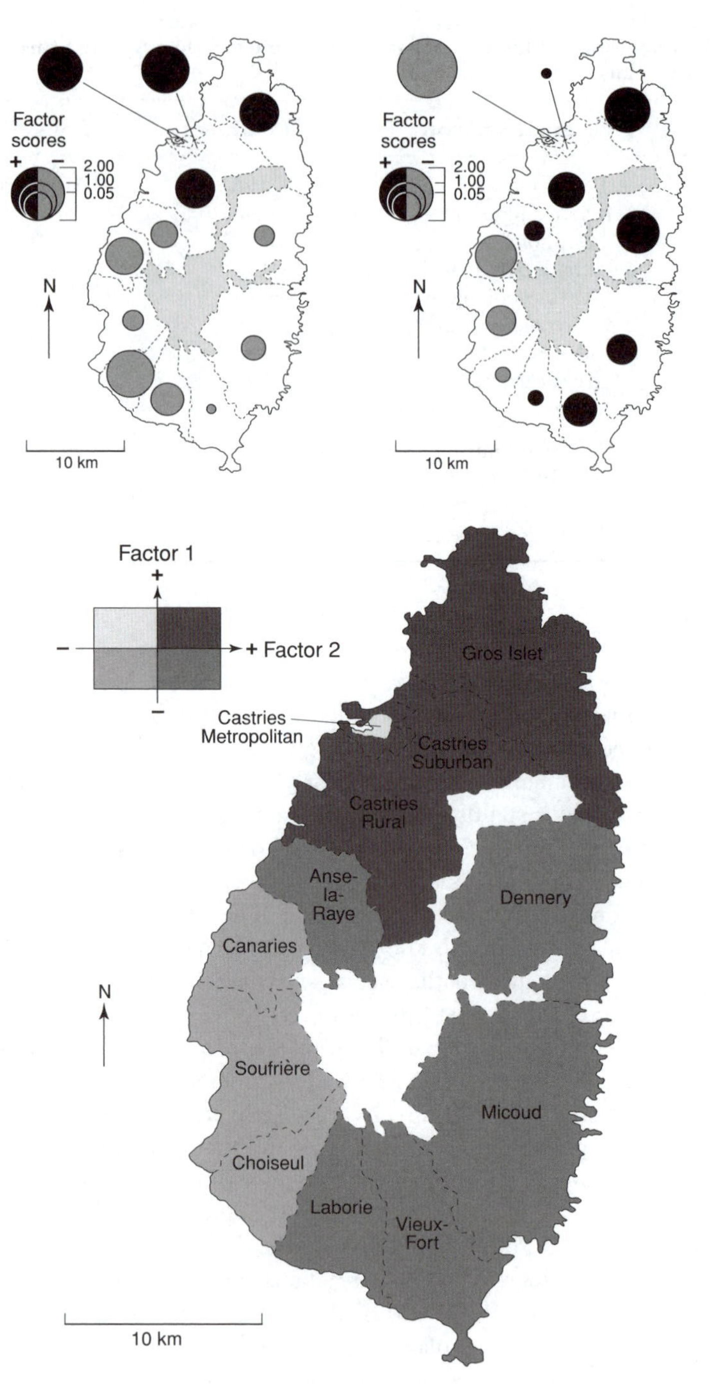

Figure 5.5 Factors 1 and 2 summarising social conditions in St Lucia, plus a final typology of social areas
Source: Potter and Jacyno (1996)

Box 5.2: The spatial incidence of relative poverty and affluence: the case of Barbados at the enumeration district level

The data collected by means of the various censuses carried out in the region provide an excellent first step towards examining social conditions in particular territories. The analysis of such data is coming to be used more and more in relation to poverty reduction strategies in the region.

These types of analysis can be carried out at the detailed elemental level of the enumeration district (or ED). In the case of Barbados, as an example, outside of the metropolitan parish of St Michael, the nation was divided into 294 enumeration districts. St Michael, the home of the capital city, Bridgetown, comprised another 183 enumeration districts, making a total of 477 finely graded areas for which data were available.

Potter *et al.* (2004) and Potter (1996, 1999) examined 36 key socio-economic, demographic and infrastructural variables at this detailed spatial level. The variables covered population, race and religion, socio-economic standing, housing and related infrastruture. These were mapped at the ED level for the entire country (see Potter *et al.* 2004b).

As expected, it was noticeable from visual inspection of the maps that many of them showed very similar patterns across the island. This visual impression was confirmed statistically, by employing the multivariate method of factor analysis to look for groups of closely associated variables. Details of this analysis are included in the *Social Atlas of Barbados* (Potter *et al.* 2004b).

This association between the census variables allowed the derivation of an *index of poverty* and an *index of affluence*. The index of poverty brought together six of the original census variables. Specifically, these were the percentage of houses built over 20 years before the census; percentage of the population over 60 years; percentage of households where all the facilities recorded in the census were not available; percentage of houses built entirely of wood; percentage using pit latrine toilets; and the percentage drawing water from a public standpipe. The average of these six percentage scores was taken for each and every ED to yield an overall measure of relative poverty.

A clear and very interesting picture of relative poverty in Barbados emerges when the scores of the enumeration districts are mapped in this fashion (Figure 5.6). The map serves to pinpoint the association of relative poverty with the rural central, northern and eastern enumeration districts of the country, where contiguous areas of poverty stretch from the north-east coast, centred on the parishes of St Andrew and St Joseph. High concentrations of poverty extend into north-western St John, northern St Phillip and eastern St George, plus much of St Thomas and St Lucy.

The research also involved calculating the reverse, that is an index of affluence (Figure 5.7). This summed together six variables that were held to be

(Box continued)

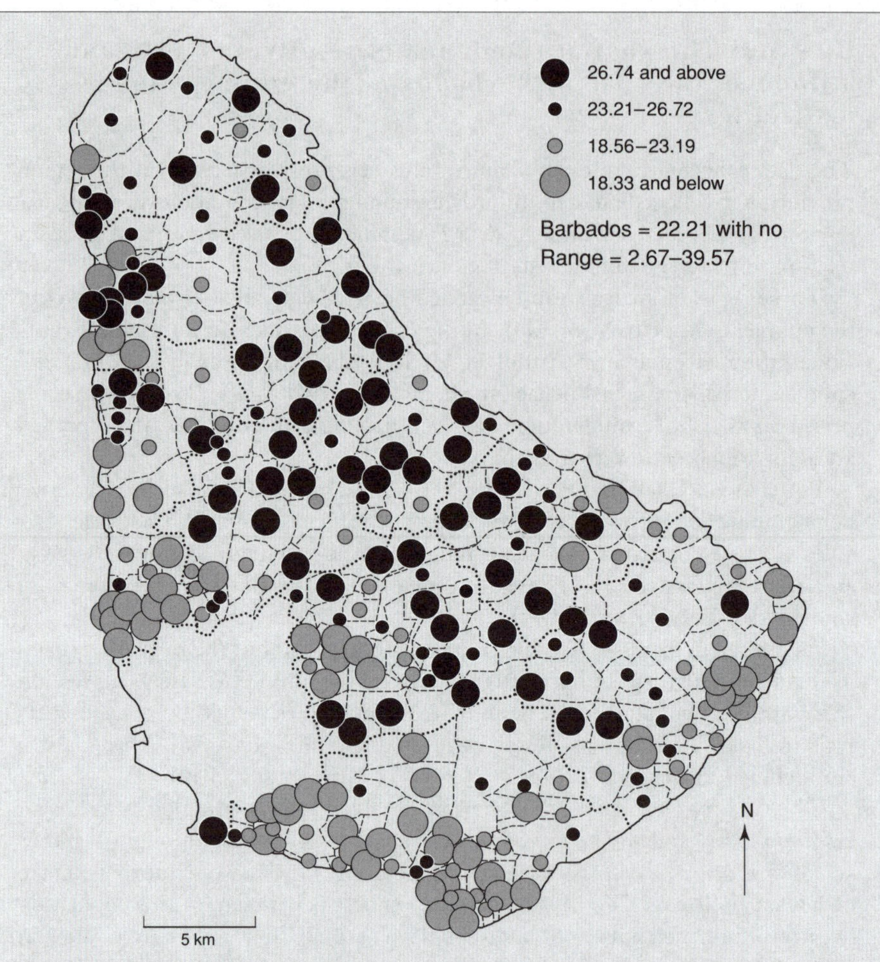

Figure 5.6 The scores of the Barbadian enumeration districts on the index of poverty
Source: Potter (1999)

highly diagnostic of relative wealth and prosperity. These were the percentage of households owning a car; percentage with solar water heating; percentage of houses with more than seven rooms; percentage of houses built entirely of concrete; percentage of households with washing machines; and the percentage of households with video recorders. Once again, the combined scores were divided by six to yield an overall index of affluence.

When mapped, the pattern is very much the complement of the one shown by the index of poverty. There is a clear concentration of enhanced levels of affluence along the south-western, western and south-western coasts. Nationally, it is apparent that the strongest concentration of wealth is to be found along the

(Box continued)

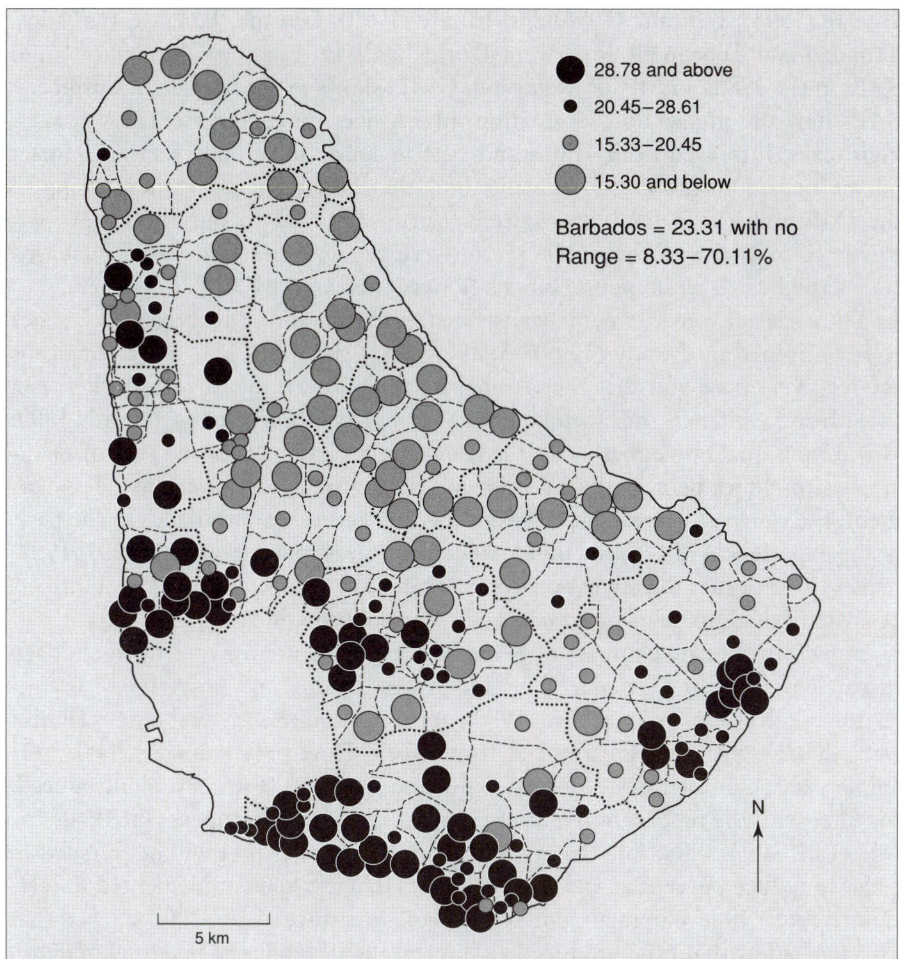

Figure 5.7 The scores of the Barbadian enumeration districts on the index of affluence
Source: Potter (1999)

western coastal areas of Christ Church, followed by the southern tracts of St James, and coastal to middle St Phillip. Another clear area of affluence is shown to be the western parts of St George. Again, the most salient feature is the strength of the geographical contrasts that are revealed. One of the very few areas where relative affluence and poverty occur relatively close together is on the north-west coast around the Speightstown urban district.

lowering wage levels, devaluation, opening up to foreign investment and abolishing subsidies.

Aspects of structural adjustment as they influenced the Caribbean have been analysed by a number of writers, including McAfee (1991), Ramsaran (1992), La

Guerre (1994), LaFranc (1994) and Lundy (1999). Guyana, Jamaica, Barbados, Trinidad and Tobago all underwent World Bank/International Monetary Fund SAPs in the 1980s and 1990s. Ramsaran (1992) shows in the context of Caribbean SAPs how the impacts of devaluation, high prices, unemployment, low wages, high taxes, increased utility rates and cuts to social and public services impact on the poor. Duncan (1994) explains how the Barbadian government going to the IMF and World Bank to secure a loan in 1991 led to the imposition of a classic SAP. Wages in the public sector were cut by 8 per cent, taxes increased across the board, while public bus fares were increased by 50 per cent.

The social and environmental consequences of structural adjustment in Jamaica were examined by Lundy (1999). At the outset, she notes that most Jamaicans are relatively poor and that there remains a high degree of correlation between 'wealth and whiteness, near-whiteness and minority ethnic groups' (Lundy 1999: viii). Using qualitative methods, it is shown that as well as the number of primary health clinics being reduced following the introduction of structural adjustment, the range of outreach services offered was severely curtailed. On the basis of the experiences of Guyana under structural adjustment, Bernard (1991, 1997) makes the plea that the education system should be protected from the ill-effects of structural adjustment programmes in Caribbean nations.

In fact, the negative impact of SAPs on the poor, and on women and children in particular, became a major focus for protest among non-government organisations such as Oxfam in the late 1990s. As a result of this general outcry (Potter *et al.* 2004a), SAPs were expunged from World Bank policy during 2000/2001 (Simon 2002). In their place, poverty reduction strategies (PRSs) were introduced, together with the preparation of poverty reduction strategy papers (PRSPs). The role of PRSPs is to outline the array of domestic policies that will be pursued in order to reduce poverty, as well as the policies that are to be implemented in relation to external development aid. World Bank literature in late 2002 showed that Guyana had by that time produced an interim PRSP, while the potential of Haiti, Dominica, St Lucia, St Vincent and the Grenadines, and Grenada for the production of PRSPs had also been identified. It is made clear in the World Bank literature that future funding from the Bank will be conditional upon the production of a satisfactory PRSP (World Bank 2003).

Given such economic vicissitudes, and the non-existence or rudimentary nature of social security systems in much of the Caribbean, we return to the theme that Caribbean families have had to be both resourceful and innovative. In short, they have had to develop effective strategies for survival (Barrow 1986). Family networks and mutual support systems have been put in place, such as mothers, grandmothers and aunts providing childcare, in some instances, where mothers work overseas. People and community groups have helped one another in systems of community self-help. The existence of mutual support networks and reciprocity as a form of glue which serves to hold members of society together has since the late 1990s been referred to as social capital (see McAslan 2001). The same is true in regard to the existence of rotating credit funds and credit unions, which allow people to save and have access to credit. The same principle has been used in the field of housing in Trinidad and is referred to as Sou Sou. Other instances of community- and self-help are reviewed in relation to the provision of low-income housing in the Caribbean. A notable example is the microbrigade system in Cuba (see Box 6.3). Indeed, the notable achievements of the socialist state of Cuba in the field of social welfare are outlined in the next section and in Chapter 7 in dealing with urban and regional patterns. With regard to regional problems, for example, a concerted effort has been made in post-revolutionary Cuba to spread infrastructure and facilities to former peripheral and rural regions of the national space.

In respect of jobs too, those facing unemployment have been forced to create their own jobs. The argument goes that unemployment is a luxury that only the relatively wealthy can afford. The resulting 'informal sector' consists of unregistered and unregulated activities, which represent the employment component of individual and community self-help. Throughout the Caribbean, in common with the developing world, the informal sector has increased in significance, along with the trend towards neo-liberalism (see Lloyd-Evans and Potter 1992, 2003; Potter and Lloyd-Evans 1998). A case study of the importance of the informal sector in Trinidad is provided in Box 5.3. Notably, this case study serves to illustrate how the informal sector is closely related to societal differences of gender and race. It also shows how closely associated the rise of the informal sector has been with debt and structural adjustment packages.

Aspects of health care and welfare

Social progress in health and welfare delivery: forwards from the 1950s to the 1980s; but backwards since the 1980s

During the post-Second World War period of decolonisation, island health care systems went from solving infant and child health concerns to addressing maternal and adult health problems. High fertility, morbidity and mortality rates among children and mothers, the continued prevalence of endemic diseases through to the 1950s – including malaria, yellow fever, hookworm, tuberculosis, smallpox and yaws – were all the legacies of underdevelopment and impoverishment under colonialism.

Health care systems needed to be developed in order to address the pressing health and nutritional problems of the poor – infant mortality, malnutrition, maternal mortality and high youth-fertility rates. In response, colonial administrations began building institutional capacity in island health systems in order to eradicate these age-old diseases, and by the 1960s many had been successfully conquered.

Independent governments in the region have continued to allocate sizeable budgets, often between 6 and 15 per cent of GDP, in tackling the nutritional and morbidity threats to their children and young people. Family planning programmes

Box 5.3: Gender and race in the Trinidadian informal sector

As noted in the main text, the informal sector has always been the refuge sector for those facing economic hardship, and has therefore always featured as a key element in household survival strategies (see Portes *et al.* 1997; Potter and Lloyd-Evans 1998; Potter *et al.* 2004a). Lloyd-Evans and Potter (2002) show how as a result of debt and structural adjustment in the Caribbean, more and more workers have had to undertake low-waged, low-skill and insecure jobs in the informal sector. This is particularly true of female workers. It is suggested by Momsen (1993) that in the Caribbean female-headed households account for 30 per cent of all households, with women constituting between 30 and 50 per cent of the formal labour force, and 40 per cent of the informal unregulated labour force. Massiah (1989) argued that if home-based work is taken into account, then the general participation of women in the labour force is increased by 10–20 per cent (see also Ellis 1986; Senior 1991).

In the case of Trinidad, the IMF structural adjustment programme implemented in 1989, and which increased in severity in 1992, served to increase unemployment sharply. Unemployment rose from 11.1 per cent in 1983 to over 20 per cent after 1990. Notably, female unemployment was running at 25.3 per cent in 1990. As already noted, Trinidad is a racially mixed society, which is divided between a large East Indian population making up 45 per cent of the total, and a slightly larger Afro-Caribbean population, amounting to 50 per cent, with small populations of European, Chinese, Syrian and Lebanese (5 per cent) (see Figure 5.2). Ethnic diversity is shown throughout society, the informal sector included, along with gender differences.

In the research, an in-depth examination of the informal sector revealed just how racialised and gendered occupational choice and constraint are within Trinidadian society (Lloyd-Evans and Potter 2002). Furthermore, these differences are spatially expressed. Thus, higglers or street vendors are generally female and of Afro-Caribbean identity. Highway trading along the main arterial routes is principally carried out by males of East Indian descent, with female members of the family assisting in a secondary role. Suitcase trading, buying at a distance and selling locally, is generally carried out in the capital Port of Spain, and is dominated by those of Afro-Caribbean descent, plus a few Surinamese, Chinese and those of mixed descent. In other words, jobs in the informal sector are both racialised and gendered. Further, the research suggested that males were more likely to run a successful business. It appeared that women were marginalised by dint of their family responsibilities, whilst in contrast, males appeared to be more likely to be able to avoid such responsibilities and, therefore, to take risks and change occupations if need be.

have proved to be highly successful in many island territories, dramatically bringing down fertility rates and slowing population growth. Equally significant to this demographic change has been the empowerment of women brought by the post-plantation transformations of island economies, from plantation agriculture to tourism and services (for the Barbados experience see Handwerker 1989).

Between 1960 and 1980, major progress was made across the region towards solving these child and maternal health problems. Chief among imperatives was increasing life expectancies, reducing unwanted pregnancies, reducing infant mortality, embracing contraceptive practices and reducing fertility rates to replacement levels in all countries, save for Haiti (total fertility rate of 4.1) and Belize (total fertility rate of 3.2) (PAHO 1989). In real terms, GDP per capita has changed only modestly (see Table 5.8), but governments have spent a measurable proportion of their revenues on health, education and social welfare services and have provided food subsidies, food stamps and other assistance to needy mothers. Although food production declined, large sums were allocated for the importation of food stuffs. At the same time, emigration provided a 'safety valve', and remittances and reduced birth rates kept per capita food availability near 'normal' nutritional requirement levels. Opportunities for secondary education, including for females rose steeply, though the degree of improvement in women's access varied markedly across the region. Most importantly, quantitative and qualitative improvements in primary health care, including environmental sanitation and health education, resulted in health delivery becoming accessible to everyone, particularly women and children (CEHI 2000; Sinha 1988).

Forty years later, today's health care systems in the Caribbean are addressing increasing incidences of the diseases of modernisation and affluence such as HIV/AIDS, heart attacks, heart failure and diabetes. The last 50 years have seen the evolution of Caribbean health delivery systems from colonial and post-colonial (European/Fabianist) public sector supported, preventive health care and labour-intensive local delivery systems to North American-style higher-cost, higher-technology medical and surgical delivery systems, encouraged and promoted by PAHO/WHO (Pan-American Health Organisation/World Health Organisation) privatisation imperatives and ideologies (PAHO 1997).

During the first 20 to 30 years of political independence, health care systems benefited from major changes in the improvement in the quantity and quality of services – especially through the training of medical personnel – and from the increase in accessibility of these systems to people – making most services free or highly subsidised. Great emphasis was placed on the development of human resources, with nurse practitioners, nurses and nurses' aides being trained to perform essential services in large numbers. In addition to traditional health workers – doctors, surgeons, operating room specialists – midwives, auxillary nurses, public health inspectors, nutrition advisors and public health nurses were also trained in large numbers. Many countries were able to build well-developed networks of health services covering the entire population through hospitals, clinics and health centres. While doctor–population ratios may be low, the nurse and nurse–

Table 5.8 Trends in human development and GDP in the Caribbean, 1960–1993/94

HDI rank – country (2000)	Human Development Index 1960	Human Development Index 1980	Human Development Index & rank 1994		GDP per capita in 1987 US$				
					1960	1970	1980	1990	1993
31. Barbados	0.678	0.856 (+26%)	0.907 (+34%)	**25**	2,290	4,282 (+87%)	5,490 (+140%)	6,002 (+162%)	5,843 (+152%)
42. Bahamas	n.a.	0.805	0.894 (+11%)	**28**	6,770	9,624 (+42%)	10,265 (+52%)	11,240 (+66%)	10,290 (+52%)
44. St Kitts and Nevis	n.a.	n.a.	0.853	**49**	n.a.	n.a.	1,782	3.085	3,577
50. Trinidad and Tobago	0.737	0.816 (+11%)	0.880 (+19%)	**40**	2,443	3,183 (+30%)	5,218 (+114%)	3,759 (+54%)	3,711 (+52%)
52. Antigua and Barbuda	n.a.	n.a.	0.892	**29**	n.a.	n.a.	2,982	5,255	5,699
58. Belize	n.a.	n.a.	0.806	**63**	750	939 (+25%)	1,592 (+112%)	1,950 (+160%)	2,123 (+183%)
61. Dominica	n.a.	n.a.	0.873	**41**	1,192	1,307 (+10%)	1,129 (–5%)	2,018 (+69%)	2,161 (+81%)
66. St Lucia	n.a.	n.a.	0.838	**56**	n.a.	n.a.	n.a.	2,722	2,665
83. Grenada	n.a.	n.a.	0.843	**54**	n.a.	n.a.	n.a.	1,960	n.a.
86. Jamaica	0.529	0.654 (+24%)	0.736 (+39%)	**83**	1,154	1,555 (+35%)	1,289 (+12%)	1,461 (+27%)	1,586 (+37%)
91. St Vincent and the Grenadines	n.a.	n.a.	0.836	**57**	743	797 (+7%)	939 (+26%)	1,620 (+103%)	1,793 (+141%)
94. Dominican Republic	0.385	0.541 (+40%)	0.718 (+86%)	**87**	386	494 (+28%)	744 (+93%)	773 (+100%)	839 (+117%)
103. Guyana	n.a.	n.a.	0.649	**104**	475	540 (+14%)	587 (+24%)	395 (–17%)	519 (+9%)
146. Haiti	0.174	0.295 (+69%)	0.338 (+94%)	**156**	386	333 (–14%)	428 (+11%)	339 (–12%)	226 (–41%)

Sources: *Human Development Reports*, 1997, 2002, United Nations Development Program and Oxford University Press, New York and Oxford, hdr.undp.org/reports/global/2002

Table 5.9 Health care indicators in the Anglophone Caribbean, 1962–1984

Country	No. hospital beds/per 1,000 pop. 1980–1984	No. of health centers 1980–1984	Graduate nurses/ 10,000 pop. 1962–1965	Graduate nurses/ 10,000 pop. 1971–1973	Graduate nurses/ 10,000 pop. 1980–1982	Pop. per physician 1993	Per cent of national budget on health services 1978–1985
Antigua and Barbuda	6.9	26	21.8	17.7	16.0	284	11.8
Bahamas	4.3	54	10.2	22.5	42.9	689	14.7
Barbados	9.0	12	16.2	20.9	30.0	874	15.1
Cayman Islands	2.4	7	7.8	6.0	26.4	n.a.	n.a.
Dominica	3.0	44	9.2	15.6	16.0	n.a.	15.0
Grenada	3.2	33	13.4	13.5	33.7	1,776	n.a.
Guyana	4.5	204	5.8	8.5	6.5	8,929	6.3
Jamaica	2.9	346	22.0	5.6	8.0	6,419	8.9
Montserrat	5.7	12	10.0	43.3	34.1	n.a.	n.a.
St Kitts and Nevis	5.6	17	11.1	27.6	56.7	n.a.	12.3
St Lucia	4.4	24	7.0	11.0	22.7	1,903	13.6
St Vincent and the Grenadines	2.4	34	9.7	11.6	10.6	2,624	14.2
Trinidad and Tobago	4.1	102	12.6	28.0	28.3	1,541	n.a.
Virgin Isles (British)	3.9	9	5.6	13.1	31.7	n.a.	11.2

Sources: 'Number of hospital beds/1,000 population, number of health centers offering maternal care, and graduate nurses/10,000 population ratios, 1962–1982' in Sinha, D. P. (1988) *Children of the Caribbean, 1945–1984: Progress in Child Survival, its Determinants and Implications*, Kingston, Jamaica, and Bridgetown, Barbados: Caribbean Food and Nutrition Institute (PAHO/WHO), in collaboration with United Nation's Children's Fund, Caribbean Area Office; 'Population per physician' in *Statistical Yearbook for Latin America and the Caribbean, 1993*, UNECLAC, Port of Spain, Trinidad and Tobago; 'Per cent of national budget allocated to health services' in *Annual Report of the Director, PAHO*, Washington

auxillary population ratios are quite high (Table 5.9). Nurses, most of whom are female, form the backbone of extensive primary health care in the contemporary Caribbean. In addition, the commitment of Caribbean governments to the development of primary health system in their countries, and their willingness to spend appreciable proportions of their revenues on this sector, remains an equally important reason for the programmatic successes they have achieved (PAHO 1989; Sinha 1988).

Since the 1980s, however, economic uncertainties have increased and debt-burdened governments in the region have had to respond to IMF/World Bank structural adjustment programmes and cut back on public sector expenditures, as discussed previously in this chapter. In doing so, they have cut back on public sector expenditure, so that the necessary financial support either to maintain or improve these health care systems is no longer assured. There are also higher expectations among the burgeoning middle classes for health services to help them solve their 'diseases of affluence' (obesity, heart disease, hypertension, diabetes, cancer, respiratory infection, bronchitis/asthma).

These non-communicable diseases have now displaced communicable ones as the region's most important health problems. Successful infant immunisation campaigns, safer drinking water supplies and higher nutritional standards have all helped to dispel the health threats of earlier years. Accessible primary health care has provided a fundamental level of socio-cultural and economic well-being, which is laudable. Today, however, other causes of mortality, such as the excessive use of illegal drugs, violence within the family and the wider community, along with increases in traffic and household fatalities, are lifestyle/behavioural issues facing island people.

Increasingly during the 1990s, a consensus emerged that 'good health' is a fundamental right of every citizen, and that citizens as well as the community have responsibility for their own health. The second part of this social contract is, of course, a notion derived from North American ideas on modernisation, and is, in large part, a reflection of the cultural penetration of US/North American 'individualist' ideology into the socio-economic realm of the Caribbean. Of course, individualism has always been a cultural trait in the Caribbean, but institutionalising this concept as a replacement for mother–child relations and responsibilities, social consciousness with regards to health care access and communal obligations, would appear to be counter-productive, if not downright regressive. Primary health care systems have proved their worth in the Caribbean, so that recent calls for substantial health system reform may not only be premature, but misplaced (PAHO 1997).

Cuba's success story

Cuba's community health care system and the high level of financial support provided by its government, amounting to 25 per cent of its total GDP, have been widely acclaimed as the best in the world. Infant mortality dropped to six per thousand live births in 2002, down from 60 per thousand before 1959. Life expectancy is now 75 years, up from 55 years before the revolution in 1959.

Because of its reliance on local, community-based care, Cuban health provision is extremely inexpensive, estimated to cost approximately US$20.00 per person per annum. Private expenditure contributes only 11 per cent of total health care expenditure in Cuba, whereas in the USA, the equivalent proportion is 55 per cent, and is growing annually. Cuba has a total of 28 medical schools, where students study medicine for free. The ratio of doctors to people is 1:189, while the equivalent in the USA is 1:358. The ratio of dentists to people in Cuba is 1:1,183; in Australia it is 1:2,500. Unlike Australia, where most people pay the full price for dental care, Cubans receive free dental treatment.

Even after the withdrawal of Soviet support, the tightened US embargo and the crisis of the Special Period, the Cuban government has continued its commitment to free, quality health care. It has also adopted alternative medical practices – for example, acupuncture – and still runs its hospitals, pays its doctors and nurses and runs its maternal health and nutritional subsidy programmes (Feinsilver 1993; Schuyler 2001). The punitive US legislation of the Cuba Democracy Act of 1992 and the Helms-Burton Act of 1996 on the importing of medical supplies and food certainly inflicted considerable damage during the Special Period, and the accompanying severe economic contraction also brought sharp declines in Cuban people's standards of living. Cubans ate less, nutritional deficiencies among pregnant women increased, mortality from diarrhoeal diseases increased, acute shortages of prescription drugs and medicines were commonplace, and tuberculosis cases almost tripled between 1990 and 1994.

Yet, Cuba's heath care system survived and even diversified in response to the challenge. For example, health workers turned to alternative medicine practices, such as acupuncture. Using its highly skilled medical professionals, Cuba intensified the growth of its pharmaceutical and biomedical industries to bring in much needed capital. Health tourism was also promoted (Schuyler 2001). Today, with over 60,000 trained physicians and medical practitioners – 54.6 per 10,000 of the population – Cuba has the highest doctor–patient ratio in the world. Cuba's 'socialised medicine' programme is still viewed as a model of primary community-level health care for the rest of the underdeveloped world (Bourne 2003).

Cuba's HIV/AIDS prevention programme has also been recognised as one of the world's best and most successful. The country has the lowest rate of HIV positive patients of any nation in the world – 4,500 out a nation of 11 million. By contrast, New York city, also with 11 million people, has 25,000 patients (Barksdale 2003; Bourne 2003). Other Caribbean 'successes' in HIV/AIDS treatment and prevention are the Bahamas and Bermuda (Camara 2001). On the other hand, HIV/AIDS has spread to become the most serious epidemic in recent times, despite many countries' efforts to limit its impact (Table 5.10). Regional

Table 5.10 HIV/AIDS cases in the Caribbean, 1994 and 2001

Country	Number of Reported AIDS cases 1994	Incidence rate per 100,000 1994	Estimate of people living with HIV/AIDS 2001	HIV/AIDS Adults and children as % of adult population 2001
Antigua and Barbuda	18	0.0	147	2.1%
Bahamas	322	123.8	6,200	2.0%
Barbados	119	45.1	2,000	1.17%
Belize	18	8.6	n.a.	n.a.
Cuba	n.a.	n.a.	4,500	n.a.
Dominica	5	6.9	n.a.	n.a.
Dominican Republic	n.a.	n.a.	130,000	2.5%
Grenada	7	7.3	n.a.	n.a.
Guadeloupe	n.a.	n.a.	n.a.	n.a.
Guyana	105	14.4	18,000	2.7%
Haiti	n.a.	n.a.	250,000	6.1%
Jamaica	359	14.3	20,000	1.2%
Martinique	n.a.	n.a.	n.a.	n.a.
St Kitts and Nevis	5	11.9	n.a.	n.a.
St Lucia	13	9.2	n.a.	n.a.
St Vincent and the Grenadines	8	7.3	n.a.	n.a.
Trinidad and Tobago	269	20.9	17,000	2.5%
United States	80,691	30.5	n.a.	n.a.

Sources: TransAfrica Forum Issue Brief (2002) and PAHO (1997)

cooperation is underway, and condom use among sex workers is actively being encouraged, both progressive signs that regional governments and NGOs are serious about tackling this latest health threat. The HIV/AIDs epidemic is expected to continue in the region, because preventive health systems cannot cure the disease and a vaccine is not available as yet (Barksdale 2003; TransAfrica Forum 2002). However, Cuba has shown that slowing the rate of its increase is attainable without any major increases in financial resources.

Final comments

The present chapter has stressed the marked social and economic contrasts that characterise contemporary Caribbean societies. Historical information has been provided to substantiate this assertion, both in relation to the social differences that characterise members of the population in terms of race, ethnicity and colour, and in respect of the spatial contrasts that are witnessed across geographical space. In the first half of the chapter, the operation of a strong colour-class system of social stratification was emphasised, this stemming from the colonial period of slavery and indentureship. The rise of a black elite population has, however, served to emphasise the growing salience of social class within present-day Caribbean societies. However, at the macro-scale of the nation, data still show that incomes are closely associated with colour.

The chapter has also served to demonstrate that Caribbean family life has evolved in response to the economic, political and social realities that have faced the region. Again, the influences of slavery and migration in relation to economic opportunities have to be recognised. Women have often taken on a disproportionate share of both familial and economic burdens. Household survival strategies have been responsive to the local needs of the population. Caribbean populations have shown extreme resourcefulness in finding solutions based on individual and community self-help and mutual assistance, contributing to what these days would be regarded as good stocks of social capital. Some of these approaches have been strongly gendered, as in the Red Thread organisation in Guyana (see Peake 1997; Nettles 2003). As emphasised by the material in this chapter, Caribbean household, family and gender divisions must not be interpreted in a wholly negative and pejorative manner.

However, Caribbean societies do remain unequal societies, and ever more so since the intensification of neo-liberal economic policies (see also Chapter 9). This was particularly well exemplified in respect of health care and social welfare provision in the wider region since the 1980s, following the wholesale improvements that were generally witnessed in the post-Second World War period. However, the contemporary health care system of Cuba has been acclaimed as one of the most progressive in the world, and in many respects a model for developing countries.

But the widening of social and spatial polarisation is not just something that is occurring at the national level, as dealt with in this chapter. Marked and sustained inequalities are also the hallmarks of Caribbean neighbourhoods and the housing that serves to make them up. Thus, the Caribbean housing system is the focus of Chapter 6. After that, the expression of inequalities in relation to national urban settlement patterns, as well as within towns, represents the substantive theme of Chapter 7.

References

Barrow, C. (1983) 'Ownership and control of resources in Barbados, 1834 to the present', *Social and Economic Studies*, **32**, 83–120.

Barrow, C. (1986) 'Finding the support: a study of strategies for survival', *Social and Economic Studies*, **35**, 131–76.

Barrow, C. (1996) *Family in the Caribbean: Theories and Perspectives*, Ian Randle, Kingston, Jamaica.

Barrow, C. (ed.) (1998) *Caribbean Portraits: Essays on Gender, Ideologies and Identities*, Ian Randle, Kingston, Jamaica.

Bernard, D. M. (1991) 'Education reform and structural adjustment: reflections on the Guyana experience', in Alexander, K. and Williams, V. (eds) *Reforming Education in a Changing World: International Persepctives*, Oxford Round Table, Oxford, pp. 76–88.

Bernard, D. M. (1997) 'Education spending and structural adjustment', ch. 2 in Williams, P. E. and Rose, J. G. (eds) *Environment and Sustainable Development*, Free Press, Georgetown, Guyana, pp. 9–26.

Booth, D. (1976) 'Cuba, colour and the revolution', *Science and Society*, **40**, 129–72.

Barksdale, B. L. (2003) *HIV/AIDS in Cuba: From Vigilante Quarantine to Vaccine Quest.* Cuba Aids Project, Maryland. http://www.cubaaidsproject.com

Bourne, P. (2003) 'Asking the right questions: lessons from the Cuban health care system', UK HEN Seminar Series presentation, London School of Economics, London: Health Equity Network. http://www.ukhen.org.uk/

Camara B. (2001) *20 Years of the HIV/AIDS Epidemic in the Caribbean: A Summary.* CAREC-SPSTI. http://www.carec.org/pdf/20-years-aids-caribbean.pdf

CEHI (2000) 'Environmental health management: building consensus, *CEHI News* (quarterly news letter of the Caribbean environmental health institute), 7.2 2/3 Quarter 2000, pp. 6–8. Caribbean Environmental Health Institute, Castries, St Lucia.

Clarke, C. (1984) 'Pluralism and plural societies: Caribbean perspectives', ch. 2 in Clarke, C., Ley. D. and Peach, C. (eds) *Geography and Ethnic Pluralism*, George Allen & Unwin, London.

Clarke, C. (1986) 'Sovereignty, dependency and social change in the Caribbean', in South *America, Central America and the Caribbean, 1986*, Europa Publications, London.

Clarke, E. (1966) *My Mother Who Fathered Me: A Study of the Family in Three Selected Communities in Jamaica*, Allen & Unwin, London.

Cross, M. (1979) *Urbanization and Urban Growth in the Caribbean*, Cambridge University Press, Cambridge.

Dann, G. and Potter, R. B. (1990) 'Yellow man in the Yellow Pages: sex and race typing in the Barbados telephone directory', *Bulletin of Eastern Caribbean Affairs*, **15**, 1–15.

Dann, G. M. S. and Potter, R. B. (1994) 'Tourism and postmodernity in a Caribbean setting', *Cahiers du Tourisme*, **Series C, 185**, 1–45.

Dann, G. (1987) *The Barbadian Male: Sexual Attitudes and Practice*, Macmillan Caribbean, London and Basingstoke.

de Albuquerque, K. and McElroy, J. (1999) 'Race, ethnicity and social stratification in three Windward Islands', *Journal of Eastern Caribbean Studies*, **24**, 1–28.

Desai, V. and Potter, R. B. (2002) *The Companion to Development Studies*, Edward Arnold, London, and Oxford University Press, New York.

Department for International Development (DFID) (2000) *Eliminating World Poverty: Making Globalisation Work for the Poor*, DFID, London.

Duncan, N. C. (1994) 'Barbados and the IMF – a case study', ch. 3 in La Guerre, J. (ed.) *Structural Adjustment*, School of Continuing Studies, University of the West Indies, St Augustine, Trinidad and Tobago, pp. 54–87.

Ellis, P. (ed.) (1986) *Women of the Caribbean*, Zed Books, London.

Feinsilver, J. M. (1993) Healing the masses: Cuban health politics at home and abroad, University of California Press, Berkeley.

Furnivall, J. S. (1948) *Colonial Policy and Practice: A Comparative Study of Burima and Netherlands,* Cambridge University Press, Cambridge.

Green, C. (1995) 'Gender, race and class in the social economy of the English-speaking Caribbean', *Social and Economic Studies*, **44**, 65–102.

Handwerker, W. P. (1989) *Women's power and social revolution: Fertility transition in the West Indies*, Sage, Newbury Park, London and New Delhi.

Harewood, J. (1971) 'Racial discrimination in employment in Trinidad and Tobago', *Social and Economic Studies*, **20**, 267–293.

Hoetink, H. (1967) *Two Variants in Caribbean Race Relations*, Oxford University Press, Oxford.

James, C. L. R. (1973) 'The middle classes', in Lowenthal, D. and Comitas, L. (eds) *Consequences of Class and Colour: West Indian Perspectives*, Doubleday, New York, pp. 79–92.

Karch, C. (1985) 'Class formation and class and race relations in the West Indies', in Johnson, D. L. (ed.) *The Middle Class in Dependent Countries*, Sage, Beverley Hills, California, pp. 107–36.

Karch, C. (1981) 'The growth of the corporate economy in Barbados: class/race factors, 1890–1977', in Craig, S. (ed.) *Contemporary Caribbean: A Sociological Reader*, The College Press, Port-of-Spain.

Keagy, T. J. (1975) 'The Redlegs of Barbados', *Americas*, **24**, 14–21.

Kowalewski, D. (1982) *Transnational Corporations and Caribbean Inequalities*, Praeger, New York.

La Guerre, J. (1994) *Structural Adjustment, Public Policy and Administration in the Caribbean*, School of Continuing Studies, University of the West Indies, St Augustine, Trinidad and Tobago.

Laguerre, M. S. (1990) *Urban Poverty in the Caribbean: French Martinique as a Social Laboratory*, Macmillan, Basingstoke.

Layne, A. (1997) 'Race, class and development in Barbados', *Caribbean Quarterly*, **25**, 120–35.

Le Franc, E. (ed.) (1994) *Consequences of Structural Adjustment: A Review of the Jamaican Experience*, Canoe Press, Jamaica.

Lewis, G. (1974) *Notes on the Puerto Rican Revolution*, Monthly Review Press, New York and London.

Lewis, G. (1994) *Notes on the Puerto Rican Revolution*, Monthly Review Press, New York and London.

Lewis, L. (1990) 'The politics of race in Barbados', *Bulletin of Eastern Caribbean Studies*, **15**, 32–45.

Lewis, L. (2001) 'The contestation of race in Barbadian society and the camouflage of conservatism', ch. 7 in Meeks, B. and Lindahl, F. (eds): *New Caribbean Thought: A Reader*, University of the West Indies Press, Jamaica, Barbados, Trinidad and Tobago, pp. 144–95.

Lloyd-Evans, S. and Potter, R. B. (1992) 'The informal sector of the economy in the Commonwealth Caribbean', *Bulletin of Eastern Caribbean Affairs*, **17**, 26–40.

Lloyd-Evans, S. and Potter, R. B. (2002) *Gender, Ethnicity and the Informal Sector in Trinidad*, Ashgate, Aldershot and Burlington.

Lundy, P. (1999) *Debt and Adjustment: Social and Environmental Consequences in Jamaica*, Ashgate, Aldershot.

Lowenthal, D. (1972) *West Indian Societies*, Oxford University Press, Oxford, and Institute for Race Relations.

Lowenthal, D. (1973) 'Black power in the Caribbean context', *Economic Geography*, **48**, 116–34.

Lowenthal, D. and Comitas, L. (1973) *Consequences of Class and Color: West Indian Perspectives*, Doubleday, New York.

Massiah, J. (1989) 'Women's lives and livelihoods; a review from the Commonwealth Caribbean', *World Development*, 17, 965–77.

McAfee, K. (1991) *Storm Signals: Structural Adjustment and Development Alternatives in the Caribbean*, Zed Books, Boston.

McAslan, E. (2001) 'Poverty reduction and social capital in Barbados', unpublished PhD thesis, University of London.

Momsen, J. (1993) *Women and Change in the Caribbean: A Pan-Caribbean Perspective*, Ian Randle, Kingston, Jamaica.

Momsen, J. (1987) 'The feminisation of agriculture in the Caribbean', in Momsen, J. and Townsend, J. (eds) *Geography and Gender in the Third World*, Hutchinson, London.

Nettles, K. (2003) 'Learning, but not always earning: the promise and problematics of women's grass-roots development in Guyana', in Pugh, J. and Potter, R. B. (eds) *Participatory Planning in the Caribbean: Lessons from Practice*, Ashgate, Aldershot.

PAHO (1989) *Primary health care and local health systems in the Caribbean*, proceedings of the workshop on primary health care and the local health systems in Tobago, November 1988, Pan American Health Organisation, Washington.

Peake, L. (1997) 'From cooperative socialism to a social housing policy? Declines and revivals in housing policy in Guyana', ch. 7 in Potter, R. B. and Conway, D. (eds) *Self-Help Housing, the Poor, and the State in the Caribbean*, University of Tennessee

Press, Knoxville, and University of the West Indies Press, Barbados, Jamaica, Trinidad and Tobago, pp. 120–40.

Portes, A., Dore-Cabral, C. and Landolt, P. (eds) (1997) *The Urban Caribbean: Transition to the New Global Economy*, Johns Hopkins University Press, Baltimore and London.

Potter, R. B. (1993) 'Urbanisation in the Caribbean and trends of global convergence-divergence', *Geographical Journal*, **159**, 1–21.

Potter, R. B. (1995) *Low-Income Housing and the State in the Eastern Caribbean*, University of the West Indies Press, Barbados, Jamaica, Trinidad and Tobago.

Potter, R. B. (1996) 'Social conditions in Barbados: aggregate analysis of the 1990 Census at the parish level', *Social and Economic Studies*, 45, 157–86.

Potter, R. B. (1999) 'The geography of relative affluence and poverty in Barbados', *Caribbean Geography*, **10**, 79–88.

Potter, R. B. (2000) *The Urban Caribbean in an Era of Global Change*, Ashgate, Aldershot, Burlington and Sydney.

Potter, R. B. and Phillips, P. (2002) 'The social dynamics of foreign-born and "young" returning nationals to the Caribbean: outline of a project', *Centre for Developing Areas Research Paper*, Royal Holloway, University of London, **37**, 30pp.

Potter, R. B. (2003a) 'Foreign-born and young returning nationals to Barbados: results of a pilot study', *Reading Geographical Paper*, **166**, 40pp.

Potter, R. B. (2003b) 'Foreign-born and young returning nationals to St Lucia: results of a pilot study', *Reading Geographical Paper*, **168**, 33pp.

Potter, R. B., Binns, J. A., Elliott, J. A. and Smith, D. (2004a) *Geographies of Development*, second edition, Pearson/Prentice Hall, London and New York.

Potter *et al.* (2004b) *Social Atlas of Barbados: Cartographic and Statistical Analysis*, Arawak Press, Kingston, Jamaica, and ISES, University of the West Indies, Barbados.

Potter, R. B. and Binns, J. A. (1988) 'Power, politics and society', ch. 7 in Pacione, M. (1988) *The Geography of the Third World: Progress and Prospects*, Routledge, London and New York, pp. 271–310.

Potter, R. B. and Conway, D. (1997) *Self-Help Housing, the Poor and the State in the Caribbean*, University of Tennessee Press, Knoxville, and University of the West Indies Press, Barbados, Jamaica, Trinidad and Tobago.

Potter, R. B. and Dann, G. (1990) 'Dependent urbanization and retail change in Barbados, West Indies', in Potter, R. B. and Salau, A. T. (eds) *Cities and Development in the Third World*, Mansell, London.

Potter, R. B. and Jacyno, J. (1996) 'Social conditions in St Lucia: aggregate analysis of the 1991 Census at the quarter level', *Journal of Eastern Caribbean Studies*, **24**, 25–50.

Potter, R. B. and Lloyd-Evans, S. (1998) *The City in the Developing World*, Prentice Hall, Harlow.

Potter, R. B. and McAslan, E. (2000) 'Urban poverty and the Urban Development Commission in Barbados', *Geography*, **85**, 263–67.

Ramsaran, R. F. (1992) *The Challenge of Structural Adjustment in the Commonwealth Caribbean*, Praeger, New York.

Roberts, G. and Sinclair, S. A. (1977) 'Family and reproduction in Jamaica', unpublished manuscript cited in Cross (1979).

Schlesinger, B. (1962) 'Family patterns in Jamaica: review and commentary', *Journal of Marriage and the Family*, **30**, 137–48.

Schuyler G. W. (2001) 'Health and neoliberalism: Venezuela and Cuba', paper presented at the 2001 Congress of the Americas, Universidad de las Americas, Puebla, Mexico. http://info.pue.udlap.mx/congress/5/papers_pdf/gws16.pdf

Senior, O. (1991) *Working Miracles: Women's Lives in the English-Speaking Caribbean*, James Currey, London.

Sinha D. P. (1988) *Children of the Caribbean, 1945–1984: Progress in Child Survival, its determinants and implications*, Caribbean Food and Nutrition Institute (PAHO/WHO), Kingston, Jamaica, and Bridgetown, Barbados, in collaboration with United Nation's Children's Fund, Caribbean area office.

Smith, M. G. (1965a) *The Plural Society in the British West Indies*, University of California Press, Berkeley.

Smith, M. G. (1965b) *Stratification in Grenada*, University of California Press, Berkeley.

Stone, C. (1980) *Democracy and Clientism in Jamaica*, Transaction Books, New Brunswick, NJ.

Thomas, C. Y. (1988) *The Poor and the Powerless: Economic Policy and Change in the Caribbean*, Latin America Bureau, London.

TransAfrica Forum (2002) 'AIDS in the Caribbean,' *TransAfrica Forum Issue Brief*, September 2002, TransAfrica Forum, Washington. http://www.transafrica.org

United Nations (2001) *Human Development Report 2001: Promoting Linkages*, Oxford University Press, Oxford, for United Nations Development Program.

West, R. C. and Augelli, J. P. (1976) *Middle America: Its Lands and Peoples*, Englewood Cliffs, New Jersey.

West Indies Royal Commission (1945) *Report*, Cmnd. 6607, HMSO, London.

Wilson, P. J. (1973) *Crab Antics: The Social Anthropology of English-Speaking Negro Societies in the Caribbean*, Yale University Press, New Haven, Connecticut.

Wilson, P. J. (1969) 'Reputation and respectability: a suggestion for Caribbean ethnology', *Man*, **4**, 70–84.

World Bank (2003) 'Poverty reduction strategy papers: good practices', pamphlet available at www.worldbank.org/poverty/strategies/review/order.htm.

Chapter 6

HOUSING IN THE CARIBBEAN

Housing in the Caribbean: an introduction

Just as social conditions in the Caribbean have always displayed such marked polarity, distinguishing between the wealthy few and the relatively impoverished majority, so housing shows a similarly marked contrast and differentiation in the Caribbean region (Potter and Conway 1997). In exemplifying this central point, Hudson (1997) relates his experience working on an archaeological dig on the Montpelier Estate near Montego Bay in Jamaica. The excavation work focused on a single former slave dwelling, and little remained of the house apart from a few stones that formed the base of the original structure.

Hudson uses this as a type of metaphor concerning everyday homes and housing in the Caribbean region through time. Thus, the entire site of the former village settlement, once home to generations of poor Jamaicans of African descent, is barely witnessed in the form of a shallow depression in what is now a cattle pasture. In contrast with these relict features representing the homes of the poor majority, the eighteenth-century Great House, built of local cut limestone, stands in full view of the former village and is still inhabited. The remains of the old sugar estate also survive nearby. This vignette is used to introduce a wider argument concerning the nature of popular housing in the Caribbean context, when it is noted how:

> . . . very little remains of the former homes of the poor majority of the population. Splendid and substantially built mansions, palaces, churches, fortresses and the like often stand for centuries, but the unprepossessing and insubstantial dwellings of the poor are usually constructed of much less durable materials and are destroyed with indifference when those in power decide to clear them away for other purposes, a theme which is of considerable relevance to the contemporary Caribbean housing situation (Hudson 1997: 14–15).

Hudson (1997) maintains that when the dwellings of the poor are considered at all in the Caribbean, they are regarded as substandard structures that are in pressing need of replacement with modern concrete-based dwellings, along with these being built as economically as possible. And in doing this, in the Caribbean as elsewhere in the developing world, the design standards, building materials and technology have frequently been imported from abroad, reflecting the ideology and building codes of the more affluent developed world. Edwards (1980) and Potter (1989, 1991, 1993; Potter and Watson 1999; Watson and Potter 2001) have bemoaned the fact that the indigenous or folk architectural systems of the Caribbean remain poorly studied. For example, in a content analysis of the journals *Social and Economic Studies* between 1953 and 1988 and the *Bulletin of Eastern Caribbean Affairs* (now the *Journal of Eastern Caribbean Studies*) from 1975 to 1983, only two out of 763 and nil out of 272 respectively (that is a total of two out of 1,025) dealt with housing (Lloyd-Evans and Potter 1992). Potter (1993) emphasises that this academic neglect is paralleled by the almost total neglect of the efforts of low-income groups to house themselves in the policy arena. He argues that this amounts to the almost total neglect of a socio-cultural resource of major importance within the Caribbean region as a whole.

The present chapter provides an overview of housing conditions and housing policies in the Caribbean region, taking as its principal theme the general neglect of the Caribbean working-class or popular dwelling. However, exceptions to this general rule, such as that provided by Cuba, are considered. In reviewing this argument, the characteristics of the vernacular architectural system of the Caribbean are outlined. This then leads to the consideration of the way in which Caribbean housing has worked as an indigenous self-help system since the days of slavery. The general neglect of the popular house form in the policy domain is then documented, before the dominance of middle- and upper-income housing is stressed in respect of the modern period.

The Caribbean popular house: architecture that works

The general neglect of Caribbean popular housing is surprising, given that as long ago as 1980, Jay Edwards proposed an interesting stage model of the way in which the Caribbean popular dwelling has evolved through time. The stages recognised by Edwards are summarised in Table 6.1.

The first stage, broadly extending from 1627 to 1675, was the period of original English settlement of islands such as Barbados and Nevis. During this period, labelled as 'antecedents', the planter-pioneers constructed what amounted to crude temporary huts. These were frequently wattled and followed quite closely the design of indigenous Arawak Indian dwellings, using forked sticks as supports. During the second stage, from 1628 to 1660 and labelled the period of 'preadaptation and the grounding of separate folk/vernacular traditions', as soon as sawn timber was available to the English planters, they began to build English-style cottages. These were built in a similar manner to the type of folk houses that were prevalent in the south of England at that time. Interestingly, these dwellings shared several basic structural characteristics with the folk houses found in the areas of West Africa from which slave labourers had been transported. In both traditions the unit was a two-roomed module and the roof generally took the form of a thatched gable.

Table 6.1 Five-stage sequence in the development of the Caribbean popular house

Stage	Description	Approximate time period
0	Antecedents	*c*. 1627–1675
1	Preadaptation and the grounding of separate folk/vernacular traditions	1628–*c*. 1660
2	Simplification: reduction of Old World variability	1642–*c*. 1800
3	Initial amalgamation and reinterpretation	*c*. 1650–1700
4	Elaboration: innovation and borrowing	*c*. 1670–present

Source: Potter (1992), (originally adapted from Edwards 1980)

In the European cottage, the door was normally placed asymmetrically on the long wall of the house.

Edwards states that at this juncture there were many British- and African-style huts, but little or no synthesis of their forms into a new regional house type. It was only as a third and distinct stage that a move towards a new synthesis occurred between 1642 and 1800 in different territories. Edwards describes this as a period of 'simplification and reduction of Old World variability', during which a process of evolutionary change occurred whereby the commonalities of the two formerly separate systems were combined to afford a new regional form. It is argued that the destruction of former houses in the region by fire and hurricanes played a vital role in advancing this process of syncretism.

Thus, it was as a distinct fourth stage, from 1650 onwards locally, that the process of initial amalgamation and reinterpretation of house forms occurred. Edwards attributes a key role in this process to the characteristics that were shared between European and African houses, in particular, the fact that they were both based on two-room cottage modules of rectangular proportions, with the length of the dwelling generally extending two times the width. This effectively became the norm for houses in the region and a process of local interpretation and development then occurred. In this respect the raising of the cottage off the ground was significant. This widespread Caribbean practice, effected either by building timber posts or by placing the house on loose-rock foundations, was carried out in order to exclude vermin and pests, improve ventilation and reduce the effects of dampness and floods (Hudson 1997). A further series of local adaptations reflected climatic influences, these including the use of wooden shingles, large windows with shutters, and careful orientations with respect to the prevailing wind.

The outcome of this gradual evolutionary process was that by 1700, true indigenous Caribbean dwellings were to be found throughout the region. In his

Plate 6.1 A 'folk' or 'popular' house in Curaçao, Netherlands Antilles
Source: Rob Potter

account, Edwards looks at the fifth stage, referred to as 'elaboration: innovation and borrowing', specifically in relation to the western Caribbean. But in Barbados, as in some other territories such as Trinidad, a strong Georgian influence can be identified, including the addition of verandahs and terraces and a strong tendency towards complete symmetry of the front façade.

The popular Caribbean house of today is, therefore, the outcome of this combination or syncretism of European and West African architectural forms. Houses are normally twice as long as they are wide, and initially, are divided into two major spaces, with the main door being placed on the long side of the dwelling (see Plate 6.1). These local vernacular architectural forms display several other distinctive elements, including jalousies, box (bell) pelmets, battened shutters and decorative fretwork. These adaptations all reflect the climatic realities of the region. It would be very naïve to interpret such houses as 'makeshift' or 'unplanned': they represent a distinct socio-cultural architectural form. However, the overall house type derived from the syncretism shows minor stylistic variations from territory to territory. Thus, the entrance is often located to one side of the main wall in the Windward Islands of Grenada, St Lucia and St Vincent and the Grenadines (Plate 6.2 and Figure 6.1). Additionally, in these territories, the shutters on the doors and windows are frequently made of solid wood.

In Barbados, the outcome of this evolution has been the 'chattel' house. A representation of a classic example of the house type is shown in Figure 6.2. The basic dwelling is often 16 feet by 8 feet, or 18 by 9, or 20 by 10. The chattel shows an entirely symmetrical front façade, with a pair of 18-inch wide doors being placed

Plate 6.2 A 'popular' home in St Vincent
Source: Rob Potter

Figure 6.1 Architectural drawings of some eastern Caribbean popular house types
Source: Potter (1995)

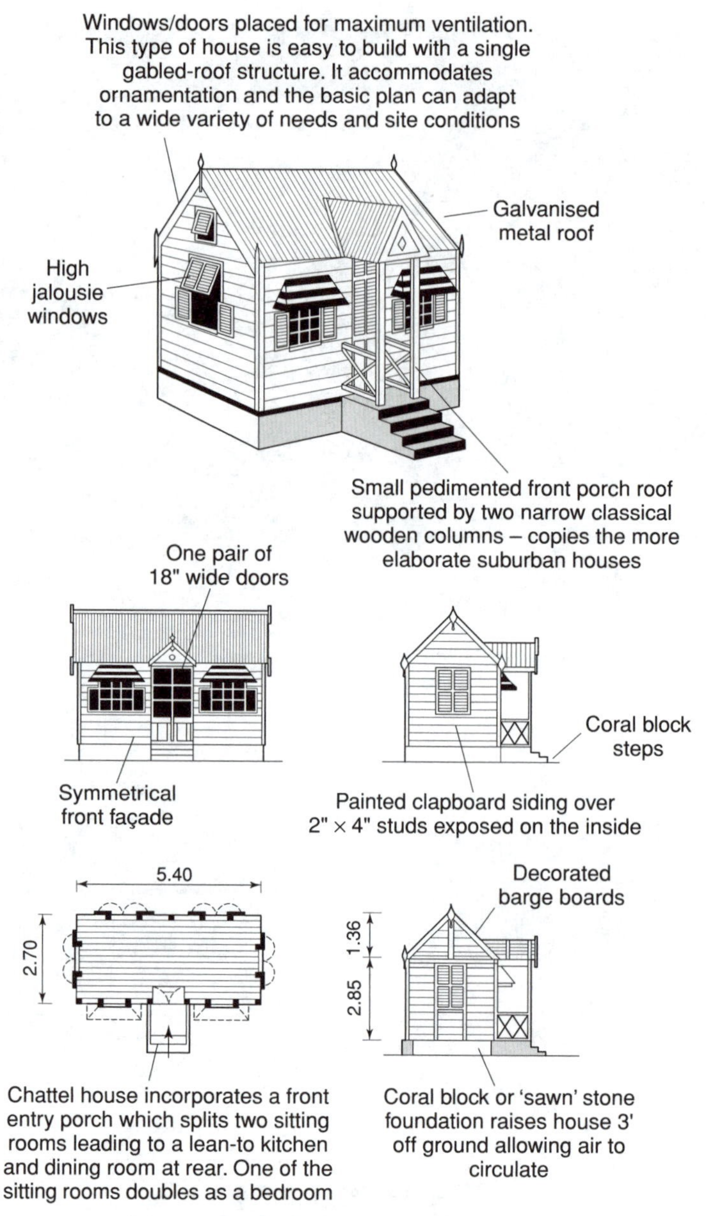

Figure 6.2 Architectural details of the Barbadian chattel house
Source: Watson and Potter (2001)

at the dead centre of the front façade. The windows and doors are placed for maximum ventilation, and wherever possible, the house is located at right angles to the prevailing trade winds. Traditionally such houses have been built on loosely consolidated foundations of broken coral stones (Figure 6.2).

Caribbean self-help housing: an indigenous system that works

Most houses in the Caribbean are built by a process of individual and extended community self-help. This is referred to as *coup-de-main* in the case of St Lucia, and friends and relatives assist in the building process, often at weekends. In return, those who help are provided with food, drink and hospitality. By such means, low-income groups have traditionally built wooden houses, not least because they do not own the land on which they have constructed their homes. This may be for a variety of reasons. It may have been because such homes had been constructed on so-called rab (or poor) land existing at the edge of plantations, as frequently happened in the case of Barbados' sugar plantations.

In the more mountainous states of the Caribbean, like St Vincent, Dominica and Grenada, there has been a greater tendency for squatters to build houses on the hillsides and steep slopes that surround the principal towns and which have not previously been developed by the formal building sector. Examples include the River Road and Grand Anse Valley areas of St George's, Grenada, and the Upper New Montrose, Camden Park and Ottley Hall areas of Kingstown, St Vincent. In yet other circumstances, squatters have built on swampy, recently reclaimed and other marginal urban lands. Examples are provided by areas such as the Conway and Four-à-Chaud residential communities in Castries, St Lucia. Four-à-Chaud is a squatter settlement located on a reclaimed area adjacent to the port (see Plate 6.3), while Conway is a longer-standing shanty area that has

Plate 6.3 Part of the Four-à-Chaud low-income area, Castries, St Lucia
Source: Rob Potter

for a long time been inhabited by fishermen. The history of the Conway squatter area is reviewed in Box 6.1.

What unites all of these areas is the lack of security of tenure. People are building homes in circumstances where they may be evicted at short notice, either by

Box 6.1: History of the Conway squatter area in Castries, St Lucia

The Conway squatter area was built on land originally owned privately by a single family and which was being developed as a rent yard. The area grew rapidly during the 1940s and 1950s, serving those low-income residents who needed to be near to the harbour and to the central commercial district. With the rapid growth of the residential area, the owners eventually felt that they faced too many difficulties in collecting rents from tenants, so they sold the land to the government early in the 1960s. At this point, the residential area effectively became a squatter settlement. At this juncture, there were more than 300 houses. Part of the area is shown in the early 1980s in Plate 6.4.

As early as the 1970s the area was scheduled for commercial redevelopment, but nothing happened for the next 15 years. Then, in the middle of the 1980s, a new duty-free tourist facility was developed across the Castries Harbour at Pointe

Plate 6.4 Part of the Conway low-income residential area photographed in the early 1980s
Source: Rob Potter

(Box continued)

Seraphine. In 1986, when this was opened, the government announced its plan to clear most of the Conway area – this representing the first intervention of its kind in St Lucia – in order to build commercial and government offices. Accordingly, the western portion of the squatter zone was cleared with some haste between April and July 1986. In the end, a total of 162 houses were relocated in three phases, and the residents were moved to an area known as Ciceron, located some 3 km to the south-west of the city centre. Those who relocated were given a EC$1,000 inducement in the form of either building materials or as a contribution to the cost of the land. The government provided free transport for the dismantled houses to be moved to their new sites.

Although two meetings were held with local residents about the clearance of these squatted lands, both took place after the relocation process had begun in earnest, so that effective public participation was minimal (Phillip 1988, 1999). Clearly the state's involvement after an extended period of quiescence reflects the interests of capital. The relocation process was embarked upon in order to release prime land adjacent to the commercial centre and the harbour. As shown by Potter (2001) the freeing up of this land can be seen as part of a process of rendering Castries Harbour entirely user-friendly to cruise ship tourism.

plantation landlords on termination of employment, or by governments seeking to repossess formerly economically inactive lands for new commercial purposes (see Box 6.1 concerning the forced removal of squatters from Castries, St Lucia). Low-income groups have thereby been pushed toward building houses of semi-permanent materials that can, if necessary, be disassembled, moved and reassembled at another location (see Plate 6.5). This is another reason why houses in the region are placed on stilts or on a base of loose rocks. The insecurity of land tenure involved has generally served to place a strict upper limit or ceiling on the types of improvement that individuals have been willing, and able, to carry out with respect to their dwellings, without the provision of land titles.

The end result has been the derivation of a distinct Caribbean indigenous vernacular architectural style. Such is the flexibility and appropriateness of the housing system that has been evolved by the people that dwellings are not only movable, they also form an expandable, modular system of self-help housing that fully subscribes to Turnerian principles of progressive improvement over time (Turner 1968; see also Potter and Lloyd-Evans 1998, chapter 7). Thus, in many of the territories, a second or third cabin can be attached at the rear of the original unit. This is shown in Plate 6.6 and Figure 6.3 in respect of the Barbadian popular housing, but essentially the same process is witnessed elsewhere. In this manner, with time, separate bedrooms, living, dining and kitchen spaces are developed. In most cases the houses are movable once disassembled and 'flat-packed' onto a suitably large lorry (Plate 6.5).

A different modular system is found in some of the non-Anglophone islands, particularly Puerto Rico, Guadeloupe and Aruba (Hudson 1997). Here a square

Plate 6.5 Barbadian chattel house being moved
Source: Rob Potter

Plate 6.6 Good example of the modular expansion of the Barbadian popular house
Source: Rob Potter

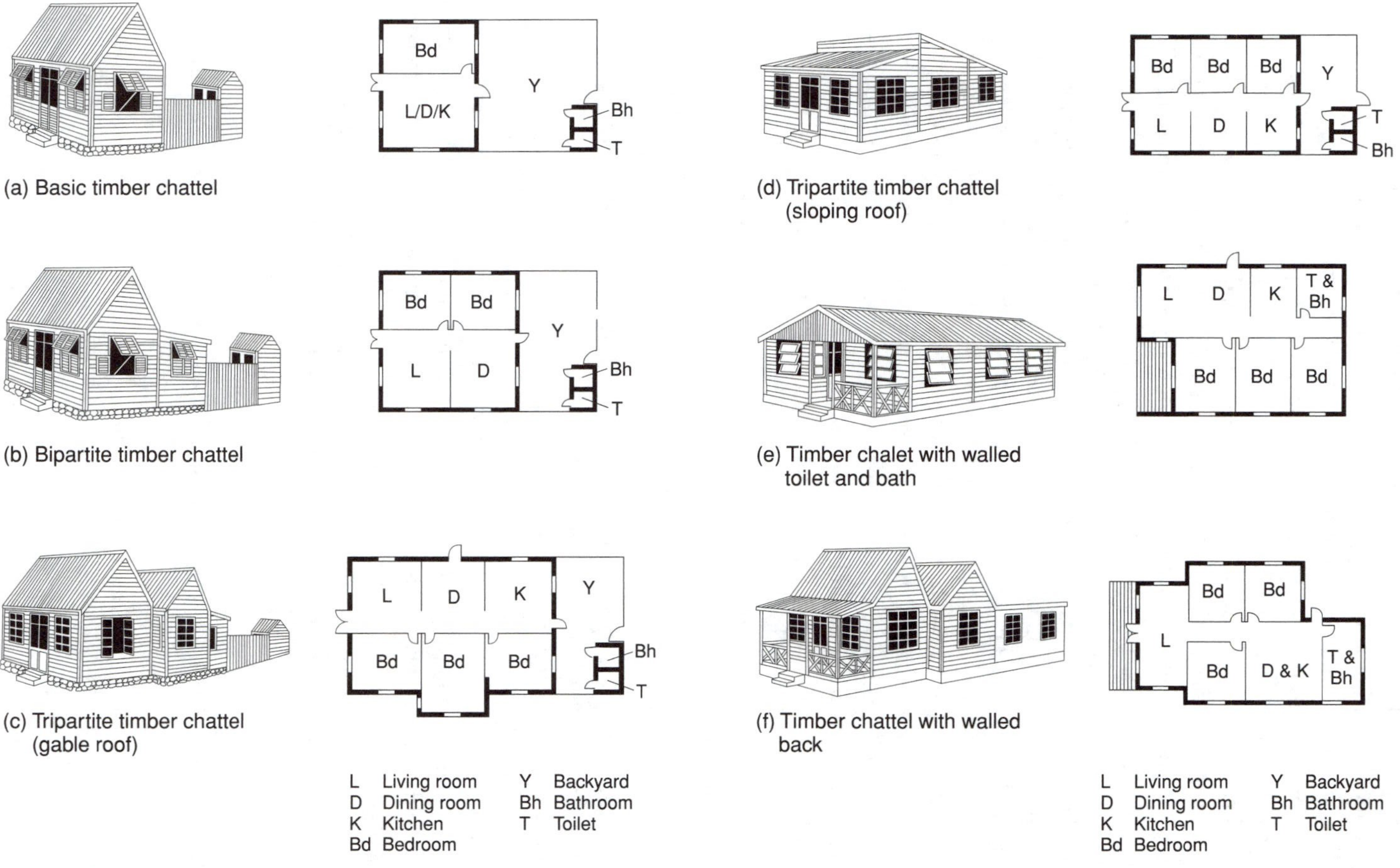

Figure 6.3 Extension and growth of the modular chattel house
Source: Potter (1992)

room, frequently 10 feet by 10 feet (3 m by 3 m) forms the base unit. The module has an opening, either a door or a window, at the middle of each side. Thus, the dwelling can be extended by placing new modules alongside existing modules. If three such units are placed together to form an L shape, then the corner can effectively form a verandah. It is thought that this system possibly originated in the former French colony of Louisiana (Berthelot and Gaume 1982). In fact, Redman (1976) suggested the implementation of such a modular system for policy-related purposes in respect of the Barbadian housing system. As Hudson (1997) observes, modular systems of construction are highly appropriate in what were essentially industrialised, slave-labour plantation economies.

But the central point is that Caribbean people constantly show ingenuity in improving their extant housing conditions in all manner of circumstances. For example, the National Housing Trust and the private building sector in Jamaica have jointly produced prefabricated units in the Portmore area. These take the form of either single-family dwellings or duplexes, the equivalent to semi-detached housing. Soon after a family moves in, the modifications begin, even though these industrially produced core units were never really designed to be modular. The house is often enlarged by adding extra rooms, an additional storey, a turret or two, a fancy façade or a porch, and some are embellished with very distinctive design features. Very quickly the urban landscape changes from uniform drabness to something quite different that can only be described as distinctly Jamaican. Portmore is the classic example of this type of progressive transformation, but the same happens in housing schemes across Jamaica, effectively ignoring planning regulations. This is functionally similar to the flexible, expandable popular housing previously described in this section, but the scale and suburban context represent a different environmental context.

Similarly, a common feature of housing among poor people in the Caribbean in both urban ghetto and squatter communities alike is multiple household occupancy and subletting. In the Jamaican context, the classic tenement yard is a kind of open compound area with several rented dwellings occupied by separate households. These households often consist of large families. A similar pattern is replicated in squatter communities and some of Eyre's (1972) early work focused on squatter landlordism in squatter areas, with people adding rooms or subdividing rooms for rent almost as soon as their initial abode is 'completed'.

Contemporary housing conditions in the Caribbean region

The virtues of the indigenous Caribbean housing system should be recognised, as a form of what Turner (1968) has described as 'architecture that works' – that is architecture by the people and for the people that slowly improves (or 'consolidates' over time). However, it is equally imperative to acknowledge what the system can and cannot be expected to do without the intervention of appropriate responses in the wider domains of social and economic policy. The remainder of this chapter considers these issues as they relate to matters of housing quality and housing policy in the contemporary Caribbean region.

For example, the frequent lack of ownership of the land on which houses have been built, as opposed to the direct ownership of the dwelling itself, has meant that a clear upper limit has been set on the degree to which individual and spontaneous improvement is likely to occur. It is also reflected in the sometimes haphazard and higgledy-piggledy layout of substantial tracts of low-income housing. Furthermore, although wood is an excellent building material in tropical climates, in the humid tropics, unless wooden houses are regularly repainted and repaired, the structures can rapidly deteriorate. Thus, without suitable policy measures, over time, such housing often comes to show an association of different types of housing stress: old houses with a high proportion of all wood dwellings are found clustered together, using basic amenities and with few signs of improvement.

An example of this is provided by the island-wide state of housing in Grenada provided by Potter (1995). This analysis showed that for Grenada as a whole, 61 per cent of the entire housing stock was constructed entirely of wood, with nearly as many dwellings making use of pit latrine toilets. Nearly half of all dwellings had been built over 20 years previously, and just over a third, 34 per cent, obtained their water from a public standpipe.

Old houses, i.e. those built more than 20 years in the past, were most prevalent in the case of the island of Carriacou, and were well above the national average for the three northern parishes of St John's, St Mark's and St Andrew's (Figure 6.4). The lowest proportion of old housing was recorded for urban St George's South, but was above the national level for the city centre. An essentially similar pattern was revealed for houses constructed entirely of wood. The highest dependence on pit latrine toilets was shown by the eastern parish of St Andrew North, and was lowest for the city centre (less than 20 per cent). In four parishes over 40 per cent of total dwellings gained their water from a public standpipe. These included the northern zones of St Mark's, St John's and St Andrew's once again (Figure 6.4).

The example provided by Grenada shows how measures of housing stress and disamenity all tend to tell the same story, pointing to areas of multiple housing deprivation. It also illuminates another salient point, namely that although some of the worst pockets of housing disamenity occur in urban areas, on average the poorest conditions are found in the rural and agricultural zones. This is well demonstrated by the analysis of housing conditions in the French West Indies in 1990 carried out by the French authorities and cited by Condon and Ogden (1997). The maps of relative housing deprivation in Guadeloupe and Martinique depicted in Figure 6.5 basically show the percentage of dwellings with a deficiency in the provision of water and/or electricity within the house itself. In the case of Martinique this accounted for nearly 14 per cent of all dwellings, and close to 20 per cent in the case of Guadeloupe. When the geographical expression of

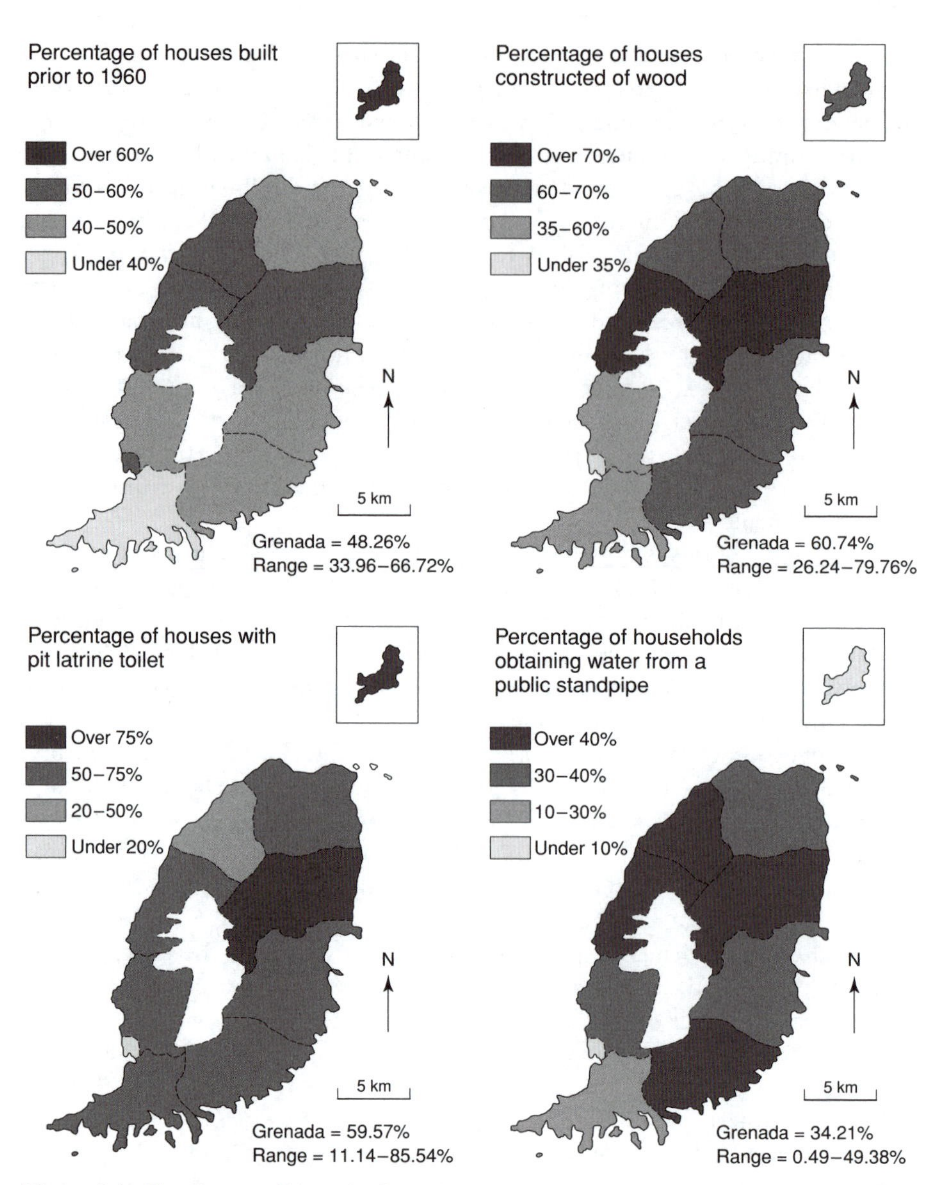

Figure 6.4 Housing conditions in Grenada
Source: Potter (1995)

such relative deprivation is considered, levels of housing stress are clearly shown to be high in the small rural farming and fishing communities of eastern and southern Martinique. This is also true of northern Martinique, together with outlying Desirade and St Martin. But although not apparent from the map, pockets of marked housing deprivation also exist in the heart of the city.

Potter (1992) employed the coincidence of different measures of housing disamenity in order to provide a composite measure of housing conditions in the

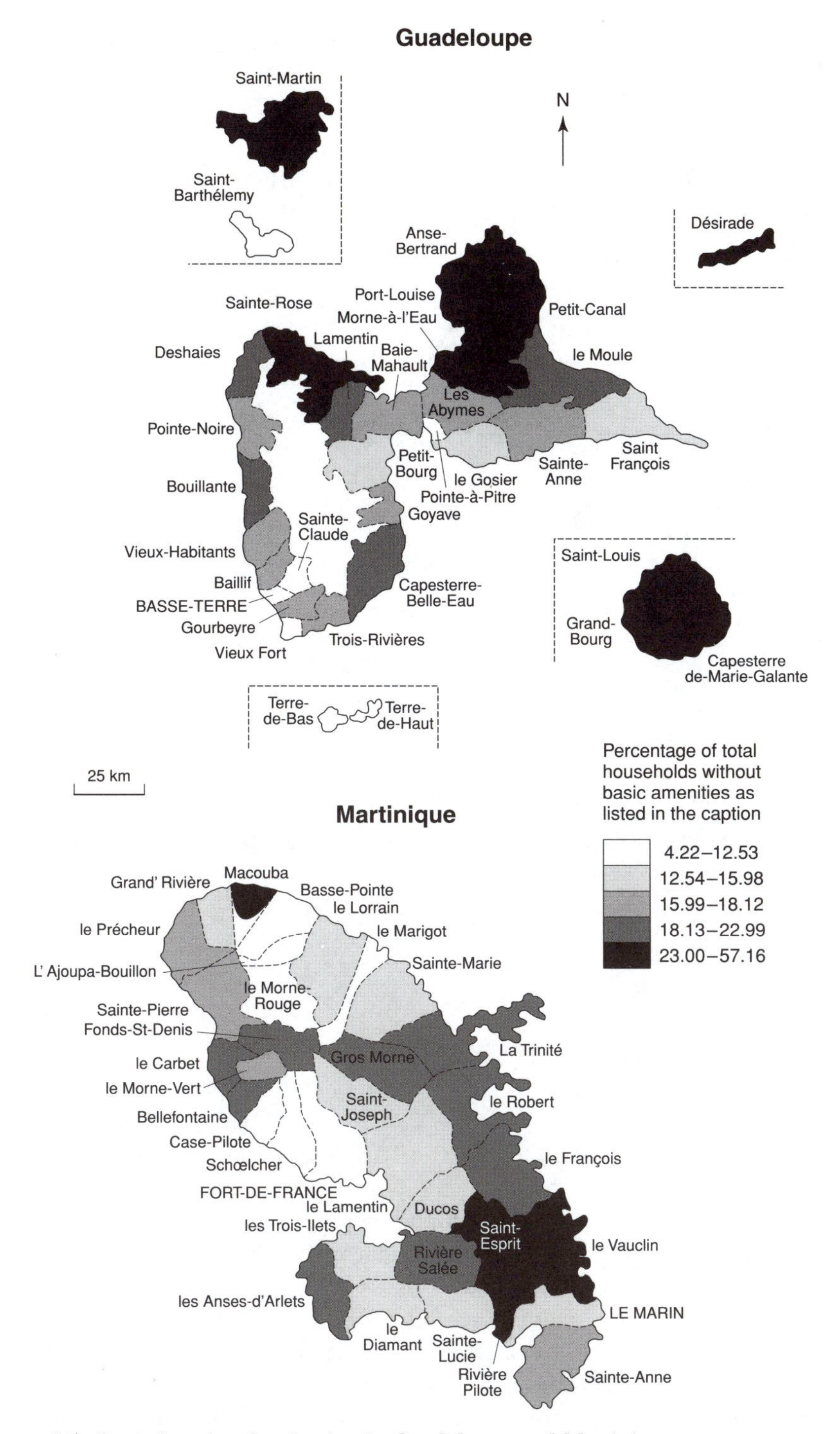

Figure 6.5 Basic housing deprivation in Guadeloupe and Martinique
Source: Condon and Ogden (1997)

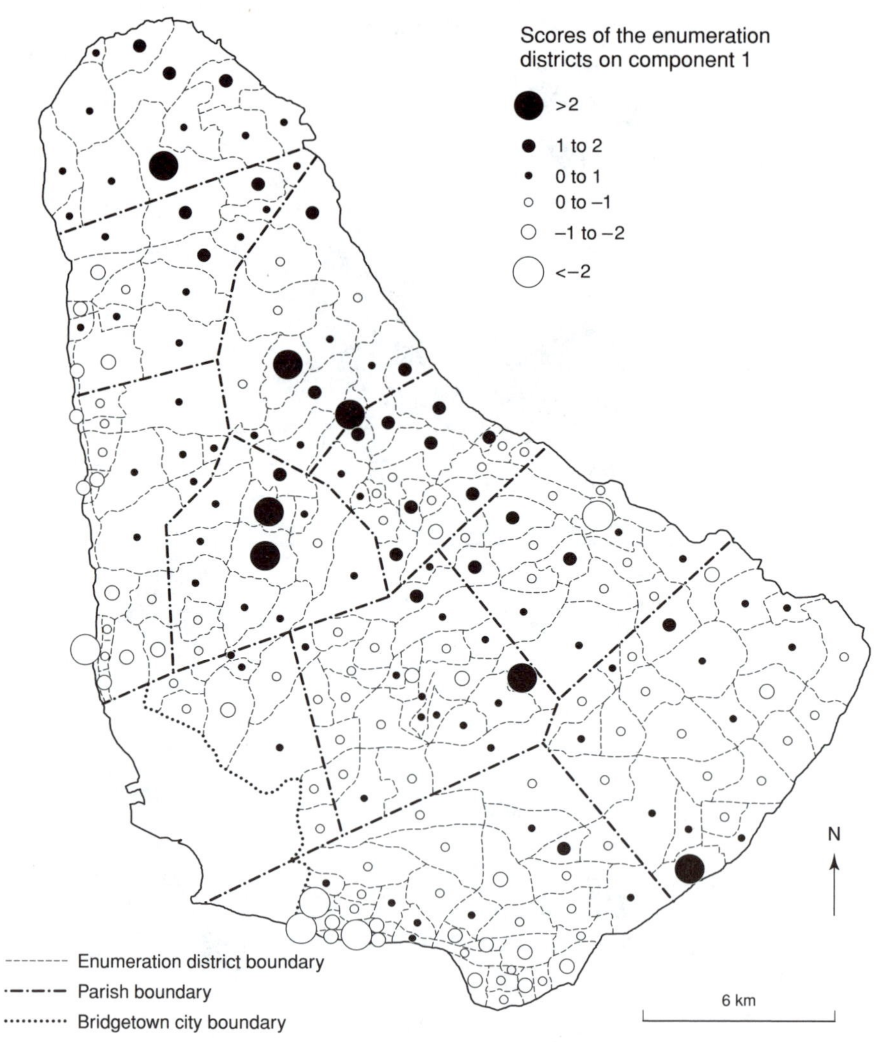

Figure 6.6 Housing disamenity in Barbados
Source: Potter (1992)

case of the eastern Caribbean territory of Barbados. Eight key housing variables were factor analysed. This basically means that patterns of association existing between the variables were searched for among the 210 census enumeration districts making up Barbados outside the Bridgetown urban area. The first major component so derived reflected houses that were constructed of wood, had pit latrine toilets and made use of public standpipes. The component was negatively associated with the availability of electric lighting and the use of gas for cooking. The scores of the original enumeration districts on this measure of housing amenity are shown in Figure 6.6. Poor housing was again shown to be typical of the entire northern and eastern rural agricultural zones, especially St Lucy, St Andrew,

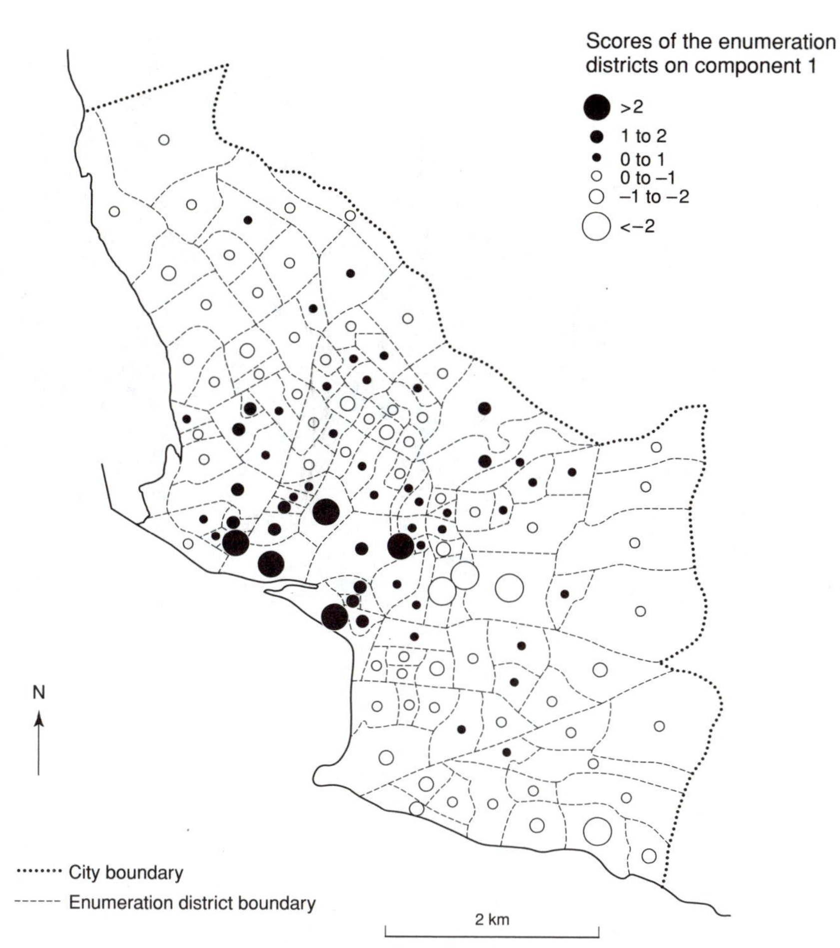

Figure 6.7 Housing disamenity in Bridgetown
Source: Potter (1992)

St Thomas and St Joseph. In the wider analysis, multiple housing deprivation was also shown to be associated with a broad inner city ring of older housing that encircles the central business district of Bridgetown (Figure 6.7) (see Potter 1992).

Each of these examples drawn for different Caribbean territories suggests that without appropriate state mechanisms of facilitation, the everyday efforts of the poor can only have so much influence on improving extant housing conditions.

Modernising housing in the contemporary Caribbean: housing and state policy in the postwar period

As Conway and Potter (1997a, 1997b) note, Caribbean housing problems reflect the existence of sprawling and largely uncontrolled informal-popular settlements,

Plate 6.7 Inner city residential area in Havana, Cuba
Source: Rob Potter

limited rental accommodation and highly unequal and distorted real estate markets. But such problems also include the existence of dilapidated and overcrowded inner city tenements. In looking at migration paths in Chapter 7 on urbanisation in the Caribbean, it is noted how in the larger urban areas, inner city tenements represent the settling grounds for many in-migrants (see Plate 6.7). Only later do such migrants tend to move out to the peripheral shanty towns. In Chapter 7, following Eyre's (1972) research, this argument is supported by the case of Montego Bay, Jamaica. Thus, in summarising housing conditions in the region, Conway and Potter (1997a) noted that:

> In common with most, if not all, developing countries, those of the Caribbean exhibit a plethora of housing problems for their burgeoning poor. Recent history has witnessed the struggle for adequate shelter in this urbanizing region. This is in a highly inequitable social context where, as early as 1964, Charles Abrams reported on the desperate plight of the poor in Trenchtown, Jamaica, the Kingston 'shanty-town' which was later to be immortalized by Bob Marley.

The account so far in this chapter has shown how during the colonial era people had no option but to get on with building for themselves. Thus, in respect of the formal sector, the region has always displayed a marked housing deficit. Notwithstanding economic problems in the early decades of the twentieth century, the provision of adequate housing was ostensibly an important objective of these peripheral capitalist, decolonising Caribbean nations. However, as noted by Conway and Potter (1997a), the record tends to suggest that in the field of housing, rhetoric – and sometimes legislation – has rarely been matched by reality and practice. In other words, most governments have talked about improving housing for the masses, but all too often little has actually been achieved on the ground.

Undoubtedly, this is partly due to the fact that these newly established state apparatuses were overburdened at independence, and the resources available for improving housing were strictly limited. Hence, Conway and Potter (1997a: 2) talk about the existence of 'either an evasive or benign neglect of the housing sector by public institutions, state enterprises, and government ministries alike . . . throughout the Caribbean, the state struggled to define its agendas, and housing more often than not remained low on the priority list, rarely warranting the attention it deserved'.

The lack of clear proactive housing policies is particularly well exemplified by the housing histories of the small and peripheral capitalist Windward Islands. With few exceptions, the state in these territories has been seen to intervene directly in the housing arena only when the demand for land for commercial purposes has suggested the efficacy of clearing low-income groups from particular residential sites. This was well illustrated in Box 6.1 dealing with the Conway squatter settlement in Castries, St Lucia. Another instance when the state has become involved is where dwellings have been threatened by the likelihood of environmental disasters such as flooding and storm damage (see below for an example). In almost all other circumstances the state has basically left low-income groups to provide for themselves (Potter 1995, 1997a). As such, housing policies have largely failed to incorporate the positive aspects of vernacular housing.

In the Windward Islands, Potter (1995) notes that the respective Ministries of Housing have had little to do with the production of houses in the direct sense. Phillip (1988: 59) comments in relation to St Lucia that the 'Ministry is therefore the least effective and active in the field of housing'. This is reflected in the minuscule proportion of the total housing stocks that is provided by the state (see Figure 6.8). This is highest in the case of St Lucia at 1.5 per cent, and as low as 0.18 per cent and 0.36 per cent for Grenada and St Vincent and the Grenadines respectively.

Even though the state has been more involved in housing in St Lucia than in the other Windward Islands, a brief review of the twists and turns of government housing policies shows how little has been achieved. In St Lucia, the state first became involved in housing when fire destroyed much of central Castries in 1948. Between 1949 and 1959, blocks containing some 357 apartments were constructed in central Castries. But very swiftly, these suffered from both subletting and persistent rent defaulting. At this juncture, the state almost entirely withdrew from the housing market. It was 12 years before the Housing and Urban Development Corporation (HUDC) was established in 1971. This was involved with two relatively small schemes in Castries. HUDC was charged with acting on the basis of cost recovery and in the end the units produced were priced at

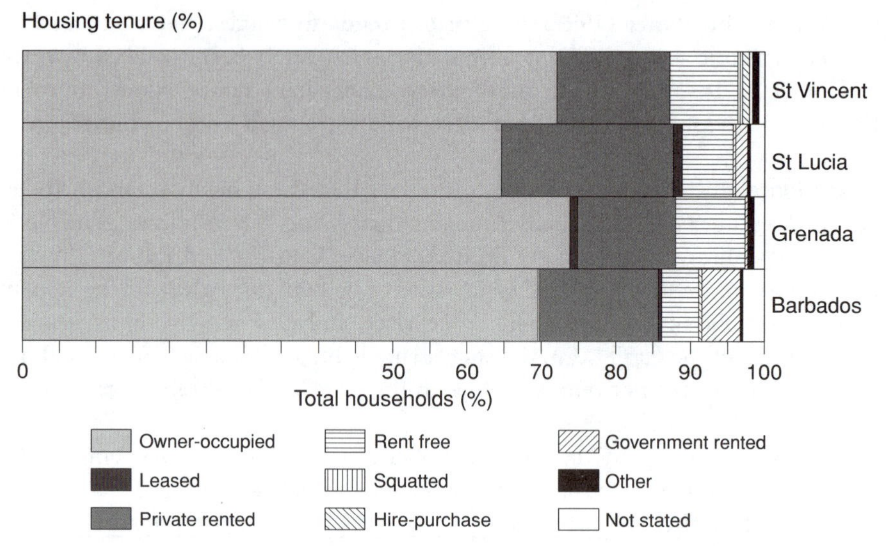

Figure 6.8 Housing tenure divide for St Vincent, St Lucia, Grenada and Barbados
Source: Potter (1995)

levels that were only affordable to middle-income earners. In terms of new building starts, HUDC was then effectively dormant from the 1980s.

In contrast to this manifestly poor performance with respect to new houses, 1980 saw the establishment of the Housing Rehabilitation Project (HRP) in St Lucia. This followed the destruction caused by Hurricane Allen in August of that year. The scheme was located within HUDC, and its aim was to provide direct assistance to the 1,940 families whose homes were devastated by the hurricane. Interestingly, many of the units provided were factory prefabricated. A local housing expert, Louis (1986) argued that with sound management and financial control, the prefabrication of homes for low-income earners could well be a feasible venture as a means of providing new dwellings for the poor. In this manner, the virtues of the Caribbean appropriate and vernacular housing system could be harnessed.

Thus, the case of St Lucia shows how easy it is for governments to get involved in the provision of dwellings at too high a specification. Having done so, it is then tempting for the state to withdraw altogether from the field of housing. The St Lucian case also illustrates how from that point onwards, the state only became involved in housing in response to specific natural disasters or the need to secure land for fresh commercial purposes. Schemes that might serve to link the indigenous housing supply systems with state involvement appear all too frequently to be neglected or ignored within the Caribbean region.

A further example of some of these trends is provided by Duany's (1997) analysis of the provision of housing in Puerto Rico between 1920 and 1950, a period associated with uncontrolled urbanisation and the rapid growth of squatter settlements in the capital city. The growth of informal housing was regarded as giving rise to problems in the form of 'social cancers'. Thus, from 1957 to 1966 some 33,500 public housing units were built, and 26,000 shanty town families were relocated throughout the island, but particularly in San Juan. But as Duany (1997) is at pains to emphasise, this programme was far from successful. The author suggests that it was associated with the disruption of those vital kin networks and mutual assistance and coping strategies that allowed people to get by day to day. In addition to this wholesale erosion of social capital, the design of the homes was disliked by many residents, and they also questioned government restrictions that prohibited the installation of relatives, renters and livestock in their apartments. Duany maintains that lower-class norms clashed with government regulations, in particular the desire to own a plot of land, however small. The increasing incidence of vandalism, theft and the erosion of male authority were also attributed to the development of such rented as opposed to owned properties (p. 204). Duany concludes:

> Planners rarely took into account the need to recreate a neighbourhood's social institutions . . . Many people resisted the move to public housing because they preferred to have their own home, no matter how humble. . . . Public housing projects tended to deteriorate over time, as opposed to shanty towns, which tended to improve (Duany 1997: 204).

This set of conclusions could be applied to the outcome of all manner of public housing schemes in the Caribbean.

During the neo-liberal era, as in so many sectors, the state has effectively withdrawn from housing altogether. As the World Bank and International Monetary Fund imposed structural adjustment packages and cost-recovery imperatives have taken hold, so at best, the state has been involved in the stimulation of the private sector in the form of the promotion of public–private partnerships (see Chapter 5). The neo-liberal global agenda has served to reinforce the state's avoidance of any responsibility to provide affordable accommodation for the poor and homeless. Where low-income housing has been provided by government agencies, as shown in the cases of Puerto Rico and St Lucia, in all too many countries this has led to the construction of a very limited number of over-specified units, frequently at a cost well beyond the means of the poor majority making up the population. And it is ironic in the extreme that this has occurred in a region where folk and architectural forms have always exhibited the capacity of modular, expandable and upgradable self-built and self-improved houses. In the words of Conway and Potter (1997a: 3) 'the poor, largely excluded from entering the legal (and fragmented) housing and land markets, have responded to their plight with self-help strategies: they have accessed the limited but growing rental markets and adapted vernacular forms of shelter to their new transitional realities, whether urban or rural'. And the folk and vernacular housing systems address the environmental realities and challenges of this tropical region, in ways that the introduced, invariably 'modern' temperate latitude designs do not.

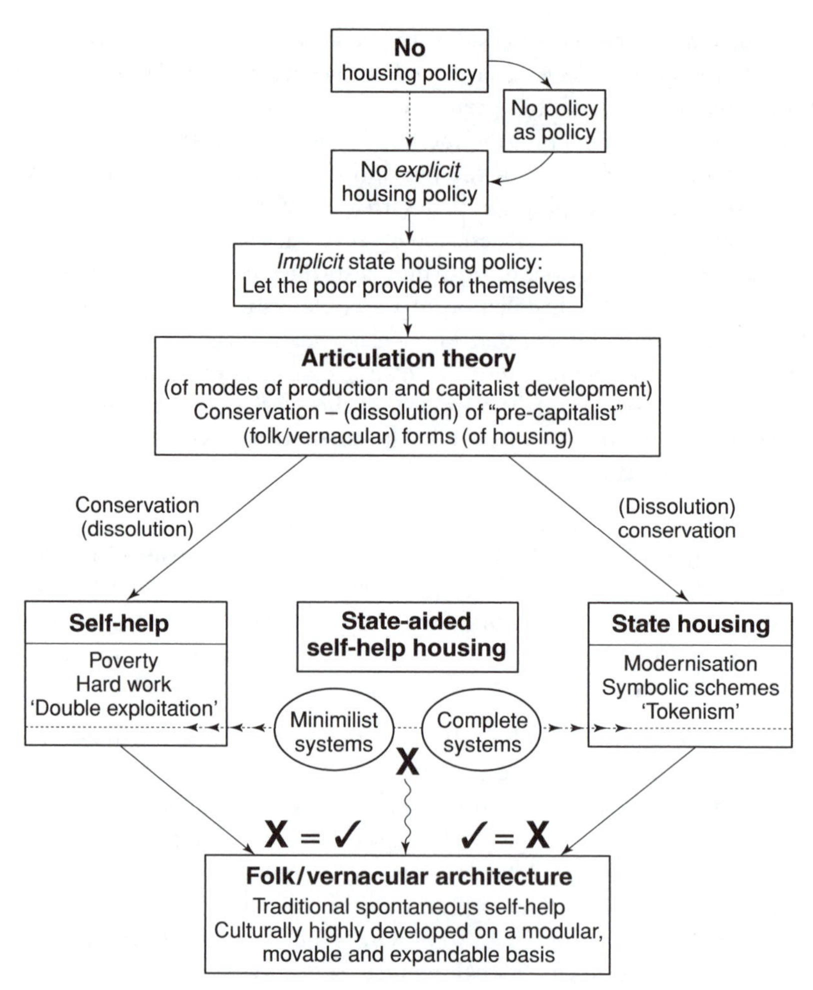

Figure 6.9 Housing policy and the state in the eastern Caribbean
Source: Potter (1995)

In fact, in the neo-liberal era, the state has in several notable instances started to turn against the self-help system where this seems to be impeding the free operation of the land and commercial markets. The clearance of the Conway squatter community in Castries, St Lucia, affords a good example of this (see Box 6.1). Similarly, Eyre (1997) stresses the long tradition of self-help housing in Jamaica and the fact that this has generally been supported by both politicians and the police. However, Eyre details a turnaround that occurred in 1994, with the state evicting squatters on a site near Montego Bay. Squatters were branded as antisocial and even immoral, with Eyre attributing this to the desire to re-incorporate commercially valuable tracts into the land market. Klak (1997) shows how in the neo-liberal era in Jamaica, the state has avoided helping the popular housing sector in any way, shape or form. Instead, it has concentrated on stimulating the provision of mortgages via the private sector.

In concluding the analysis of housing conditions and housing policies in the eastern Caribbean, Potter (1995) presented a model framework. This related to what is known as 'articulation theory', as shown in Figure 6.9. The failure to produce explicit housing policies can be seen as an implicit state policy of allowing the poor to fend for themselves. Thus, no policy has to be seen as a clear housing policy, and this is elaborated in the upper part of Figure 6.9. This can be linked to ideas of articulation theory, concerning the deployment of modes of production under capitalism. The basic argument is that capitalism as a system tends to 'conserve' (that is retain) traditional pre-capitalist forms that work in its favour, but 'dissolve' (dispense with) those which do not, or where a profit can be made. Self-help housing can be seen as keeping wage demands low and therefore reducing the cost of labour. Thus, the stance of governments in the Caribbean can be interpreted as one of seeking primarily to 'conserve' the pre-capitalist or informal housing system in its most basic of forms. This is shown on the bottom left of the diagram. On the bottom right, the state's brief flirtations with modern, complete systems is graphically represented. But such schemes seem doomed to failure and lead to the ultimate conservation of the local form, when the state withdraws almost entirely from the housing scene (Potter 1995).

In contrast to this rather negative perspective, writers on the regional housing scene including Hudson (1997), Potter (1992, 1995), and Conway and Potter (1997a) have argued that notwithstanding the neo-liberal era, there remains an urgent need for the state to be involved in the housing arena, even though it should not be involved in providing houses directly itself. Thus, this involvement must be at an appropriate level. State planning machineries should encompass principles of self-help and modularity. They should involve the provision of self-help housing manuals and schemes, along with retrofitting to meet local weather conditions and the operation of forms of advocacy planning. At a higher-order level, the state needs to facilitate the orderly operation of the informal sector. This may be by encouraging schemes of upgrading and improvement to indigenous housing schemes.

There are regional examples that show what can be achieved when the state apparatus acts in ways that facilitate the local self-build imperatives of the people. In this vein, the Barbadian state actively supported the upgrading of the historic plantation tenantry system, as detailed in Box 6.2. This is a good illustration of what can be achieved if the vernacular system is harnessed. But the box also shows how later schemes outside the plantation system effectively ignored the success of the tenantries scheme. The effective operation of such programmes often necessitates the state getting involved in the establishment of systems of land banking and land pooling – the purchase and holding of tracts of land that are subsequently made available to low-income self-builders and improvers. This approach subscribes to the important argument that the so-called 'housing problem' is all too often a 'land problem'. By such means, the state can assist people

Box 6.2: Housing in Barbados with particular reference to the tenantries programme

A National Housing Board was established in Barbados in 1939, two years after the appointment of the West Indian Royal Commission to examine the causes of social unrest in the Caribbean region. After the passage of Hurricane Janet in 1955, the board was transformed into the Barbados National Housing Corporation (NHC). The long-term goal of the NHC was to improve the housing stock, but the corporation fell short of its targets from the start (Watson and Potter 1997). Redman (1976) argued that at this stage finance that was badly needed for housing purposes was redirected to the promotion of tourism by the government of Barbados. In contrast, housing was seen as non-productive. By the mid-1970s, the NHC had provided a total of 4,282 housing units, 3,316 of which were rental units, accounting for just 4.94 per cent of the national housing stock. The row of houses shown in Plate 6.8 are typical examples of the type built by the NHC during this period.

A very interesting feature of the housing history of Barbados is that in 1963, the doyen of self-help housing, Charles Abrams, prepared a consultancy report on the national housing situation at the request of the Barbadian government

Plate 6.8 NHC row housing in part of Bridgetown, Barbados
Source: Rob Potter

(Box continued)

(Abrams 1963). Abrams was basically quite critical of government policy and urged the establishment of a mortgage finance system and the use of core housing systems to capitalise directly on the indigenous self-help system. However, the evidence suggests that the government of Barbados rejected Abram's advice wholesale. As noted above, throughout the 1970s, state housing took the form of finished concrete units for sale or rental.

But an exception to this marked tendency of stressing the modern came about in 1980 when the then Prime Minister, Tom Adams, announced the Tenantries Freehold Purchase Act and the Tenantries Development Act (see Potter 1986, 1992; Watson and Potter 1993, 1997, 2001; Potter and Watson 1999). Plantation tenantries are the localities where plantation workers built their chattel-style houses, which they could move if fired from their jobs. The first act created the right for individuals who had lived on the tenantries for five years or more to purchase the freeholds of their lots, at $1 per square metre. The second act provided the framework for the parallel physical upgrading of the infrastructure of these areas.

The number of lots conveyed under the act ran at between 220 and 380 per annum during the 1980s. A survey of 150 plantation tenantries carried out in the mid-1990s showed that significant levels of upgrading and modernisation had occurred since the 1980s (see Watson and Potter 2001). Nearly 20 per cent of the homes surveyed had undergone the process of wood to wall conversion over the twelve-year period surveyed. Nearly 30 per cent of houses were constructed of permanent walling materials. Statistical evidence showed that the overall quality of residential construction varied quite markedly from tenantry to tenantry in a patchy manner. But, overall, upgrading activity seemed to be higher in the north than in the south, where housing conditions are generally better. Such a patterning tends to suggest that under the legislation, the northern tenantries are acting as real vehicles of housing improvement (see Figure 6.10).

Some have argued that the tenantries programme represented a 'political sop', suggesting that although it has been labelled as a social revolution, there has been little direct investment by the state. The low level of infrastructural improvement that has occurred, other than the provision of slippered roads, is cited in this connection. At the same time, some commentators point to the fact that the tenantries programme has brought a new section of the poor into the land taxation system (see Watson and Potter 2001: 214–15).

There is another linked argument of importance in this connection. Watson and Potter (2001), in the only ever evaluation of the tenantries programme, express surprise that the same philosophy has not been harnessed as part of wider Barbadian housing policy outside the rural tenantries. In housing plans unveiled as part of the 1986 National Housing Plan, proposals were made for the provision of wet cores, kitchens and toilet units, alongside which pre-existing wooden chattel houses could be placed (Figure 6.11(a)). The provision of the same type of wet core, but connected with concrete 'starter homes', was also illustrated (see the connected starter house in Figure 6.11(b)). But in the final analysis the option of incorporating chattel houses into the equation was never taken up in earnest.

(Box continued)

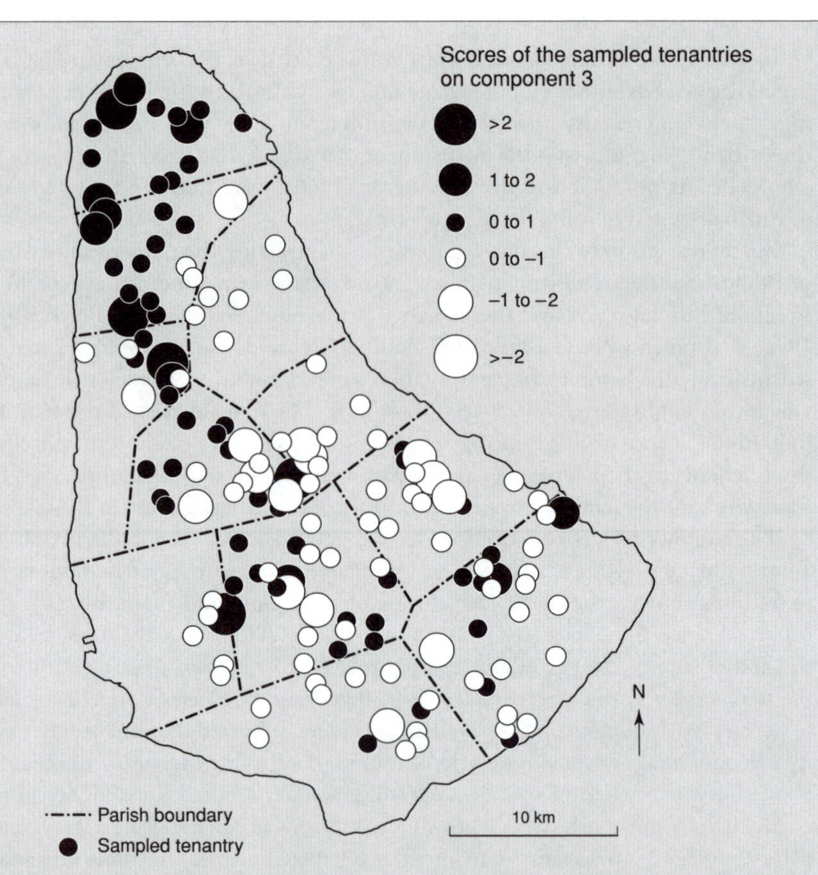

Figure 6.10 Overall level of upgrading on the plantation tenantries of Barbados
Source: Watson and Potter (2001)

Rather, the NHC built two prototype starter houses (see Plate 6.9). Both were built entirely of concrete, and cost Bds$35,000 and Bds$42,000 to purchase, way beyond the means of the poor. At the same time, the government made it clear that it was strongly advocating private sector involvement in the housing market.

As Watson and Potter (2001) were writing the conclusions to their assessment of the overall success of the rural plantation tenantries scheme and calling for the wider imposition of its generic principles, the government of Barbados under Owen Arthur talked about applying the principles to the urban situation (see Watson and Potter 2001). The proposed extension of the upgrading scheme to the smaller urban tenantries involved some very progressive social elements, including the provision of rent-free houses for selected needy single pensioners, and the recognition of the specific needs of those on the lowest incomes, the disabled and battered women. As yet, the wider scheme of funding that would launch such a progressive social platform does not seem to have been obtained.

(Box continued)

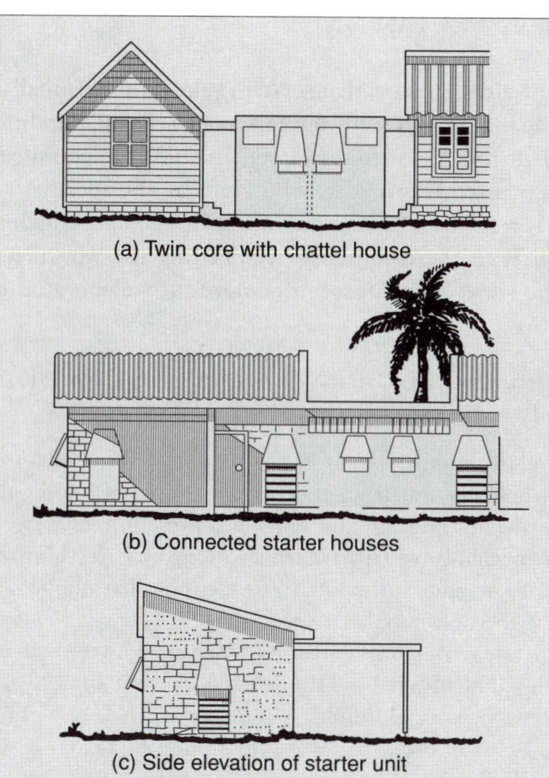

Figure 6.11 Proposed twin core and starter housing development
Source: Potter (1992)

Plate 6.9 NHC prototype starter house in Bridgetown, Barbados
Source: Rob Potter

to provide for themselves, in ways that serve to valorise and capitalise on the appropriate nature of the indigenous self-help folk and vernacular architectural systems that have existed in the region for so long. In this connection, the encouragement of workplace-based teams of self-builders by the socialist state of Cuba is often cited. This is referred to as the microbrigade scheme, which sees the linking of the self-builder with the community, enabled by the workplace and officially encouraged by the state. The approach is more fully elaborated in Box 6.3.

Box 6.3: Housing in Cuba: the contribution of the microbrigade system

Before the arrival of the Spanish, most ordinary houses in Cuba were self-built. The traditional hut was constructed of organic materials. After the revolution in 1959, state-provided housing and self-build programmes were the two main housing supply mechanisms. But a third and very innovative mixed alternative currently needs to be added: the microbrigade, one that might be of relevance in the context of the non-socialist world.

The microbrigade movement represents a mixture between self-help and state-provided housing. The idea behind the scheme was first aired by Castro in 1970 and the basic concept is very simple. A number of workers within a factory or office can be released in order to form into a building team or 'microbrigade'. The colleagues who remain in the workplace guarantee to maintain the previous level of productivity. The dwellings constructed are then distributed among the workers at the base workplace, according to need.

By 1978 more than 1,100 microbrigade teams were in operation, incorporating a total of 30,000 workers. They had completed 82,000 dwellings, and were strongest in the capital Havana (Segre 1984; Mathey 1997). The product was often four- and five-storey walk-up apartment blocks, similar to those shown in the foreground in Plate 6.10. By such means, Mathey notes that entire satellite townships were created around Havana.

In the late 1970s there were criticisms of the quality of the units produced by the microbrigade system, and also concerning the selective nature of the process of distribution of the units produced. As a result, the system went into eclipse. But in June 1986, Fidel Castro proposed a revitalisation of the principle and sang the virtues of the earlier scheme. It was argued that the system worked particularly well where factories were experiencing intermittent work stoppages. It was also proposed that microbrigades should build for general needs as well as those of the workplace-based building team. With four months of Castro's pronouncement, 75 new microbrigades had been formed, composed of 2,400 workers. By November 1988, 38,000 were employed in as many as 10,000 schemes. Some 60 per cent of the dwellings created were allocated for members and 40 per cent were for distribution within the community at large. From late 1987, a new twist occurred. Microbrigades were also formulated

(Box continued)

Plate 6.10 A variety of high-rise apartment blocks on the periphery of Havana
Source: Rob Potter

for the purpose of neighbourhood regeneration and improvement, and these schemes were referred to as 'social microbrigades'. Mathey (1997) concludes by arguing that self-help can work in the socialist state. Furthermore, it is argued that the concept of the social microbrigade could well be employed in the context of non-socialist states.

Middle- and upper-class housing in the Caribbean

As noted at the beginning of this account on housing, via the views of Hudson (1997), just as with social structure generally, the residential mosaic of the Caribbean consists of the everyday homes of the overwhelming majority on the one hand, and the imposing, grand and permanent houses of the wealthy few on the other. Such inequalities, however, are not merely the outcome of the past polarity between the plantocracy and a broad lumpenproletariat. It is cogently witnessed today in the occurrence of middle- and upper-income housing in Caribbean urban areas – including affluent-looking town houses or terraced units in Kingston, Jamaica, replete with Volvos in the parking spaces outside, as portrayed in Plate 6.11. It is also signified by the growth of 'terraces', 'heights' and 'parks', often perched on one of the uplifted marine terraces, in the context of the increasing middle-class character of Barbadian society (Watson and Potter 1997).

Plate 6.11 Town houses or terraced housing units in Kingston, Jamaica
Source: Rob Potter

For example, housing conditions as a whole have shown a marked improvement in Barbados over the past 25 years (Plate 6.12). A major change has been the progressive reduction in the proportion of all-wood chattel-style houses. This had fallen to 39.89 per cent of all dwellings in 1990 from 75.25 per cent in 1970. On the other hand, the proportion of houses constructed of concrete increased from 9.64 per cent in 1970 to 35.47 per cent in 1990. Just as saliently, houses constructed of wood plus concrete accounted for 21.3 per cent of the total in 1990, having increased from 4.41 per cent in 1970. The provision of electric lighting increased from 83.02 in 1980 to 92.58 per cent of all dwellings in 1990.

Broadening the argument somewhat, Conway and Potter (1997a: 4) note that 'competing for scarce resources and housing and land are the middle and upper classes of these transitional, peripheral capitalist societies'. The same authors go on to note that the widespread occurrence of uncontrolled or mismanaged land subdivisions and low-density residential developments is one clear expression of this situation. Such developments are frequently associated with middle-class housing schemes, normally of a private orientation, but sometimes publicly sponsored.

A good example of this is provided by the subdivision of plots of land for private sector housing development in Barbados (see Nurse 1983; Potter 1997b; Potter and Watson 1999) (see Plate 6.13). On the one hand, some commentators

Plate 6.12 A 'modern' house in Bridgetown, constructed in the early 1980s
Source: Rob Potter

Plate 6.13 Vacant land subdivisions in the parish of Christ Church, Barbados
Source: Rob Potter

have bemoaned the conversion of rural to urban land, which is contingent upon the 'one family, one home' mentality that is engenderd by the indigenous Barbadian housing system. On the other, there are very extensive tracts of land that have been subdivided for housing construction, but which have not yet been built upon.

The Town and Country Planning Office in its 1988 Physical Development Plan noted that between 1965 and 1977 new private residential subdivisions with more than ten lots created a total of 16,196 lots. The same survey showed that by 1977, only 2,590 (or 14 per cent) of these had been developed. Thus, 13,606 lots or some 86 per cent had not been developed. The data also showed that 56 per cent of all the lots under construction were sold during the period under study, and that this level of land sales from the creation of subdivisions was enough to make the operation financially feasible for the developers concerned. This gives some idea as to how far the land and housing markets are dominated by the needs of the relatively wealthy few.

Estimates made in 1993 by the Town and Country Development Planning Office, using land tax valuation records, suggest that the number of vacant and undeveloped lots on private subdivisions in Barbados is as high as 17,702. In order to give some idea of the magnitude of this phenomenon, the distribution of such vacant residential subdivisions by parish area and by lot size is given in Table 6.2. Geographically, a clear pattern emerges, with the greatest number of vacant subdivisions occurring in the urban and suburban parishes of Christ Church (5,805), St Philip (4,691), St Michael (2,838) and St James (1,625), as shown in the table. This spatial patterning indicates the close association of the process with middle- and upper-income subdivisions. In direct contrast, there were only six vacant lots in predominantly rural St Andrew, 75 in St Joseph and 98 in St John. With regard to the size of the lots, the feature that stands out is how large they are. The data show that 6,359 of the 17,702 vacant lots, by far and away the largest number, were in the top 10,000 square feet and above category (Table 6.2).

Further exemplifying the magnitude of the issue posed by vacant lots, a survey conducted by the Division of Housing in 1995 showed that there were some 1,174 vacant lots in Greater Bridgetown, comprising a total area of 300 acres. It was noted that these unbuilt lots reflected land speculation, and absentee and deceased land owners. The survey also identified just over 400 abandoned or ruined structures covering about 100 acres of land. The suggestion can be made therefore that there may be as many as 20,000 idle land subdivisions within the existing residential fabric of Barbados. In a background consultancy report prepared by Potter (1997b) in connection with the Third Physical Development Plan, it was noted that this unused land bank was sufficient to provide land to meet the housing needs of the nation for the next eight to ten years. Residential densities have been very low, following an essentially middle-class and suburban model of detached and extensive house construction throughout the recent period. For example, a survey of 326 residential subdivisions of ten or more lots approved between 1965 and 1977 indicted that 52 per cent of all the lots were larger than

Table 6.2 Vacant subdivisions in Barbados by area and plot size

Parish	Number of lots	Number of vacant lots	Average lot size (square feet)				
			Under 4,000	4,000–5,999	6,000–7,999	8,000–9,999	Over 10,000
St Michael	17,328	2,838	705	651	372	262	848
St James	4,429	1,625	82	198	329	330	686
St George	644	732	46	82	137	121	346
St Thomas	1,030	667	22	145	127	118	255
St Joseph	166	75	8	13	8	3	43
St John	81	98	21	15	10	19	33
St Peter	1,213	781	79	66	100	152	384
St Lucy	614	384	40	48	160	36	100
St Andrew	10	6	–	–	–	–	6
St Philip	3,838	4,691	139	1,208	908	561	1,875
Christ Church	9,815	5,805	391	1,400	1,367	864	1,783
TOTAL	39,168	17,702	1,533	3,826	3,518	2,466	6,359

Source: from Town and Country Development Planning Office, based on data from the Land Tax Department (Valuation Section), Land Tax Roll 1992–1993

800 square meters. While some of these parcels of land have been purchased by Barbadians living abroad, it is clear that the majority are owned by resident Barbadians who have purchased such plots for speculative reasons, as well as to provide for their children in the future.

Conway and Potter (1997a, 1997b) also contend that some of the land market distortions discussed serve to encourage property speculation and overseas investment. They argue that the state's role is again challenged and all too often found wanting. The openness of Caribbean economies, along with poorly managed tourism entrepreneurship, remittance investments and externally controlled flows of capital, are all seen as contributing to the essentially anarchic state of housing and land markets in the Caribbean region. They conclude that it is not so much a case of mismanagement by the state as the state's powerlessness and/or functional limitations within these markets.

A case illustrating some of the associations between elite residential areas, tourism and the provision of sports, recreational and other activities is provided by the high-status residential zones of St Lucia. Both of the premier residential districts of Castries, the Cap Estate right at the north of the island, and the Rodney Heights area, located midway between central Castries and Cap Estate, are closely linked with recreational, tourist and expatriate activities. Thus, the premier Cap Estate area is also home to the St Lucia Racquet Club and Golf Course, plus the Club St Lucia resort area and the Great House restaurant. The Rodney Heights district is located to the east of the main road between Castries and Cap Estate. The new detached houses clustered on the hillside directly overlook the Rodney Bay Marina (Plate 6.14) and Reduit Beach tourist zone with its restaurants, hotels, bars and clubs.

Concluding comments

In the Caribbean region, in common with other regions of the 'global South', the struggle between the relatively poor and powerless and the socially and economically more powerful middle and upper classes is a highly unequal one (Conway and Potter 1997a). The body politic has almost inevitably acceded to the needs and desires of the more powerful and wealthy, and frequently this set of imperatives has been turned to in the name of so-called 'modernisation' and 'development'. This particular trend is well exemplified in the case of housing in the Caribbean region. Nevertheless, the political–economic realities of the nation state should not be so retrogressive. The acknowledged energy of autonomous, community-based efforts by the Caribbean poor to find their own means of access to adequate shelter suggest that the situation is far from hopeless. As this chapter has shown, self-help strategies have frequently been employed by the urban and rural lower classes in the region. The burgeoning shanty towns, a reliance on urban squatting as an alternative to legal residential processes, reliance on information networks to finance accommodation alternatives, dependence upon informal activities in the

Plate 6.14 New middle-income housing overlooking Rodney Bay Marina, St Lucia
Source: Rob Potter

construction sector, and varied land access strategies developed in response to different forms of land tenure all characterise the Caribbean region. In every sense of the phrase, the efforts of the poor can be seen as classic examples of community empowerment. But it has to be recognised that this largely represents a form of empowerment necessitated by pressing external circumstances, and in this sense largely reflects the relative 'unfreedoms' (Sen 2000; Potter *et al.* 2004) that remain to be faced by the region's poor and relatively powerless.

References

Abrams, C. (1963) *Housing in the Modern World: Man's Struggle for Shelter in an Urbanizing World*, Faber and Faber, London.

Berthelot, J. and Gaume, M. (1982) *Kaz Antiye (Caribbean Popular Dwelling)*, Editions Perspectives Creoles, Pointe-a-Pitre, Guadeloupe.

Condon, S. A. and Ogden, P. E. (1997) 'Housing and the state in the French Caribbean', ch. 11 in Potter, R. B. and Conway, D. (eds) *Self-Help Housing, the Poor, and the State in the Caribbean*, University of Tennessee Press, Knoxville, and University of the West Indies Press, Jamaica, Barbados and Trinidad and Tobago, pp. 217–42.

Conway, D. and Potter, R. B. (1997a) 'Caribbean housing, the state, and self-help; an overview', ch. 1 in Potter, R. B. and Conway, D. (eds) *Self-Help Housing, the Poor, and the State in the Caribbean*, University of Tennessee Press, Knoxville, and

University of the West Indies Press, Jamaica, Barbados and Trinidad and Tobago, pp. 1–13.

Conway, D. and Potter, R. B. (1997b): 'Caribbean housing futures: building communities for sustainability, ch. 12 in Potter, R. B. and Conway, D. (eds) *Self-Help Housing, the Poor, and the State in the Caribbean*, University of Tennessee Press, Knoxville, and University of the West Indies Press, Jamaica, Barbados and Trinidad and Tobago, pp. 242–59.

Duany, J. (1997) 'From the *Bohio* to the *Caserio*: urban housing conditions in Puerto Rico', ch. 10 in Potter, R. B. and Conway, D. (eds) *Self-Help Housing, the Poor, and the State in the Caribbean*, University of Tennessee Press, Knoxville, and University of the West Indies Press, Jamaica, Barbados and Trinidad and Tobago, pp. 188–216.

Edwards, J. D. (1980) 'The evolution of vernacular architecture in the western Caribbean', in Wilkersenn, S. J. K. (ed.) *Cultural Traditions and Caribbean Identity: the Question of Patrimony*, Center for Latin American Studies, University of Florida, Gainsville, pp. 291–339.

Eyre, L. A. (1972) 'The shantytowns of Montego Bay, Jamaica', *Geographical Review*, **62**, 394–412.

Eyre, L. A. (1997) 'Self-help housing in Jamaica', ch. 5 in Potter, R. B. and Conway, D. (eds) *Self-Help Housing, the Poor, and the State in the Caribbean*, University of Tennessee Press, Knoxville, and University of the West Indies Press, Jamaica, Barbados and Trinidad and Tobago, pp. 75–101.

Hudson, B. J. (1997) 'Houses in the Caribbean: homes and heritage', ch. 2 in Potter, R. B. and Conway, D. (eds) *Self-Help Housing, the Poor, and the State in the Caribbean*, University of Tennessee Press, Knoxville, and University of the West Indies Press, Jamaica, Barbados and Trinidad and Tobago, pp. 14–29.

Klak, T. (1997) 'Obstacles to low-income housing assistance in the capitalist periphery: the case of Jamaica', ch. 6 in Potter, R. B. and Conway, D. (eds) *Self-Help Housing, the Poor, and the State in the Caribbean*, University of Tennessee Press, Knoxville, and University of the West Indies Press, Jamaica, Barbados and Trinidad and Tobago, pp. 102–119.

Lloyd-Evans, S. and Potter, R. B. (1992) 'The informal sector of the economy in the Commonwealth Caribbean: an overview', *Bulletin of Eastern Caribbean Affairs*, **17**, 26–40.

Louis, E. L. (1986) 'A critical analysis of low-income housing in St Lucia', unpublished MSc dissertation, St Augustine campus, University of the West Indies.

Mathey, K. (1997) 'Self-help housing in Cuba: an alternative to conventional wisdom?' ch. 9 in Potter, R. B. and Conway, D. (eds) *Self-Help Housing, the Poor, and the State in the Caribbean*, University of Tennessee Press, Knoxville, and University of the West Indies Press, Jamaica, Barbados and Trinidad and Tobago, pp. 164–87.

Nurse, L. (1983) 'Residential Subdivisions of Barbados', Occasional Paper number 14, Cave Hill, Institute of Social and Economic Research, University of the West Indies.

Phillip, M. P. (1988) 'Urban low income housing in St Lucia: an analysis of the formal and informal sectors', unpublished MPhil thesis, University of London.

Phillip, M. P. (1999) 'Low-income housing, the environment and the state: the case of St Lucia', unpublished PhD thesis, University of London.

Potter, R. B. (1986) 'Housing upgrading in Barbados: the tenantries programme', *Geography*, **71**, 255–7.

Potter, R. B. (1989) 'Urban housing in Barbados, West Indies', *Geographical Journal*, 155, 81–93.

Potter, R. B. (1991) 'An analysis of housing in Grenada, St Lucia and St Vincent and the Grenadines', *Caribbean Geography*, **3**, 106–125.

Potter, R. B. (1992) *Housing Conditions in Barbados: A Geographical Analysis*, Institute of Social and Economic Research, University of the West Indies, Mona, Jamaica.

Potter, R. B. (1993) 'The neglect of Caribbean vernacular architecture', *Bahamas Journal of Science*, **1**, 46–51.

Potter, R. B. (1995) *Low-Income Housing and State Policy in the Eastern Caribbean*, University of the West Indies Press, Barbados, Jamaica, Trinidad and Tobago.

Potter, R. B. (1997a) 'Housing and the state in the eastern Caribbean', ch. 4 in Potter, R. B. and Conway, D. (eds) *Self-Help Housing, the Poor, and the State in the Caribbean*, University of Tennessee Press, Knoxville, and University of the West Indies Press, Jamaica, Barbados and Trinidad and Tobago, pp. 527–74.

Potter, R. B. (1997b) 'Housing conditions background report', Environmental Management and Land Use Planning for Sustainable Development, Ministry of Health and Environment, Government of Barbados, Bridgetown.

Potter, R. B. (2001) 'Urban Castries, St Lucia revisited: global forces and local responses', *Geography*, **86**, 329–36.

Potter, R. B. and Conway, D. (eds) (1997) *Self-Help Housing, the Poor, and the State in the Caribbean*, University of Tennessee Press, Knoxville, and University of the West Indies Press, Jamaica, Barbados and Trindad and Tobago.

Potter, R. B. and Lloyd-Evans, S. (1998) *The City in the Developing World*, Pearson/Prentice Hall, London and New York.

Potter, R. B. and O'Flaherty, P. (1995) 'An analysis of housing conditions in Trinidad and Tobago', *Social and Economic Studies*, **44**, 165–83.

Potter, R. B. and Watson, M. R. (1999) 'Current housing policy issues in Barbados: with particular reference to vacant subdivisions', *Third World Planning Review*, **21**, 237–60.

Potter, R. B., Binns, J. A., Elliott, J. A. and Smith, D. (2004) *Geographies of Development*, 2nd edition, Pearson/Prentice Hall, London and New York.

Redman, L. (1976) 'Proposals for rural-scale housing in Barbados', paper presented at the 27th International Conference on Housing, Planning and Building, Rotterdam.

Seagre, R. (1984) 'Architecture in the revolution', in Hatch, C. R. (ed.) *The Scope of Social Architecture*, Van Nostrand, New York, pp. 348–60.

Sen, A. (2000) *Development as Freedom: Human Capability and Global Need*, Anchor Books, New York.

Turner, J. F. C. (1968) 'The squatter settlement: an architecture that works', *Architectural Design*, **38**, 355–60.

Watson, M. R. and Potter, R. B. (1993) 'Housing and housing policy in Barbados: the relevance of the chattel house', *Third World Planning Review*, **15**, 373–95.

Watson, M. R. and Potter, R. B. (1997) 'Housing conditions, vernacular architecture, and state policy in Barbados', ch. 3 in Potter, R. B. and Conway, D. (eds) *Self-Help Housing, the Poor, and the State in the Caribbean*, University of Tennessee Press, Knoxville, and University of the West Indies Press, Jamaica, Barbados and Trinidad and Tobago, pp. 301–51.

Watson, M. R. and Potter, R. B. (2001) *Low-Cost Housing in Barbados: Evolution or Social Revolution?*, University of the West Indies Press, Barbados, Jamaica and Trinidad and Tobago.

Chapter 7

URBAN DYNAMICS AND TOWNSCAPES

Introduction

There are many aspects that contribute to the overall image of the Caribbean, as attested by the earlier chapters of this book. One strong set of images undoubtedly relates to the pervasive influence of agriculture in association with colonial plantations and slavery. As stressed in Chapter 3, this aspect of the history of the Caribbean gives rise to the strong rural standing of the region. However, the contemporary Caribbean is also strongly associated in the minds of millions of people with steel bands, reggae, ska, soca and dancehell. Although the imperatives for such music have as often been rural as urban, nonetheless, a strong link with urban ghettos and 'downtown' recording studios exists for many people, linked, for example, to Kingston, Jamaica, Port of Spain, Trinidad and Tobago, San Juan, Puerto Rico and Port-au-Prince, Haiti.

The theme of the present chapter on urban conditions and the dynamics of townscapes in the Caribbean is essentially similar to the one put forward with respect to music. Just as the foundations of the Caribbean region lie in agriculture and demanding rural lives, so it is to be recognised that the same forces also gave rise to distinctly urban settlement structures and livelihood patterns at the macro-spatial scale of the nation and the region. Indeed, the importance of rural-to-urban migration flows in the region has already been emphasised in Chapter 2. The present chapter initially fleshes out this broad argument in a historical context, and in so doing, presents a number of conceptual models, which are then empirically verified.

Following this evolutionary overview, statistics establishing contemporary levels of urban development in the Caribbean region are presented in detail. In both this and the foregoing account, considerable emphasis is placed on the occurrence of what is referred to as 'urban primacy' and the associated conditions of urban bias in development and the maintenance of spatial polarisation and regional inequalities in even the smallest island territories. Linking directly to Chapters 8 to 12 of the present book, the chapter proceeds to show that contemporary, post-war developments in the fields of industrialisation and the development of tourism associated with the concomitant process of globalisation are further serving to concentrate people, activities and capital in urban regions. The consequences of these macro-structural processes are viewed in terms of the development of new urban structures and forms in the present-day Caribbean. In this account, waterfront redevelopment projects in association with modern and post-modern forms of development and change are highlighted. Some of these issues are also considered in Chapter 11 on tourism.

Before turning to consider these substantive topics, it is necessary to define the basic terms that are needed in order to be able to discuss such issues. The expression 'urban development' has already been employed, and this is normally used to refer to the totality of urban structure and character making up a particular area. The word 'urbanisation' has a very strict and clear meaning in the statistical sense, and is employed by countries all around the world for census purposes. Urbanisation refers to the proportion of the population of a nation or region that is to be found living in those settlements that are designated as urban, and is normally expressed as a percentage figure. Distinct from this, the word 'urban growth' is also frequently used, to refer to either the physical expansion of cities on the ground, or the growth in their overall population. The basic point is that throughout history, urban growth and urbanisation have generally occurred together. But there is no absolute and invariant association between the two processes, for individual cities can continue to grow in both population and area even when the level of urbanisation flattens out or decreases. Finally, when referring to changes in the degree to which given populations experience what may be deemed to be an urban way of life, the word 'urbanism' is customarily employed in the literature (Wirth 1938). Armed with these basic definitions, we now turn to consider the reasons why as a primarily rural-agricultural zone of economic production, the Caribbean, developed a strongly urban-based settlement fabric from the time of its earliest settlement and colonisation.

The historical evolution of settlement systems in the Caribbean

The customary framework employed in order to explain the evolution and character of settlement systems is referred to as central place theory. A German geographer called Walter Christaller derived the first principles of central place theory. His work was empirically tested in the context of southern Germany, and his book, *Central Places in Southern Germany*, first appeared in 1933, based on his doctoral research. The work was translated into English and published in 1966 (listed in references), whereupon it gained international currency.

Put in the simplest terms, Christaller argued that all other things being equal, the economics of marketing will lead to a triangular lattice of 'central' places supplying goods and services to the surrounding population. If factors such as population density, purchasing power and consumer behaviour vary from person to person and place to place, then it was accepted that the outcome would be a less even and more diverse distribution of towns and cities. Christaller also demonstrated that if efficient transport or public administration were more important than maximising the efficiency of marketing, then the outcome would be a somewhat different arrangement of central places. However, in all cases, the overall outcome would be a relatively uniform distribution of settlements over a given territory.

Consideration of the way in which Christaller's central place theory was developed has led to the argument that it is basically a model of the evolution of settlement under the sorts of conditions that pertained in Europe during the feudal period. In other words, it describes what happens in a relatively isolated area, which does not trade either locally or internationally, so that development essentially comes *from within*. But the reality is that since Columbus set sail in 1492, global settlement patterns have developed in direct response to the demands of long-distance trade. Indeed, Christaller was fully aware of this distinction, and in his original work he started by drawing the clear distinction between non-central places (such as transport-based places and special resource-oriented places) on the one hand, and central places on the other.

Empirical observation in many world regions suggests the occurrence of settlement patterns that are far less regular and even than those suggested by pure central place theory (Potter 2000). In the first place, the settlement patterns of whole regions are strongly coastal in orientation. A further observation is that where settlements extend inland, they frequently exhibit very pronounced patterns of linearity. This first observation is particularly true of the Caribbean region, where settlements are frequently aligned along coasts, especially those to the leeward.

In 1970, an American geographer, J. E. Vance (1970), argued that in the seventeenth and eighteenth centuries mercantile entrepreneurs started to look outwards from Europe. This was due to the long history of parochial trade and what he referred to as 'the confining honeycomb of Christaller cells that had grown up with feudalism' (Vance 1970: 48). With overseas colonialism, merchants were for the first time confronted by an unorganised landmass, whereas in Europe the spatial framework for economic activities was already established in the pre-mercantile period.

Vance thereby argued that with the rise of mercantile societies, settlement systems globally started to evolve along far more complex lines than previously. The main development came with colonialism when continued economic growth necessitated greater land resources. During this period, coastal ports came to dominate both colonies and colonial powers alike. In colonies, once established, ports acted as gateways to the interior lands. Subsequent evolutionary change witnessed increasing spatial concentration at certain nodes, and, later on, lateral interconnection of the coastal gateways and the establishment of new inland areas for expansion. At the same time, the settlement patterns of the colonial power underwent considerable change, with the original feudal-type central place system experiencing considerable polarisation with respect to the principal ports.

These historical facets of trade articulation led Vance to put forward in graphical terms what he regarded as a corrective to the historical *naïveté* of the central place model. Vance's framework, referred to as the mercantile model, is firmly based on the history of exploration, colonialism and mercantilism (Vance 1970; Potter and Lloyd-Evans 1998; Potter 2000; Potter *et al.* 2004). Before looking at the Vance framework in some detail, however, it should be recalled that Christaller specifically excluded the consideration of transport- and special resource-based places from his basic models, although he recognised their importance in reality. Christaller's perspective was very much that of the conventional economist employing the partial equilibrium approach, albeit in a spatial context.

A simplified version of Vance's mercantile model is shown as Figure 7.1. The graphical model itself is divided into five stages, each of which depicts a key stage in the evolution of globalised settlements, in both the colony and the colonial power. The first stage represents the initial search phase of mercantilism, which basically involved the search for economic information on the part of the prospective colonial power. The second stage saw the testing of productivity and the harvest of the pre-existing natural storage, with the periodic harvesting of staples such as fish, furs and timber. At this stage, no permanent settlement was established in the colony. The planting of settlers who produce staple agricultural items and consume the manufactures of the home country represents a third distinct phase in the mercantile model. At this juncture, the settlement system of the colony is established, via a point of attachment. The development of a symbiotic relationship between the colony and the colonial power is witnessed by a sharp reduction in the effective distance separating them. At this stage, the major port in the homeland becomes pre-eminent. The fourth stage is characterised by the introduction of internal trade and manufacturing in the colony. At this point, inland penetration occurs from the major gateways in the colony, based on staple production. There is rapid growth of manufacturing in the homeland to supply both the overseas and home markets. Ports become even more significant. The fifth and final stage sees the establishment of a mercantile settlement pattern with central place infilling occurring in the colony and the emergence of a central place-type system with a mercantile overlay in the homeland.

As is clearly attested by Figure 7.1, the hallmark of the mercantile model is the remarkable linearity of the basic settlement pattern, first along coasts, especially in colonies, and secondly, along the routes that develop between the coastal point of attachment and the staple-producing interiors. These two alignments are given direct expression in several other locational models, including Taaffe *et al.*'s (1963) conceptualisation of transport expansion in less developed countries, based on the histories of Brazil, Malaya, East Africa, Nigeria and Ghana. It is more than evident, however, that such a framework holds very high explanatory power in the context of the Caribbean.

In the Caribbean region, however, what may be regarded as a local historical variant of the mercantile settlement system can be recognised. This framework,

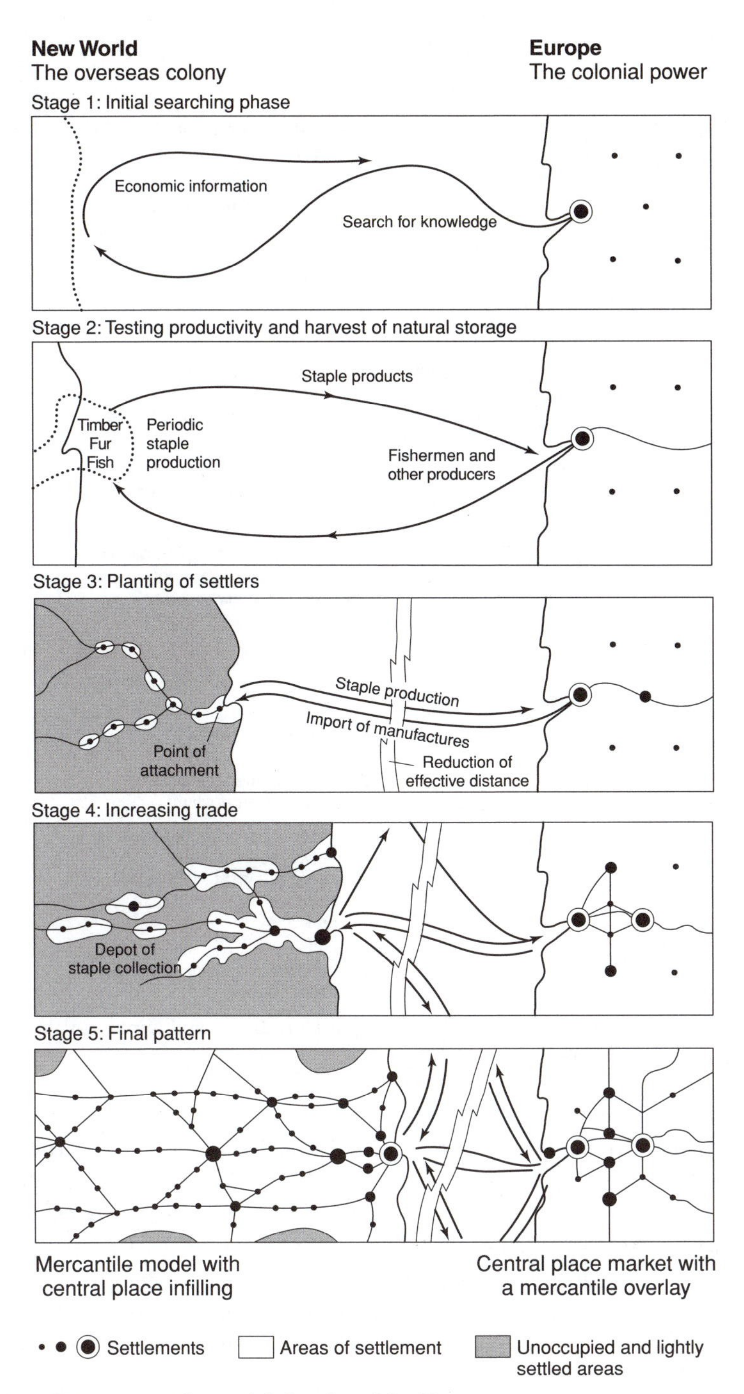

Figure 7.1 The mercantile model developed by Vance
Source: Adapted from Vance (1970)

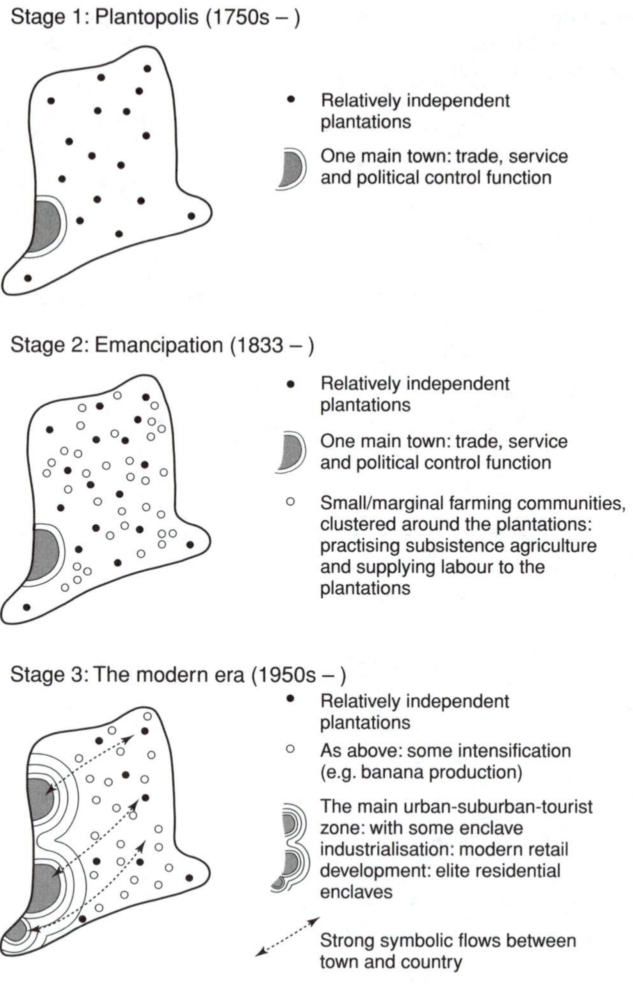

Figure 7.2 The three-stage plantopolis model of Caribbean settlement development
Source: Potter (1995, 2000)

which is produced here as Figure 7.2, specifically acknowledges the foundations of the Caribbean space economy in the globally-oriented plantation system, and maps this into the evolution of the settlement system. This framework is referred to as the plantopolis model. The first two stages of the plantopolis model shown in Figure 7.2 are based on the account provided by Rojas (1989), who originated the framework. On the other hand, Potter (1995, 1997) effected the graphical depiction of the sequence and its extension into the so-called 'modern era'.

In the first stage of the framework, plantopolis (1750 onwards), it is emphasised that the plantations formed self-contained bases for the settlement pattern of the region, such that only one main town performing trade, service and political control functions was required in addition to the basic agricultural points of control. This stage of the model can be adapted to take into account the development of more than one commercial centre, if applied, for example, to the Greater Antilles. Further, the birth of one centre and the subsequent ascendancy of another can also be built into the model in a relatively straightforward manner.

Following Emancipation (1834 onwards), small marginal farming communities, which were clustered around the plantations, and which practised subsistence agriculture and supplied labour to the plantations, added a third layer to the settlement system (see stage 2 in Figure 7.2). Once again it can be appreciated that the distribution of these would, of course, be modified according to agrarian structure and physical geographical features, especially difficult relief, in order to fit variant circumstances.

Potter (1995, 2000) sought to extend the model into the modern era (post-1950), as a third distinct stage in the framework (Figure 7.2). It was suggested that the modern period has seen the *extension* of the highly polarised pattern of development and change. The emphasis is placed on extension, for the changes being outlined may not amount in all cases to spatial intensification *per se*. In other words, the modern era may well have seen the spatial extension of an increasingly complex and differentiated urban region. The outcome has been the development of what may be referred to as mini-metropolitan regions. These are small urban constellations, but ones which nevertheless show all the sharpness of differentiation exhibited by large metropolises such as London and New York. This type of patterning has been referred to in the Caribbean by both Sahr (1998) and Potter (1995, 1998, 2000) and is the focus of further attention in the second half of this chapter. Such urban areas are multi-centred, and multinational capital is busy creating new nodes of activity within them, often in relation to tourism and recreational facilities and functions. Larger urban agglomerations of this type are referred to as 'extended metropolitan regions', and frequently urban and rural activities are intermixed in a complex manner within such areas (see Potter *et al.* 2004, chapter 9, for a summary).

The virtues of the mercantile and plantopolis models are many. Principally, however, they both serve to stress that the development of settlement systems in most developing countries amounts to a form of dependent urbanisation. Certainly, both frameworks serve to remind that the high degree of concentration in the principal urban areas and the strong littoral orientation of settlement fabrics in the Caribbean, as elsewhere in the developing world, is the direct product of colonialism, and not some chance or aberrant case. The case study offered by Guyana is reviewed in Box 7.1.

Thus, a pattern of spatially uneven or polarised growth emerged as characteristic in the Caribbean region several hundred years ago, with the development of this strong symbiotic relationship between the colony and the colonial power. As a consequence, the overarching suggestion is that due to the dictates of the global economy, far greater levels of spatial concentration and inequality exist within the region's settlement fabric than may be socially desirable.

Box 7.1: Patterns of settlement and economic activity in Guyana

Guyana, located on the mainland of South America, recorded a population of 739,553 in 1992. In fact, its population has been in decline, having been recorded at 801,000 at the time of the 1980 Census. Extending over an area of 214,970 km^2, Guyana is similar in size to Britain. The overall population density is low at around 3.73 persons per km^2. However, this statistic is very deceptive, because it hides the degree to which the population is heavily concentrated on the coastal belt. This zone, made up of alluvial mud, varies between 15 and 65 km in width. Three major rivers, the Essequibo, Demerara and Berbice, serve to divide the coastal plain into four distinct parts, many areas of which are below sea level.

At various times in the sixteenth century, fortune hunters from Europe explored the rivers of South America north of the equator. The Dutch began to settle the area in 1616. They established plantations along the banks of such rivers as the Essequibo, Demerara, Berbice, Corentyne, Coppename and Suriname. Export trade in tobacco, sugar cane, cocoa, cotton and coffee developed. However, these riverside settlements did not last very long. From this point onwards, the settlers began to develop the coastlands, despite the enormous scale of the task involved. In order to reclaim the land, a sea wall had to be built, plus canals to drain the swamps left behind.

From that time onwards, settlement has focused strongly on the coast. Today, 90 per cent of the total national population lives on the coastal plain, a situation which is discernible from the accompanying figure. Today, 50 per cent of the population is classified as living in urban settlements, although this is employing a broad definition of what constitutes a town. Georgetown, together with its suburbs, consists of around 200,000 people, which is about one-quarter of the national population. It grew first as a small Dutch port protecting the settlements established along the Demerara River, and became a small township when the riverside estates were abandoned in favour of the coast and sugar cane became the dominant crop. A capital city and chief port was needed and Georgetown fitted the bill admirably. Today, Georgetown handles practically all of the country's trade, except the export of bauxite and some of the sugar.

Other settlements on the coast vary in size from small villages to townships of several thousand people. New Amsterdam (*c.* 20,000), the country's third largest town, occupies a similar position to Georgetown, in this case in relation to the Berbice River. In contrast, there are only two settlements of any size in the interior. One is the mining and industrial centre of Linden (*c.* 27,000). The other is Bartica, with a population of around 3,500, which acts as the focus of most overland transport routes into the interior and as a collecting centre for timber and gold (see Figure 7.3).

(Box continued)

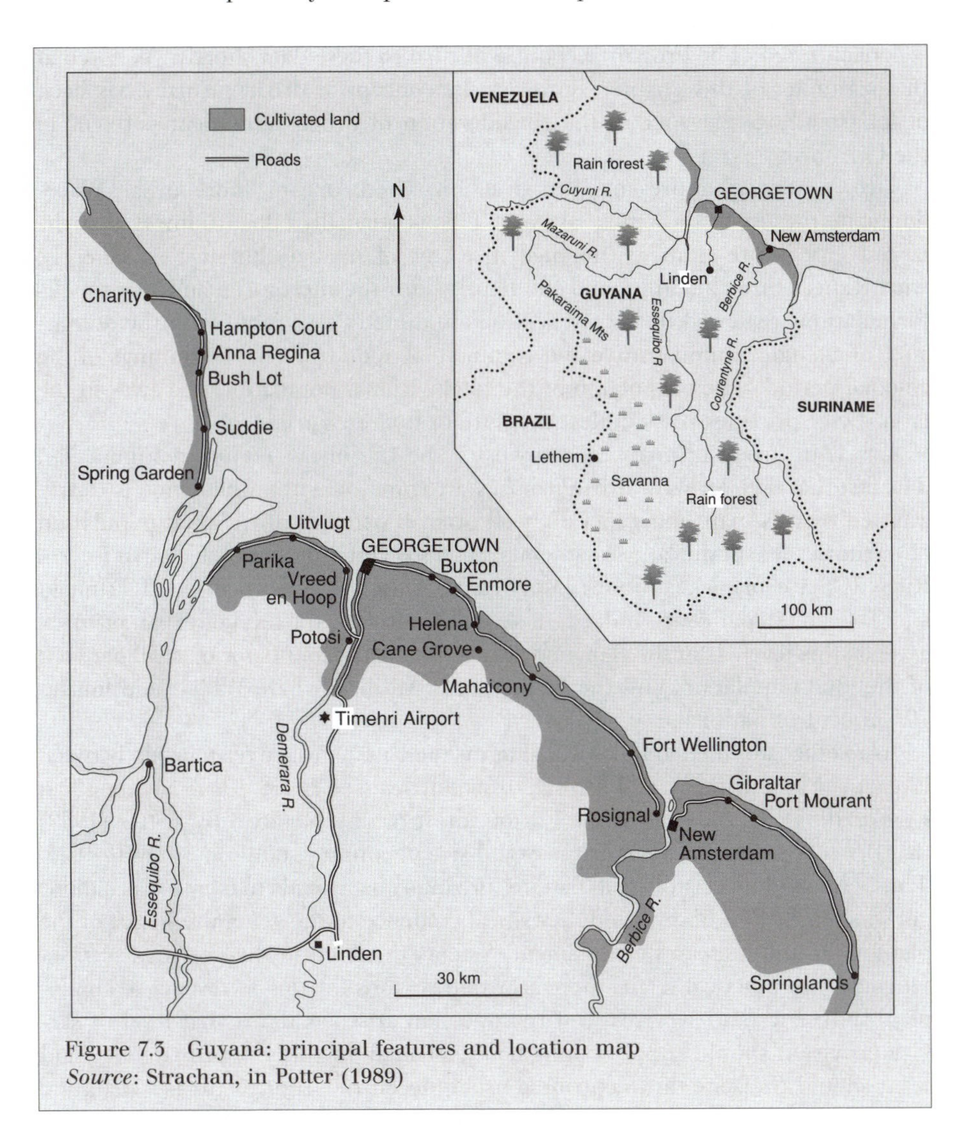

Figure 7.3 Guyana: principal features and location map
Source: Strachan, in Potter (1989)

Urban primacy and polarised development in the Caribbean: empirical verification

Urban primacy viewed at the regional scale in the Caribbean

A good deal of visual and statistical evidence exists showing that the basic settlement fabrics of the Caribbean region are highly concentrated. In the paper by Rojas (1989) concerning the plantopolis settlement system, extensive data were published illustrating the remarkable degree to which population in the Caribbean is concentrated. The present account will turn to these data shortly. As noted at the beginning of this chapter, however, the concept of urban primacy has been of longstanding relevance in the consideration of urban settlement patterns in the Caribbean region.

Urban primacy represents the situation where one or more urban places dominate the urban settlement system. This is normally taken to imply that the largest city is pre-eminent, not just in terms of its size, but just as saliently, with respect to its standing and influence. Both the mercantile and plantopolis formulations reviewed in the last section are directly associated with the dominance of a single administrative and commercial node from the beginnings of the colonial period. The proportion of the total national population that lives in the largest city provides the simplest measure of first city primacy.

Levels of urban primacy are shown for the Caribbean region in Figure 7.4. The fact that high levels of urban primacy are common within the region is clearly attested by this map, and generally well over 30 per cent of the total population of territories live in the largest settlement. This is certainly so in the case of Puerto Rico (31 per cent), Martinique (33 per cent), Antigua (33 per cent), and Trinidad and Tobago (40 per cent). Indeed, in several territories, the level of urban primacy exceeds this level, as in the Bahamas, where Nassau accounts for over 50 per cent of the total population, and the Netherlands Antilles, where Willemstad houses 70 per cent of nationals.

As in other world regions, in overall terms there is an inverse relationship between levels of urban primacy and the size of countries (see Berry 1961; Mehta 1964; Linsky 1965). In the case of the Caribbean, it has been shown by Potter (1989) that the correlation between primacy and size of country stands at around −0.25. Thus, even within a set of small nations, urban primacy tends to increase as nations get smaller. Such a finding adds statistical credence to the general salience of the plantopolis and mercantile settlement systems reviewed in the previous section. Of course, in practical terms there are good reasons for this in respect of single-plant industries and services. But the question arises as to at what level of service provision should supply points be established outside the primate city, and what should the state do to encourage such decentralisation. This is a planning issue of some concern in regions like the Caribbean.

But one form of urban dominance included in Figure 7.4 has not been discussed thus far, and that is the dominance of the American city of Miami within the wider Caribbean urban system. In chronicling the emergence of Miami as a global or world city in the 1970s and 1980s, Grosfoguel (1995: 162) refers to it as the 'capital of the Caribbean'. From the 1960s, Miami grew as a centre for tourism and a retirement haven. Grosfoguel argues that illegal drugs have now taken over as the city's number one industry. However, the city's growth in status as an international banking and trade centre for the entire Caribbean Basin has been very rapid and reflects the relative decentralisation of such activities from New York and other large northern cities since the early 1970s. Grosfoguel also suggests that San Juan, Puerto Rico, has become a semi-peripheral world city,

Table 7.1 Population distribution by size class of settlement for Antigua, Dominica and St Lucia

Country	Antigua				Dominica				St Lucia			
Size range	Number of settlements	%	Population	%	Number of settlements	%	Population	%	Number of settlements	%	Population	%
0–200	1	3.23	91	0.13	25	31.25	8,516	11.39	45	26.16	5,585	4.76
201–400	5	16.13	1,343	1.94	9	11.25	2,628	3.51	54	31.40	15,570	13.28
401–600	4	12.90	2,034	2.94	14	17.50	6,175	8.26	29	16.86	13,864	11.82
601–800	2	6.45	1,420	2.05	12	15.00	8,560	11.45	19	11.05	12,992	11.08
801–1,000	5	16.13	4,691	6.77	3	3.75	2,577	3.45	8	4.65	6,937	5.91
1,001–1,200	4	12.90	4,409	6.36	5	6.25	5,292	7.08	2	1.16	2,119	1.81
1,201–2,500	9	29.03	14,615	21.09	8	10.00	14,646	19.58	12	6.98	20,179	17.20
2,501–4,000	0	0.00	0	0.00	3	3.75	9,276	12.40	1	0.58	3,705	3.16
4,001–6,000	0	0.00	0	0.00	0	0.00	0	0.00	1	0.58	4,065	3.47
6,001–8,000	0	0.00	0	0.00	0	0.00	0	0.00	0	0.00	0	0.00
8,001–12,000	0	0.00	0	0.00	0	0.00	0	0.00	0	0.00	0	0.00
12,001–20,000	0	0.00	0	0.00	1	1.25	17,115	22.89	0	0.00	0	0.00
20,001+	1	3.23	40,687	58.72	0	0.00	0	0.00	1	0.58	32,272	27.52
Total	31	100	69,290	100	80	100	74,785	100	172	100	11,728	100

Source: Adapted from a table reprinted from *Cities: the international quarterly on urban policy*, Vol. 6, Rojas, E., (1989), '*Human settlement of the Eastern Caribbean: development problems and policy options*', pp. 243–258, Copyright © 1989 Butterworth Scientific, (Elsevier), with permission from Elsevier.

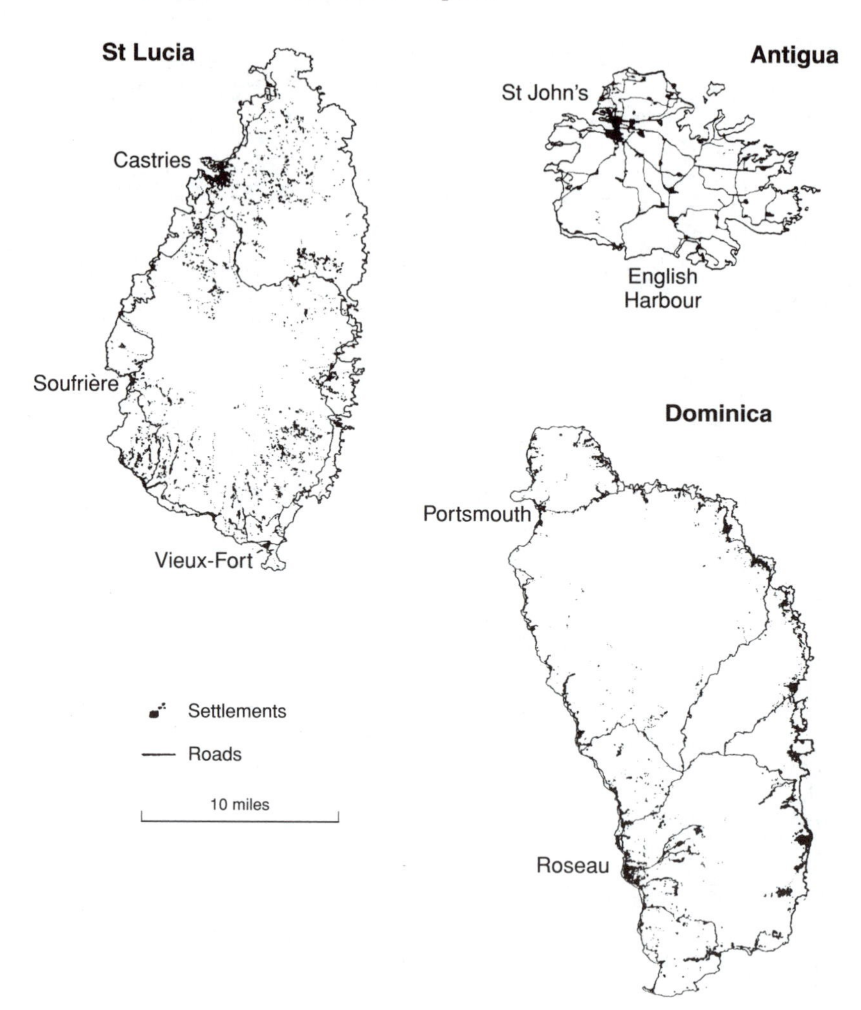

Figure 7.5 The distribution of settlements in St Lucia, Antigua and Dominica
Source: Adapted from Rojas (1989)

countries. The high level of spatial polarisation which is characteristic of Caribbean territories is further demonstrated if the distribution of urban services by settlement size category is considered. In the case of St Lucia, Rojas (1989) looked at the incidence of a total of 78 urban services and functions. Only the primate urban area, Castries, contained the entire set of urban functions (Table 7.2). The two second-order settlements had only a very small proportion of the urban functions found in the main urban area, with Vieux Fort containing only half the urban functions present in the capital.

While larger territories in the region, such as Jamaica, have some equivalent medium-sized towns – those in the 8,000–25,000 population range – in overall terms such settlements are still under-represented in the urban hierarchy. However, it is worth noting that some of these medium-sized towns are currently showing reasonably fast growth rates.

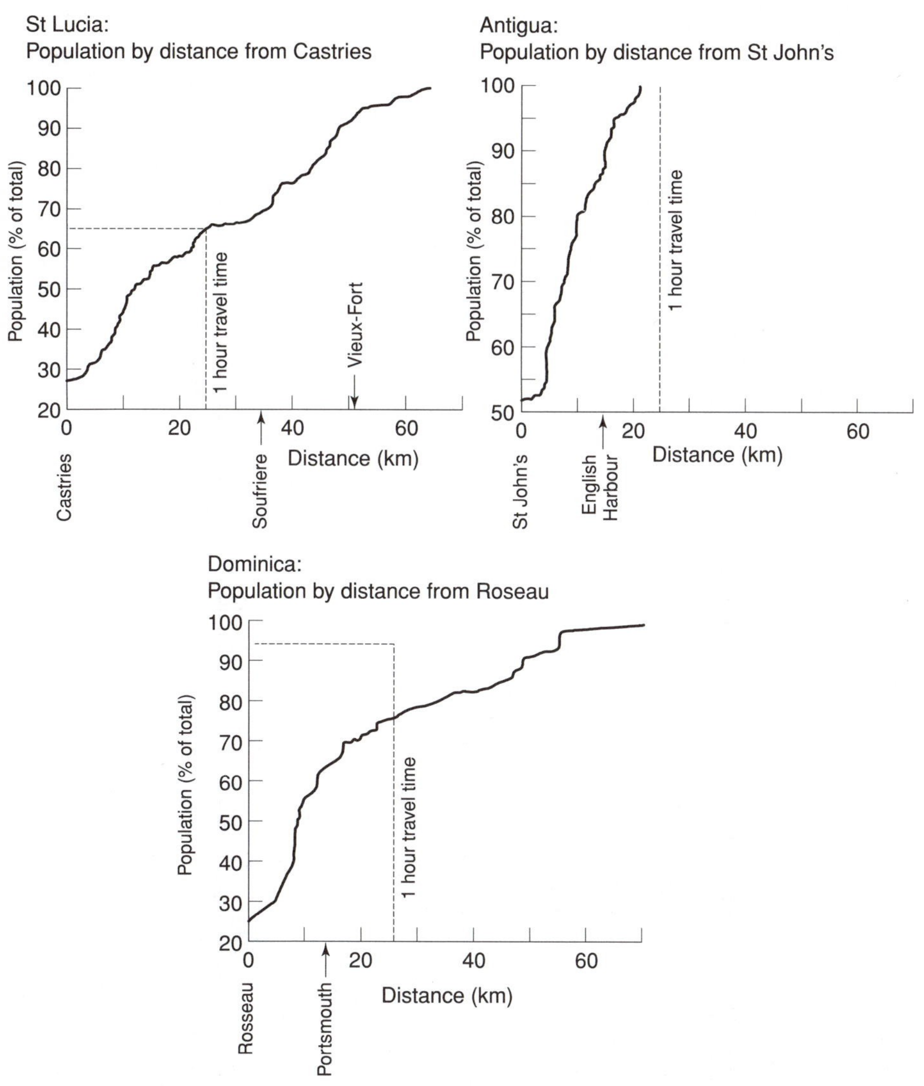

Figure 7.6 The distribution of national population by distance from the capital in St Lucia, Antigua and Dominica
Source: Reprinted from *Cities: the international quarterly on urban policy*, Vol. 6, Rojas, E., (1989), '*Human settlement of the Eastern Caribbean: development problems and policy options*', pp. 243–258, Copyright © 1989 Butterworth Scientific, (Elsevier), with permission from Elsevier.

Urban primacy in the Caribbean and neo-liberalism

There is an argument that the recent strictures of structural adjustment programmes, austerity packages and the seemingly inexorable march of neo-liberalism and free trade (see Chapters 5, 8 and 10) have all taken their toll on patterns of service

Table 7.2 The occurrence of urban activities by size groupings of settlements in St Lucia

Size range	Number of settlements	Public utilities	Transport	Communications	Education	Health	Government	Commerce	Tourism	Finance	Industry	Professions	Recreation	Community services
0–200	45	2	0	0	0	0	0	0	0	0	0	0	1	0
201–400	54	3	0	1	0	0	0	1	0	0	0	0	1	0
401–600	29	3	0	1	0	0	0	1	0	0	0	0	2	1
601–800	19	3	0	1	1	0	0	1	1	0	0	0	2	1
801–1,000	8	3	0	1	1	1	0	1	0	0	0	0	2	2
1,001–1,200	2	3	1	2	1	1	0	1	0	0	1	0	3	4
1,201–2,500	12	3	1	2	1	1	1	1	0	0	1	0	3	2
2,501–4,000	1	5	7	3	3	2	3	8	3	1	2	0	6	4
4,001–6,000	1	4	5	3	3	1	3	5	3	2	2	0	6	4
6,001–8,000	0													
8,001–12,000	0													
12,001–20,000	0													
20,000+	1	6	7	6	6	7	7	10	5	4	2	5	9	4

Source: Adapted from a table reprinted from *Cities: the international quarterly on urban policy*, Vol. 6, Rojas, E., (1989), '*Human settlement of the Eastern Caribbean: development problems and policy options*', pp. 243–258, Copyright © 1989 Butterworth Scientific, (Elsevier), with permission from Elsevier.

Table 7.3 Private and public sector service provision in the urban and rural areas of Barbados

Sector/service	Numerical provision in		
	Central Bridgetown	Greater Bridgetown	Rest of Barbados
Private Sector			
Retail floor space (m^2)	69,254	85,140	20,003
Lawyers	76	77	0
Banks	17	32	16
Medical doctors	95	52	19
Public Sector			
Polyclinics	0	4	4
Libraries	1	2	6
Secondary schools	4	11	10
Police stations	2	3	11
Fire stations	1	2	3
Post offices	1	5	11

Source: Potter and Wilson (1989: 121)

provision in the Caribbean region. In the case of Barbados, Potter and Wilson (1989) have argued that whilst state planning policies have seen some decentralisation of public sector urban facilities, this has not been true of private sector services. If the national distribution of lawyers, banks and general practitioners in Barbados is scrutinised, their paucity outside non-metropolitan Bridgetown is striking. As shown in Table 7.3, whilst 76 lawyers were to be found in the central city and 77 in the greater urban area, none were to be found in the remaining areas of the nation. Years of closely fought elections and the evenly balanced two-party system have ensured that basic forms of rural development have been attended to since the first national Physical Development Plan was produced in 1970 (Government of Barbados 1970; Potter and Hunte 1979; Potter 1986; Pugh and Potter 2000). Thus polyclinics, branch libraries, community centres and schools are relatively equitably distributed across the island.

In conclusion, what does all of this mean for Caribbean patterns of urban change and development? Here the account takes us to the consideration of overall development patterns. The exogenous realities for Caribbean nations remain relative dependency and aspects of emulation in the guise of imported foods, manufactures, new technologies, along with key traits of consumerism, especially in association with tourism, enclave manufacturing and data processing. Internally, the geographical pull of accessible and previously highly developed sites with good infrastructural facilities for industry, and of scenic and safe beaches for tourism, have served to further skew recent development to those very same leeward coastal tracts that centuries earlier had first attracted mercantile capitalism (Potter 1985, 1989).

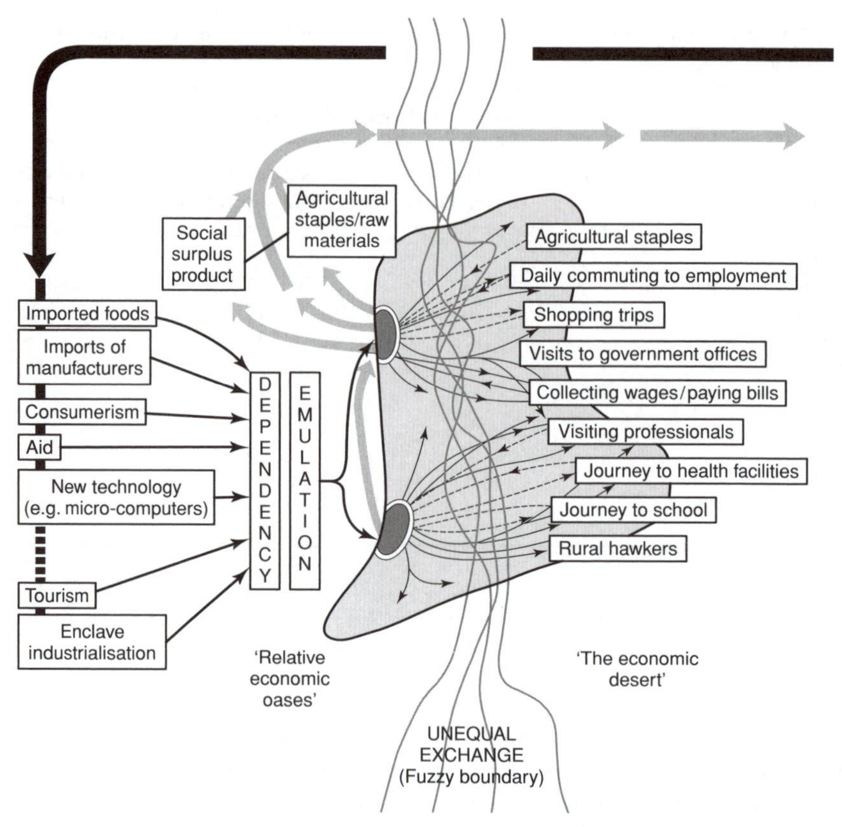

Figure 7.7 A model of Caribbean small island spatial development
Source: Potter (2000)

The outcome of these trends is given graphical expression in Figure 7.7. The channelling of new developments into what may be referred to as 'relative economic oases' might be considered by some to have left 'relative economic deserts' in their wake (see also Potter 1989). The socio-economic contrasts between these two area types are real and impinge directly on the daily lives of Caribbean citizens. In particular, such contrasts remain associated with markedly unequal exchange, as well as being predicated on substantial rural-to-urban flows. As shown previously, in the past, relative inaccessibility and, more recently, government policies have led to patterns of deconcentrated decentralisation within the primate core–coastal littoral region. However, long trips to visit professionals, government offices and even to collect wages are still to be faced by the denizens of rural areas, on a monthly, if not a weekly or, indeed, a daily basis. It is in this manner that models of settlement location and overall development patterns in the Caribbean still exhibit strong affinity with longstanding polarised models such as the core–periphery framework (Potter 1995).

All of this stands in marked contrast to the arguments presented by Portes *et al.* (1997a), that the dictates of structural adjustment and neo-liberalism are leading to reductions in first city levels of primacy. In fact, their hypothesis was based on earlier observations made in Latin American cities. On applying their arguments to Haiti, Jamaica, the Dominican Republic, Costa Rica and Guatemala, perhaps not too surprisingly, evidence was not generally supportive. Only in the case of Jamaica were they able to discern a clear reduction in levels of urban primacy. Indeed, evidence showed quite clearly that in the cases of both Haiti and Guatemala, urban primacy had, in fact, increased significantly (see Potter 1999, 2000). In the wider context, the results serve to demonstrate, with some force, that it is dangerous to assume that trends discerned in one region can be applied directly to another in a simple and straightforward manner, however proximate the two regions may happen to be. Specifically, Caribbean cities are not to be included when making blanket generalisations concerning the 'Latin American' city, as is erroneously implied by the Portes *et al.* work.

Urban primacy and national urban development strategies

At some juncture, most Caribbean countries have made public declarations of their intent to tackle the consequences of urban primacy, urban bias in development and gross spatial inequalities. For example, the first Physical Development Plan for Barbados produced in 1970 (Government of Barbados 1970), presented what can only be described as an extremely ambitious programme for the decentralisation of employment, population and government offices into an almost perfect Christaller-type central place hierarchy (Potter and Hunte 1979; Potter 1985). Although this first espoused national settlement strategy was immensely idealistic, and certainly never implementable, the second and third national Physical Development Plans, produced in 1986 and 1998 respectively, envisaged spatially more focused decentralisation along the linear urban corridor (Government of Barbados 1983, 1998). In this way, the local east–west impact of Speightstown, Holetown, Oistin and Six Cross Roads as service centres was to be enhanced (see Potter 1987; Pugh and Potter 2000).

Another example of state awareness of the issue of regional imbalance is provided by the Physical Development Plan produced for Trinidad and Tobago in the early 1980s (Town and Country Planning Division 1978). The significance of the plan is that it suggested and evaluated four alternative strategies, which could be employed to reduce the strong urban concentration that had developed from the colonial days. The first, referred to as trends development, envisaged the continued growth of the existing urban areas of Port-of-Spain and San Fernando-Couva. As a second strategy, that of planned concentration, future growth would be directed east along the main urban corridor, towards Sangre Grande. In the third strategy, complete dispersion was envisaged, with new growth being spread throughout the island, but in the rural south and east in particular. The fourth compromise solution was to combine dispersion with concentration. Growth was to be concentrated in a number of localities, such as Sangre Grande and eastern Port-of-Spain, but was also to be spread to key centres in the rural east and south of the country. In the final analysis, it was this mini–max solution that was selected to apply to the problem of urban primacy and concentration in Trinidad and Tobago (Potter 1985: 133–9).

However, as argued by Potter and Lloyd-Evans (1998), it is entirely possible for the state to be involved in making a set of public pronouncements whilst doing little or nothing, or indeed something quite different, in the policy arena. Potter and Pugh (2001) look at this type of scenario in the case of physical development planning in St Lucia. Specifically, the state can talk about promoting regional equality whilst doing very little to actually achieve this. Such a situation has not been the case in the socialist state of Cuba, however, and it is generally in such collectivist states that action has been taken to reduce urban dominance.

Prior to the revolution in 1959, Havana stood as the classic primate capital. Cuba showed very high rates of rural–urban migration and the rapid growth of spontaneous settlements in Havana. The Census of 1953 indicated that 55 per cent of all urban housing was either unsanitary or inadequate.

After 1963, Havana was actively discriminated against, being regarded as imperialist, privileged and corrupt. The freezing of metropolitan Havana in order to enable the rest of the nation to catch up is frequently referred to as the 'Havana strategy'. In contrast, at the inter-regional scale, the growth of provincial towns in the 20,000 to 200,000 size range was stimulated in an effort to counterbalance the primacy of Havana (see Figure 7.8). At the next level down, the regrouping of villages into 'rural towns' was effected (Hall 1989). In the fields of both health and education, massive efforts have been made to develop new secondary schools along with primary, secondary and tertiary health care facilities in what are predominantly rural areas. But the point is that the policy had been designed to reduce urban–rural imbalances (see also Mathey 1997; Colantonio and Potter 2003). The price has, of course, been state involvement, so that planning in Cuba is often described as both dictatorial and technocratic.

A statistical overview of Caribbean urbanisation

As noted elsewhere (Potter 2000), in common with other regions of the developing world, rapid urbanisation in the Caribbean has primarily occurred in the period since the Second World War. Although as noted previously, Caribbean towns and cities had their origins in the colonial administration and control of territories and the development of plantation agriculture, their rapid growth has resulted from the high rates of rural-to-urban migration that have been experienced since 1945 (see Chapter 2). But this movement of population towards the opportunities offered by urban areas and urban labour markets has also served to swell rates of natural increase in urban populations. The twin 'push' of rural poverty and the 'pull' of socio-economic opportunities in the urban areas – both real and perceived – have thereby been causal.

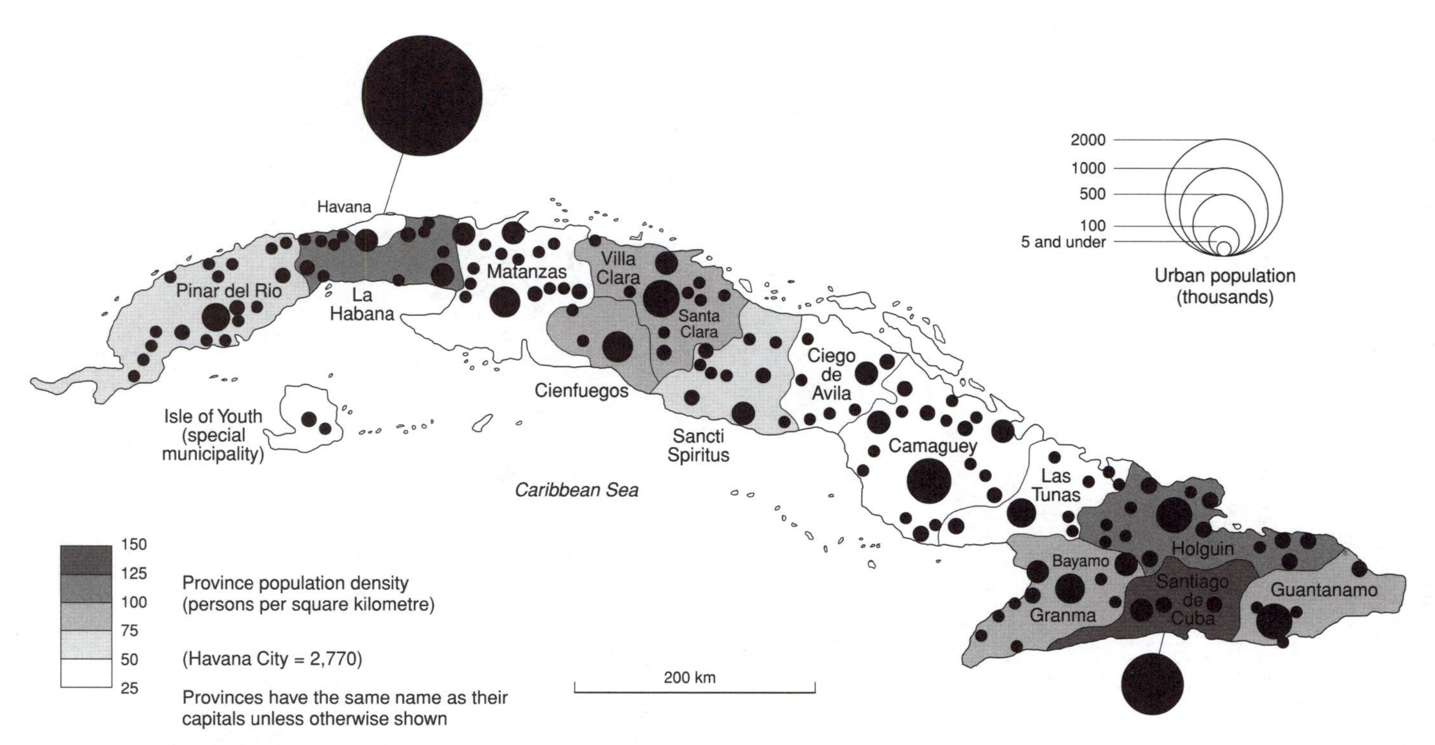

Figure 7.8 The distribution of towns and cities and population densities by province in Cuba
Source: Hall, in Potter (1989)

Table 7.4 Total population living in towns and cities and level of urbanisation in the Caribbean, 1960–2000

Date	Total population of the Caribbean living in towns and cities (millions)	Percentage of total population living in urban areas	
		Caribbean	World
1960	7.7	38.2	33.9
1970	11.1	45.1	37.5
1980	15.7	52.2	41.3
1990	21.6	58.7	45.9
2000	28.8	64.6	51.3

Source: United Nations, 1980, in Potter (2000)

Interestingly, data published by the United Nations (1980) show that the contemporary Caribbean is not only considerably more highly urbanised than the developing world taken as a whole; in fact, the region is more highly urbanised than the world in aggregate. This generalisation was true in 1960, when just in excess of one-third of the total population of the Caribbean region was classified as urban. It is just as applicable to the urbanisation level of 64.6 per cent that applied in 2000 (Table 7.4). The urban proportion is projected to rise to over three-quarters by the end of the first quarter of the twenty-first century (Table 7.5).

The impact of cities on the socio-economic landscape of the Caribbean region has been dramatic. Thus, West and Augelli (1976) observed that in 1950, the region supported only seven cities with populations of 100,000 or more, three of these being in Cuba. By 1970, the number had risen to 12. Calculations by the present

Table 7.5 Projected levels of urbanisation in the Caribbean, 2005–2025

Year	Percentage of total population living in urban areas
2005	67.9
2010	70.0
2015	72.1
2020	74.1
2025	75.9

Source: United Nations, 1989, in Potter (2000)

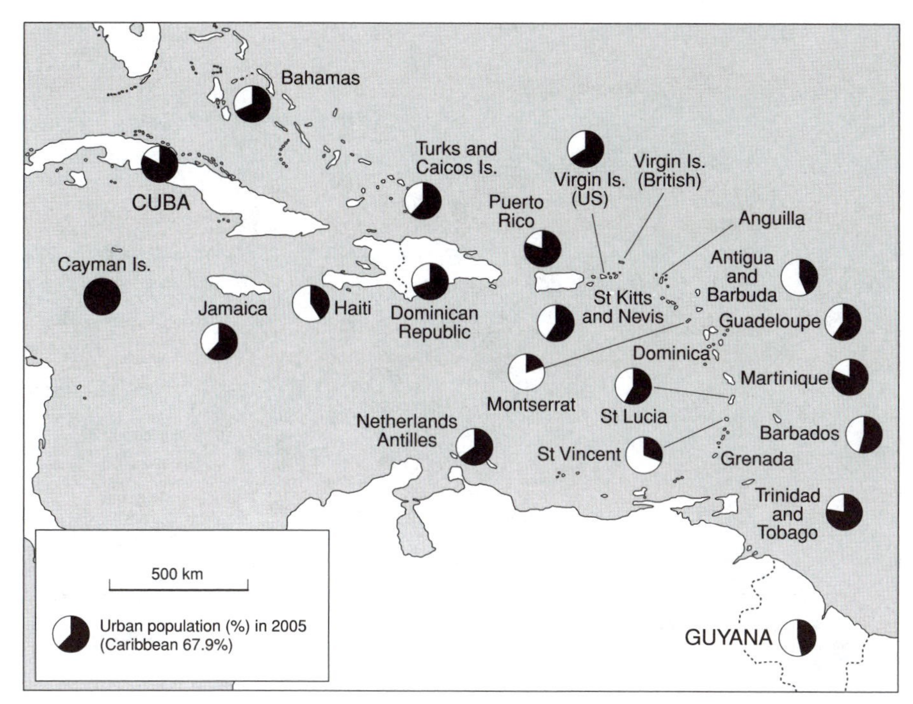

Figure 7.9 Levels of urbanisation for Caribbean territories in 2005
Source: Potter (2000)

author indicated that, in the early to mid-1980s, there were at least 24 urban places of 100,000 or more inhabitants in the Caribbean (Potter 1989, 2000). Indeed, three of these located in the Hispanic Caribbean, Havana, Santo Domingo and San Juan, had exceeded the million population level by that date. Data tabulated by the present author for the 1990s show that the number of large cities has continued to increase within the Caribbean (Potter 2000).

Statistics concerning current levels of urbanisation and rates of urban growth for the Caribbean region are shown mapped in Figures 7.9 and 7.10. Projected overall levels of urbanisation in 2005 are shown for Caribbean territories in Figure 7.9. The general pattern is one of relatively high levels of urbanisation, averaging 67.9 per cent for the Caribbean as a regional entity. Levels of urbanisation in excess of two-thirds of the total population apply in the case of the Bahamas, Cayman Islands, Cuba, Dominican Republic, Martinique, Puerto Rico, Trinidad and Tobago, and the United States Virgin Islands.

At the same time, Caribbean states are showing relatively high rates of urban population growth, with a regional average of 2.14 per cent per annum pertaining for the period 1995–2000 (Figure 7.10). Notably, Antigua and Barbuda, the Dominican Republic, Haiti, Montserrat (before the volcanic eruption), St Lucia, St Vincent and Guyana are all characterised by urban growth rates in excess

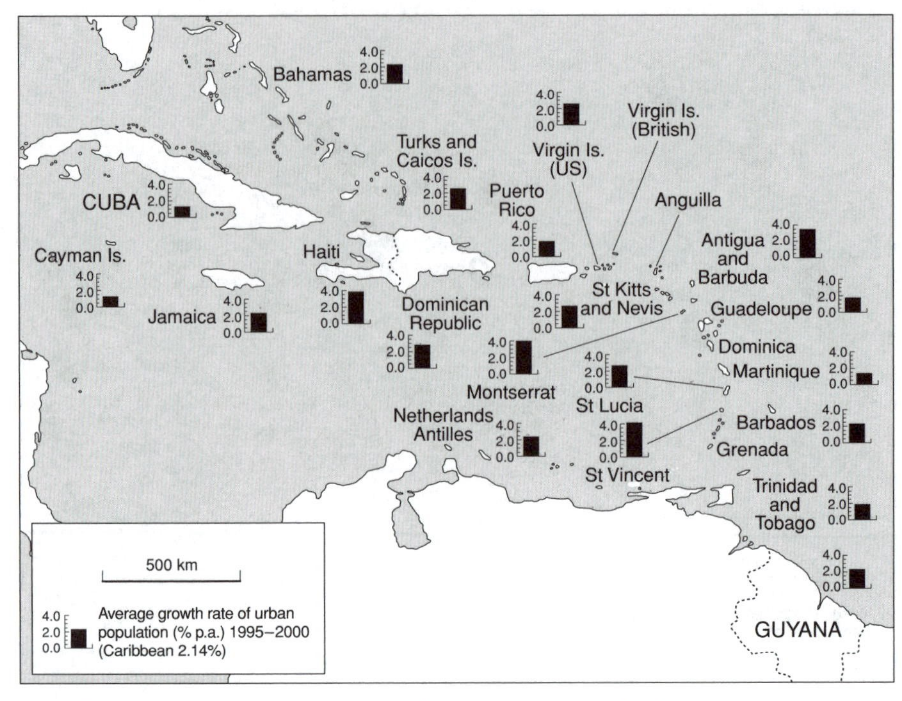

Figure 7.10 Average growth rates of urban population for Caribbean territories, 1995–2000
Source: Potter (2001)

of 2.75 per cent per annum. In addition, as already commented on extensively, levels of urban primacy are also high in the region, with generally 40–50 per cent of the total population of nations living in the capital city region.

The overall level of urbanisation estimated for the Caribbean region between 1950 and 2025 by the United Nations is shown in Figure 7.11. In this graph, the urbanisation curve for the Caribbean region is shown juxtaposed with the same curves relating to the more developed and less developed regions of the world taken as a whole. The convergence of the Caribbean region's aggregate level of urbanisation on that of the more developed areas of the world is clearly apparent from this figure. Although this means that the average growth rate of urban population in the Caribbean region has fallen below 2 per cent per annum since 2000, the level of urban growth will still stand at 1.22 per cent per annum at the end of the first quarter of the twenty-first century.

This short section has summarised the current and future likely pattern of urbanisation in the Caribbean to the year 2025. In so doing, two major points made in the previous sections of this chapter have been exemplified and illustrated. The first is the fact stressed throughout the first part of the chapter that, despite the region's foundation in plantation agriculture, in overall terms, the contemporary Caribbean is strongly urban in character. Secondly, the statistical

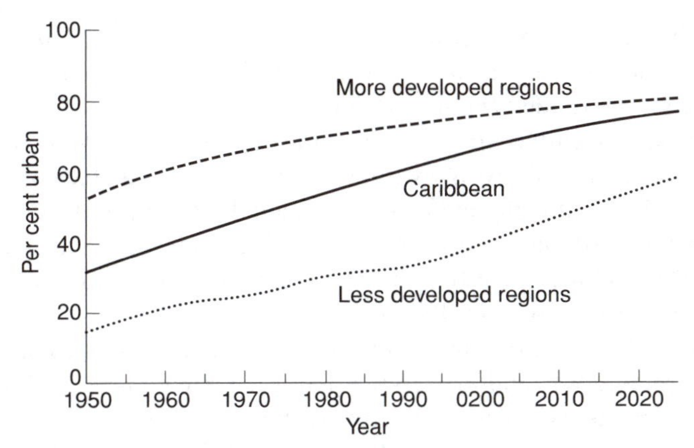

Figure 7.11 Urbanisation curve for the Caribbean region against those for the more developed and less developed regions of the world taken in aggregate
Source: Potter (2000)

account has emphasised that this strong urban polarity will continue in the future. This statistical reality is the outcome of continuing urban-based developments in the fields of commerce, manufacturing and informatics, together with the further development of coastal-based tourism throughout the Caribbean region. Indeed, as noted in respect of the third stage of the plantopolis model, tourism must be conceptualised as a powerful agent in the promotion of urbanisation in the contemporary Caribbean region (see Weaver 1988, 1993; Potter 1997, 2000; Sahr 1998). This theme is addressed in further detail in Chapter 11 when dealing with urban tourism.

Urban internal structure and urban dynamics in the Caribbean

Settlement location and early development

The earlier sections of this chapter explained the way in which Caribbean cities have developed from their mercantile origins into complex coastal-oriented settlement systems. This account also served to stress the relatively long history of urban development in the region. The purpose of this second half of the chapter is to consider the internal structure and dynamics of Caribbean urban places.

Obviously, the mercantile and plantopolis-based origins of settlements were reflected in the initial location of Caribbean urban places. Indeed, most of the towns and cities established in the region by the European colonists after 1500 had direct access to tidewater (Hudson 1989, 1998). Hudson notes that some of the earliest ports, developed on the larger islands by the Spanish, were located a short distance inland on tidal inlets. It can be assumed that such locations were initially preferred in order to avoid the unhealthy low-lying coastal swamps. Examples include the settlements of Trinidad in Cuba, Spanish Town in Jamaica and St Joseph in Trinidad (Hudson 1989).

From the seventeenth century onwards, the development of West Indian trade, principally in sugar and slaves, saw urban expansion on suitable coastal sites, as envisaged by the plantopolis and mercantile models. The importance of the early inland settlements of the Spanish colonists declined in favour of the seaport cities (Hudson 1989). Thus, Spanish Town was eclipsed by Kingston in Jamaica and St Joseph by Port-of-Spain in Trinidad.

Hudson (1998) notes that in the case of the smaller islands colonised by the French and the English, and to a lesser extent the Dutch, Danes and Swedes, the original settlements were based directly on the coasts and these have remained the dominant urban force as envisaged in the plantopolis framework. These settlements classically combined the functions of chief port and capital city with strong administrative and control functions and linkages to the European metropole. In this sense, the plantopolis and mercantile models can be related directly to the dependency and core–periphery models (see Chapter 8) (see also Potter and Lloyd-Evans 1998; Potter *et al.* 2004). As summarised by Hudson:

> For a community dependent on sea-borne trade the most obvious site for a settlement is one with direct access to (a) safe, sheltered harbour, preferably with sufficient level or gently sloping land for the requisite buildings (Hudson 1998: 77).

As Potter (1989) notes, the early gateway origins of urban centres within the Caribbean have not just resulted in strong urban primacy and spatial polarisation at the national level. Together with the topography of Caribbean territories, these mercantile origins have also contributed to the cramped sites that many towns and cities in the region exhibit today at the intra-urban scale. In a similar vein to Hudson, West and Augelli (1976: 120) record how the chief consideration in the origin of most Caribbean cities was their function as sheltered maritime ports that could be defended with relative ease. Thus, the typical urban location was on 'the shore of an embayment protected at the land-ward side by commanding hills on which forts overlooking the sea approaches were built'. Typical examples are Kingstown, St Vincent and Castries, St Lucia, as shown by Plates 7.1 to 7.3. It is these strategic origins that give many contemporary Caribbean capitals their somewhat huddled and cramped character today.

The site, situation and contemporary land-use pattern of Castries is shown in Figure 7.12. Castries is based on a natural east–west running harbour. However, the amount of natural flat land is very limited, and most of the present-day central area stands on reclaimed land (see Figure 7.14). Inland to the south, south-east and east there is higher land. Indeed, as shown in Figure 7.14, the only easy direction for further expansion, including that of tourism, is along the coastal strip towards the north-east (Potter 1985; Sahr 1998; Potter 1999).

Plate 7.1 The centre of Kingstown, St Vincent
Source: Rob Potter

Plate 7.2 The centre of Castries, St Lucia
Source: Rob Potter

Plate 7.3 Settlement on the steep hills rising from the centre of Castries
Source: Rob Potter

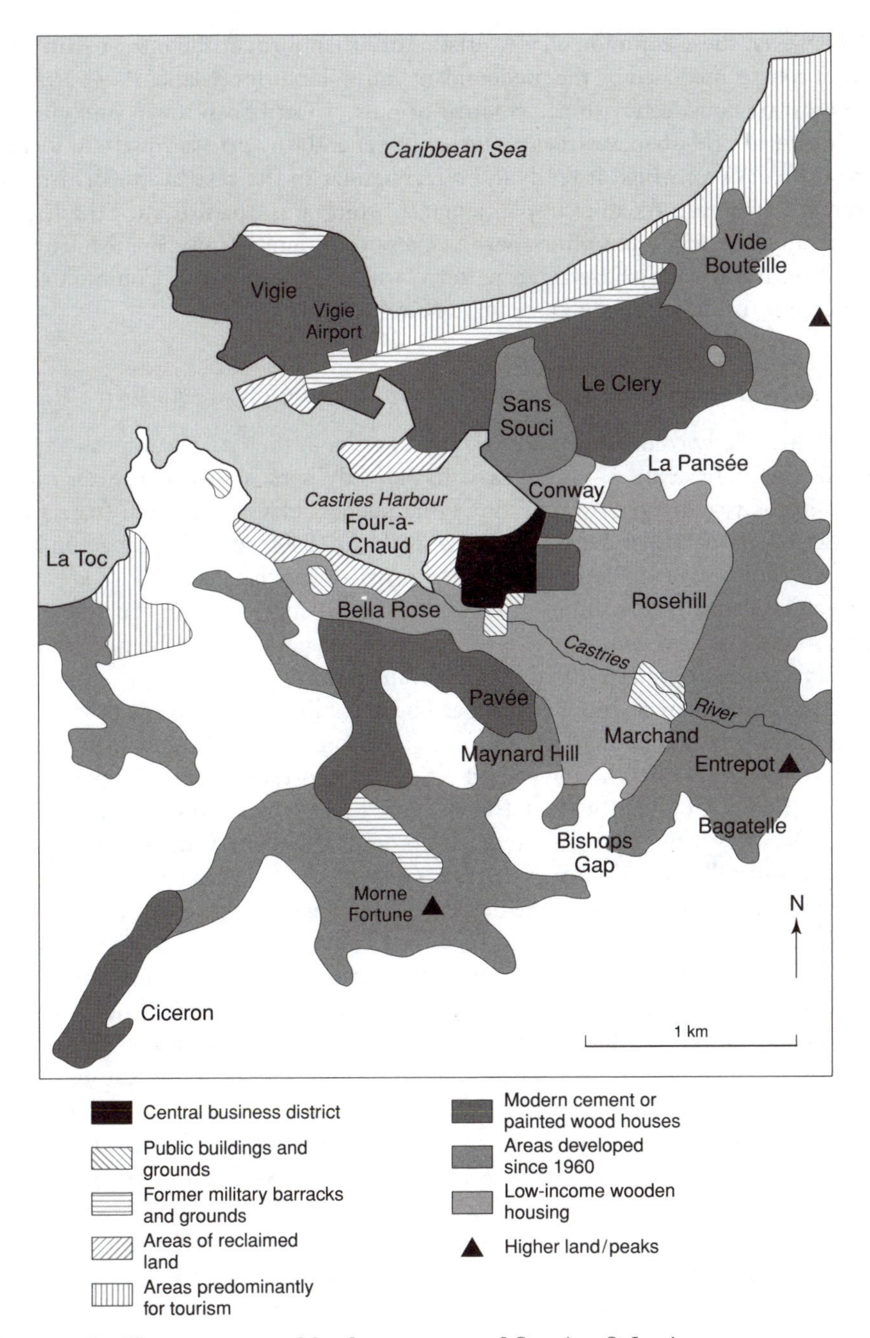

Figure 7.12 The structure and land-use pattern of Castries, St Lucia
Source: Potter (1985)

However, these common characteristics of urban form at the intra-urban level are, of course, mediated by the predominant cultural imprint (Clarke 1974). Perhaps the most obvious legacy of the colonial origins of Caribbean towns and cities is their central grid-shaped street patterns. In general, the grid is square in former Spanish colonies, whilst it tends to be rectangular in the case of the British and French Caribbean. As an example, Jamaica, which was Spanish until the British occupation of 1655, exemplifies both of these traditions. Thus, Spanish Town is based on the square grid-shaped pattern, whilst Kingston as a Commonwealth settlement displays a rectangular form (Clarke 1974).

The evolution and structure of contemporary towns and cities

From these early colonial beginnings, urban areas in the Caribbean grew rapidly in the twentieth century. However, it has already been shown that even capital cities in the region vary from million cities such as Havana and San Juan, to far smaller functional urban regions in the 20,000 to 40,000 population range. In addition, in the shadow of these primate cities, whatever their size, there are many very small urban places, often in the 3,000–6,000 size range. Given this essential diversity of size, is it possible to derive any general notions concerning the structure and social patterning of Caribbean towns and cities?

Like all urban areas, those of the Caribbean exhibit clear patterns of zonation, reflecting land-use competition and specialisation. At the functional if not the geometrical heart of these towns and cities, the central business district (CBD) is to be found. Frequently, the CBD is located just inland from the coast, along which lie the principal harbour or docks and attendant warehouses. It is in the CBD, or following the American terminology, the 'downtown', that the principal department stores, duty-free outlets, commercial and professional offices and transport termini are to be found. In some Caribbean cities, the old downtown area has come to be associated with pollution, crime, poor access and congestion. For example, Kingston, Jamaica, really has two CBDs, with different functions. New Kingston houses the high-rise financial sector offices, banks and insurance companies, plus recently developed night-time entertainment and eating places. The old downtown is predominantly commercial and retail, along with huge produce markets. It also acts as the main transport hub for the island, and is home to a good deal of informal sector activity. Downtown is much busier than New Kingston. Furthermore, many government offices are still located downtown. Such developments are giving rise to a distinctly multi-nodal character within the city.

The broad area immediately surrounding the CBD may be referred to as the inner city. It is in this area, where commerce and retailing give way to housing, that the residences of the poor are primarily concentrated, frequently in close proximity to informal sector shops, rum houses, workplaces and workshops. Such areas manifest one of the principal outcomes of competition within the dominant capitalist world system. For example, Laguerre (1990) has considered what he regards as the reproduction of urban poverty in two central neighbourhoods of Fort-de-France, Martinique. He argued that this represents a structural system reflecting the capitalist mode of accumulation. The housing in such areas is generally of two types. In older larger cities such as Havana, Cuba, former large residences have been split into smaller apartments for rent. Elsewhere in this zone, some of the oldest informal settlements are sited, some of them having experienced a great deal of consolidation and upgrading over the years. Good examples of the latter are provided by the squatter settlements of Four à Chaud and Conway in central Castries (Figure 7.12) (see also Chapter 6).

Of course, it would be wrong to associate the inner cities of the larger Caribbean territories with universally poor and static conditions. For example, the central areas of large cities in the Hispanic Caribbean date back to the earliest colonial times. Reflecting this, UNESCO has designated Santo Domingo in the Dominican Republic a World Cultural Heritage site, and in the old part of the city, there are many fine early sixteenth-century buildings. The same is true of Havana, Cuba, which is also a UNESCO World Heritage Site. This designation is leading to substantial efforts to refurbish the old city, which dates from 1515. More than 900 of Old Havana's buildings are regarded as being of historical importance. Furthermore, it is estimated that 144 buildings date from the sixteenth and seventeenth centuries, so that the city has been described as four hundred years deep. San Juan, Puerto Rico, founded in 1510, still houses the Old Walled City, together with the adjacent narrow city streets. The earliest buildings of Willemstad, the capital of the Netherlands Antilles, were exact copies of Dutch mid-seventeenth-century buildings, and were built high and close together in order to save both time and money. The only acknowledgement of their new tropical location was being painted in bright pastel colours. It was not until considerably later that the Dutch adapted their buildings to their new environment.

It can also be seen that Caribbean towns and cities, whether large or small, seemingly display another common characteristic – the movement out from the inner core of those who are relatively wealthy. As in cities of the developed world, high-income groups have been 'pushed' out by the often poor, polluted and congested nature of the centre, as well as being 'pulled' by cheaper and more extensive tracts of land on the urban periphery. As envisaged by the American land economist Homer Hoyt (1937) in connection with the fast-growing North American city of the early 1900s, the progressive movement outwards of the relatively wealthy has occurred in certain preferential directions. The resultant high-income zone is a characteristic wedge or sectoral shape. For example, there are many parallels between the physical growth of Kingston, Jamaica, and European and North American cities, including the expansion of Victorian middle-class suburbs along lines and sectors defined initially by horse-drawn and then electric tramcars. Later in the twentieth century, specifically post-1950, Kingston's middle classes moved out farther, and Victorian suburbs became urban ghettos and working-class high-density residential areas.

In a number of Caribbean territories, this process of outward sectoral migration of the high-income quarter has been quite sudden, being predicated mainly on the increasing availability and use of the automobile. Thus, the emergence of a new middle class and an American-style flight to the suburbs in the 1960s and 1970s has been documented in the case of San Juan (de Albuquerque and McElroy 1989), whilst a similar process has been noted for Nassau, Bahamas (Boswell and Briggs 1989). The progress of urbanisation and urban structuring in the case of San Juan, Puerto Rico, is treated in Box 7.2.

Box 7.2: Urbanisation and urban structure of San Juan, Puerto Rico

During the 'Operation Bootstrap' period from 1950 to 1970 (see Chapter 4), the rate of urban absorption experienced by Puerto Rico was second only to that recorded in the case of the Netherlands Antilles. Puerto Rico single-mindedly followed the development model known as 'industrialisation by invitation', seeking to attract mainland USA capital by means of tax concessions, cheap labour, rapid access to the US market and the provision of subsidised infrastructure (de Alburquerque and McElroy 1989).

As a result, between 1950 and 1970, the urban population of Puerto Rico almost doubled, despite some efforts on the part of the government to promote rural development. However, this was not enough to offset the economies of scale experienced in the urban areas, plus the grinding effects of rural poverty. For example, it is estimated that during the 1960s, rural unemployment rates ran at between 22 and 47 per cent.

Today, around 67 per cent of the population live in urban areas, and San Juan, the capital, has joined the ranks of cities with a million or more inhabitants. By the end of the eighteenth century, San Juan had developed a Spanish-style pattern of residential differentiation (Duany 1997), with the best residences, those constructed of stone or brick walls with wooden roofs, being located near to the city centre. On the other hand, thatched local houses with dirt floors (*bohios*) dominated the outer parts of the city. This colonial or pre-industrial pattern meant that, on average, the farther one lived from the urban centre, the lower one's standing was in the social class hierarchy (Duany 1997).

Following the American occupation of Puerto Rico in 1898, large-scale migration from the rural to the urban areas was experienced, reflected in overcrowding and poor urban housing. Duany argues that living conditions deteriorated for many workers in the period between 1898 and 1920. In the growing towns, which by this stage included Ponce, Guyama and Mayaguez as well as San Juan, workers frequently built their own houses, whilst others rented rooms in Old San Juan and Santurce.

Since 1950, the Spanish colonial pattern of residential differentiation, with the rich living near to the centre, has broken down. In the 1960s, upper-status groups moved to the peripheries of the cities, whilst lower-income groups have remained much closer to the urban core. Thus, Duany (1997: 197) concludes

(Box continued)

that 'Modern San Juan looks more and more like a typical American city in its physical and sociological features: the upper-class flight to the suburbs, the marginalisation of old downtown areas, the concentration of capital resources and services in a financial district, and the residential segregation of rich and poor'.

Since 1970, these trends have continued, and suburbanisation has become a major feature of the urban process (de Albuquerque and McElroy 1989; Duany 1997). Indeed, since the 1970s, the core area of San Juan has actually lost population. However, the five municipal districts immediately outside San Juan (Toa Baja, Bayamon, Guaynabo, Trujillo Alto and Carolina) have all increased their populations very rapidly. Together, their population increased by 54 per cent between 1970 and 1980, totaling 546,171 in 1980, whilst San Juan itself stood at 433,901. As a result, Greater San Juan amounted to just under 1 million people by 1980, and as previously noted, now well and truly exceeds this mark.

Even in the case of a medium-sized urban area such as Port-of-Spain, the structure and development of the city has been described as a history of sectoral expansion (Conway 1981, 1989). Within the city, a pronounced high-density low-income residential sector has developed, radiating from the CBD outwards towards the eastern periphery. This zone includes residential areas such as Belmont, Success Village, Eastern Quarry/Prizgar Lands, Troumacaque and Chinapoo, as shown in Figure 7.13. This low-income area really consists of three zones: a central tenement area, a ring of early established spontaneous settlements and a newer periphery of uncontrolled settlements such as Morvant (see Conway 1989; Potter 1993; Potter and O'Flaherty 1995). In the meantime, wealthy urbanites have relocated farther away from the commercial core. To the east, they have leapfrogged across the eastern inner city. In addition, high-status inner city residential areas have been developed in the northern and north-western valleys, such as in the Diego Martin and Maraval areas (Figure 7.13).

Similar processes and outcomes can be seen in the contemporary social spaces of other Caribbean towns and cities. Portes *et al.* (1997) present a map of residential strata and irregular settlements in Santo Domingo, in the Dominican Republic. An adapted version of this map is shown in Figure 7.14. Two clearly prescribed sectoral zones running out from the CBD and Old City can be recognised. The first runs towards the north-east of the city and the second towards the south-west. Together these two areas represent the most noticeable features of the map. The poorest areas are located in the central sector of the city. The authors note how a recent dynamic has been the development of a strong middle-class sector to the east of the city. Locally, the process leading to the growth of this area has been referred to as 'crossing the bridge' (that is, across the Rio Ozama). The other major feature of the social space of Santo Domingo is the spread of irregular or squatter settlements throughout the entire urban area, although there is a heavy concentration of such communities north and east of the downtown (Figure 7.14).

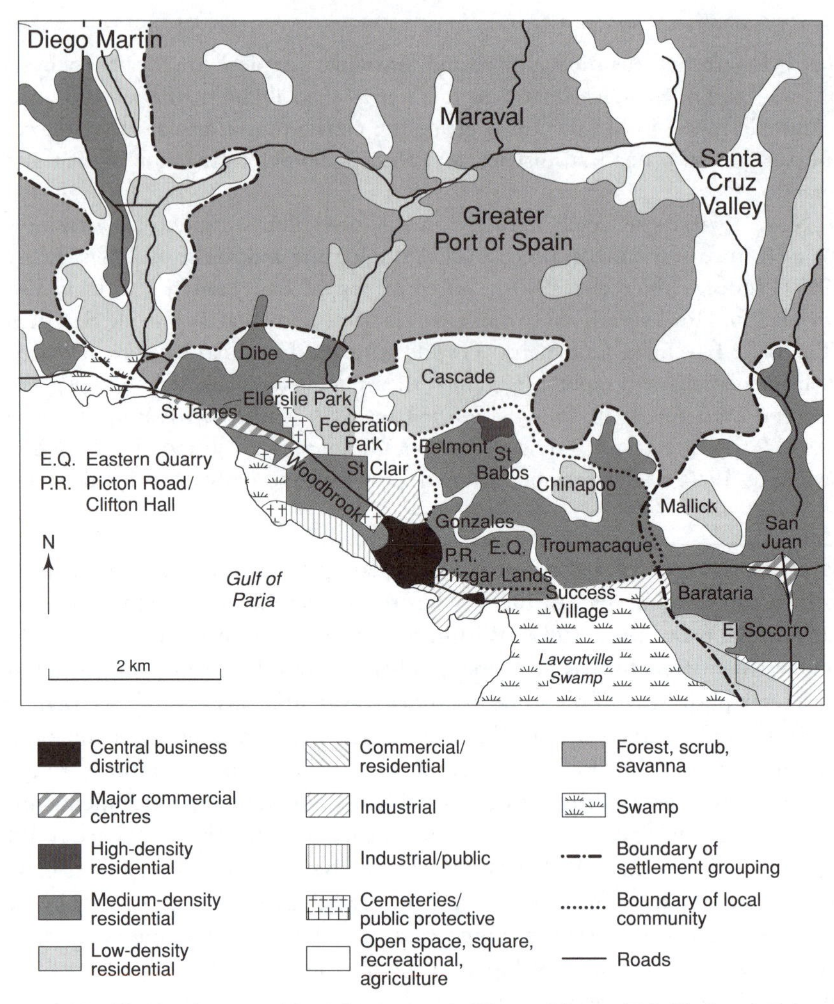

Figure 7.13 The land use and social structure of Port of Spain, Trinidad and Tobago
Source: Potter (1985)

In an essentially similar manner, what can be described as a banded sectoral patterning of social classes has emerged in Kingston, Jamaica (Figure 7.15). Clarke (1989) argues that the polarisation of white and black populations in the city is expressed in both social and spatial terms. Clarke maintains that the predominantly white population is confined to the low-density elite suburbs of the north-eastern sector of the city (Figure 7.15). It is argued that the predominant black population is concentrated at high densities in West Kingston and East Kingston, forming a sector that runs right across the city. Clarke argues that people of mixed race are located socially and geographically between the black and white populations and merge with these other areas (Figure 7.15). However, as Clarke (1989) notes, since independence, but especially since the economic

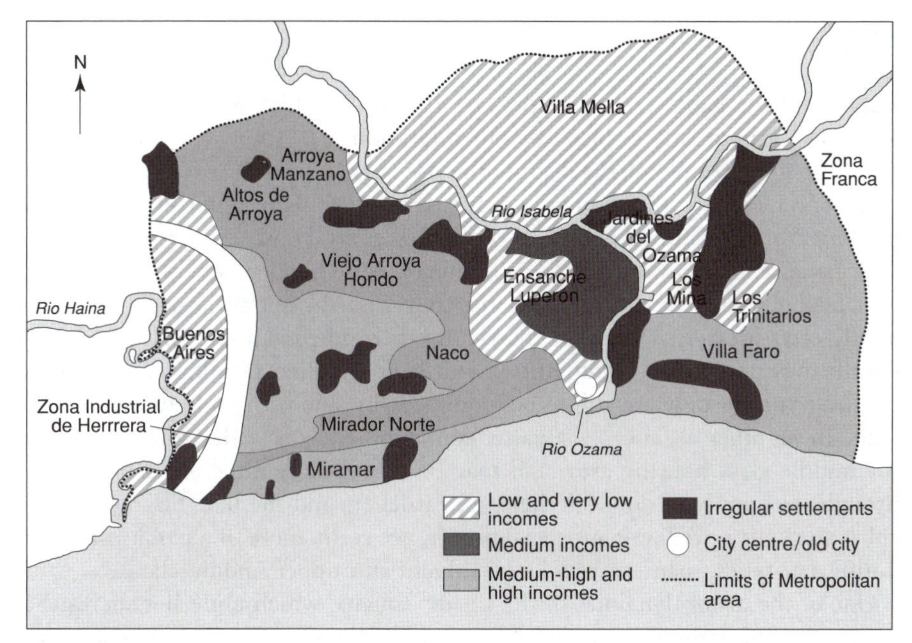

Figure 7.14 The social areas of Santo Domingo, Dominican Republic
Source: Portes *et al.* (1997b)

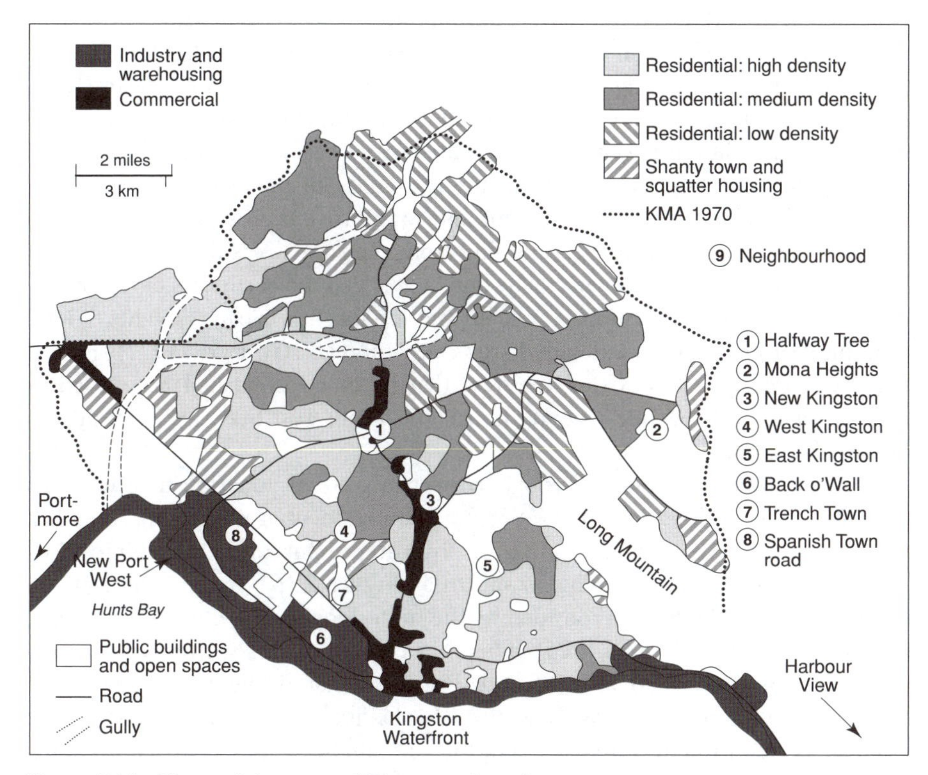

Figure 7.15 The social spaces of Kingston, Jamaica
Source: Clarke (1989), in Potter (1989)

decline and the gang warfare of the late 1970s, the white and Chinese populations have been greatly reduced by out-migration.

Whilst some people might dispute the use of these racially exclusive terms in the Jamaican context, there is no doubting that uptown Kingston communities are different from downtown Kingston residential areas. However, upwardly mobile black Jamaicans live uptown in large numbers, as do higglers and other working-class people who have acquired money but who do not necessarily aspire to middle-class lifestyles. Many of these uptown residential areas might better be described as peri-urban communities, and have engulfed farming communities, and these rural people are still to be found side by side with their affluent neighbours. In yet other instances, a squatter community may have developed next door to a middle-class housing area, and may be functionally interconnected with it through the supply of domestic helpers, gardeners and the like. So, these areas, whilst showing a predominant social profile, are really more of a patchwork quilt of different types, many of which are affluent and upper middle-class.

One of the major dynamics of the Caribbean city, which underlies such social differentiation, is the movement of members of the majority low-income group in their quest to earn a living. In a string of papers, Alan Eyre (1972, 1979, 1984, 1997) has reported on some of the most detailed work ever carried out on the dynamics of squatters and squatter communities in the Caribbean. All of this research was carried out in the context of Jamaica. Fundamentally, Eyre's work serves to show that the principal flow of migrants is not to the shanty towns (Eyre, 1972). The main flow is of rural migrants to the tenement slums of the inner city. Only when they are established urbanites with a secure foothold in the urban economy do they tend to move from the inner city to the shanty towns, in an effort to reduce their payments on rent and to improve their general living conditions. Other migratory flows, including from the rural areas direct to informal settlements, are of far less salience. The outcome is that the occupants of the shanty towns are not rootless recent migrants but well-established urban residents of long standing. For example, in a detailed study of Montego Bay, Eyre (1972) showed that over three-quarters of the squatters that he interviewed had been born in the city itself, and that on average, household heads had been urban residents for over 11 years.

The observations presented above, concerning social zonation in Caribbean cities, can be combined with Eyre's findings on residential migration. This has been attempted by the present author in order to present a generalised model of Caribbean urban structure. This is shown as Figure 7.16. At the centre of the city is the CBD, which is surrounded by the high-density, low-income inner city, or 'transition zone' (Burgess 1925). This may contain older squatter settlements, which have been consolidated. Beyond this, the social gradient rises, reflecting the out-migration of the relatively wealthy. In the figure, the middle-income zone is shown as a continuous circular band, but in reality this may be discontinuous. From the middle-class residential areas, a number of high-income sectoral zones wedge out towards the periphery. Thus, between the high-status zones and the

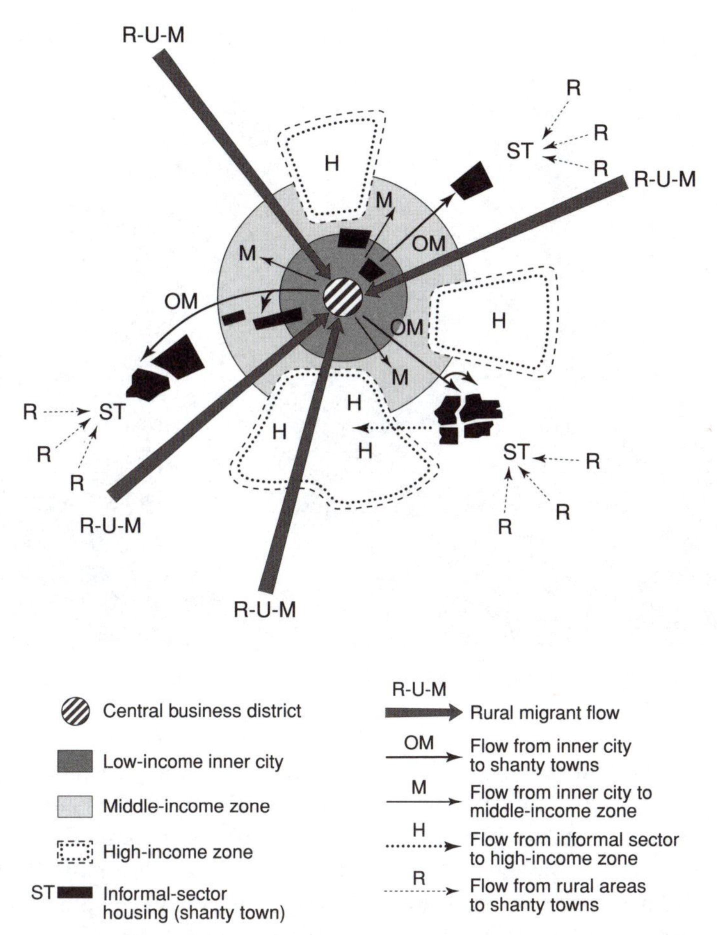

Figure 7.16 A model of Caribbean urban structure and population dynamics

low-income areas, there exist middle-income 'buffer zones'. Beyond the city proper there may be squatter settlements on vacant lands in the largest and fastest-growing cities (see Figure 7.16).

One final point worthy of consideration is that many of the smaller urban places of the Caribbean are relatively low-rise, with the majority of buildings not extending above two or three storeys. This is certainly true of Bridgetown, Castries, St George's, Roseau and Kingstown. This may well be an interesting area of urban change in the future, given the small size and limited land availability that is characteristic of such countries. Indeed, there are some indications of incipient high-rise development in these territories, with respect to symbolic high-rise government office blocks such as the twin towers of Port of Spain, the government and commercial office blocks adjacent to Castries Harbour in St Lucia, and the new government office block located in Kingstown, St Vincent. However, tourism together with Americanisation and the demonstration effect seem to be

Plate 7.4 High-rise townscape in the Condado area of San Juan, Puerto Rico
Source: Rob Potter

the key influences in relation to the occurrence of high-rise urban developments, as exemplified by the Condado area of San Juan, Puerto Rico (Plate 7.4).

As an example of the dynamics involved in the establishment of high-rise properties, the recently prepared third National Physical Development Plan for Barbados (Government of Barbados 1998) envisages the development of urban-based tourism in the capital Bridgetown in order to complement long-established beach-based tourism. In turn, this is linked to the development of the urban waterfront and a number of relatively high-rise modern buildings. Interestingly, the Port St Charles development on the west coast near Speightstown is a three- to four-storey development aimed primarily at wealthy overseas visitors and expatriates (Potter and Phillips 2004). Developed in the mid-1990s, by 2000 an exclusive waterfront home in the development, comprising 3,500 square feet, was reported to cost US$6 million, whilst condominiums were priced at around US$2.5 million. At a more affordable level, Marlin Quay overlooking Rodney Bay in St Lucia offers apartments and villas for shared ownership or outright purchase, based on what can be described as a three-storey mock-Spanish/Moorish template.

Urban redevelopment and the future of Caribbean urban regions

This brings us to the concluding section of this chapter. What forces are currently shaping the Caribbean urban area and what forces may be expected to exert a powerful influence in the future?

In the last section, it was stressed how the restricted sites occupied by many West Indian towns and cities led to the need for land reclamation schemes. An aspect of this change, which has not been discussed thus far, is the development of deepwater harbours. As Hudson (1998) notes, port improvements to facilitate larger shipping and loading equipment have frequently gone hand in hand with this process, involving waterfront landfill to create deepwater frontage, thereby allowing ships to lie alongside the wharf. This has facilitated both increasing cargo and cruise ship traffic.

Hudson (1998) describes in detail the extensive land reclamation schemes that were undertaken in Port Royal, Jamaica, from 1870 onwards. Port of Spain's town council took similar steps in the early nineteenth century, thereby both enhancing the port and providing much needed additional land for urban development at a time of rapid urban growth. Another interesting example is afforded by the case mentioned earlier, Castries, where as pointed out in relation to Figure 7.14, much of the site around the enclosed natural harbour at the foot of steep slopes is largely artificial (Hudson 1989). Further landfill reclamation has occurred on the Castries waterfront in modern times, providing land that is now used for low-income and squatter housing at Four-à-Chaud (see Figure 7.12), and for the recently completed Point Seraphine tourist development, where deepwater berths for cruise ships are situated in close proximity to duty-free shopping facilities and restaurants (Hudson 1998; Sahr 1998; Potter 2001) (Figure 7.17).

Thus, another change that is occurring within the urban envelope is that, as mentioned in the last section, the urban waterfront is coming to be increasingly positively valued. While a large number of long-distance holidaymakers appear to have remained satisfied with the sand, sea and sun formula, more specialised niche markets are emerging. These are frequently more culturally oriented and tend to stress more indigenous settings, including those that are to be found in urban areas.

Indeed, since the mid-1980s, the entire urban geography of Castries has been redesigned in accordance with the needs of tourism, especially cruise ships. From 1987, the informal housing area of Conway, which fronted directly onto Castries Harbour, was rapidly relocated to the south-west periphery of the city at Ciceron (see also Chapter 6). Government offices, including those of the prime minister, were created along with those of commercial banks. The market area has been completely redeveloped, and the Point Seraphine duty-free zone built. The view from the harhour might now cynically be described as 'entirely cruise-ship friendly'. At the same time, poorer housing is generally only visible as patches ascending the surrounding steep hills (Figure 7.12) (see Potter 1985, 1995; Sahr 1998). Potter (2001) has recently documented this tourism- and cruise ship-oriented urban redevelopment programme in some detail.

Likewise, the development of the waterfront in St John's, Antigua, has been interpreted by one analyst as the 'gentrification of paradise' (Thomas 1991). Since 1980, a number of nineteenth-century buildings along the harbour front have been restored and transformed into boutiques, restaurants, offices and a disco. The development is known as Redcliffe Quay. Thomas (1991) draws strong parallels between what he sees as gentrification in Antigua and the process as it

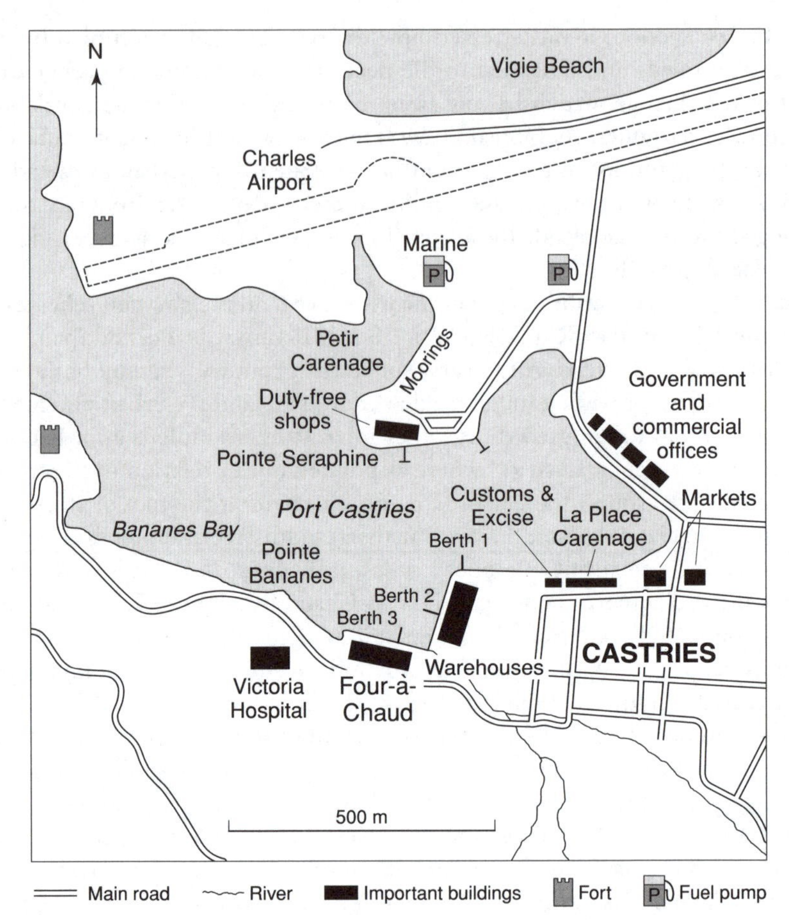

Figure 7.17 The principal features of the redevelopment of the waterfront in Castries, St Lucia
Source: Potter (2001)

occurs in developed countries, involving the upper-income use of buildings that are of architectural value, and the associated displacement of low-income groups (see also Chapter 11).

There are undoubtedly aspects of the post-modern condition here, with the old and the new being fused together, along with juxtapositions of the cultural and the functional, as well as the aesthetic and the commercial. Perhaps we should not be entirely surprised at such seemingly aberrant admixtures. In a slightly different urban context, Eyre (1979) referred to '*quasi-urban melange*' settlements in Caribbean towns. In these, unplanned rapid urban growth gives rise to a mosaic of juxtaposed sub-units of settlement, which display widely diverse social and physical characteristics (see also Conway (1989) and de Albuqueque *et al.* (1980) on so-called 'uncontrolled urbanisation' in the Caribbean). Eyre defined such urban zones as exhibiting little uniformity in terms of religious or political affiliations, or the use of facilities and utilities. Some commentators might even be tempted to argue that such fluidity is typical of Caribbean cities. Indeed, in respect of urban change and development in Kingston, Jamaica, Austin-Bross (1995) maintains that the residents of the city have always been essentially 'postmodern'. This is because they have long expressed scepticism concerning modernity and the metropolitan world, principally as expressed in the attempted modernisation of the city that occurred in the 1970s.

This brings the chapter to a final comment. Ultimately, despite the differences created by informal settlements, tourism and waterfront redevelopment projects, Caribbean urban places do show striking similarities in respect of their overall structure, as exemplified in this chapter. This is principally because in growing rapidly from their colonial origins, most have followed a strong path of modernisation along Western capitalist lines. Of course, as noted earlier in this chapter, there have been exceptions, such as that offered by Cuban cities like Havana. However, in the light of the Special Period, it is perhaps tempting to see this as a holding operation, with the market having taken over since 1989 (Mathey 1997).

Both Sahr (1997, 1998) and Potter (1998) have described Caribbean towns and cities as mini-metropolises or micro-metropolises. The salient point is that, in several instances, the structure of most contemporary Caribbean cities resembles very closely that found in the larger metropolitan centres of developed countries. As shown in the present chapter, they are markedly socially and functionally differentiated, and are frequently distinctly multi-nuclear. They exhibit rapid suburbanisation and the development of suburban fringe zones, with a predisposition towards sectoral residential areas, as well as linear or strip developments. In short, despite some major contrasts, especially those relating to low-income housing and squatter settlements, in these respects Caribbean cities appear like scaled-down versions of the extended metropolises of the Western world. This, should not be seen as surprising, since like those found in other world regions, Caribbean urban areas reflect the meeting of the global and the local, in the name of late capitalism.

References

Austin-Bross, D. J. (1995) 'Gay nights and Kingston Town: representations of Kingston, Jamaica', in Watson, S. and Gibson, K. (eds) *Postmodern Cities and Spaces*, Blackwell, Oxford, pp. 149–64.

Berry, B. J. L. (1961) 'City size distributions and economic development', *Economic Development and Cultural Change*, **9**, 573–87.

Boswell, T. and Briggs (1989) 'Bahamas', in Potter, R. B. (ed.) *Urbanization, Planning and Development in the Caribbean*, Mansell, London and New York.

Burgess, E. (1925) in Park, R. E., Burgess, E. W. and McKenzie, R. D. (eds) *The City*, University of Chicago Press.

Christaller, W. (1933) *Central Places in Southern Germany* (translated by Baskin, C. W.) Prentice Hall, Englewood Cliffs, NJ.

Clarke, C. (1974) 'Urbanization in the Caribbean', *Geography*, **59**, 223–32.

Clarke, C. (1989) 'Jamaica', ch. 2 in Potter, R. B. (ed.) *Urbanisation, Planning and Development in the Caribbean*, Mansell, London and New York, pp. 21–48.

Colantonio, A. and Potter, R. B. (2003) 'Participatory and collaborative planning in "Special Period" Cuba', in Pugh, J. and Potter, R. B. (eds) *Participatory Planning in the Caribbean: Lessons from Practice*, Ashgate, Aldershot and Burlington.

Conway, D. (1981) 'Fact or opinion on uncontrolled peripheral settlements in Trinidad: or how different conclusions arise from the same data', *Ekistics*, **286**, 37–43.

Conway, C. (1989) 'Trinidad and Tobago', ch. 3 in Potter, R. B. (ed.) *Urbanisation, Planning and Development in the Caribbean*, Mansell, London and New York, pp. 49–76.

de Albuquerque, K., van Riel, W. and Taylor, J. M. (1980) 'Uncontrolled urbanization in the developing world: a Jamaican case study', *Journal of Developing Areas*, **14**, 361–86.

de Albuquerque, K. and McElroy, J. (1989) 'Puerto Rico and the United States Virgin Islands', ch. 12 in Potter, R. B. (ed.) *Urbanisation, Planning and Development in the Caribbean*, Mansell, London and New York, pp. 285–315.

Duany, J. (1997) 'From the Bohio to the Caserio: urban housing conditions in Puerto Rico', ch. 10 in Potter, R. B. and Conway, D. (eds) *Self-help Housing, the Poor, and the State in the Caribbean,* University of Tennessee Press, Knoxville, and Jamaica, Barbados, Trinidad and Tobago, pp. 189–216.

Eyre, L. A. (1972) 'The shanty towns of Montego Bay, Jamaica', *Geographical Review*, **62**, 394–412.

Eyre, L. A. (1979) 'Quasi-urban mélange settlements', *Caribbean Review*, **8**, 32–5.

Eyre, L. A. (1984) 'Political violence and urban geography in Kingston, Jamaica', *Geographical Review*, **74**, 24–37.

Eyre, A. (1997) 'Self-help housing in Jamaica' ch. 5 in Potter, R. B. and Conway, D. (eds) *Self-Help Housing, the State and the Poor in the Caribbean*, University of Tennessee Press, Knoxville, and University of the West Indies Press, Jamaica, Barbados, Trinidad and Tobago, pp. 75–101.

Government of Barbados (1970) *Physical Development Plan for Barbados*, Government of Barbados, Bridgetown, Barbados.

Government of Barbados (1983) *Barbados Physical Development Plan Amended 1983*, Government of Barbados, Bridgetown, Barbados.

Government of Barbados (1998) *Draft National Physical Development Plan*, Government of Barbados, Bridgetown, Barbados.

Grosfoguel, R. (1995) 'Global logistics in the Caribbean city system: the case of Miami', ch. 8 in Knox, P. L. and Taylor, P. J. (eds) *World Cities in a World-System*, Cambridge University Press, Cambridge, pp. 156–70.

Hall, D. (1989) 'Cuba', ch. 4 in Potter, R. B. (ed.) *Urbanisation, Planning and Development in the Caribbean*, Mansell, London and New York, pp. 77–113.

Hoyt, H. (1937) *The Structure and Growth of Residential Neighborhoods in American Cities*, Federal Housing Administration, Washington.

Hudson, B. (1989) 'The Commonwealth eastern Caribbean', ch. 8 in Potter, R. B. (ed.) *Urbanisation, Planning and Development in the Caribbean*, Mansell, London and New York, pp. 181–211.

Hudson, B. (1998) *Urban Waterfront Development*, Mansell, London and New York.

Laguerre, M. S. (1990) *Urban Poverty in the Caribbean: French Martinique as a Social Laboratory*, Macmillan, Basingstoke.

Linsky, A. S. (1965) 'Some generalizations concerning primate cities', *Annals of the Association of American Geographers*, **55**, 506–13.

Mathey, K. (1997) 'Self-help housing strategies in Cuba: an alternative to conventional wisdom?' ch. 9 in Potter, R. B. and Conway, D. (eds) *Self-Help Housing, the State and the Poor in the Caribbean*, University of Tennessee Press, Knoxville, and University of the West Indies Press, Jamaica, Barbados, Trinidad and Tobago, pp. 164–87.

Mehta, S. K. (1964) 'Some demographic and economic correlates of primate cities: a case for revaluation', *Demography*, **1**, 136–47.

Portes, A. Dore-Cabral, C. and Landolt, P. (1997a) *The Urban Caribbean: Transition to the New Global Economy*, Johns Hopkins University Press, Baltimore.

Portes, A. Itzigsohn, J. and Dore-Cabral (eds) (1997b) 'Urbanization in the Caribbean Basin: social change during the years of crisis', ch. 2 in Portes, A. Dore-Cabral, C. and Landolt, P. (eds) *The Urban Caribbean: Transition to the New Global Economy*, Johns Hopkins University Press, Baltimore, pp. 16–54.

Potter, R. B. (1985) *Urbanisation and Planning in the Third World: Spatial Perceptions and Public Participation*, Croom Helm, Beckenham, and St Martin's Press, New York.

Potter, R. B. (1986) 'Physical development or spatial land use planning in Barbados: retrospect and prospect', *Bulletin of Eastern Caribbean Affairs*, **12**, 24–32.

Potter, R. B. (ed.) (1989) *Urbanisation, Planning and Development in the Caribbean*, Mansell, London and New York.

Potter, R. B. (1993) 'Urbanization in the Caribbean and trends of global convergence-divergence', *Geographical Journal*, **159**, 1–21.

Potter, R. B. (1995) 'Urbanisation and development in the Caribbean', *Geography*, **80**, 334–41.

Potter, R. B. (1997) 'Third world urbanization in a global context', *Geography Review*, **10**, 2–6.

Potter, R. B. (1998) 'From plantopolis to mini-metropolis in the eastern Caribbean; reflections on urban sustainability', in McGregor, D. F. M., Barker, D. and Lloyd-Evans, S. (eds) *Resources, Sustainbability and Caribbean Development*, University of the West Indies Press, Jamaica, Barbados, Trinidad and Tobago.

Potter, R. B. (1999) 'Review of *The Urban Caribbean: Transition to the New Global Economy*', *Urban Studies*, **36**, 402–4.

Potter, R. B. (2000) *The Urban Caribbean in an Era of Global Change*, Ashgate, Aldershot, Burlington and Sydney.

Potter, R. B. (2001) 'Urban Castries revisited: global forces and local responses', *Geography*, **86**, 329–36.

Potter, R. B. and Hunte, M. (1979) 'Recent developments in planning the settlement hierarchy in Barbados: implications concerning the debate on urban primacy', *Geoforum*, **10**, 335–62.

Potter, R. B. and Wilson, M. (1989) 'Barbados', ch. 5 in Potter, R. B. (ed.) *Urbanisation, Planning and Development in the Caribbean*, Mansell, London and New York, pp. 114–39.

Potter, R. B. and Pugh, J. (2001) 'Planning without plans and the neoliberal state: the case of St Lucia', *Third World Planning Review*, **23**, 323–40.

Potter, R. B. and Lloyd-Evans, S. (1998) *The City in the Developing World*, Pearson/Prentice Hall, Harlow, UK and New York.

Potter, R. B., Binns, J. A., Elliott, J. A. and Smith, D. (2004) *Geographies of Development*, second edition, Pearson/Prentice Hall, Harlow, UK and New York.

Potter, R. B. and O'Flaherty, P. (1995) 'An analysis of housing conditions in Trinidad and Tobago', *Social and Economic Studies*, **44**, 165–83.

Potter, R. B. and Phillips, J. (2004) 'The rejuvenation of tourism in Barbados', (in press) *Geography* **89**(3).

Pugh, J. and Potter, R. B. (2000) 'Rolling back the state and physical development planning: the case of Barbados', *Singapore Journal of Tropical Geography*, **21**, 175–91.

Rojas, E. (1989) 'Human settlement of the eastern Caribbean; development problems and policy options', *Cities*, **6**, 243–58.

Sahr, W.-D. (1998) 'Micro-metropolis in the eastern Caribbean: the example of St Lucia', in McGregor, D., Barker, D. and Lloyd-Evans, S. (eds) *Resources, Sustainbability and Caribbean Development*, University of the West Indies Press, Jamaica, Barbados, Trinidad and Tobago.

Taaffe, E. J. Morrill, R. L. and Gould, P. (1963) 'Transport expansion in underdeveloped countries: a comparative analysis', *Geographical Review*, **53**, 503–29.

Thomas, G. A. (1991) 'The gentrification of paradise: St John's, Antigua', *Urban Geography*, **12**, 469–87.

Town and Country Planning Division (1978), *National Physical Development Plan*, Government of Trinidad and Tobago, Port of Spain.

United Nations (1980) *Patterns of Urban and Rural Population Growth*, United Nations, New York.

Vance, J. E. (1970) *The Merchant's World: the Geography of Wholesaling*, Prentice Hall, Englewood Cliffs, NJ.

Weaver, D. (1988) 'The evolution of a plantation-tourism landscape on the Caribbean island of Antigua', *Tijdschrift voor Economische en Sociale Geografie*, **79**, 319–31.

Weaver, D. (1993) 'Ecotourism in the small island Caribbean', *Geojournal*, **31**, 457–65.

West, R. C. and Augelli, J. P. (1976) *Middle America: its Land and Peoples*, Prentice Hall, Englewood Cliffs, NJ.

Wirth, L. (1938) 'Urbanism as a way of life', *American Journal of Sociology'*, **44**, 1–24.

Part III

GLOBAL RESTRUCTURING AND THE CARIBBEAN: INDUSTRY, GLOBALISATION, TOURISM AND POLITICS

Chapter 8

INDUSTRIALISATION, DEVELOPMENT AND ECONOMIC CHANGE

This chapter presents the history and contemporary features of Caribbean development and industrialisation. The account is divided into eight sections. First, we review the broad patterns of social and economic development that characterise the region. Second, we offer a brief tour of the first four centuries of Caribbean development, aiming to show how the Caribbean was from the outset of colonialism a globalised region, dependent on and shaped by outside interests and agendas. Third, we note the Caribbean's principal natural resource endowments. Fourth, we present the region's two main industrial development models from the mid-twentieth century (industrialisation by invitation and Operation Bootstrap) and suggest how both perpetuated the region's historical reliance on outside powers. Fifth, we briefly review radical Caribbean versions of development theory and note how adherents attempted to influence state policy. Sixth, we describe the connections between foreign debt, neo-liberal reforms and export-oriented manufacturing within development policies since the 1980s. Seventh, we review the results of Caribbean experiments with socialist development. Finally, we conclude the chapter with a discussion of the region's ongoing trade dependence, particularly in relation to the United States.

Regional patterns of social and economic development

An overview of the basic economic and social indicators for the Caribbean reveals much diversity, as shown in Table 8.1. In an attempt to identify the main patterns, it is useful to think of Caribbean territories as falling into four groups. The territories can be grouped by their political status (dependent or independent), and by the two most commonly used indicators of economic and social development, per capita income, measured as purchasing power parity, and the infant mortality rate (IMR).

Group 1: the political dependencies

Twelve of the Caribbean's 28 territories are still overseas dependencies of various types (see Preface Table 1, 'Countries and territories of the Caribbean region',

Table 8.1 Social and economic indicators for Caribbean countries and territories

	Grouping (*see text for explanation*)	Area (square miles)	Population 2002 (000s)	Infant mortality rate, 2002	External debt in $1,000s (most recent year)	Per capita GDP, (PPP US$) (most recent year)	Human Development Index, 2000[2]
Anguilla	1	35	12	24	9 (98)	8,600 (01)	n.a.
Antigua and Barbuda	3	171	66	21	231 (99)	10,000 (01)	52
Aruba	1	75	70	6	285 (99)	28,000 (01)	n.a.
Bahamas	2	5,382	300	17	382 (00)	16,000 (00)	41
Barbados	2	166	276	12	425 (00)	14,500 (01)	31
Belize	3	8,867	262	24	500 (00)	3,250 (01)	58
Cayman Islands	1	102	36	10	70 (96)	30,000 (99)	n.a.
Cuba	4	42,804	11,224	7	11,000 (00) 20,000 (01)[1]	2,300 (01)	55
Dominica	3	290	70	16	150 (00)	3,700 (98)	61
Dominican Republic	3	18,704	8,721	33	5 (01)	5,800 (01)	94
French Guiana	1	33,399	182	13	1,200 (88)	6,000 (98)	n.a.
Grenada	3	133	89	15	196 (00)	4,750 (01)	83
Guadeloupe	1	687	436	9	n.a.	9,000 (97)	n.a.
Guyana	3	83,000	698	38	1,100 (00)	3,600 (99)	103
Haiti	4	10,597	7,064	93	1,200 (99)	1,700 (01)	146
Jamaica	3	4,244	2,680	14	5,200 (01)	3,700 (01)	86
Martinique	1	421	422	8	n.a.	11,000 (97)	n.a.
Montserrat	1	40	8	8	9 (97)	2,400 (99)	n.a.
Netherlands Antilles	1	308	214	11	1,350 (96)	11,400 (00)	n.a.
Puerto Rico	1	3,515	3,958	9	n.a.	11,200 (01)	n.a.
St Kitts and Nevis	3	104	39	16	140 (00)	8,700 (01)	44
St Lucia	3	238	160	15	214 (00)	4,400 (01)	66
St Vincent and the Grenadines	3	150	116	16	167 (00)	2,900 (01)	91
Suriname	3	63,251	436	23	512 (00)	3,500 (00)	74
Trinidad and Tobago	3	1,978	1,164	24	2,200 (00)	9,000 (01)	50
Turks and Caicos Islands	1	193	19	17	n.a.	7,300 (99)	n.a.
Virgin Islands (British)	1	59	21	20	36 (97)	16,000 (01)	n.a.
Virgin Islands (US)	1	136	123	9	n.a.	15,000 (00)	n.a.

[1] Russian debt

[2] Rank out of 173 countries

Sources: *The Human Development Report*, hdr.undp.org/reports/global/2002/en/pdf/backone.pdf; *The World Factbook 2002*, www.cia.gov/cia/publications/factbook/countrylisting.html

Plate 8.1 Scene from the main drag in Fort-de-France, Martinique, illustrating the consumer wealth found among political dependencies
Source: Tom Klak

for the political status of each). The dependencies sacrifice political autonomy, sovereignty and territorial integrity in as much as their laws are written in the far-off metropolitan capitals. At the same time, however, the dependencies obtain a range of benefits from their connections to core countries. Benefits vary among the 12, according to decisions made by the specific 'mother country' (Ramos and Rivera 2001). Typically, benefits include special investment laws and provisions, social subsidies, a guaranteed market for exports, and other development aid (Plate 8.1). There are also important indirect benefits from the stability that investors perceive in these territories, because they fly the flag and operate under the laws of a core country. Payoffs include considerable foreign investment and income flows from tourism (Chapter 11) and offshore banking (Chapter 9).[1]

Among the political dependencies there are two groups. The first consists of four of the smaller Caribbean territories: Aruba, the Cayman Islands, and the US and British Virgin Islands. Three of these territories have impressive social and economic indicators: IMRs of 10 or less (the British Virgin Islands is the exception with an IMR of 21), and all four have per capita income levels of $15,000 or more. These are very good indicators compared with global averages. They are, in fact, similar to those of the wealthy core countries to which these islands continue to be politically attached (United Nations 2000).

The remaining eight dependencies are Anguilla, French Guiana, Guadeloupe, Martinique, Monserrat, the Netherlands Antilles, Puerto Rico, and the Turks and Caicos Islands. The indicators for these dependencies are not as impressive as the four above, but are still very good by regional and global standards. Their IMRs range from 8 to 25, and per capita income levels range from $6,000 to $12,000.

Group 2: impressive Barbados and the Bahamas

These two independent states have managed to translate sizeable foreign exchange revenues, mainly from tourism and offshore services, into development gains (Brohman 1995). They have lowered their IMRs to below 20, and have reached per capita income levels above $10,000. These statistics place them well above the norm for the developing world. Indeed, these indicators are comparable with those achieved after decades of industrial export development in East Asia (United Nations 2000). The Bahamas has benefited from proximity to the USA and its tourists (1.4 million arrivals in 2002), cruise ships (2.8 million visitors in 2002) and offshore capital flows (mainly banking, insurance, and business and ship registration; CTO 2003). It must also be noted that the trafficking of drugs into the USA also enriches a Bahamian minority (Ferguson 2003).

Barbados' success is more domestic in origin, and is additionally impressive when one considers that its population density is 26 times that of the Bahamas (1,560 people per square mile compared with 58). Much of the credit in Barbados can be attributed to sound public administration. Successive governments from both major parties since independence have been relatively non-ideological, technocratic, pragmatic, rational and committed to social welfare (Conway 1998). These positive contributions from the Bajan state to political and economic stability and to enhancing social capital have, in turn, attracted substantial foreign investment and strengthened the manufacturing sector (Downes 2001).

Group 3: 'archetypal' Caribbean countries

This group consists of 12 of the 16 independent Caribbean territories that arguably represent the Caribbean norm for development conditions and problems. They represent a great range in territorial size. The group includes the three relatively large mainland countries of Suriname, Guyana and Belize, as well as the Dominican Republic, the largest island territory after Cuba. The other eight members are smaller, ranging in size from Jamaica and Trinidad and Tobago, to the six eastern Caribbean micro-states of Antigua and Barbuda, Dominica, Grenada, St Kitts and Nevis, St Lucia, and St Vincent and the Grenadines. These 12 countries have per capita income levels in the $2,500–10,000 range. Their IMRs range from 15 to 40. These figures are near the mid-level for the world as a whole – that is, in the middle of what has become a widening gulf separating the rich core countries and the impoverished global periphery (Wade 2001).

Group 4: the outliers of Haiti and Cuba

Lastly, the Caribbean has two outliers of different sorts – Haiti and Cuba. Both have per capita incomes of $1,000–2,000, low by world standards. These income levels are more in line with those of sub-Saharan Africa and South Asia than with their Latin American and Caribbean neighbors. Beyond income, these two countries diverge. Haiti's destitute social conditions categorise it with the world's least developed countries. Its IMR of around 100 is also similar to that of sub-Saharan Africa. In contrast, Cuba boasts an IMR of 7. This is well below the OECD average of 12, and similar to that of the United States and most of Western Europe (United Nations 2000: 186–9). Even World Bank president James Wolfensohn has acknowledged that 'Cuba has done a great job on education and health' (Lobe 2001).

Notwithstanding this apparent economic and social diversity in the Caribbean, however, lie many similar territorial and structural problems. The region is characterised by territorial fragmentation and isolation. The Caribbean lacks land-based access to major markets. Development policies are hampered when they must incorporate small, non-contiguous territories separated by large stretches of sea (see Figure 1.1).

Many of the 12 dependencies in the region are themselves territorially fragmented. This fragmentation is suggested by names that are pluralised (e.g. the Cayman Island*s*, or the Netherlands Antille*s*) or that include multiple entities (e.g. Turks *and* Caicos). Eight of the independent states also consist of several islands. This is again suggested by their pluralised names (e.g. the Bahamas Island*s*) or by those including 'and' (Antigua *and* Barbuda, or St Kitts *and* Nevis).

By world standards, the countries of the Caribbean also suffer from an array of economic structural weaknesses. Their economies place undue emphasis on a few sectors such as bauxite, bananas or beaches (mass tourism). Their manufacturing sectors are also not diverse. Furthermore, they are highly vulnerable to external decision making and change, heavily reliant on international trade, and burdened by large foreign debts (Table 8.1). They also have wide income disparities. Other macro-economic indicators signal additional long-term problems. Traditionally the Caribbean has relied on primary product exports. But these have lost their export vitality. Non-traditional exports such as garments or vegetables have been inadequate substitutes in terms of earnings, employment and linkages. In addition, after a decade of net capital export, Caribbean countries continue to be burdened by foreign debt servicing that drains a large share of national resources. For many Caribbean countries, total debt has continued to increase, and is now at or near record levels (Table 8.1). The region also has a growing trade deficit whereby imports exceed exports by about 50 per cent, with the difference made up primarily by tourism and remittances. Work forces grow while employment stagnates or even declines in terms of quantity and quality. Further, the World Bank's structural adjustment dictates forced cuts in education, health care and other basic needs.

This chapter tries to account for how this underdevelopment has come about. The chapter reviews the key historical factors as a backdrop for an analysis principally focused on the contemporary industrial development policies in the region. Only in the decades following the Second World War did Caribbean governments, like their counterparts across the 'global South', try to cultivate a manufacturing sector that both produces from the Caribbean and serves Caribbean consumers. Since the foreign debt crisis of the 1980s, the Caribbean has needed to shift towards export-oriented industrialisation. This policy shift involves the use of public funds to build and maintain export processing zones (EPZs), and to subsidise foreign investment through tax breaks and infrastructural provision.

A brief sketch of the first 400 years of Caribbean development: the making of a globalised, peripheral region

The exploitation of Caribbean natural resources and labour extends back to the sixteenth century. Since the 1500s the Caribbean has also been a globalised region – that is, a region shaped by the interests, priorities and agendas of outside powers within a world system of trade and profit making. The Caribbean islands were Europe's first overseas colonies, and for centuries they were among the principal sources of Europe's profits.

In search of gold and silver, the Spanish and other Europeans scoured and pillaged the Caribbean islands and, later, the Latin American mainland. As a result of this first round of resource extraction, during the sixteenth century, Europe's supply of gold increased by 20 per cent, while its silver increased by 300 per cent. These precious metals began a several centuries-long process whereby the Caribbean fuelled the growth of a Europe-centred, global capitalism. The precious metals were used to buy out and marginalise Europe's feudal land-owning elite, and to strengthen the process of urban industrialisation and capitalist development there (Blaut 1992).

Then in the 1640s, the English, mindful of Portugal's tremendous profit making from its Brazilian sugar plantations, began growing the first Caribbean sugar cane on Barbados. Sugar has been the Caribbean's principal export ever since. By the late eighteenth century, profits primarily from sugar, but also other primary products, made Haiti the world's most profitable colonial possession, more valuable than all of continental Latin America combined. Other Caribbean islands ranked close behind Haiti as overseas sources of profit. When Haitian slaves revolted and sought independence from 1791 to 1804, imperial France not only sought to suppress the uprising, but it also had to contend with Spain and England, as all three vied for the territory. Independence was finally achieved, but Haiti's profitable infrastructure and landscape were ruined. Then Cuba and many other Caribbean colonies rose to supplant Haiti as the world's major sugar producers. During the nineteenth century, Caribbean sugar accounted for a third or more of the world's total and its profits were a major source of surplus that helped to fuel Europe's industrial revolution (Wallerstein 1979). Since the late nineteenth century, however, competition from sugar beets grown in mid-latitudes and other factors have steadily reduced the global significance of Caribbean sugar. Today it accounts for a mere 5 per cent of the total (Galloway 2000).

A few additional points about the political economy of sugar production during the colonial period will further demonstrate the historical process of Caribbean underdevelopment. Sugar cane requires local refining. The sugar content of cane falls dramatically within days of harvest. There is insufficient time to ship cut cane all the way to Europe. In addition, cane is very bulky compared to granular sugar. So European colonial interests needed to put sugar mills throughout the Caribbean. Why, then, did not the Caribbean accumulate great wealth from growing, processing and exporting vast quantities of that most sought-after commodity? Sugar cane has four derivatives: brown and white granular sugar, molasses and bagasse (the leftover waste material). Europeans declared brown sugar to be only semi-refined. White refined sugar was considered the final product. Only white sugar could be marketed across Europe. The Caribbean mills' output of brown sugar was therefore worth little on the market. The colonies were not allowed to fully refine sugar themselves. This ensured sugar profits would accumulate in Europe, and not where the sugar cane was grown and partially processed (Richardson 1992: 44).

The fate of the 'vacuum pan technology' further illustrates the workings of a capitalist world system in which core countries restrict development initiatives in peripheral regions such as the Caribbean. The vacuum pan was first used in British Guiana in 1833. It was an important innovation because it yielded more of the valuable sugar juice and less of the molasses by-product from the sugar cane boiling process. Great Britain's response to this Caribbean-based increase in productivity was not favourable. On vacuum pan-produced sugar, the colonial power slapped a 'punitive import duty so as to reduce competition faced by Britain's domestic sugar-refining industry'. British Guiana's governor observed in 1852 that without that punitive duty, all Caribbean plantations would have adopted the vacuum pan (Richardson 1992: 49).

The lessons from the vacuum pan technology example are clear and important: investors and policy makers from the core applaud innovations and productivity gains only when they serve their interests. Colonial powers' domination of the Caribbean brought them development and systematically inhibited development in the islands. Through core regions' initial economic advantages, and mercantilist policies by core states to protect their domestic economic interests, core regions and peripheral regions are path dependent, meaning that they tend to stay in similar positions in the world system over time (Gwynne *et al.* 2003).

The main point of reviewing this history of precious metals and sugar exports is to emphasise how the globalised, dependent nature of Caribbean development has drained resources and value out of the region. This is not simply due to the fact that the Caribbean has focused on primary product exports. More important than *what* is produced is the issue of *how* it is produced, and *to the benefit of whom*. In fact, it is argued that the dispersal of sugar milling factories throughout the Caribbean beginning in the seventeenth century meant that the region was industrialised before Europe (Richardson 1992). Like the garment industries spread throughout the Caribbean and other peripheral regions today, however, it matters less where the factories are located than who controls them and how they are run. Historically the Caribbean has been controlled by outside powers, based economically on imported labour, cleared to create monocrop landscapes, and reliant on the import of virtually everything else needed to sustain local populations.

Because all dimensions of Caribbean society were exogenously constructed and transplanted there, the culture of Afro-Caribbean people from the outset has been detached from its historical and geographical roots, and therefore from early on could be considered economically modern and global. Indeed, historical globalisation has not been *entirely* bad for the Caribbean. It has made the region culturally diverse, rich, open-minded, flexible and adaptable. Caribbean culture blends a wide range of African regional traditions, Asian traditions, West Indian adaptations, and many other foreign influences introduced by Europeans or picked up during circular migration abroad (see Chapter 2 and Richardson 1992; Mintz 1996).

This section of the chapter has thus far focused on the historical relationship between the Caribbean and Europe. Little has been said about the impacts of all this history on the direct producers – slaves and then emancipated farmers and other workers. The colonial-era development of mines and then plantations established the societal role for the region's workforce. The development of 'human capital', so much in vogue today, clearly was not a consideration during the era of slavery, and has not changed much since. The treatment of labour has not been conducive to either economic or social development. Economic historian Hla Myint summarised the labour problem associated with economies based on mines and plantations:

> Their low wage policy induced them to use labor extravagantly, merely as an undifferentiated mass of 'cheap' or 'expendable' brawn-power. So, through the vicious cycle of low wages and low productivity, the productivity of the indigenous labour . . . was fossilized at its very low initial level (Myint 1964: 56–7; quoted in Mandle 1996: 30).

Richardson (1992) echoes this characterisation of the role of labour when he describes the typical way in which plantation owners dealt with declining soil fertility and sugar yields. They simply imported more slaves to work the land harder and, in the process, overpopulated the fragile islands. This attitude among employers that Caribbean labour is expendable continued through the twentieth century. For example, Mandle (1996: 57) notes that the working-class revolts in the mid-1930s across the Caribbean were caused by labourers being 'regarded merely as suppliers of cheap labor'. Unfortunately, this portrayal rings true for contemporary factory conditions as well, including those in the neo-liberal export-processing zones and the data-processing sectors, as we discuss later in this chapter (see Carter 1997; Klak 1998).

This chapter has briefly reviewed Caribbean development history. The full story on agriculture is provided in Chapter 3. The main point here is that the

Caribbean's historical economic basis in agriculture and other primary products has left many unanswered needs and desires among Caribbean people, to which contemporary industrial policy must try to respond. Agriculture has been unable to provide the basis for prosperity and inclusive development in the Caribbean. Problems began with the region's limited land and soil resources, which were unable to support large post-colonial populations. The mal-distribution of land has worsened these problems. A highly unequal distribution of land has remained since colonial times as a barrier to agricultural and overall development (Rampersad *et al.* 1997). Also, the colonial governments limited the expansion of non-plantation agriculture, which in turn restricted the emergence of a class of productive small farmers and other entrepreneurs that could promote economic development (Mandle 1996: 31). Throughout the Caribbean, rural peasants and small farmers have been unable to sizeably expand production and income levels. The colonial or post-independence state has not implemented serious land reform accompanied by appropriate infrastructure and extension services. The limitations of Caribbean agriculture lead to two conclusions. First, it highlights the overall weakness of state development policy making, a weakness that carries over into industrial policy. Second, it underscores the urgent need to find alternatives to agriculture as the region's foundation for modern development and prosperity.

Natural resources

By international standards the Caribbean is not well endowed with natural resources. Table 8.2 lists the natural resources that make important contributions to domestic economies or to exports. These resources generate employment, foreign investment and/or foreign exchange. As Table 8.2 suggests, the resource distribution is geographically uneven. The natural resource endowments of many Caribbean territories amount to little more than sandy beaches to lure tourists. Indeed it is sometimes said that the region's most important and ubiquitous natural resource is its tropical island setting, which attract tourists from the global North (Pattullo 1996). While a good number of Caribbean territories possess some valuable resources, three countries hold the greatest concentrations of internationally valuable minerals: Cuba, Jamaica and Trinidad and Tobago.

While Cuban nickel (and associated cobalt) and petroleum were exploited during the Soviet era, they have become much more economically important since 1991. Cuba's economic rebound since the low point in its post-Soviet transition (1993) is in large part due to the productivity growth associated with foreign investment in tourism, oil and nickel. Cuba's nickel reserves are the world's fourth largest and yearly earnings grew to around $400 million by the late 1990s. Low international prices have limited Cuba's earnings since. The expansion in the oil sector is most dramatic. Oil companies from Europe, Canada and Brazil are tapping enough fuel from their new finds to be able to cover 90 per cent of Cuba's

Table 8.2 Caribbean national resources by country

Country or territory	Significant natural resources*
Anguilla	Lobsters
Antigua and Barbuda	Beaches
Aruba	Beaches
Bahamas	Salt, aragonite, timber, lobsters
Barbados	Petroleum, natural gas, fish
Belize	Timber, fish, shrimp, limestone
Cayman Islands	Fish, beaches
Cuba	Petroleum, nickel and cobalt, iron ore, copper, manganese, salt, timber, silica
Dominica	Timber, hydropower
Dominican Republic	Nickel, bauxite, gold, silver, iron, limestone, copper, gypsum, mercury, salt and sulphur; marble, travertine and onyx for the local construction industry
French Guiana	Shrimp, timber, gold
Grenada	Timber, deepwater harbours, tropical fruit
Guadeloupe	Beaches
Guyana	Bauxite, gold, timber, shrimp
Haiti	Gold, bauxite, copper, marble, calcium carbonate, hydropower
Jamaica	Bauxite, gypsum, limestone
Martinique	Beaches
Montserrat	Negligible
Netherlands Antilles	Phosphates (Curaçao), salt (Bonaire)
Puerto Rico	Copper, nickel, potential onshore and offshore oil
St Kitts and Nevis	Negligible
St Lucia	Minerals (pumice), mineral springs, geothermal potential, forests, beaches
St Vincent and the Grenadines	Hydropower
Suriname	Bauxite, gold, oil, timber, shrimp, fish
Trinidad and Tobago	Petroleum, natural gas, asphalt
Turks and Caicos	Lobsters, conches
Virgin Islands (British)	Negligible
Virgin Islands (US)	Beaches

* Defined as resources that currently have important domestic economic value and/or earn foreign exchange. In addition to the natural resources listed, virtually all of these countries or territories have at least 1 per cent of their land surface dedicated to farming. The two exceptions to this rule are Anguilla and the Cayman Islands. For some countries or territories, farming accounts for a much greater share of land and therefore of products.

Sources: World Eagle (1997), Sealey (2001), and the following websites: www.immigration-usa.com/wfb/antigua_and_barbuda_geography.html; www.google.com/search?hl=en&lr=&ie=UTF-8&oe=UTF-8&q=anguilla+-+natural+resources; www.villadawn.com/st_croix/info/usviinfo.htm; www.odci.gov/cia/publications/factbook/index.html; lcweb2.loc.gov/frd/cs/cshome.html

electricity needs by 2001 (Fletcher 2000). Foreign oil companies continue prospecting for what are predicted to be even larger offshore fields.

Jamaica's bauxite and alumina industry began in 1952–53 when Reynolds Aluminum, Kaiser Aluminum and Alcan (Aluminum Company of Canada) began mining and exporting the ores. (Alumina is partially processed bauxite whereby volume is reduced by 50 per cent.) Foreign companies bought huge tracts of the island very cheaply in the 1950s and 1960s. Jamaica rose quickly to become the world's second-largest supplier of bauxite and alumina after Australia, contributing 18 per cent of world total by 1973 (Manley 1987). Jamaica's industry has since declined owing to depleted reserves and alternative suppliers. Still, in 1998, it was the world's third-largest producer, contributing 10 per cent and 7 per cent of the world's total of bauxite and alumina, respectively. Within CARICOM, Jamaica's bauxite and alumina are second only to Trinidad's petroleum as the leading export, earning $2 billion in 1998 (Sealey 2001).

Because the refining of bauxite into its final product (aluminium) is highly energy-intensive, and because North American firms have dominated Jamaica's bauxite industry, bauxite and alumina have always been exported northwards for final processing. This arrangement leaves Jamaica with a small share of the value added, with limited influence over the industry, and vulnerable to exogenous events. In 1999, for example, an explosion at the Gramercy Louisiana Alumina refinery, which buys more than 60 per cent of Jamaica's bauxite, reduced Jamaica's annual earnings by $10 million (Sealey 2001). Then, in 2001, Alcan announced it was pulling out of Jamaica and selling its assets to Swiss-based Glencore. Alcan was itself bought by another Swiss company in 2000 and has subsequently been restructuring its global operations. Global trading of this sort has created many uncertainties for the Jamaican economy and for policy makers.

Trinidad and Tobago's considerable petroleum reserves make it the Caribbean's most well-endowed territory in terms of resources. Using its petroleum both as an energy source and as a refinable product, Trinidad produces the region's widest range of heavy industrial products. Trinidad's oil and especially natural gas industry has become the basis for a vibrant petrochemical industry. Trinidad's industrial products include asphalt, ammonia and urea fertilisers, methanol, iron and steel. Its petroleum and natural gas industries alone contribute 25 per cent of GDP and are the region's largest export, earning over $3 billion in 1998 (Sealey 2001). The principal player in Trinidad's oil industry is the giant state-owned firm Petrotrin, which was incorporated in 1993 through a series of buy-outs of the local assets of foreign oil companies. Trinidad also produces an expanding range of household products and consumer items such as detergents and soft drinks that have become widely available throughout the CARICOM region since the 1990s. Despite its considerable natural resource endowments, however, Trinidad has been challenged to convert them into sustained development gains. Its exports have suffered from protectionism in the global North and competition from alternative producers. In addition, the enclave nature of its heavy industries has meant low labour absorption, tax yields and domestic linkages (Thomas 1988).

The environmental impacts of the Caribbean's natural resource-based heavy industries must also be considered and have often been severe. In Guyana, for example, cyanide spills are associated with gold mining. In Jamaica, bauxite has been the main agent of deforestation, and waste disposal from alumina produces red mud lakes that pollute groundwater aquifers. Industrial effluent flows from Trinidad's Point Lisas industrial complex have poisoned the Gulf of Paria.

Natural resources have formed the basis of a certain form of Caribbean industrialisation, one of an enclave variety that is highly dependent on outside technologies, TNCs and markets that have considerable influence over the industry. Indeed, at a global level, natural resource-rich countries have underperformed resource-poor countries. This owes to a variety of factors, including the 'Dutch disease', whereby an export commodity boom shrinks domestic agriculture and manufacturing while it encourages imports (Auty 2002).

Postwar industrial strategies

Efforts to develop manufacturing activities in the Caribbean go back only about a half-century. Through the first few decades of the twentieth century, 'manufacturing was not even conceived of as an option for the region' (Mandle 1996: 51). Attitudes among the economic and political elite changed after the Second World War as it became increasingly clear that traditional agriculture could not provide sufficient exports and employment. The region therefore began to diversify both into a wider range of agricultural products, and into manufacturing. On the agricultural front, although the region has moved slowly away from its historical basis in sugar cane, it has not been enough to achieve global competitiveness and decent living standards for the majority of the population. Postwar Caribbean states and the region's large land owners have tried to reinvigorate the profitability of their monocrop plantation systems and associated export sectors by replacing an outmoded crop with another that appeared to have more promise. For example, various Caribbean countries substituted bananas, citrus or coffee for sugar, arrowroot or cocoa. However, such limited adjustments to the countries' agrarian production did not bring back the profitability of the colonial economies (Conway 1998).

How has industrialisation fared? This section reviews the region's two main industrial development models. These 'models' have been held up as either the analytical blueprints or the envied empirical examples to which Caribbean policy makers have aspired. The industrial models are similar to one another in that they have reinforced and extended ties with the North Atlantic region. In contrast, Caribbean experiments in socialist development have attempted to reconfigure the region's global relations, but these experiments (save for Cuba's) have been fleeting. Socialistic regimes are discussed in a separate section later in this chapter.

Plate 8.2 Gravestone of Nobel Prize-winning development economist Sir Arthur Lewis in St Lucia
Source: Tom Klak

Industrialisation-by-invitation: the contribution and legacy of Caribbean economist Sir Arthur Lewis

Sir Arthur Lewis of St Lucia is the Caribbean's best-known and most influential development economist (Plate 8.2). His recommendations that the Caribbean should aggressively pursue policies of economic diversification and industrialisation were in direct response to the failures of British colonial administrations until then (Lewis 1950). Arthur Lewis is virtually synonymous with Caribbean industrialisation strategies during the twentieth century. His work was a significant Caribbean contribution both to development theories and to their application to promote development across the global periphery.

As often happens when policy prescriptions are put into practice, however, the applied version of Lewis's ideas was simplistic compared with the original (Conway 1998). Lewis's policy recommendations came to be known as 'industrialisation-by-invitation' (here we will refer to it as 'the Lewis strategy'). Based on the writings of Lewis and kindred Caribbean economic analysts, policy makers came to believe that foreign capitalists held the key to the region's industrialisation. Industrial growth would occur only following 'a considerable inflow of foreign capital and capitalists and a period of wooing and fawning upon such people' (Lewis 1950). The assessment was that 'in the absence of the nurturing of a West Indian business sector, all the region could do was hope that overseas business people would flock to the West Indies' (Mandle 1996: 67). Although Lewis had the greatest influence on the region's English-speaking governments, similar policies were enacted in postwar Haiti and the Dominican Republic (Thoumi 1989; Tirado de Alonso 1992).

At the same time that the Caribbean began pursuing the Lewis strategy, researchers at the United Nations Economic Commission for Latin America (ECLA) in Chile were working in parallel fashion. They articulated a somewhat more autonomous development strategy called import substitution industrialisation (ISI). The thinking from ECLA was as follows. ISI involved:

> policies of protectionism, exchange controls, attracting foreign investment to home industries, encouraging national investment and raising wages in order to provide effective demand, [which] were all aimed at stimulating the production of goods locally to replace imports, that is import substitution or IS (Cubitt 1995: 36).

Compared with the Lewis strategy, ISI was less driven by the actions and interests of foreign investors. ISI placed greater emphasis on stimulating domestic industrial entrepreneurship and increasing wage levels.

A fundamental problem with the Lewis strategy for the Caribbean is that despite all of the 'wooing and fawning' of foreign investors (i.e. subsidies, protections and other state support), they have never been convinced to put much money into the region. To cite a typical example, in Jamaica foreign capital was only 5 per cent of the total in the late 1980s (Rampersad *et al.* 1997). Foreign industrialists have always perceived the Caribbean as unstable, and its (predominantly male) industrial workers as rebellious and undisciplined (Segal 1989; SRI 1992). Therefore, after two decades of policy efforts based on the Lewis strategy, the Caribbean attracted little (mostly enclave-oriented) foreign investment. Most of what did come was of the assembly plant type, with few linkages to the rest of the economy. The Lewis strategy largely involved US MNCs relocating production facilities within the Caribbean to serve customers within the region. Only 4 per cent of Caribbean-manufactured goods were exported as of 1970 (Deere *et al.* 1990). It yielded no major expansion of domestically owned industries. Much of the new industry was capital-intensive as well, notwithstanding the huge and growing demand for decent working-class employment.

Lewis persuasively argued that the Caribbean could not attempt to industrialise autonomously, but must instead rely on attracting outside capital, technology and industry. The broader lesson from the Lewis strategy is that historical dependency has bred new forms in both theory and practice. Caribbean countries have historically relied on imported industrial and commercial goods and on foreign capital and expertise to feed people, service industry and finance internal capital expansion (Grugel 1995). Lewis extended this foreign dependency and vulnerability, by establishing an externally driven approach to industrialisation that continues until the present.

Operation Bootstrap

For a half-century, Caribbean political leaders have viewed Puerto Rico's Operation Bootstrap as an industrial development model. Arthur Lewis and his economist colleagues were similarly impressed (Farrell 1980). Upon closer inspection, however, Operation Bootstrap can be seen as a model that (1) cannot be replicated beyond Puerto Rico's unique political circumstances, (2) excludes key domestic interests from its benefits, and (3) deepens rather than resolves the region's long-term problem of external dependence and vulnerability.

Puerto Rican industrialisation is attributable to the island's unique commonwealth relationship with the United States. Operation Bootstrap began in 1947 when US tax law amendments gave corporations multi-year tax exemptions on any net income generated on the island and then deposited in banks there. The typical flow of this income is that it is generated in manufacturing and then invested in tourism and other real estate activities. For their part, Puerto Rican officials encouraged US investors by promising them industrial infrastructure and low-wage, non-union and, by Caribbean standards, skilled labour (Thomas 1988; Cordero-Guzman 1993). Puerto Rico's export shift away from primary products and towards manufactured products has been dramatic. Sugar exports to the USA have declined precipitously since the 1960s, to the point now in which they are insignificant (Galloway 2000). Industrial exports have more than made up for the decline.

The first industries to relocate to Puerto Rico included clothing, shoes and glassware. Then came heavier and more environmentally damaging industries of petroleum refining, petrochemicals and pharmaceuticals. They contributed to impressive annual economic growth rates of 6 per cent in the 1950s, 5 per cent in the 1960s and 4 per cent in the 1970s. All totalled, Puerto Rico's real GDP growth rate 1950–1990 was the eighth highest of the 74 countries worldwide for which data are available, and the highest in the hemisphere (Baumol and Wolff 1996). In 1979 Puerto Rico had the world's highest per capita level of US imports, modern transportation facilities, and 34 per cent of all US foreign direct investment. The 22 US drug companies manufacturing in Puerto Rico during the 1980s saved $8.5 billion in income tax exemptions over the decade (Freudenheim 1992). The various subsidies from the US mainland to Puerto Rico amount to around $9 billion annually, almost as much as all of the US aid to the rest of the world combined (de Blij and Muller 1998). Cuba's $5–6 billion annual subsidy from the USSR during the 1980s makes an interesting comparison, and is discussed in a later section of this chapter.

The geopolitical price of Operation Bootstrap is great dependency on US subsidies, capital and trade. The US Department of Commerce has acknowledged that US firms have not incorporated Puerto Rican ones into their chains of production (Paus 1988). Local manufacturers account for only 20 per cent of factory jobs, and a mere 2 per cent of industrial profits (Susman 1990). In 1996 the US Congress began a ten-year phase-out of Puerto Rico's corporate tax exemption, leaving the island entrenched in a bankrupt model of dependent industrialisation (Pantojas-Garcia 1990). Puerto Rican people have made the most of their special political status by migrating to and circulating back from the US mainland in great numbers, and by relying on the federally mandated social welfare net (Conway 1998). The human face of Puerto Rico's industrial expansion includes the fact that real per capita income has not grown relative to that of the mainland USA during Operation Bootstrap, and that unemployment has often been in excess of 20 per cent since the 1970s (Cordero-Guzman 1993; Grugel 1995). Puerto Rico's Operation Bootstrap illustrates the double-edged sword of growth/prosperity and dependency/vulnerability associated with relying on special trade preferences from the core capitalist countries.

New world dependency theory

It was already clear to many Caribbean social scientists by the 1960s that Arthur Lewis's prescription for industrialisation and development for the region was unviable. The Lewis industrialisation strategy replicated the historical over-reliance on foreign capital and failed in its promise to provide an engine for sustained economic growth. New world dependency theory (NWDT) was conceived in part in opposition to Lewis. NWDT was an alternative to modernisation theory-based prescriptions for development. Articulated in the 1960s and 1970s in the English-speaking West Indies, NWDT was a distinctively Caribbean version of Marxism-inspired development theory (Demas 1980; Thomas 1988). Political circumstances also engendered NWDT. Beginning with Jamaica and Trinidad and Tobago in 1962, the British colonies were gaining independence. This provided additional motivation to set the region on a new course for social and economic development and self-reliance.

Yet another stimulus for NWDT came from Latin America. There, beginning in the 1950s, scholars first at ECLA, and then across Spanish-speaking Latin America, were interpreting the region's underdevelopment as a result of unequal trade between their primary products and the manufactured goods from Europe and North America (Conway and Heynen 2002). Caribbean intellectuals were informed by the writings of their Latin American counterparts. They were also inspired to define a uniquely Caribbean critical economic history based on the plantation experience and to create policies based on autonomous industrialisation, social justice and third world solidarity.

Never before or since has there been so much intellectual fervour in the English-speaking Caribbean and even extending beyond it. Scholars formed new world groups to debate and devise development theory in various locales from Guyana, St Kitts and Anguilla to Montreal and Washington (Marshall 2002). The most vibrant locale was Kingston, Jamaica, where new world scholars at the University of the West Indies (UWI) advised Michael Manley's administration during the 1970s in its pursuit of democratic socialism. From the dialogue between UWI faculty and leaders of the People's National Party (PNP), a range of policy priorities emerged, such as investing in land reform for small farmers, imposing a levy on the bauxite exported by foreign companies, empowering community councils, and committing to social welfare, redistribution and equity (Manley 1987; Thomas 1988).

By the 1980s, however, NWDT was history. Why did it collapse? Observers have identified a variety of contributing factors:

- Much of the dependency theory writing in the Caribbean, Latin America, and beyond was flawed, in that it tended to be rigidly deterministic, overly economistic and tautological (dependency was presented as both cause and effect; Booth 1985).
- Careerism among new world intellectuals diverted their attention away from developing sound theory and policy, and spawned 'a culture marked by keynote address, cocktail attendance and doctoral authority' (Marshall 2002: 105).
- In the case of Jamaica, the PNP was said to be more democratic socialist in rhetoric than in practice. It therefore could not withstand attack by right-wing forces inside the country and from the United States. It then lost the 1980 election to the Jamaican Labour Party's Edward Seaga, who allied himself with Ronald Reagan and Margaret Thatcher and pursued the Puerto Rican industrialisation model (Payne 1982; Thomas 1988).
- Beginning in the early 1980s, the debt crisis forced a priority shift in the Caribbean and the global South as a whole away from autonomous development and towards attracting investment from core regions (Klak 1995).

Is NWDT of any relevance to contemporary development theory and practice? There are two valid answers that, at first glance, may seem contradictory, but upon closer inspection are not. One is that Caribbean dependency theory, like its Latin American counterpart, is dead. Gone are the fertile and optimistic intellectual exchanges in which regional history and contemporary development options were challenged and re-theorised. Gone too is the direct application of dependency theory to the conceptualisation and implementation of social and economic policy by Caribbean governments. Instead, most Caribbean policy makers and scholars argue for greater insertion into the international market and reliance on foreign capital as the engine of economic growth. For example, the contributors to an edited volume on Caribbean policy choices are typical of the enthusiasm for relying on foreign investors. They are especially clear in their pro-capitalist priorities for the region, arguing that one of the keys to the Caribbean's future prosperity is that governments 'shape more attractive business environments' (Dookeran 1996: 127). Indeed, compared with the dependency theory era, today's development studies scholarship and policy applications seem like an intellectual desert that do little more than parrot the free-market dogma emanating out of Washington (Klak and Das 1999; Marshall 2002).

Another answer is more sanguine about NWDT's legacy. It finds continuity and extensions from NWDT in certain scholarship since the 1980s (Beckford and Girvan 1989; Dupuy 1997; Klak 1998; Girvan 1999; Meeks 2000). Such work in the political economy tradition improves on NWDT in several ways. It is less deterministic and economistic, it is informed by past policy failures, and it takes account of the new realities emanating from the evolving conditions in the global political economy. The remaining challenges for such scholars are to consolidate the intellectual legacies from NWDT and to find ways to employ the theory to influence policy. Undoubtedly, since the 1970s, alternatives to mainstream thinking have been far less central to Caribbean theory and practice. Most Caribbean scholars are finding it difficult to imagine an alternative to development reliant on foreign investors (Dupuy 1991; Bourdieu 1998). However, the region's ongoing development problems indicate the continued need for such alternatives (Klak and Das 1999), as the next section suggests.

The debt crisis and the neo-liberal solution

The origins and ongoing impacts of foreign debt

If we wish to understand why development policies today in the Caribbean and elsewhere in the global South focus on attracting foreign investors to generate new exports, we need to trace them back to the debt crisis of the 1980s. The foreign loan taking itself occurred primarily during the 1970s. In the 1970s, international banks were awash in deposits, and therefore eager to make loans to both public and private concerns in Latin America and the Caribbean (Corbridge 1993). Most of these loans carried floating interest rates that, at that time, were below the rate of inflation. The loans were therefore hard to pass up, especially for countries seeking ways to stave off deteriorating terms of trade for their primary products, and to pump-prime their development efforts (Klak 1995).

The foreign loan-based development strategy in the global South was foiled by US monetary policy. President Reagan raised interest rates markedly in 1980–1981 to attract global capital to the USA, in part as a way to finance his administration's deficit spending (Reifer and Sudler 1996). The US interest rate hikes increased the real rate of interest on loans that were made earlier and that carried floating rates (Figure 8.1). This interest rate increase swelled third world debt and helped to push the world economy into recession. The interest rate hikes also contributed to making the foreign debts of most developing countries, including those of the Caribbean, unserviceable. Debt arrears accumulated. In turn, global recession meant that the vulnerable primary product and tourism sectors of the Caribbean lost many of their customers. Economic decline was precipitous regionwide. By the end of 1983, the economies of Barbados, Dominican Republic,

Back to the Future

The effective federal funds rate is at its lowest point since the Kennedy administration.

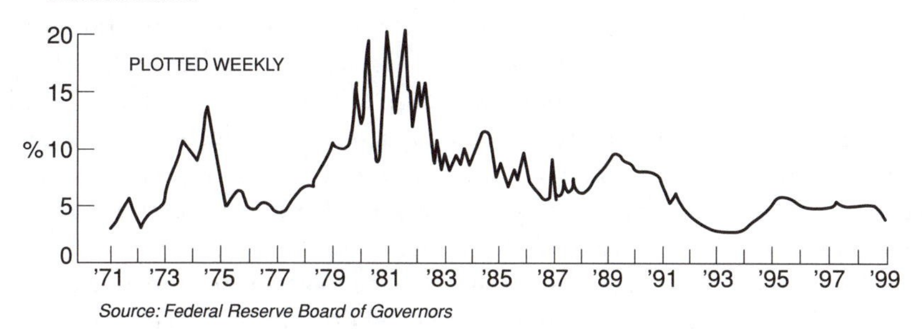

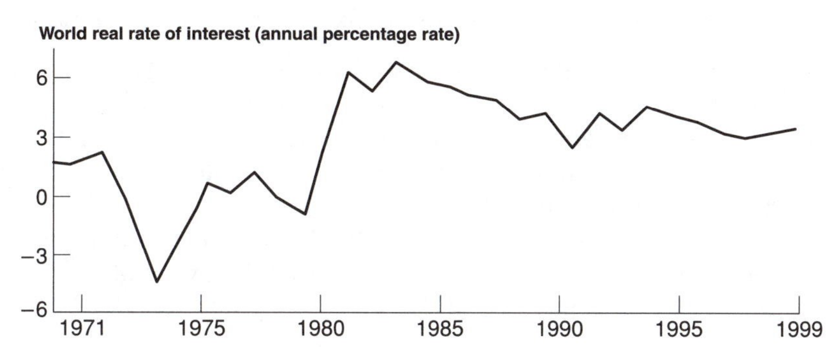

* 'real' = nominal or face value interest rate minus inflation rate

Source: From graph from *100 ways of seeing an unequal world*, Sutcliffe, Robert B., Zed Books (2001), by permission of Zed Books.

Figure 8.1 Graphs showing (a) US federal rates and (b) world real interest rates to show how (a) causes (b)

Guyana, Haiti, Jamaica and Trinidad and Tobago shrank an average of 17 per cent in real terms compared with 1980 (Conway 1998). Taken together, an array of related factors, from poor agricultural yields and prices to wasted windfall loans and political thinking that seeks industrialisation through foreign investment, led to governmental insolvency by the 1980s (Klak 1995). It took only a few more years for socialist Cuba to enter a crisis parallel to the capitalist Caribbean brought on by declining trade and agricultural production, and by mounting foreign debt. The case of Cuba is reviewed later in the chapter.

The many interrelated problems that we have outlined have provided strong motivation for Caribbean governments to approach the international development agencies seeking debt rescheduling, additional loans and, in the words of its leaders, the IMF's 'blessing' (McBain 1990; Killick and Malik 1992). Since the 1980s, Caribbean governments have needed to sign onto a series of agreements with the IMF and World Bank in exchange for financial bailouts. The bailouts come with strings attached. The international agencies seek to implement a new vision of how to develop the global economy and its member countries, which has come to be known as neo-liberalism. This vision includes fiscal conservatism, privatisation of public assets, currency devaluation, economic opening, and aggressive promotion of foreign investment and non-traditional exports.

Haiti provided an early case study of these neo-liberal transformations. It has for many years been the poorest country in Latin America (Table 8.1). For the vast majority of Haitians, conditions are grim: a high infant mortality rate, low life expectancy, low birth weights, high maternal deaths, inadequate caloric intake, squalid housing conditions and a lack of basic services such as potable water and sewer lines. The country is also very economically dependent, importing around 70 per cent of all that it consumes. It was in this context that the World Bank in 1985 offered advice to the Haitian government in a report intended to have restricted distribution, but which became public. The report, entitled 'Haiti: policy proposals for growth', suggested the following reorientation of national priorities:

> The development strategy must be export-oriented . . . [Domestic] consumption . . . will have to be markedly restrained in order to shift the required share of output increases to exports . . . More emphasis will have to be put on development projects that support the expansion of private enterprises in agriculture, industry and services. Private projects with high economic returns should be strongly supported with accordingly less relative emphasis on public expenditures in the social services (Wilentz 1989: 272–3).

Given how reliant Haiti is on foreign assistance and investment, World Bank recommendations such as these carry a lot of weight in government circles.

The main point to take from this discussion is that the third world debt crisis ushered in a new global development paradigm. Since the 1980s, the representatives of the multilateral aid agencies have advocated the neo-liberal paradigm during their frequent visits to the capital cities of the developing world, and thereby have moved the neo-liberal transition slowly but irreversibly along (Stiglitz 2000). Policy makers, in the Caribbean and in other peripheral regions, have placed great emphasis on attracting foreign investors to produce for export. As states compete for the same foreign investment, investors are empowered to shape the negotiations to their advantage.

Neo-liberal reforms have had the effect of homogenising government policies across the Caribbean and indeed all of the third world. Distinctive development strategies, socialist or otherwise, are history. Developing countries are likely to continue along the homogeneous neo-liberal track in the future, in part because of the ongoing constraints of foreign debt. In 1996 the total foreign debt of Latin America and the Caribbean stood at $622 billion, or $189 billion more than in 1987, when bankers and first world politicians considered that the crisis was at its worst (IDB 1997: 256; Rosen 1997). Measured against population size or against various economic indicators such as foreign exchange earnings, the foreign debts of four Caribbean Basin countries (Guyana, Nicaragua, Honduras and Jamaica) are now the hemisphere's worst and among the most burdensome in the world.

Non-traditional export promotion as development policy

For the small countries of Latin America and the Caribbean, neo-liberal reforms have placed great emphasis on non-traditional export promotion (NEP). The IMF, the World Bank and USAID have encouraged the governments of the smaller countries to promote non-traditional agricultural and industrial exports as a way out of the crisis of foreign debt and the collapse of traditional, primary product exports. For Caribbean countries, NEP has focused more on industrial than agricultural products, and so this section of the chapter also emphasises manufacturing.

In 1983, the US government provided some partial policy support for a conversion of the economies of the Caribbean and Central America towards NEP through the Caribbean Basin Iniative (CBI). Note that the CBI was initially a disguised US *military aid* policy against leftists in Central America. As an *economic* development policy, the CBI primarily involved a further opening of Caribbean markets to US products, and a relocation of US garment factories to the Caribbean to take advantage of cheap labour and subsidised factory space (Deere *et al.* 1990). The CBI virtually ignored tourism, and thereby forwent its potential contribution to sustainable development (Chapter 10). The CBI's opening of the internal markets of Caribbean Basin countries to US products has had a pronounced effect. For many years, the United States has had a large and growing export surplus with the Caribbean, which along with the rest of Latin America, is the only world region with which the US has had a trade surplus.

NEP has also had a significant impact on Caribbean exports. As a result of both the decline in traditional exports and the expansion of new ones, a significant shift in the composition of Caribbean exports has indeed occurred. For many Caribbean countries, such non-traditional industrial exports were contributing one-quarter to one-half of total exports to the USA by the 1990s (Schoepfle and Pérez-López 1992). If measured by growth in foreign exchange earnings and employment during the 1990s, the Caribbean region's two most important economic sectors were manufacturing for export and tourism. As we will argue, however, upon closer inspection Caribbean non-traditional exports display a number of significant weaknesses.

For the host countries of the Caribbean, NEP involves a wide range of public expenditures and incentives designed to encourage investment: governments have established export-processing zones. States have also offered investors duty and tax holidays of 10–15 years or more, and the unrestricted repatriation of profits (Deere *et al.* 1990; World Bank 1992). Other public costs associated with the promotion of non-traditional exports include expenditures for staffing public agencies, advertising the free zones abroad, preparing, operating and maintaining the industrial sites, entertaining prospective investors, training workers, and assisting foreign managers to establish operations and new residences and to hire desirable workers. Incentives and associated expenditures such as these have become standard offerings in many peripheral countries. As a result, they have come to serve as expensive antes into the international game of investment promotion, rather than as targeted and cost-effective means to expand exports (Klak and Myers 1997). As Dicken (1994: 112) put it, governments 'are locked in competitive struggles' over investment.

For all the efforts to conform to IMF dictates, the Caribbean has little progress to show for it. Jamaica is a case in point, having gone through more structural adjustment agreements over a longer time frame than virtually any other developing country. For Jamaica these began in the late 1970s when democratic socialist Prime Minister Michael Manley got into trouble with domestic investors, the USA and the international development agencies. Jamaica has for long done what the World Bank and IMF have requested but its exports have not increased (Mandle 1996), nor has it shown signs of being on a path towards development (Dookeran 1997; Meeks 2000).

A smaller state, or a different kind of state?

Under neo-liberalism, the state's role has shifted away from direct ownership of economic assets, production activities and the provision of social services, and towards attracting and subsidising export-oriented investors. The state's new role is often portrayed as one of downsizing (see Weiss 1997). However, it is more accurately viewed as a qualitatively different relation between the state, investors and workers. For some neo-liberal activities, such as promoting exports and competing for investment, the state's role has actually considerably expanded (see Box 8.1 on St Lucia' s National Development Corporation).

Export-processing zones as development policy

Export-processing zones (EPZs) are labour-intensive manufacturing centres that involve the import of raw materials and the export of factory products. EPZs (more

Box 8.1: Promoting export-oriented manufacturing: the example of St Lucia's National Development Corporation

A central player in St Lucia's efforts to create new sources of foreign exchange is the National Development Corporation, a parastatal agency (i.e. governmental but with considerable autonomy from the traditional state). Within its array of duties, NDC seeks to promote foreign investment and to expand non-traditional exports, ranging from garments and electronics to processed agricultural goods and data. NDC's mission and approach reflects the general outline of the role assigned to peripheral states in the neo-liberal transition. This role is usually

(Box continued)

carried out by newly created parastatal agencies that were initially funded in whole or part by the US Agency for International Development (Clark 1997).

NDC's stated objectives are to use product differentiation and niche marketing to revive and reorient the country's manufacturing sector, and to increase its contribution to national development (MFSN 1997: 55). NDC pursues these broad objectives by planning and executing industrial site development on the vast land holdings under its authority. NDC also leases, maintains and eventually, after the foreign loans for construction are repaid, tries to sell the factory shells (thus far unsuccessfully). In accordance with national development plans that encourage investment in manufacturing for export, NDC charges rents for factory space that are priced at 'very reasonable concessionary rates', as the Permanent Secretary in the Ministry of Commerce, Industry and Consumer Affairs phrased it (Richardson 1997). Fifteen-year tax holidays (which are usually extended when they near their end), tariff-free import of industrial inputs, unlimited repatriation of capital and profits, and worker training programmes round off the St Lucian government's offerings to qualified investors (NDC 1997a).

Agencies such as NDC that promote investment and exports cannot expect to recover their operational costs through the income they receive from the rental or sale of land and factory shells. On the contrary, proffering significant subsidies to (primarily foreign) investors has become a mandatory component of this development strategy, as the following comments by the head of NDC illustrate. In November 1996 Jacki Emmanuel Albertini became the general manager of NDC after years of work in the private sector. At a meeting soon after taking office with her more senior counterparts from Barbados, Jamaica and other Caribbean countries, Albertini rather naïvely asked when and under what conditions an agency such as NDC can be expected to move into the financial black. Their response, as she put it, was, 'child, it will never be'. Short of the seemingly impossible goal of financial solvency, Albertini is instead more modestly 'trying to keep her head above water' regarding the scale of NDC's financial deficit (Albertini 1997). Furthermore, NDC has limited funds to deliver quality service to its investor clients. The agency's burdensome expenses associated with investment promotion, site development, maintenance, management and unpaid factory rental bills have left it with inadequate cash flows, limiting its capacity to service its factory shells and to respond to tenant requests for basic assistance. NDC's inadequate site servicing and responses to its customers undercuts the broader national development policy of using state agencies and national resources to create pro-investment conditions.

In sum, NDC is a significant user of foreign exchange. No one knows how much net income it generates for St Lucia because a full cost accounting of the EPZ strategy has never been done. The agency is also financially strapped, overburdened and overextended in terms of responsibilities and projects. A ramification of these constraints is that NDC has little room for forward thinking and planning that could identify and aggressively pursue emerging market niches in which St Lucia could be competitive.

often called 'free zones' by their supporters) are extraterritorial sites in which manufacturing can proceed outside local regulations. EPZs have been created to streamline importing and exporting and to sidestep national laws that privilege some domestic social groups over foreign industrial exporters (Gereffi and Korzeniewicz 1994; Klak 1995). Most of the new export-oriented manufacturing occurs in state-designated free zones. In some cases, such as Barbados, the state has not designated free zones proper, but nonetheless offers a package of infrastructure and incentives to encourage foreign investment for export manufacturing which parallels that found in countries with official EPZs (Schoepfle and Perez-Lopez 1989).

More than 90 countries worldwide now offer export-processing zones. This global homogenisation of development policy emerged from a confluence of three exogenously controlled factors (Mody and Wheeler 1987; Klak and Rulli 1993): (1) massive foreign exchange shortfalls in less developed countries (that is, the debt crisis described earlier in this chapter); (2) the search by international investors for cost-saving components in manufacturing; and (3) the spread of neo-liberal ideas that encourage open economies, foreign investment and non-traditional exports (Williamson 1990, 1993).

EPZs primarily attract manufacturers seeking bargain-priced and compliant labour as a cost-saving component of global commodity chains. Many researchers have raised concerns about the overemphasis placed in NEP on the availability of low-wage, unorganised and low-skill employees (Warr 1989; Safa 1992; Klak 1998). In relation, other researchers argue that NEP is a self-contradictory development policy, inasmuch as the competitive advantage rests on labour cost minimisation rather than on more development-oriented criteria such as the productivity/wage ratio (Deere *et al.* 1990). Furthermore, investors are often prepared to relocate if provoked by social instability or by exogenous changes in trade policies or production methods (Gereffi and Korzeniewicz 1994; Klak 1995). This degree of exogenous control over export processing suggests that Caribbean countries have only limited latitude for capturing benefits (Kaplinsky 1993). EPZs are emblematic of the broader neo-liberal agenda, inasmuch as both emphasise open economies, foreign investment, non-traditional exports, privatisation and a withdrawal of the state to the role of 'market facilitator' (Williamson 1990; Krueger 1993).

Caribbean governments have aggressively promoted and subsidised non-traditional exports, including garments and high-value agricultural commodities such as pepper sauces and other condiments (Klak 1998). These efforts can be likened to a 'shatter-shot' approach to searching out export market niches, defined as new, high-value products that the Caribbean and other parts of the global South could sell competitively abroad. Even in the smallest countries, policy makers have been actively promoting investment in a host of non-traditional activities. In tiny Dominica, for example, these range from tourism, assembly operations and data processing to vegetables, fruits, seafood and cut flowers (Wiley 1998). Such experimentation raises the question of whether Caribbean countries can identify product niches with considerable promise or whether they are replacing monocrop and single market dependence with a new form of vulnerability. In other words, are Caribbean countries trying to do many things at once while not doing any of them well?

The non-traditional export activities that employed the most people and were most widespread in the Caribbean at their peak in the mid-1990s were the *maquiladoras* or assembly factories. Virtually every Caribbean state created export-processing zones and took loans to build and service factory shells within them. Garments have been by far the most common product, although electronics, plastic goods and shoes have also been produced. At their peak, tens of thousands of mainly young females were employed in such factories in Jamaica and Haiti. St Lucia and Grenada had more than a thousand workers each. The Dominican Republic had attracted the most garment assembly factories by far. Employment reached over 160,000 factory workers, whose products were worth $4.1 billion, or 82 per cent of the country's exports, in 1998 (Sealey 2001). Even there, however, the assembly plants have been low-paying economic enclaves with minimal positive impacts on the local economy. Since NAFTA in 1994, many foreign factory owners have left the Caribbean in favour of much lower-wage sites in Mexico and especially in China (see Box 8.2).

Although it is widely claimed that the assembly plants are net economic gains for Caribbean countries, that assertion would need to be validated through a thorough cost accounting, which appears never to have been undertaken (Klak 1995, 1998). The gross income from low factory wages and subsidised rents and utilities would need to be weighed against the public costs of promoting the country as an investment site, building, operating and maintaining the factories and related infrastructure, and training workers (see Box 8.1). The main beneficiaries of the low-cost products made in Caribbean assembly plants are the corporate

Box 8.2: Footloose investors and underpaid workers: export industries in St Lucia

In 1997, St Lucia's formal sector workforce included 3,565 manufacturing employees, who worked in 70 different factories (NDC 1997b). This represented 9 per cent of the formal sector workforce, or an average of 51 workers per factory. Since then there has been a precipitous decline in export-oriented manufacturing in St Lucia. The total number of factory workers has declined primarily because a number of garment and electronics factories have left the island. Garments have contracted the most, declining in terms of both output and value in each of the last six years for which data are available (MFSN 1997: 58). Because the industrial sites offered by the National Development Corporation (NDC) provide the majority of the island's total factory space, the pullout has worsened the agency's financial disequilibrium. Twenty-seven per cent of NDC's available factory space was vacant as of September 1997. In

(Box continued)

addition, firms fleeing as of 1996 owed over US$1 million to NDC in rents, and over US$4 million more in social security contributions (*St Lucia Business Focus*, 1996). These declines and losses have left most St Lucians, from policy makers to people on the street, less than sanguine about foreign manufacturers on their soil. As one official in the Ministry of Commerce, Industry and Consumer Affairs, which oversees the promotion of foreign industrial investment, laments, 'many establishments that came here in the early days were what are called "footloose" ' (Richardson 1997). Amid these absolute declines in the industrial sector, the national workforce continues to grow by about 3,000 per year, thereby making the task of providing sufficient numbers of decent jobs ever more difficult.

Over 90 per cent of workers in the existing export industries are women. This statistic is largely explained by their willingness to work for lower wages and to be more tolerant than men. As in other Caribbean Basin countries, the unemployment rate for women is higher than for men, and is highest among young women, many of whom are also heads of households. Young women are therefore quite desperate for work. NDC can take credit for providing jobs for some of them, although the wage rates for most factory jobs are inadequate. Ironically, one of main reasons that factories have been closing down in St Lucia is that industrial wages, which are usually between US$1 and $2 per hour, are higher than in many countries competing for the same investment. Wages are relatively high in St Lucia because they are buoyed up by the high cost of labour reproduction. Labour costs are relatively high in St Lucia for at least two reasons. Firstly, the standard of living in St Lucia is higher than in virtually every other Caribbean Basin country offering lower wage rates. Secondly, the cost of living in St Lucia is relatively high by regional standards. Because less than a third of the island's arable land is devoted to growing food other than bananas, and because of St Lucians' partiality for foreign products, much food is imported (Barrow 1992). In other words, both the small domestic food supply and a large import component make local prices relatively high.

However, to say that wages in St Lucia are relatively high should not be taken to mean that they are adequate. Indeed, the combination of relatively high wages and expenses creates an especially difficult situation for the St Lucian women working in the export factories. For the typical female factory worker, the pay level is less than what is required to reproduce her own labour power. These women have been forced to resort to a wide range of additional and creative survival strategies. For many St Lucian women these strategies include some combination of wage work, informal sector activities, homework, backyard gardening, food provision by relatives and neighbours, monthly support payments from one or more fathers of the children in the household, and rural living to minimise the cost of living. The factory pay rates are not high enough to allow many of the workers to live in an urban area, where public services and other amenities are better, but rents are higher and subsistence food production is not possible (Klak 1998).

brand names they carry, the department stores where they are sold, and US consumers who buy the low-priced products.

The threat and the allure of NAFTA

NAFTA (the North American Free Trade Association), which took effect on 1 January 1994, has decreased the Caribbean's economic importance in the global economy, and has relocated manufacturing investment to Mexico. Mexico's foreign investment and exports have grown relative to the Caribbean since 1994. Members of CARICOM (the Caribbean Community and Common Market) have reacted to NAFTA by altering their trade and investment policies to better match US preferences. As Demas (1997: 20) laments for CARICOM, 'We have already conceded far too much, far too quickly, by way of trade liberalisation'. CARICOM countries have also been pressing the United States for trade concessions. The most urgent recent aim from the Caribbean is parity with Mexico within NAFTA, so that the region will no longer be comparatively disadvantaged as an export platform. After years of Caribbean lobbying efforts, then US President Bill Clinton signed into law a modest trade bill compared with Caribbean demands. It is called 'NAFTA for the Caribbean and Central America', although it is also referred to as an extension of a Reagan-era bill, the Caribbean Basin Initiative. US garment retailers and producers predicted a great increase in production from Central America and the Caribbean as a result of this bill, but it has not been enough to overcome the forces producing a decline in the region's maquiladoras since 1994.

There is more hard evidence of the impacts of NAFTA on Mexico since it came into effect in 1994. Note that Caribbean and Central American countries seek NAFTA membership despite their competitiveness *vis-à-vis* the USA for little beyond low-wage labour. And they do this despite all the hardship in Mexico over NAFTA's initial years (Anderson *et al.* 1994; Schrieberg 1997). For example, open unemployment in Mexico increased from 3.4 to 7.3 per cent during NAFTA's first two years, while real median salaries fell by 50 per cent and interest rates and foreign debt soared. Mexico's per capita GDP was lower in 1997 than in 1982. Somewhere between 45 and 75 per cent of Mexico's nearly 100 million people live in poverty (Otero 2000). The distribution of NAFTA's benefits has also been economically and spatially unequal. Only 2.2 per cent of Mexican companies account for 80 per cent of the country's non-oil exports (Gayle 1998). Furthermore, benefits have been concentrated in Mexico's northern states. In southern states such as Chiapas impacts are largely negative and have fuelled the Zapatista rebellion.

The fact that Caribbean policy makers want to make their countries like Mexico suggests their *comprador* class positions. They are able to benefit personally from international ties that may not be in the interest of local residents (Cardoso and Faletto 1979). Indeed, investment promotion agencies have expanded throughout the Caribbean under neo-liberalism. The Caribbean continues to engage in what has been called a 'race to the bottom', whereby countries of the global

South compete with each other by offering foreign investors more subsidies and cheaper and more compliant labour (Green 1995). This race to the bottom suggests neo-liberalism's hegemony as a global development policy. It also serves to underscore the Caribbean's abject dependency and limited development options.

The data-processing industry: technopoles or electronic sweatshops?

The export market niche strategy that has become popular during neo-liberalism may be conceptually appealing, but the Caribbean's experience with it shows that it is fraught with difficulties. The problems are well illustrated by the data-processing sector, which Caribbean policy makers have hailed as a globally competitive industry that bodes well for the region's future. In order to contribute to Caribbean development, the data-processing sector would need to incorporate many local workers and to nurture local firms to take advantage of expanding opportunities in the global data-processing industry, which earns $1 trillion yearly (Plate 8.3).

Plate 8.3 The Montego Bay Free Zone in Jamaica specialises in data processing. The female workers prefer the more pleasant computerised work environment to that of textile factories. Unfortunately, the number of jobs available and the pay levels in either data processing or the textile industry have been far below the workers' needs
Source: Tom Klak

Mullings (1995, 1998) draws on a detailed analysis of the rise and fall of information services in Jamaica to explain why this industry, which has real potential (however narrow) for growth in employment, wages, managerial expertise and backward linkages into the local economy, has stagnated in terms of all these criteria. She identifies four factors that have hampered the industry:

- inadequate state support for local firms;
- continued policy steering and dampening by a traditional, non-dynamic private elite;
- investment fear on the part of foreigners;
- an extremely narrow role allotted to Jamaican firms and workers by US outsourcing firms.

Rather than propelling Jamaica to a heightened position in the international division of labour as the neo-liberal model predicts, the information services sector has slumped and entrenched the gender, class and international inequalities that have long characterised this peripheral capitalist country.

Despite Jamaica's negative experience, eastern Caribbean policy makers have in recent years been promoting the data-processing industry. With the garment sector collapsing, the World Bank has deemed it a viable new export industry for the eastern Caribbean (NDC 1998). Ironically, St Lucian policy makers view Jamaica's data-processing sector as a model they can emulate. The St Lucian state's own evidence of obstacles to the expansion of the data-processing sector would seem to be enough to discredit the idea. It hired two outside consulting firms to evaluate the prospects. The joint report by the Chicago Group and Wolf, Arnold, and Cardoso notes obstacles such as the following:

> Voice grade services cost US$1.85 per minute from St Lucia, while similar services with volume discounts cost US$.22 per minute from the Jamaica Digiport International [at the Montego Bay Free Zone] and as little as US$.10 per minute in the United States (*Diagnostic Evaluation of the Enabling Environment for Informatics in St Lucia* 1995: 19).

The negotiations of Caribbean states with the British firm Cable and Wireless, which has for decades commanded a telecommunications monopoly in Jamaica and elsewhere (Dunn 1995), were unable to open the telecommunications sector to competition or to bring local rates down to competitive levels through the 1990s (Vitalis 1998). Jamaican officials, followed by their eastern Caribbean counterparts, were finally able to loosen Cable and Wireless's monopoly after 2001. However, the question remains as to whether the Caribbean can overcome its long-term telecommunication disadvantages and now quickly rise to be internationally competitive for telecommunication-based industries. Despite the many reasons for scepticism, Caribbean governments are aggressively pursuing telecommunication services, as this policy position from St Lucia's National Development Corporation illustrates:

> Information services is without a doubt the fastest growing industry in the world today. The NDC believes that St Lucia is well placed to benefit from this business development both in terms of the export of information services and the technology infusion within all sectors of the economy. . . . The informatics companies . . . require more attractive work spaces which more closely resemble commercial office space . . . [Thus the existing vacant factory shells are unsuitable and new shells are being built] (NDC 1997c: 13–15).

Caribbean experiments with socialist development: narrow manoeuvring room *vis-à-vis* the USA

Global capitalism has not been generous to Caribbean people, some of whom have therefore experimented with an alternative model of development and societal organisation. Cuba has had a socialist regime since 1959. Three other Caribbean territories have had briefer and more modest socialist experiences: Guyana in 1961–1964 under Cheddi Jagan; Jamaica in 1972–1980 under Michael Manley; and Grenada in 1979–1983 under Maurice Bishop.

The essential challenge for the Caribbean's socialist regimes has been to harness an underindustrialised and unequal peripheral economy and to rapidly expand the productive forces and redistribute resources more fairly. This is in itself a tall order, made more difficult because it runs against the tide of global capitalism and US hegemony. Such departures raise broad questions such as 'How can a Caribbean country, in the US "backyard", pursue a non-capitalist path without the trade, capital and blessing of the United States and the rest of the largely capitalist world?' 'How can internal resources be progressively redistributed when there are few to go around, and when societies are poor and unequal and therefore prone to internal divisions, unrealisable pent-up popular expectations, and patronage systems?' The challenges are many and the successes not surprisingly are few.

Shutting down the socialist development option

Although this chapter has focused on economic development issues, they can never be divorced from geopolitics. The economic power of the United States in the region allows it to express its geopolitical power. Since the 1950s, direct or covert intervention by the United States in several Caribbean countries (e.g. Cuba, Guyana, the Dominican Republic, Grenada and Haiti) attests to the importance the superpower places on pursuing and maintaining its interests in the region (Blum 1986; Dupuy 1997; Walker 1997). Grenada felt Washington's political pressure in the early 1980s (see Box 8.3). The US economic embargo against Cuba is over four decades old and was only strengthened with the passage of the Helms–Burton Amendment in 1996 (LeoGrande 1997).

Box 8.3: Grenada, from 'Not for Sale' to 'For Sale': US opposition to its socialist experiment

'We are not in anybody's backyard, and we are definitely not for sale'. Maurice Bishop spoke these words in 1979, one month after taking power in Grenada's New Jewel Movement revolution (Conway 1983: 3). (How ironic his words were!) The Reagan administration was not amused by such defiance, and soon after taking office in January 1981 began work towards toppling Bishop's regime. At an OAS meeting in Washington in 1981, Reagan made an assessment and a warning that everyone knew was directed at Grenada:

> In the Caribbean we above all seek to protect those values and principles that shape the proud heritage of this hemisphere. Some, however, have turned from their American neighbors and their heritage. Let them return to the traditions and common values of this hemisphere and we all will welcome them. The choice is theirs (Hersh 1991).

Then the United States chose to distance itself diplomatically from the tiny island and conducted a mock invasion exercise on the Puerto Rican island of Vieques that was clearly aimed at Grenada. When Bishop traveled to the US capital in 1983 to assuage concerns about Grenada's international relations and development priorities, he got the cold shoulder. Upon his return home from cajoling Washington, the radical fringe of the New Jewel Movement arrested him and murdered him amid the ensuing chaos and struggle for power. In the words of the US ambassador to the eastern Caribbean at that time, Sally Shelton Colby, that murder and power struggle provided the United States with 'an enormous piece of luck' that it could use to justify the invasion it had long been planning (Hersh 1991). A predictable sequence of events followed: an election of the US-chosen presidential candidate, an inflow of modest US aid (including funds to complete the airstrip Reagan insisted Bishop was building for Soviet warplanes), and a pro-US policy. Although these events have perhaps left it less sanguine than many of its neighbours about neo-liberal development policy (Wiley 1999), the Grenada government now joins the chorus by selling its country and people to foreign investors in non-traditional export sectors. In March 2003, key planners of the US invasion, including Oliver North and former CIA operative Dana Rohrabacher, participated in a well-publicised cruise to and tour of the island under the theme of 'Celebrating the 20th Anniversary of the Liberation of Grenada' (www.freedomalliance.org/). (For further analysis of Grenada's brief socialist experiment, see Ambursley 1983; Conway 1983, 1998; Clark 1987; Brierley 1985; Hudson 1989; McAfee 1991).

As Grenada's disastrous power struggle in 1983 illustrates, the Caribbean's economic and geopolitical dependency on the United States is a constant factor weighing on the region's policy makers. This is in contrast to the relative decision-making autonomy enjoyed by the American hemisphere's larger countries, especially those of South America (Gwynne and Kay 2004). Caribbean countries must always take account of how the United States will react to their policy proposals. For Cuba's socialist regime, US animosity has been a constant preoccupation since the 1959 revolution.

Enemies from within and without (mainly the USA) have worked systematically against socialist regimes in the Caribbean. Included in the active enemies of Caribbean socialist experiments is the news media, as illustrated by the scare tactics employed by Jamaica's *Daily Gleaner* and the *New York Times* from the USA. These opponents capitalised on any of the inevitable errors in leadership, and they sabotaged the Caribbean's socialist experiments before their developmental capabilities could be ascertained (Blum 1986; Sunshine 1988; Booth and Walker 1993). While Cuba retains elements of socialism, Cuba is following the rest of the Caribbean region by courting foreign capital with special labour codes and incentives, and by relying on international tourism as its economic mainstay (see Chapter 10 and Pattullo 1996).

Economic development policies in Cuba: the Soviet era

Socialist Cuba's recent economic problems are not unrelated to those of the capitalist countries of the region. Since the nineteenth century Cuba has essentially had a one-crop (sugar cane) exporting economy, with the concomitant vulnerabilities of output and price fluctuations and deteriorating terms of trade (see Chapter 3 on agriculture). In part, Cuba's crisis is unique to its experience in CMEA (Council for Mutual Economic Assistance), the now defunct trade alliance among state socialist countries led by the USSR. Rather than embarking on a major post-revolutionary economic diversification effort from the 1960s through the 1980s, Cuba was encouraged to specialise primarily in sugar exports and to import most other requirements from other CMEA members. A one-for-one trading deal with Russia, sugar in exchange for petroleum, encouraged Cuba's specialisation and yielded a $5 billion annual subsidy compared with open market prices (note however that little sugar actually trades on the open market). The USSR also availed to Cuba a generous line of credit that has left an outstanding debt of $20 billion, in addition to the island's $11 billion of debt to Western creditors (Table 8.1).

Cuba's CMEA trade relationship delivered a fair assortment of industrial inputs and consumer goods until the unexpected events of the late 1980s. With the collapse of the Soviet bloc in 1989–1991, Cuba lost 75 per cent of its trade flows and imports, and aid worth 22 per cent of national income (Marshall 1998). Cuban imports fell from $8.1 billion in 1989 to only $2 billion in 1993, and its economy went into a free fall. Independent of a country's political-economic organisation, it is hard to imagine any country enduring such vast economic losses, hardships and required transitions without a major internal upheaval. Cubans have sombrely belt-tightened, coped, and adjusted to their suddenly transformed reality.

Economic development policies in Cuba today: socialist island amid global capitalism

Castro coined the euphemism 'The special period in the time of peace' to describe the post-Soviet era of massive shortages, daily hardships and new policies that encourage private initiative and court foreign investors. Cuba's domestic policies now allow farmers to sell their surplus directly to consumers, private vendors to offer many other consumer goods, and all citizens to use the US dollar as legal tender. New production policies include encouraging foreign investment by allowing (at least partial) foreign ownership of enterprises in Cuba, promoting tourism, reorienting state investment to enhance product development for global markets, and reconfiguring large state farms into autonomous and productive cooperatives (Susman 1998).

Figuring out creative ways to obtain food and other daily necessities now preoccupies most Cubans. Habaneros have been left especially vulnerable since CMEA's collapse because they are the most isolated from rural food production. The shortages of items of basic needs have encouraged the theft of state resources and black marketeering. For example, nearly 30 per cent of all tobacco produced on state farms is stolen for illegal sale, Cuban authorities admit (Chauvin 1998). Cuban officials have been forced to legalise a range of other capitalistic practices (Plate 8.4). The foreign tourism sector has acted like a magnet, drawing Cubans

Plate 8.4 Illegal informal vending at Varadero Beach, Cuba. Since this photo was taken in 1993, the Cuban government has legalised the US dollar and has offered licences to informal vendors. The government taxes and restricts the activities of small businesspeople, fearing that they may accumulate inordinate wealth and contribute to social inequalities. As a result, the number of licence holders for small businesses peaked at 208,000 in 1995, but fell to 113,000 by 2000
Source: Tom Klak

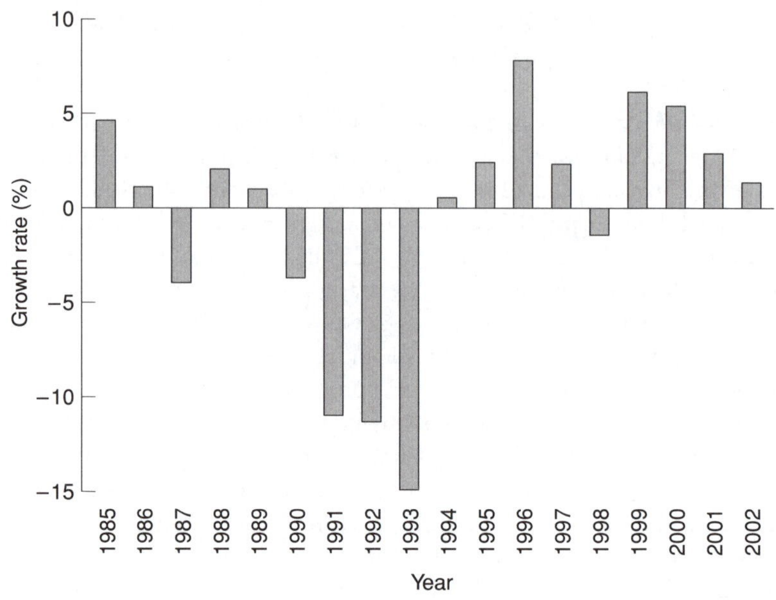

Figure 8.2 Cuban economic growth rates, 1985–2002
Note: Figures for 1985–1989 are GSP (global social product), while those for 1990–2002 are the more standard GDP (gross domestic product); for an analysis of their differences, see Zimbalist and Brundenius (1989)
Source: *Anuario Estatistico de America Latina y el* Caribe, United Nations, Santiago, Chile, various years; *Latin America Press, Inter-Press Service*, and others

of all stripes away from skilled professions and towards the dollar-earning possibilities. A highly trained professional now earns much more driving a taxi cab or mixing drinks for tourists than at her/his formal sector job (Segre *et al.* 1997). A result is an 'inverted income pyramid', whereby low-skill jobs pay much more than those requiring higher education.

The Cuban economy bottomed out in 1993 and has since haltingly recovered due to expanded domestic supplies of petroleum and foreign exchange earnings from tourism, nickel and cobalt (Figure 8.2). On top of that, in October 2000, President Hugo Chavez of Venezuela, which has more petroleum reserves than any country outside the Middle East, offered Castro the Caracas Energy Accord. It provides oil to Cuba and other Caribbean countries at reduced prices, with low interest and with the option of payment in kind. Many economic problems remain in Cuba, however. The 1998 sugar harvest of around 3 millions tonnes was the lowest since 1959. Cuba has therefore had little alternative but to join most other Caribbean islands that are members of the Caribbean Tourism Association, and to rely on foreign tourism as its main economic sector (see Chapter 10). Cuba's tourism receipts reached $1.5 billion by 1997, six times higher than in 1990.

While the United States continues to try to isolate Cuba, other countries such as Canada, Mexico and Spain are taking advantage of the profitable opportunities in tourism, as well as in mining, utilities, biotechnology and many other sectors availed by the lack of competition from US firms. Pressure on the US government from domestic farmers and industrialists to open trade with Cuba continues to mount. This pressure has thus far been checked by adept lobbying by the Cuban American National Foundation (CANF), and more broadly by the inertia of 40 years of policy attempting to isolate the socialist renegade 90 miles from Florida (Alzugaray 1999).

Conclusion: where do all these development schemes leave the contemporary Caribbean?

After a half-millennium of development schemes, the Caribbean today is economically vulnerable and dependent on the United States. Some trade statistics will provide a sense of the scale of the problem (Table 8.3). The first column of data expresses trade dependence as a ratio of exports and imports divided by GDP. The lower the value, the greater the share of a country's economic activity that involves domestic suppliers, producers and consumers. In other words, lower values mean more economic autonomy. The data indicate that, by and large, Caribbean countries are considerably more trade-dependent than the larger and more industrialised countries represented in the table. Values of well over 100 for three of the Caribbean countries indicate extreme trade dependency.

Note the contrasts with mainland Latin American countries. Caribbean countries are significantly more trade-dependent than Guatemala, the least dependent on trade of the Central American countries. Guatemala has a value similar to that of Chile, the most trade-dependent of the South American countries shown. Indeed, Chile's pronounced primary product-based export orientation has been the subject of considerable attention and debate in scholarly and policy circles (Gwynne and Kay 2004). However, Chile has a diversity of trading partners and is not only considerably less dependent on imports from and exports to the United States, but its exports there are now surpassed by those going to the EU (and to Japan). The UK is similarly trade-dependent, but with a diversity of trade partners.

The patterns just described for Caribbean trade dependency are extended by the data for imports to and exports from the USA in Table 8.3. For virtually every country shown, the United States is both the largest importer and exporter. At the same time, note that the USA is the least trade-dependent country in the table (it has a value of 13 in the middle column). This is testimony to the profound economic power of the USA in the American hemisphere. The only exception to the rule of US trade dependency among the Caribbean countries shown is St Lucia, whose banana shipments to a guaranteed UK market dominate its exports. Dominica and St Vincent display similar trade patterns. Unfortunately for these eastern Caribbean islands, their lack of trade dependency on the United States has only created a bigger problem. The USA has successfully argued to the World Trade Organization that the guaranteed market in the UK is illegal and must be terminated.

Table 8.3 Trade dependency: selected Caribbean countries in comparative perspective

Region	Exports+Imports/ GDP %	Exports to USA %	Imports from USA %
The Caribbean			
Bahamas	130	51	55
Barbados	117	13	36
Dominican Republic	48	48	60
Jamaica	82	47	54
St Lucia	166	22*	34
Trinidad and Tobago	96	48	48
Central America			
Guatemala	35	30	44
Honduras	59	53	50
Panama	77	39	40
South America and Mexico			
Argentina	20	9	21
Brazil	15	17*	23
Chile	38	15*	25
Colombia	28	39	36
Mexico	22	85	69
Industrialised countries			
United States	13	n.a.	n.a.
United Kingdom	48	13*	12*

* The USA is the largest export outlet and import source for all countries listed except those marked with an asterisk. For St Lucia, 56 per cent of exports go to the UK; for Brazil, 28 per cent of exports go to the EU; for Chile, 25 per cent of exports go to the EU; for the UK, 13 per cent of exports go to Germany and 15 per cent of imports come from Germany.
Sources: Exports+Imports/GDP: Baumol and Wolff (1996: 877); the data are for 1950–1990 with slight differences for Bahamas, Barbados and Jamaica; exports to and imports from the USA: *The World Factbook 1997*, www.odci.gov/cia/publications/factbook/indexgeo.html; the data are for 1993, 1994 or 1995, depending on the country

Besides Chile and St Lucia, the other exception to the rule of US trade dependency in Table 8.3 is Brazil, which also stands out for its exceptionally low level of trade dependency. Brazil, together with Argentina, stand out as the most economically autonomous of the Latin American countries in Table 8.3. Note that the USA heavily dominates the international trade of Mexico. However, the fact that imports and exports are only 22 per cent of Mexico's GDP indicates more economic autonomy than most other countries, especially those of the Caribbean.

Trade dependency is a key economic characteristic for all of the Caribbean, but countries differ in terms of level. The smaller islands of the eastern Caribbean are more trade-dependent than the larger islands. Despite the variations between

Caribbean countries, however, there is much regional commonality. On average, the Caribbean is more dependent on trade, and is less industrialised, than mainland Latin America. The Caribbean's economic dependence on the United States is also an important element of regional commonality, as distinct from the situation in South America. The Caribbean's relatively high levels of trade dependency distinguishes how neo-liberalism is applied there compared with most of South America. Neo-liberalism puts pressure on already highly trade-dependent Caribbean countries to export more. The Caribbean's relatively low levels of output and industrialisation also contribute to its peripheral status in the global economy.

The Caribbean region has endured many economic cycles and policy experiments over the decades. Despite these economic and political twists and turns, the Caribbean remains on a similar course. The Caribbean has been and continues to be a peripheral region in the global economy. Today, as the Caribbean travels further along the neo-liberal free-market path, the region appears to have few available policy options. Fortunately, the Caribbean people's historically refined and tested attributes of adaptability, creativity and ability to cope with external change and adversity will provide them with some protection during the neo-liberal era.

Note

1 These direct and indirect benefits associated with a politically dependent status in part explain why a slight majority of Jamaicans polled immediately before the 40th anniversary of the island's independence in August 2002 felt the island was better off under British colonial rule.

Useful websites

www.acs-aec.org/ is the website for the Association of Caribbean States, with a wide range of up-to-date economic and trade data.

www.planningcaribbean.org.uk/ is the website maintained by Dr Jonathan Pugh devoted to participatory planning in the Caribbean and related topics.

www.giraldilla.com is the website that reports policy trends in the USA toward Cuba and in Cuba itself that are relevant to US investors.

References

Albertini, J. E. (general manager of NDC) (1997) Interview with author, 18 September, Castries.

Alzugaray, C. (1999) 'Is normalization possible in Cuban–US relations after 100 years of history? paper presented at the International Studies Association Congress in Washington, 16–20 February, available from the author calzugaray@minrex.gov.cu.

Ambursely, F. (1983) 'Grenada: the new jewel revolution', in F. Ambursely and R. Cohen (eds) *Crisis in the Caribbean*, Heineman, London, pp. 191–222.

Anderson, S., Cavanagh, J., Ranney, D. and Schwalb, P. (eds) (1994) *Nafta's First Year: Lessons for the Hemisphere*, Institute for Policy Studies, Washington, 4 December.

Auty, R. (2002) 'The "resource curse" in developing countries', In Desai, V. and Potter, R. (eds) *The Companion to Development Studies*, Edward Arnold, London, pp. 224–9.

Barrow, C. (1992) *Family Land and Development in St Lucia*, Institute of Social and Economic Research, University of the West Indies, Cave Hill, Barbados.

Baumol, W. and Wolff, E. (1996) 'Catching up in the postwar period: Puerto Rico as the fifth "Tiger"?' *World Development*, **24**, 5, 869–85.

Beckford, G. and Girvan, N. (eds) (1989) *Development in Suspense; Selected Papers and Proceedings of the First Conference of Caribbean Economists*, Friedrich Ebert Stiftung/Association of Caribbean Economists, Kingston, Jamaica.

Blaut, J. M. (1992) 'Fourteen ninety-two', *Annals of the Association of American Geographers*, **11**, 4, 355–85.

Blum, W. (1986) *The CIA: A Forgotten History*, Zed Books, London.

Booth, D. (1985) 'Marxism and development sociology: interpreting the impasse', *World Development*, **13**, 761–87.

Booth, J. and Walker, T. (1993) *Understanding Central America*, 2nd edition, Westview, Boulder, Colorado.

Bourdieu, P. (1998) 'A reasoned utopia and economic fatalism', *New Left Review*, **227**, Jan/Feb, 125–30.

Brierley, J. (1985) 'A review of development strategies and programmes of the People's Revolutionary Government in Grenada', *Geographical Journal*, **151**, pp. 40–52.

Brohman, J. (1995) 'New directions in tourism for third world development', *Annals of Tourism Research*, **22**, 4.

Cardoso, F. H. and Faletto, E. (1979) *Dependency and Development in Latin America*. University of California Press, Berkeley.

Carter, K. (1997) *Why Workers Won't Work: The Worker in a Developing Economy: a Case Study of Jamaica*, Macmillan Caribbean, London.

Chauvin, L. (1998) 'Smoking economy', *Latin America Press* (Lima, Peru), 5 Feb, 4–5.

Clark, S. (1987) 'The second assassination of Maurice Bishop', *New International*, **6**, 11–96.

Clark, M. A. (1997) 'Transnational alliances and development policy in Latin America: nontraditional export promotion in Costa Rica', *Latin American Research Review*, **32**, 2, 71–97.

Conway, D. (1983) 'Grenada–United States relations part I, 1979–1983: a prelude to invasion', *University Field Staff International Reports*, **39**, Hanover, NH.

Conway, D. (1998) 'Misguided directions, mismanaged models, or missed paths?' in Klak, T. (ed.) *Globalization and Neoliberalism: The Caribbean Context*, Rowman & Littlefield, Lanham, Maryland, pp. 29–50.

Conway, D. and Heynen, N. (2002) 'Classical dependency theories: from ECLA to Andre Gunder Frank', in Desai V. and Potter, R. (eds) *The Companion to Development Studies*, Edward Arnold, London, pp. 97–101.

Corbridge, S. (1993) *Debt and Development*, Blackwell, Oxford.

Cordero-Guzman, H. (1993) 'Lessons from Operation Bootstrap', *NACLA Report on the Americas* **27**, 3, 7–10.

CTO (Caribbean Tourism Organization) (2003) *Tourist and Cruise Arrivals in 2002*, available at www.onecaribbean.org/information/documentview.php?rowid=262.

Cubitt, T. (1995) *Latin American Society*, 2nd edition, John Wiley, New York.

de Blij, H. and Muller, P. (1998) 'Puerto Rico's clouded future', in de Blij, H. and Muller, P. (eds) *Geography: Realms, Regions, and Concepts*, 8th edition, John Wiley, New York.

Deere, C. *et al.* (1990). *In the Shadows of the Sun: Caribbean Development Alternatives and U.S. Policy*, Westview, Boulder, Colorado.

Demas, W. (1980) 'Arthur Lewis and his last development policy model for the Caribbean', *Social and Economic Studies*, **29**, 85–94.

Demas, W. (1997) *West Indian Development and the Deepening and Widening of the Caribbean Community*, Ian Randle, Kingston, Jamaica.

Dicken, P. (1994) 'Global–local tensions: firms and states in the global space-economy', *Economic Geography*, **70**, 2, 101–28.

Dookeran, W. (ed.) (1996) *Choices And Change: Reflections On The Caribbean*, Interamerican Development Bank, Washington.

Downes, A. (2001) 'Economic growth and development in Barbados during the twentieth century', *Integration and Trade Journal*, **15**, 145–76.

Dunn, H. (1995) 'Caribbean telecommunication policy: fashioned by debt, dependency, and underdevelopment', *Media, Culture and Society*, **17**, 201–22.

Dupuy, A. (1991) 'Political intellectuals in the third world: the Caribbean case', in Lemert, C. (ed.) *Intellectuals and Politics: Social Theory in a Changing World*, Sage, Newbury Park, California, pp. 74–93.

Dupuy, A. (1997) *Haiti in the New World Order: The Limits of the Democratic Revolution*, Westview, Boulder, Colorado.

Farrell, T. (1980) 'Arthur Lewis and the case for Caribbean industrialization', *Social and Economic Studies*, **29**, 52–75.

Ferguson, J. (2003) 'Bahamas', *The New Internationalist*, **355**, 36.

Fletcher, P. (2000) 'Cuba reduces dependence on oil imports', *Calgary Herald*, **August 22**, D2.

Freudenheim, M. (1992) 'Tax credits of $8.5 billion received by 22 drug makers', *New York Times*, **15 May**, C3.

Galloway, J. (2000) 'Decline of a staple: the Caribbean sugar industry in the 20th century', in Munting, R. and Szmrecsanyi, T. (eds) *Competing for the Sugar Bowl*, Scripta Mecaturae Verlag, St Katharinen, Germany.

Gayle, D. (1998) 'Trade policies and the hemispheric integration process', in Klak, T. (ed.) *Globalization and Neoliberalism*, Rowman & Littlefield.

Gereffi, G. and Korzeniewicz, M. (eds) (1994) *Commodity Chains and Global Capitalism*, Praeger, Westport, Connecticut.

Green, D. (1995) *Silent Revolution: The Rise of Market Economics in Latin America*, Cassell, London.

Grugel, J. (1995) *Politics and Development in the Caribbean Basin: Central America and the Caribbean in the New World Order*, Indiana University Press, Bloomington.

Girvan, N. (1999) 'Globalisation and counter-globalisation: the Caribbean in the context of the South,' a paper presented at a seminar at the University of the West Indies, Mona, entitled 'Globalisation: A strategic response from the south', February 1–2. View online at www.acs-aec.org/SG/G15.htm.

Gwynne, R. and Kay, C. (2004) *Latin America Transformed: Globalization and Modernity*, 2nd edition, Edward Arnold, London.

Gwynne, R., Klak, T. and Shaw, D. (2003) *Alternative Capitalisms: Geographies of 'Emerging Regions'*. Edward Arnold, London, and Oxford University Press, New York.

Hudson, B. (1989) 'The Commonwealth eastern Caribbean', in Potter, R. B. (ed.) *Urbanization, Planning and Development in the Caribbean*, Mansell, London and New York.

Hersh, S. (1991) 'Operation Urgent Fury', *Frontline* television documentary, The Corporation for Public Broadcasting, aired 29 January.

IDB (Inter-American Development Bank) (1997) *Latin America After a Decade of Reforms: Economic and Social Progress, 1997 Report*, Johns Hopkins University Press, Baltimore.

Kaplinsky, R. (1993) 'Export processing zones in the Dominican Republic: transforming manufactures into commodities', *World Development*, **21**, 11, 1851–65.

Killick, T. and Malik, M. (1992) 'Country experience with IMF programmes in the 1980s', *The World Economy*, **15**, 599–632.

Klak, T. (1995) 'A framework for studying Caribbean industrial policy', *Economic Geography*, **71**, 3 (July), pp. 297–317.

Klak, T. (1998) 'Is the neoliberal industrial export model working? An assessment from the eastern Caribbean', *European Review of Latin American and Caribbean Studies*, **65**, December, 5 and 67–90.

Klak, T. and Rulli, J. (1993) 'Regimes of accumulation, the Caribbean Basin Initiative, and export-processing zones: scales of influence on Caribbean development', in Goetz, E. and Clarke, S. (eds) *The New Localism*, Sage Publications, Beverly Hills, pp. 117–50.

Klak, T. and Myers, G. (1997) 'The discursive tactics of neoliberal development in small third world countries', *Geoforum*, **28**, 2, 133–49.

Klak, T. and Das, R. (1999) 'The underdevelopment of the Caribbean and its scholarship', *Latin American Research Review*, **34**, 3, 209–24.

Krueger, A. O. (1993) *Political Economy of Policy Reform in Developing Countries*, MIT Press, Cambridge, Massachusetts.

LeoGrande, W. (1997) 'Enemies evermore: US policy towards Cuba after Helms-Burton', *Journal of Latin American Studies*, **29**, 211–21.

Lewis, A. (1950) 'The industrialization of the British West Indies', *Caribbean Economic Review*, **2**, 1–39.

Lobe, J. (2001) 'Learn from Cuba, says World Bank', *Inter Press Service*, 1 May.

Mandle, J. (1996) *Persistent Underdevelopment: Change And Economic Modernization In The West Indies*, Gordon and Breach, Amsterdam.

Manley, M. (1987) *Up the Down Escalator: Development and the International Economy: A Jamaican Case Study*. Howard University Press, Washington.

Marshall, D. (2002) 'The new world group of dependency scholars: reflections on a Caribbean avant-garde movement', in Desai, V. and Potter, R. (eds) *The Companion to Development Studies*, Edward Arnold, London, pp. 102–7.

Marshall, J. (1998) 'The political viability of free market experimentation in Cuba: evidence from *Los Mercados Agropecuarios*', *World Development*, **26**, 2, 277–88.

McAfee, K., (1991) *Storm Signals: Structural Adjustment and Development Alternatives in the Caribbean*. South End Press, Boston, and Oxfam America.

McBain, H. (1990) 'Government financing of economic growth and development in Jamaica: problems and prospects', *Social and Economic Studies*, **39**, 179–212.

Meeks, B. (2000) *Narratives of Resistance: Jamaica, Trinidad, the Caribbean*, University of West Indies Press, Mona, Jamaica.

MFSN (The Ministry of Finance, Statistics and Negotiating, Government of St Lucia) (1997) *Economic and Social Review 1996*, MFSN, Castries.

Mintz, S. 1996 'Enduring substances, trying theories: the Caribbean region as *oikoumene*', *Journal of the Royal Anthropological Society*, **2**, 2, 289–311.

Mody, A. and Wheeler, D. (1987) 'Towards a vanishing middle: competition in the world garment industry', *World Development*, **15**, 1269–84.

Mullings, B. (1995) 'Telecommunications restructuring and the development of export information processing services in Jamaica', in Dunn, H. (ed.) *Globalization, Communications and Caribbean Identity*, Ian Randle, Kingston, Jamaica, pp. 174–91.

Mullings, B. (1998) 'Jamaica's information processing services: neoliberal niche or structural limitation?' in Klak, T. (ed.) *Globalization and Neoliberalism: The Caribbean Context*, Rowman & Littlefield, Lanham, Maryland.

Myint, H. (1964) *The Economics of Developing Countries*, Praeger, New York.

NDC (National Development Corporation of St Lucia) (1997a) 'Trade and investment incentives', available from NDC's website: www.stluciandc.com.

NDC (National Development Corporation of St Lucia) (1997b) 'Manufacturers' list – St Lucia', unpublished database current as of May 2003.

NDC (National Development Corporation of St Lucia) (1997c) 'New strategic directions', NDC document.

NDC (National Development Corporation of St Lucia) (1998) 'Information services industry', available from NDC's website: www.stluciandc.com/info.htm

Pantojas-García, E. (1990) *Development Strategies as Ideology: Puerto Rico's Export-Led Industrialization Experience*, Lynne Rienner, Boulder, Colorado.

Pattullo, P. (1996) *Last Resorts: The Cost of Tourism in the Caribbean*, Monthly Review Press, New York.

Paus, E. (ed.) (1988) *Struggle Against Dependency: Nontraditional Export Growth in Central America and the Caribbean*, Westview, Boulder, Colorado.

Payne, A. (1982) 'Politics and political economy in contemporary Jamaica', University of London Institute of Commonwealth Studies, *Collected Seminar Papers,* **29**, 72–7.

Ramos, A. G. and Rivera, A. I. (2001) *Islands at the Crossroads: Politics in the Non-Independent Caribbean,* Ian Randle, Kingston, Jamaica.

Rampersad, F. (with Stewart, T., Rampersad, G. and Rampersad, R. (1997)) *The New World Order: Uruguay Round Agreements and Implications for Caricom States*, Ian Randle, Kingston, Jamaica.

Reifer, T. and Sudler, J. (1996) 'The interstate system', in Hopkins, T. K. and Wallerstein, I. (eds) *The Age of Transition: Trajectory of the World-System 1945–2025*, Atlantic Highlands, London, and Zed Books, New Jersey.

Richardson, B. (1992) *The Caribbean in the Wider World, 1492–1992: A Regional Geography*, Cambridge University Press, New York.

Richardson, C. (Permanent Secretary in the Ministry of Commerce, Industry and Consumer Affairs) (1997) Interview with author, 16 September, Castries.

Rosen, F. (1997) 'Back on the agenda: ten years after the debt crisis', *NACLA Report on the Americas*, **3**, 3, 21–24.

Safa, H. (1992) 'Women and industrialization in the Caribbean', in Stichter, S. and Parpart, J. (eds) *Women, Employment and the Family in the International Division of Labour*, Temple University Press, pp. 72–97.

St Lucia Business Focus (1996) 'NDC – a new investment focus' *St Lucia Business Focus*, **2** (Oct/Nov), 6.

Schoepfle, G. K. and Perez-Lopez, J. F. (1989) 'Export assembly operations in Mexico and the Caribbean', *Journal of Interamerican Studies*, **31**, 4, 131–61.

Schrieberg, D. (1997) 'Dateline Latin America: the growing fury', *Foreign Policy* 106: 161–175.

Sealey, N. (editorial advisor) (2001) *Caribbean Certificate Atlas*, 3rd edition, MacMillan Education, London.

Segal, G. (1989) 'The state of 807/CBI', *Bobbin Magazine*, **31**, 56–83.

Segre, R., Coyula, M. and Scarpaci, J. (1997) *Havana: Two Faces of the Antillean Metropolis*, John Wiley, New York.

SRI Consultants (1992) 'Free zone survey: final report', prepared for the Port Authority of Jamaica and paid for by USAID, Arlington, Virginia.

Stiglitz, J. (2000) 'What I learned at the world economic crisis', *The New Republic*, 17 April, available at www.thenewrepublic.com/041700/stiglitz041700.html.

Sunshine, C. A. (1988) *The Caribbean: Survival, Struggle and Sovereignty*, Ecumenical Program on Central America and the Caribbean, Washington.

Susman, P. (1990) 'Losing ground in the Caribbean', paper read at the Annual Meeting of the Association of American Geographers, Toronto.

Susman, P. (1998) 'Cuban socialism in crisis: a neoliberal solution?' in Klak, T. (ed.) *Globalization and Neoliberalism*, pp. 179–208.

Sutcliffe, B. (2001) *101 Ways of Seeing on Unequal World*, Zed Books, London.

Thomas, C. (1988) *The Poor and Powerless: Economic Policy and Change in the Caribbean*, Monthly Review Press, New York.

Thoumi, F. E. (1989) 'Thwarted comparative advantage, economic policy and industrialization in the Dominican Republic and Trinidad and Tobago', *Journal of Interamerican Studies and World Affairs*, **31**, 147–68.

Tirado de Alonso, I. (ed.) (1992) *Trade Issues in the Caribbean*, Gordon and Breach, Philadelphia.

United Nations (2000) *Human Development Report 2000*, Oxford University Press, New York.

Vitalis, D. (1998) 'Competition for Cable and Wireless: is Government ready for the fight?' *St Lucian Mirror*, Friday, 6 November, **5**, 10.

Wade, R. (2001) 'Winners and losers', *The Economist*, 26 April.

Walker, T. (ed.) (1997) *Nicaragua Without Illusions: Regime Transition and Structural Adjustment in the 1990s*, Scholarly Resources, Wilmington, Delaware.

Wallerstein, I. (1979) *The Capitalist World-Economy*, Cambridge University Press, Cambridge.

Warr, P. G. (1989) 'Export processing zones and trade policy', *Finance and Development*, June, 34–6.

Weiss, L. (1997) 'Globalization and the myth of the powerless state', *New Left Review*, **225**, Sept/Oct, 3–27.

Wiley, J. (1998) 'Dominica's economic diversification: microstates in a neoliberal era?' in Klak, T. (ed.) *Globalization and Neoliberalism*, pp. 155–78.

Wiley, J. (1999) 'Dominica, Grenada, and the NTEA imperative', *CLAG Yearbook 1999*, Journal of the Conference of Latin Americanist Geographers.

Williamson, J. (1990) 'The progress of policy reform in Latin America', in Williamson, J. (ed.) *Latin American Adjustment: How Much has Happened?* Institute for International Economics, Washington, pp. 353–420.

Williamson, J. (1993) 'Democracy and the "Washington consensus"', *World Development*, **21**, 1329–36.

Wilentz, A. (1989) *In the Rainy Season: Haiti Since Duvalier*, Touchstone, New York.

World Eagle Inc. (1997) *Latin America Today: A Reproductive Atlas*, World Eagle Publishers.

World Bank (1992) *Export Processing Zones*, Policy and Research Series, 20, World Bank, Washington.

Chapter 9

OFFSHORE SERVICES

Introduction

> The [OECD] members are finding difficulty adapting to the new global environment, of which they have been the chief architects (Owen Arthur, Prime Minister of Barbados and Chairperson of the CARICOM Heads of Government, responding to OECD attacks on offshore banking centres for allegedly harbouring tax evaders and money launderers, March 2001).

As the quote from Owen Arthur suggests, there has recently been considerable international sabre rattling over the growing role of offshore financial centres in the global economy. It is an intriguing conflict for several reasons. It pits the economically dominant OECD core countries against some of the world's smallest and otherwise most peripheral jurisdictions, from Barbados, Cayman Islands and Grenada in the Caribbean, to Tonga, Marshall Islands and Nauru in the South Pacific. Wedding these two types of state is increasingly mobile capital. As the shady financial practices of the bankrupt energy giant Enron reveal (Box 9.1), much of the capital moving offshore is based in, and/or flows through, the corporations and banks of the same OECD states pointing fingers at the offshore banks. The core states claim that offshore banking hurts them by harbouring illegal activities. Non-core states defend their gains, legality and sovereign powers. Who is right? Who wins and loses, and at what costs? Do offshore services contribute to a global decentralisation of financial benefits and to non-core (specifically Caribbean) development? Such questions underlie this chapter about the growing role of offshore banking and other transnational services in the Caribbean region.

The use of the term 'offshore services' in this chapter is admittedly based on a perspective grounded in the core countries. This is intentional. The term captures the fact that processes originating in the core propel the services of interest in this chapter – such as data processing, telephone call centres, banking, insurance and e-commerce. Much of the growth in these services is generated from the global core: either by the demand on the part of TNCs and banks from core countries for less regulated or lower-wage environments, or by the surplus purchasing power of affluent consumers from core countries. In either case, the

Box 9.1: Enron and its Caribbean subsidiaries

At the height of its power and influence in the first half of 2001, Enron was the seventh largest US corporation and the world's 16th largest. It had $100 billion in annual sales, stock valued at $90 per share and 20,000 employees. By early 2002 the energy trading company had become the largest firm ever to declare bankruptcy. Enron's stock had fallen to 26 cents a share, its workforce had been cut by thousands, and many thousands more employees and investors had lost billions of dollars from their retirement and other financial accounts. To US Treasury Secretary Paul O'Neill, Enron's rapid rise and fall illustrates 'the genius of capitalism'. To most others, however, the events are deeply troubling, particularly because Enron executives received millions of dollars of compensation right up to the company's bankruptcy. As the investigation into the causes of Enron's collapse unfolded, it became clear that Enron engaged in a number of questionable business practices. One of Enron's practices among the many pushing the boundary of legality was the use of offshore subsidiaries to move money offshore and then back onshore. Enron had 881 offshore subsidiaries. The Cayman Islands hosted more than 600 of them, while the Turks and Caicos Islands hosted another 100 or so. Through these offshore affiliates Enron was able to use creative accounting methods to save itself hundreds of millions of dollars in taxes. The essence of these sophisticated accounting methods, arranged by well-known US firms such as Chase Manhattan and Arthur Anderson and by tax consultants reportedly earning $1,000 per hour, is really quite simple. Enron moved taxable income offshore to accounts that are outside US tax authority. The offshore subsidiaries then returned the money to Enron in the USA, but in a form of capital that was not taxable. As a result, Enron paid no taxes at all during four of the five years prior to bankruptcy. Creative accounting methods, including income averaging, had in fact managed to arrange for the US government to owe Enron $183 million in tax credits. While Enron is the most notorious of the major firms that avoid taxation in these ways, it is certainly not alone. US tax codes have been revised in recent years to allow for such corporate practices. As a result, corporate taxes as a share of federal tax revenue have fallen, with the difference made up by individual taxes and by deficit spending.

economic orientation remains firmly core-focused. The economic development issue that this chapter addresses is can non-core sites such as those of the Caribbean benefit substantially from the relatively recent and rapid decentralisation of these services? Caribbean leaders are hopeful in this regard. When launching a new policy effort towards offshore services, Grenada's Prime Minister Keith Mitchell, for example, argued that 'information and communication technologies render the size of the economy less important than its knowledge and skill competitiveness' (Lee 2001). Do services offer the Caribbean real advantages for development gains?

Figure 9.1 Technical skills required for key transnational services
* APS – Advanced Producer Services (see Sassen 2000; Taylor *et al.* 2001)

In a wide-ranging overview of global service industries, Peter Daniels (1993) found that three main factors account for their recent expansion:

- technological change through advances in telecommunications;
- the growing role of TNCs specialising in the provision of services;
- changes in the regulatory environment at both the national and international levels.

By distinguishing between the contributions of technology, TNCs and public policies, Daniels provides a useful organisational framework for this chapter. The Arthur–OECD clash is but one of the examples explored later in this chapter of the political negotiations over how to regulate international capital flows and the associated services. The chapter reviews the Caribbean experience with three broad categories of offshore services:

- labour-intensive telecommunication-based services (e.g. data processing, back-office services, call centres and telephone operators, and telemarketing);
- e-commerce, involving non-core commodities, firms and sites (e.g. business-to-business sales, business-to-consumer sales, domestic versus export markets, online gambling);
- offshore financial services (e.g. banking, insurance, shipping registration, subsidiary incorporation, trusts).

These and other transnational services can be arrayed along a continuum ranging from the relatively simple tasks associated with information processing to the production of knowledge capital. Moving from left to right on the continuum requires more technical skills (Figure 9.1).

Notably, the most common transnational services in the Caribbean are the four less knowledge-intensive types on the left side of the continuum (Gwynne *et al.* 2003). The Caribbean's focus on relatively low-skill services puts the region at a disadvantage in international bargaining over the distribution of benefits. Excluded from Figure 9.1 are a variety of other services that have become prominent in recent decades. Transportation, construction, education, real estate and other services are not treated here because they are primarily domestically oriented (but see Selya 1999; Illeris 1996). Here we focus on the developmental prospects of international services that are most affected by global time–space compression and that have relocated to the Caribbean (Harvey 1989).

Labour-intensive telecommunication-based services

Various types of lower-skill and labour-intensive services have moved offshore in recent decades. Such decentralisation occurs in those aspects of services that involve routine information processing. This includes data handling for insurance firms, publishers, physicians, lawyers and airlines, telemarketing, and dealing with telephone enquiries. The main core country providing investors, capital and customers for offshore telecommunication-based services is the United States. The work is therefore subject to English language constraints. Ireland and the Caribbean have been among primary recipients of the offshore work. Many of these decentralised lower-order services are of the 'back office' type. These are routine clerical and labour-intensive tasks performed on computer keyboards by low- or semi-skilled women (Figure 9.2; Warf 1999).

There are many examples of US firms outsourcing labour-intensive services to the Caribbean. For years American Airlines has outsourced labour-intensive data-entry work there. Operating through a subsidiary called Caribbean Data Services (CDS) it created in 1983, American once had a major presence in Barbados, where it processed airline tickets and health insurance claims. By the late 1980s, CDS's Barbados activity had been scaled back and shifted to the relatively low-wage Dominican Republic.

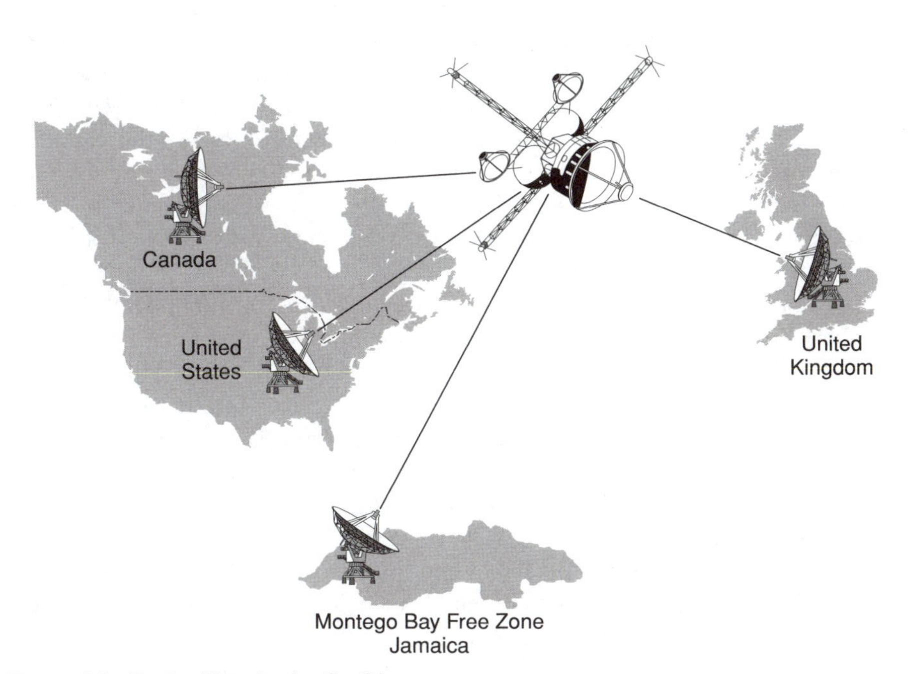

Figure 9.2 Back offices in the Caribbean
Source: *The Jamaican Advantage: An Economic Alternative*, Jamaica Digiport International Ltd. (n.d.)

Caribbean policy makers have had great expectations for the offshore data-processing industry (Klak 1998a). However, Jamaica's disappointing experience with data processing offers a lesson on the difficulties associated with pursuing economic development through offshore services of this kind. At first glance, the development prospects of the offshore data-processing sector are seemingly large. They entail incorporating local labour and nurturing local firms to take advantage of expanding opportunities in the global data-processing industry, which earns over $1 trillion yearly, of which labour-intensive computer-based services is a component. However, Beverley Mullings has drawn on a detailed analysis of the rise and fall of information services in Jamaica to explain why this industry, with its real potential for growth in employment, wages, managerial expertise and backward linkages into the local economy, has stagnated on all of these criteria. A combination of factors explain the stagnation: inadequate state support for local firms; continued policy steering and dampening by a traditional, non-dynamic private sector elite; investment fear on the part of foreigners; and an extremely narrow role allotted to Jamaican firms and workers by US outsourcing firms (Mullings 1995, 1998). Rather than propelling Jamaica to a heightened position in the international division of labour, the information services sector has slumped and perpetuated the gender, class and international inequalities that have long characterised Jamaica (cf. Klak 1998b).

Labour-intensive telephone services that are now outsourced offshore take several forms. One such service is the handling of customer queries about the products or services of core TNCs. Another is known as pay-per-call information services. In most cases, money is made when callers from the USA charge their credit cards or bill their long-distance account to designated 1–900 or 1–976 phone numbers. Sex talk is popular, and US telephone companies say the countries that are most often called for 'dial-a-porn' include the Dominican Republic and Guyana in the Caribbean region, and elsewhere Portugal, Moldova and Sao Tome. Customers often find out later that that their half-hour calls to Sao Tome cost about US$100 (Stroud 1996).

Yet another offshore telephone service is telemarketing. Calls are placed to 1–800 numbers but the operator is sometimes outside of the United States. Major US telecommunications firms such as AT&T, Sprint and TRICO (a Motorola subsidiary) use the Dominican Republic, the country's major Internet relay station, and its bilingual workers to telemarket to the growing US Latino population, which now exceeds that of African Americans (Pantojas-Garcia 2001).

The Dominican Republic offers other offshore telephone services. Television and newspaper ads for 'psychic consultants' now appear throughout the western hemisphere. Calls connect to 'psychic' wage labourers in the Dominican Republic. These sometimes lengthy phone calls have been profitable for US telecommunication and telemarketing firms, and for the US-based celebrities who franchise their names for the psychic services. Benefits are fewer for the low-wage Caribbean workers and for the state coffers from this tax-free component of a service commodity chain (Pantojas-Garcia 2001).

E-commerce involving Caribbean commodities, firms and sites

Trends in offshore e-commerce

Some people in the Caribbean are exuberant about e-commerce. They believe that creating an e-commerce web page will attract 'hordes of visitors to the site to buy everything and anything that's available on it, making the owners multi-millionaires in no time' (Panton 2003). Is there any validity to such claims? Unfortunately there are many limitations to e-commerce in the Caribbean, as the following discussion suggests.

What exactly is electronic commerce? The United Nations says it is 'any form of commercial or administrative transaction or information exchange that is transmitted by electronic means. This could be by phone, fax, television, EDI, Internet, and so on, although much of current policy discussion centres on transactions undertaken over the Internet' (UNCTAD 2000: 40). The Caribbean joins many non-core countries that now have websites selling consumer products such as food and other household items. Until recently the target of this e-commerce has primarily been domestic shoppers. Even though the goal may be to sell online to core consumers and thereby earn foreign exchange revenue, a domestic emphasis in the early stages is not inappropriate. This would parallel the experience with developing a domestic manufacturing sector in East Asia (Wade 1996), South America (Bromley and Bromley 1988) and the Caribbean islands (Klak 1998a). In all these cases, import substitution industrialisation (ISI) was used to develop and strengthen the firms and their products, which then successfully export many of those same products. For e-commerce, the next logical market after the domestic one is the Caribbean diaspora community, which is large and wealthy by standards back home.

The greater challenge, however, is how to sell to non-Caribbean customers in core regions. The idea of vending products over the Internet to core consumers immediately suggests a variety of advantages and disadvantages. The advantages include that the Internet opens the possibility for Caribbean-based vendors to offer their products directly to business or household consumers in the core countries, a large share of whom regularly access the Web from home and office. The Internet effectively reduces the distance and many other international barriers between vendor and consumer. Intermediaries and their rents are also reduced, leaving more income for the vendors. Small wonder, then, that some Caribbean policy makers have placed e-commerce high on their national priority lists.

There seem to be many more disadvantages, however. Internet technologies and security regulations are so much further developed in core countries that vendors there are likely to usurp many of the non-core's vending opportunities. Vendors from core countries can (and do) simply appropriate products from the Caribbean and sell them themselves to core consumers. The actual physical location of the operator is a minor detail of any Internet site or in the suffix of its website address.

And that assumes customers are aware of, or care about, where the vending originates. If customers do care, they are likely to favour core vendors. Concerns about the security of credit card information over the Internet only strengthens the core-based vendors. As one example, US credit cards have recently been refusing to pay out to offshore sites that are billing US customers for 1–900 number calls to the Dominican Republic (Pantojas-Garcia 2001). Core vendors are also advantaged by their lower cost and faster shipping to core buyers, and the greater ease of tracking down the vendor in the event that there are problems with the product or billing. In sum, while e-commerce in core countries continues to make rapid advances in terms of technology, security and markets, Internet vending outside the core continues to be saddled by a host of constraints.

Viewing e-commerce in its broader context, one can see that the difficulties Caribbean vendors face reflect the electronic circular and cumulative causation highlighted by Paul Krugman (cited in James 1999): technological advantages accrue more rapidly to leaders than to laggards. These widening technological gaps are among the processes behind the growing disparity between rich and poor in the world over recent decades (Wade 2001). Despite these limitations, one form of offshore e-commerce that has attracted considerable attention in the wider Caribbean region is online gambling, to which we now turn.

Online gambling

From early on Caribbean and Central American countries saw their locational advantages for hosting Internet gambling sites. Advantages include the region's proximity to the world's largest gambling market (the USA), legal restrictions on electronic gambling within the USA, the region's 'tropical getaway' atmosphere, which is attractive to North American operators (some of these entrepreneurs have boasted of making a fortune while wearing shorts and flip-flops!), a time zone within one hour of that of the eastern USA, and (for the Caribbean) convenient telephone area codes as in the USA. To exploit these locational advantages, Central American and Caribbean countries have established some of the world's first offshore regulatory environments for Internet gambling (Martin 2001).

As of May 2001, there were about 1,400 offshore gambling sites operated by about 250 companies. All of these online gambling sites are located outside of the United States, and most are in Central America and the Caribbean (Table 9.1). Costa Rica hosts about 15 per cent of all gambling sites. Its startup fees of less than US$10,000 are far below some Caribbean sites, which charge as much as US$250,000. Host countries also charge annual renewal fees. In Costa Rica, about 3,000 workers, mostly college students and foreigners staying on after teaching English, earn US$4–5 per hour taking bets or answering customer queries over the phone. Staff include speakers of at least nine languages, from English, Spanish and German to Japanese, Italian and Portuguese (Delude 2000). Wherever a gambling site is registered (and this is usually difficult for the public to discern), customers can log on from home or office to play virtual card games, craps, roulette,

Table 9.1 Forty-five offshore banking or gambling jurisdictions under OECD scrutiny

Jurisdictions accused of harbouring tax dodgers (except*)	**Region[1]**	**Historically part of the British Empire – now independent (I) or dependent (D)**	**Blacklisted for harbouring money launderers 2000–2002[2]**	**Licensed online gambling centres[3]**
Antigua and Barbuda	C	I		X
Anguilla	C	D		
Aruba	C			X
Bahamas	C	I	XOO	
Barbados	C	I		
Belize	C	I		X
Cayman Islands	C	D	XOO	
Dominica	C	I	XXX	X
Grenada	C	I	OOX	X
Guatemala*	C		OXX	
Montserrat	C	D		
Netherlands Antilles	C			X
Panama*	C		XOO	
St Lucia	C	I		
St Kitts and Nevis	C	I	XXO	X
St Vincent and the Grenadines	C	I	XXX	
Turks and Caicos Islands	C	D		
Virgin Islands (British)	C	D		
Virgin Islands (US)	C			
Andorra	E	I		
Guernsey, Sark, Alderney	E	D		
Gibraltar	E	D		X
Hungary*	E		OXO	
Isle of Man	E	D		
Jersey	E	D		
Liechtenstein	E		XOO	
Monaco	E			

Table 9.1 (*continued*)

Jurisdictions accused of harbouring tax dodgers (except*)	Region[1]	Historically part of the British Empire – now independent (I) or dependent (D)	Blacklisted for harbouring money launderers 2000–2002[2]	Licensed online gambling centres[3]
Russia*	E		XXX	
Cook Islands	P	D	XXX	
Indonesia*	P		OXX	
Maldives	P	I		
Marshall Islands	P		XXX	
Burma (Myanmar)*	P	I	OXX	
Nauru	P	I	XXX	
Niue	P	I	XXX	
Philippines	P		XXX	
Seychelles	P	I		
Tonga	P	I		
Vanuatu	P	I		
Western Samoa	P			
Bahrain	ME	I		
Egypt	ME	I	OXX	
Israel*	ME	I	XXO	
Lebanon*	ME		XXO	
Nigeria*	SSA	I	OXX	

* All jurisdictions in this table (except those marked *) the OECD claimed were uncooperative and harboured tax evaders as of 2001. Liberia joined the 2002 uncooperative tax haven list, which otherwise included only those six jurisdictions in bold.

[1] The offshore banking jurisdictions under OECD scrutiny come from five world regions: 19 from the Caribbean (C), nine from Europe (E), 12 from the Pacific/Indian Ocean (P), four from the Middle East (ME), and two from sub-Saharan Africa (SSA).

[2] Each June, the FATF issues its annual report, which blacklists jurisdictions for 'non-cooperation' in fighting money laundering. XXX on 2000, 2001 and 2002 list; XOO on 2000 list only; XXO on 2000 and 2001 list but not on 2002 list, etc.

[3] Other jurisdictions that have licensed Internet gambling but are not under OECD investigation for offshore banking practices include Australia, Costa Rica, Dominican Republic, French Guiana, Guyana, Honduras, Jamaica, Nicaragua, South Africa, Suriname, Trinidad and Venezuela. The list continues to grow.

Source: Compiled by the authors from a range of Internet sources.

slot machines, bingo or the lottery, or bet on professional sports or races. Extrapolating from recent trends, offshore gamblers have increased to about 43 million people by 2001. Besides the United States, gamblers are concentrated in Canada and Asia. The industry made at least $1.6 billion in 2000 and is projected to take in $5 billion by 2003.

Note that no US federal law explicitly prohibits gambling over the Internet. What is invoked is a 1961 US law called the 'Wire Wage Act', which prohibits gambling operations to use interstate telephone lines. The prohibition is generally understood to include the Internet. In addition, some US states have passed laws explicitly prohibiting Internet gambling. Most states, legal experts say, have older laws that could be interpreted that way.

Big US casinos have not sat idly by as gambling revenues move offshore, however. They have got a foot in the door by offering online gambling in which one can win prizes, but not win or lose money. The casinos are also lobbying the US government to legalise online gambling inside the country. This would reverse the legislative tide: in recent years the US Congress has considered legislation that would criminalise online gambling in two ways. The legislation would explicitly prohibit Internet gambling operations and customers, and it would force Internet service providers to attempt what computer specialists say is not technologically feasible: to block access to offshore gambling sites. For now, and so long as US law restricts Internet gambling within its borders, Caribbean countries that host offshore gambling sites can earn some foreign exchange and create some jobs for foreign language proficient residents. But this is a highly tenuous offshore services sector, because its advantages would be eliminated if the USA were to legalise Internet gambling.

Offshore financial services

> Money is an electron somewhere on somebody's hard drive and . . . it can bounce from here to there faster than any legal system can ever figure out where it went (Jack Blum, UN Consultant on Financial Corruption, quoted in Masters *et al.* 2001).

As the above quote implies, offshore financial activity is today a fast-paced, global enterprise. The Caribbean has emerged as a very significant player in offshore finance, but it is not possible to examine the capital flows and regulatory efforts in the Caribbean in isolation from developments in other world regions.

For more than a decade, offshore banking has been big business for many non-core countries, particularly in the Caribbean region, and a big concern for the OECD worried about the exodus of capital and tax revenues. This section reviews the global geographies and core–periphery clashes associated with offshore banking. Most of the coverage pertains to the situation before 11 September 2001, during which time the Bush administration, in contrast to its OECD allies, opposed the regulation of offshore banks. Prior to September 11th, neither the OECD nor the USA seemed particularly concerned about the financing of terrorists. Only in the fine print did the OECD's *Financial Action Task Force Annual Report* issued on 22 June 2001 broach the topic of international terrorism. After September 11th the global controversy over offshore banking shifted radically. Bush declared that global terrorists were financing their operations through offshore banks and that global regulation is required. The new terrain of controversy that is unfolding in the post-9/11 world is introduced in the final subsection.

Trends behind offshore banking

The confluence of four processes associated with the global distribution of capital has created a huge demand for Caribbean offshore banking since the 1960s.

First, from the 1960s on, Britain encouraged its overseas dependencies to create legislation to become offshore finance centres (OFCs). The British Foreign Office pursued this policy to help generate new sources of tax and other revenues in the dependencies and thereby reduce their reliance on subsidies from the UK (Hampton and Christensen 2002). Since the 1960s, many British dependencies achieved independence but continued to pursue the offshore banking sector. Today, the Caribbean region includes five of the world's few remaining British dependencies. The Cayman Islands, Anguilla, Montserrat, the Turks and Caicos Islands, and the British Virgin Islands have all become important OFCs and have attracted international capital, particularly from the United States. The British Virgin Islands now has 400,000 registered international business corporations (IBCs) along with thousands of other foreign depositors, but only 20 regulators. These numbers should convey a sense of both the scale of offshore financial activity and the limited authority many OFCs have over it.

Second, there is the legal structure in core countries that regulates and constrains capital. Core investors control most of the world's wealth but face quite thoroughly regulated banking environments at home. There are fairly strict banking regulations and oversight systems in most core countries. Further, deposits are open to public scrutiny and subject to seizure in the event of legal challenge.

Third, there is the growing global electronic mobility of capital thanks to the technical revolution in computer electronics, and the political revolution of neo-liberal policies promoting open markets, privatisation and less regulation (see Chapter 8). A trillion and a half dollars now circles the globe each day in currency trading. Public stock markets are open in various world regions, which allows for 24-hour trading. Money can be moved 'virtually' anywhere with a keystroke.

Fourth, there were two rapid increases in global oil prices in the mid- to late 1970s. They created a surge in the volume of eurodollars (dollars traded outside the USA) and other internationally tradable currencies in need of deposit (Corbridge 1993). As a result of these trends, today there is indeed much money deposited offshore, over five trillion dollars according to the US government. Even more striking are a variety of estimates that places 50–60 per cent or more of the world's wealth offshore. And the proportion of wealth offshore is growing. The scale of money involved has provoked aggressive policy actions to snare a share of it by Caribbean and core governments alike.

Where in the world are the offshore banks?

> The problem . . . for those who would control capital flight or crime through stopping up the money-laundering channels is that offshore resembles a balloon: pinch it in here, and it will expand over there. All tax havens have to be stopped at once if anything is to be achieved. The OECD and its fellow multilaterals have attacked 20 or 30 jurisdictions in 2000, but there are nearly a hundred of them, with more opening up all the time. There are a lot more tropical islands than there are bureaucrats at the OECD! (Robinson (2001) of *Tax-news.com*, a website devoted to 'international transactions and tax-minimisation techniques both on- and offshore').

As this quote from some of the facilitators of offshore capital mobility suggests, it would be difficult to generate a complete list of the offshore banking jurisdictions, let alone to police them. If nearly 100 states can claim to be OFCs, then a very large share of the world's governments are seeking to attract mobile capital. Caribbean leaders have been faced with the fact that most of their traditional as well as contemporary economic development pursuits have proved to be weak and inadequate. They have therefore viewed offshore banking and other online services as enterprises worth pursuing vigorously.

The availability of large amounts of eurodollars in search of interest-bearing deposits since the 1970s accounts for why many offshore banking centres originated in the 1980s (Table 9.2). They tend to be among the world's smaller jurisdictions (both independent and dependent). The 43 offshore banking jurisdictions in existence as of the early 1990s, what might be called the 'first wave' of offshore banking centres, cluster into four regional groups:

- the Caribbean (15 of the total)
- Europe (12)
- South-west Asia (5)
- two clusters in the South Asia-Pacific region (9).

The remaining two OFCs are African – Liberia and Mauritius.

The divisions between core and periphery in the offshore banking controversy are pretty clear, but they are not total. This is because not all of what core governments have defined as offshore banking actually occurs offshore. Many OFCs are in Europe. Luxembourg, for example, hosts some 220 foreign banks, and attracts deposits in part from its low tax rates (by European standards) and by its strict account secrecy laws. Other EU members are pressing Luxembourg to harmonise its banking laws to EU standards and to renounce account secrecy but Luxembourg is resisting this.

Table 9.2 Caribbean offshore banking centres: key features

Country	First activities	Conditions for setting up a company[1]			Compulsory bookkeeping	Tax %			No. of banks	No. of companies	Laws enacted
		Costs in US$	Time needed	Yearly costs in US$		Income	Corporation	Withholding			
Anguilla	Early 80s	–	24 hours	*	no	no	no	no	37	–	Trust 1994, IBC 1994, Company 1994, Limited Partnership 1994
Antigua	Early 80s	700–900	IBC: 24 hours	IBC: 250 banks: 5,250	yes	25	40	40	12	–	Trust 1994
Bahamas	Late 50s	1,000–2,500	2–3 days	750–50,000	yes	no	no	no	>400	>15,000 IBC, 50 IC	Company 1992
Barbados	Early 80s	2,500	2 weeks	100–15,000	yes	25–40	40	15	19	250 CIC, 1,171 IBC, 926 FSC	IBC 1992
BVI	Mid 80s	300–1,000	3–5 days	1,180	yes	3–20	15	no	6	>80,000 IBC, 1,500 IC	Trustee 1993, Insurance 1991
Cayman Islands	Mid 60s	–	3–4 days, banks 6–8 weeks	900–1,800 CI$, banks 6,000–32,400 CI$	yes	no	no	no	>500	23,000, 350 CIC	Limited Partnership 1991, Limited Duration 1993
Montserrat	Mid 80s	300–3,000	1–2 days	banks 8,000–12,000	yes	5–20	20	20	21	22,000	Trust 1992, Insurance 1992, Banking 1991
Nevis	Mid 80s	1,000	–	300	no	no	40	10	–	–	–
NL-Antilles	Mid 60s	250–4,000 NAF	a few weeks	1,000–3,000 NAF	yes	5–40	7–34	no	73	>19,000	–
TCI	Early 80s	400–15,000	1–2 days	500–10,000	yes	no	no	no	7	11 Trusts, 500 IC, 10,500 others	Trustee 1992, Limited Partnership 1992, Limited Life 1993

[1] Here the conditions for setting up a company according to the different forms of companies and banks are given. The costs of forming a company and the yearly costs vary accordingly and the largest range is given. Normally there are specific conditions for each form of company. The following abbreviations are used: IBC = International Business Companies; IC = Insurance Companies; CIC = Captive Insurance Companies.

* Costs for companies CI$ 900–1,800, banks pay an additional sum of $6,000.

Source: Possekel (1996)

A 'second wave' of OFCs appeared in the 1990s, and these too are concentrated in the Caribbean. They include Belize, Dominica, Grenada, St Lucia, and St Kitts and Nevis. At least 20 Caribbean jurisdictions now offer offshore banking services. Many of these 'second wave' OFCs have very quickly come under OECD criticism for allegedly harbouring tax evaders and/or money launderers (Table 9.1).

Benefits of OFCs for mobile capital

OFCs typically offer depositors a range of benefits compared with banking in most core countries. These include:

- banking secrecy laws;
- asset protection against creditors or lawsuits originating outside the host country;
- the ability to maximise corporate returns by setting up holding companies and subsidiaries and by using transfer pricing (see Box 9.1);
- lower income tax rates on the earnings from bank deposits (some jurisdictions even offer tax-free investment categories);
- higher interest rates paid on tax-free bank accounts (this is possible because of lower capital reserve requirements, and therefore the ability to lend out a larger share of deposits).

Such benefits are regularly invoked on the many websites of firms that provide the service of facilitating the movement of corporate and individual assets offshore (e.g. Lakeway International Equities Ltd 1998). Lucrative and secretive bank accounts are only part of the offerings of offshore banking service centres. Others include insurance, shipping registration, subsidiary incorporation, trusts and even citizenship (Evans 1993; *The Economist* 2000).

Core countries attack; offshore banking jurisdictions counterattack

Core governments have in recent years expressed three concerns about offshore deposits. First, and most obviously, core governments are worried about tax evasion. Money is moved out of the core countries to escape taxation. The OECD has termed this offshore banking provision 'harmful tax practices'. Second, core governments are worried about other unfair economic impacts from offshore banking. Offshore banks are generally characterised by less regulation, oversight and taxation of accounts. The offshore banks and their depositors can therefore engage in unfair competition against financial institutions and their depositors that operate within the core countries' more conservative regulatory and lending rules and taxation regimes. Third and most troubling are concerns about money laundering, which occurs when illegally obtained funds pass through one of the many offshore banks that do not keep records as to where the funds originated. Such banks effectively 'launder' those dirty funds. Money laundering is associated with a host of illegal activities worth billions of dollars, such as the trafficking of drugs, the sale of natural resources such as diamonds for weapons, and the theft of public assets.

In light of these concerns and the growing volume of money deposited in offshore accounts, in the late 1980s core governments initiated a range of actions aimed at reigning in offshore banking (see Box 9.2). The efforts of core governments to impinge upon OFC sovereignty and impose banking regulations follow the recent precedent from the 'ship rider' controversy. This refers literally to US government officials boarding ships in Caribbean territorial waters. The ship rider controversy was over whether or not the US Coast Guard should be allowed to enter Caribbean jurisdictions in pursuit of drug traffickers. An estimated 40 per cent of the illegal drugs entering the USA pass through the Caribbean (Barnes 2000). From the US government's perspective, Caribbean states do not have sufficient policing capacity to combat drug trafficking, and therefore they need the US agents' help. In the ship rider case, following many rounds of discussion, most Caribbean governments in 1996 granted the US government authorisation to engage in such counter-narcotics policing. Critics warn that this authorisation represents a significant breach of Caribbean sovereignty in favour of the United States. Despite these additional efforts to reduce drug trafficking through the Caribbean, local authorities believe that the flow of drugs has continued to increase since 1996 (Barnes 2000).

The rich countries may have even less success curbing money laundering than they have had with drug trafficking. The OECD countries' united front against offshore banking began to fray when the Bush administration came to power in January 2001. The ideological basis of the dispute is in the contrast between the social democrats who predominate in Western European countries, and therefore in the OECD, and the more conservative free marketers in the Bush camp. The Bush team publicised that it would like international banking reform limited to information exchange between core countries and offshore banking centres. The reforms would be limited to promoting 'cooperation when investigating specific cases of wrongdoing' (Armey 2000). The Bush team rejected the idea of trying to extend and enforce OECD banking rules on foreign soil, something the OECD criticised the Bush team over in its June 2001 report. Bush and company also rejected international tax harmonisation because it could raise US tax rates, the anathema of the Republican Party's platform (Armey 2000).

Many OFCs, including Caribbean ones, saw this intra-OECD disagreement as a window of opportunity to resist compromising and to counterattack in the negotiations and in the associated public relations. To defend their interests in the face of attack by the OECD, CARICOM in 2000 created the Caribbean Association of Regulators of International Business (CARIB) to formulate and implement strategies. Then, early in 2001, the OFCs created the International Tax and Investment Organisation (ITIO) to represent their interests against the OECD. The ITIO represents a disparate group of 12 offshore banking jurisdictions (Anguilla, Antigua and Barbuda, Bahamas, Barbados, British Virgin Islands,

Box 9.2: Offshore banking – a dodgy business?

OECD governments have take a variety of actions in recent years to investigate the legality and legitimacy of Caribbean OFCs. Britain has been particularly interested in curbing illegal activities associated with banks in its remaining overseas possessions. UK and US drug enforcement and tax authorities labelled one sting operation in 1993 'Operation Dinero'. They created a fake Anguillan bank and routed all of its mail and e-mail to their Atlanta headquarters. The sting yielded nine tonnes of South American cocaine, a shipment of arms bound for Croatia, and US$50 million in assets. US$382,000 of the recovered funds went to Anguillan authorities for their cooperation (Ferguson 1997; *The Light of Anguilla* nd).

In 2000, the OECD took concerted and publicised action against offshore banking. In an annual report distributed each June, the OECD, acting through a specialised operation called the FATF (the Financial Action Task Force on Money Laundering), blacklisted 15 offshore banking centres judged uncooperative in fighting money launderers. In its 2001 report, the FATF removed four of these jurisdictions while adding another six (Table 9.1). The OECD has demanded changes in their laws and practices so that it can identify, combat and prosecute money launderers. The OECD has also demanded access to these jurisdictions' banking systems and records so that it can judge for itself whether or not compliance has been achieved and money laundering thwarted (FATF 2001).

Then, through the 'Harmful Tax Competition Initiative', the OECD further widened its net. It identified 35 jurisdictions suspected of harbouring tax dodgers (Table 9.1). Nineteen of these are in the Caribbean, attesting to the region's extensive efforts to attract mobile capital (legal or not), especially from the USA. All but seven of these 35 jurisdictions under scrutiny have roots in the British Empire, suggesting the attractiveness of British law and English language to offshore capital. The OECD is attempting to force these jurisdictions to implement more rigorous banking rules. These would include stricter banking procedures, a more transparent banking environment including less account secrecy, and 'harmonised' tax rates for certain activities that would apply to all countries. Based on the national laws of its member countries, the OECD would also like to create international tax regulations to which all jurisdictions would comply. The OECD warns the offshore centres that the penalty for non-compliance with these regulatory initiatives will be punitive economic sanctions, thus far unspecified. Given that investors' perception of stability is fundamental to offshore banking, and given the trade imbalances and economic dependency that characterise the Caribbean (see Chapter 7), this is not an idle threat.

Cayman Islands, Cook Islands, Dominica, Malaysia, St Kitts and Nevis, Turks and Caicos, and Vanuatu).

As recounted at the start of this chapter, the prime minister of Barbados, Owen Arthur, has been among the most vocal leaders of the counterattack. Arthur has served as the chairperson of the CARICOM Heads of Government and the ITIO.

He has attempted to turn the tables on the OECD by pointing out its own flaws. Arthur has highlighted the central role of core countries in illicit activities by arguing that 'if the developed world wishes seriously to confront the problems associated with money laundering, they should begin in London and New York'. Arthur further implicated the OECD, and particularly the United States, when suggesting that the investigation consider

> the impacts on the Caribbean and other developing countries, [of] issues such as 'harmful drug consumption practices', 'harmful violent cinematic practices', 'harmful gun control practices', and 'harmful criminal deportation practices' (*Executive Time Magazine* 2001).

The political leaders of small peripheral jurisdictions realise that their negotiating power exceeds their economic power. Consider that CARICOM accounts for a mere 0.2 per cent of western hemispheric GDP. Eastern Caribbean leaders learned of their diplomatic power during the recent 'banana wars'. They successfully dangled the threat of turning banana farmers into marijuana growers and traffickers to get the USA to retreat from its original aim to immediately eliminate the Lomé convention. In the future, the OECD prefers to have the peripheral islands cooperate in its global agenda, including the fight against arms traffickers, computer hackers and spreaders of viruses, be they biological or electronic. The challenge for Caribbean leaders is to use their narrow negotiating leverage to strike deals that are to their jurisdictions' longer-term developmental advantage.

At the same time that many political leaders of peripheral states are challenging OECD authority, others are acting unilaterally and actively placating the OECD. The publication of the OECD annual offshore blacklist provides a strong incentive to OFCs to demonstrate that they are cooperating and cleaning up their operations. Since the 2000 report, many OFCs have enacted legislation and pursued international public relations in the hope of removing themselves from the list (Table 9.1). The Cayman Islands and the Bahamas, for example, which cleared the blacklist in June 2001, took it as a cause for celebration and self-promotion.

Geographies of Caribbean offshore banking

Many Caribbean governments have tried to attract offshore banks and insurance companies by advertising themselves and offering investment incentives, but only a few have succeeded at a large scale. The resulting geographical patterns are pretty clear, and follow the logic that 'success breeds success'. The places that are winners and losers in the 'casino capitalism' associated with the quest for hypermobile international finance capital have their own set of pre-existing attributes and consequent problems.

The Caribbean islands most successful at attracting offshore banking deposits, corporate subsidiaries and associated activities are relatively small territories with small populations, even by regional standards. And most of these 'winners' are those that continue to fly the Union flag: the Cayman Islands (now the world's fifth largest banking centre), Bermuda, the British Virgin Islands, the Turks and Caicos Islands, and Montserrat, until volcanic eruptions that began in 1997 made two-thirds of the island uninhabitable. These British dependencies confer the highest level of political stability and investor confidence. Their attractiveness for offshore banking extends from the proven record of British offshore territories closer to home, particularly the Isle of Man and the Channel Islands (Cobb 2002). These successful jurisdictions are still politically dependent on core states (i.e. Britain), and while they are geographically 'offshore', they are not really part of the global periphery. They receive subsidies and special export market access unavailable to the independent Caribbean. While the legal structures of these British overseas possessions are deemed trustworthy by publicly traded corporations, some of those same firms use offshore sites as a way to reduce their tax obligations to core states (Box 9.1).

Besides the British dependencies, other Caribbean islands notable for their ability to attract the financial holdings of thousands of foreign firms include the Netherlands Antilles and the Bahamas (which is politically independent but heavily dependent in economic terms on the USA, accounting for more than half of its imports and exports). In some places, such as the Caymans and the British Virgin Islands, offshore finance has become such a major economic component as to dramatically alter domestic conditions. Real estate prices have spiralled, and there is a growing reliance on foreign capital, management and technical expertise (Evans 1993; Roberts 1994, 1995; Hampton and Christensen 2002).

Unfortunately, a Caribbean jurisdiction's ability to attract legitimate investment capital into an offshore finance sector is negatively correlated with its need for new sources of foreign exchange. Most of the Caribbean is now comprised of independent countries with many features viewed as unattractive to finance capital, including weak and unstable economies, poverty, high unemployment and social tensions. These and other neo-colonial legacies leave many places isolated and uninteresting for capital investment. Then again, one thing that the smaller Caribbean OFCs have going for them is mobile capital's desire to diversify and spread risk among OFCs and even across regions. Attracting even relatively small amounts of capital by global standards can generate significant income for a small jurisdiction.

However, the money that the economically weaker, independent islands have been able to attract is often of questionable repute. Hot money is drawn to the most under-regulated banking environments, as Grenada illustrates. With over 3,000 registered banks, Grenada has about one for every 30 people on the island. The OECD has expressed consternation with Grenada's offshore banking sector. This prompted a major investigation by local authorities. A senior Grenadian financial investigator characterised the situation as one of 'shady dealing that seems to have pervaded the island recently'. The Grenada government has taken some regulatory action. It closed down 17 banks, some of which have been repeated targets of fraud or corruption allegations (Lashmar and Manneh 2001). It also pledged to consider dropping its banking secrecy laws.

The impacts of September 11th on offshore banking

> To follow the money is a trail to the terrorists (George W. Bush, 24 September 2001).

The US government's opposition to regulating global financial flows ended on 11 September 2001. Within days of the attack the Bush administration reversed its earlier position and publicly advocated the use of strong sanctions 'to pressure countries with dangerously loose banking regulations to adopt and enforce stricter rules' (*New York Times* 2001). This regulatory offensive, coming on the heels of Bush's declaration that 'either you are with us, or you are with the terrorists', muted much of the public opposition to US policies throughout the world. Even the Cuban government felt it necessary to remove its prominent billboard in Havana declaring, 'Senior Imperialists, we are not at all afraid!' (Months later the billboard was redisplayed, although in a less central location.) Fear of landing on the enemy list became widespread. A previously defiant CARICOM under Owen Arthur's leadership issued a supportive declaration:

> We undertake, as part of our contribution to the international coalition against terrorism, to redouble our efforts to prevent the use and abuse of our financial services sectors by fully cooperating with the United Nations and the international community in the tracing and freezing of the assets of terrorists, their agents and supporters (CARICOM 2001).

Note, however, that although CARICOM commits to the anti-terrorism effort, it strategically places authority in the hands of the UN rather than the USA.

The Bush administration is taking aggressive actions to identify and cut off international sources of financing for terrorist actions against the United States. A major component of these efforts is directed at disclosing sources of money laundering. They include blacklisting suspected individuals and organisations, the freezing of assets and investigating the internal records of US banks and their foreign affiliates. The Bush team is also pressuring OFCs with bank secrecy laws to cooperate with the FBI and other US government agencies by providing information on depositors. OFCs are shifting their agendas and enhancing their efforts in this regard to avoid being blacklisted.

It is too soon to know the full impacts of this US-led campaign against the international financing of terrorist activities on the efforts of Caribbean governments to benefit from offshore banking. Developments so far suggest that the US government is casting a very wide net with its anti-terrorism rhetoric, surveillance and interventions, military or regulatory. This US campaign is likely to serve as a check on the further advance of Caribbean efforts to reach a favourable settlement with the OECD. With the USA no longer dissenting, the OECD is stronger and more unified in its prosecution of weakly regulated OFCs. Further, the more regulatory oversight of OFCs by core countries, the fewer advantages OFCs have that can attract mobile capital, criminal or not. The result of the ongoing investigation of Enron's use of overseas subsidiaries (Box 9.1) is also likely to increase the international scrutiny of OFCs. It is therefore probable that one longer-term result of all of this will be a reduction of capital flows through, and therefore revenues going to, OFCs, particularly the Caribbean's smaller, poorer and less regulated ones.

Summary and conclusions

This has been a wide-ranging chapter devoted to identifying factors behind the recent growth in three types of offshore services in the Caribbean. A principal interest has been to assess the potential of these services to provide development gains for the host countries. This assessment of offshore services has involved examining the associated technological impacts and obstacles, geopolitical and legal struggles, the global geographies, and the distribution of benefits.

Has the technology-based reduction in the 'friction of distance' helped Caribbean offshore service providers overcome global economic imbalances? Electronic telecommunication has contributed much to time–space compression. We now have a *global* economy, capable of operating as a unit in real time. Economic activities in distant places are integrated electronically like never before. One result is a decentralisation of various offshore services from banking and insurance, to data processing and call centres.

We began by observing the geopolitical clash between core and periphery over the growing amount of capital flowing through OFCs. After reviewing recent trends in banking and other offshore services, we find it difficult to blame Owen Arthur for his defiance. Peripheral jurisdictions have responded to rapid, exogenous changes and found a few niches in the global economy in forms such as offshore banking and various labour-intensive services. Small wonder that Caribbean countries, with few alternative income sources, take offence when the architects of neo-liberal free-market capitalism attack some of their limited gains. Insult is added to injury when the Caribbean jurisdictions consider the fact that most of the offshore capital is based in and/or flows through the corporations and banks of those same core states pointing fingers at the OFCs.

Much of our evidence suggests that offshore services tend to replicate many of the problems the Caribbean has encountered over recent decades in its free trade zones for manufacturing (Chapter 8). These problems include low wages and low value-added contributions, and TNC control of the commodity chains. Problems also include vulnerability owing to the reliance on technologies, capital and discretionary income from core regions and to protectionist or capricious policies from core governments.

It is also notable that offshore services and the associated capital flows are very unevenly distributed outside the core regions. They create ever more complicated geographies of service specialisation, inequality, differential access and exclusion. We are therefore led to agree with Warf when he concludes that 'telecommunications . . . systems in fact produce new rounds of unevenness, forming new geographies that are imposed upon the rules of the past' (Warf 1999: 63).

Electronic telecommunications and other time–space compressing technologies make possible greater Caribbean *participation*, but this does not mean the region has managed to attain *control over* the associated technologies, legal structures and capital flows.

The post-September 11th international campaign reveals how global priorities are set. In the realm of international finance, the campaign broadly targets jurisdictions alleged to have loose financial regulations, banking secrecy or money laundering activities. The campaign provides insight into the structure of power in the contemporary world system. Many of the earlier priority issues for people and governments outside (and indeed inside) the core regions, whether they concerned financial flows or broader themes of equity, access or ecology, have been muted. The anti-terrorism campaign demonstrates how the USA and its core allies can quickly and decisively shift and focus global priorities, marginalise and effectively discredit other concerns, and dominate the global agenda.

Why do offshore services have so many limitations? The simplest answer is that the limitations emanate from the organisation of the capitalist world system, which spawns the demand for offshore services in the first place. Economic, technological and political processes originating and largely controlled in core regions propel offshore services. Despite the global decentralisation of a wide range of services, their economic orientation remains firmly core-focused. In this political-economic environment, most Caribbean countries are likely to benefit only marginally from offshore services.

Useful websites

Gambling Online Magazine (www.gamblingonlinemagazine.com) and Online Casino News (www.onlinecasinonews.com) are two of the electronic gambling industry's magazines.

Offshore Watch (visar.csustan.edu/aaba/jerseypage.htm) is a frequently updated website managed by offshore banking researcher, John Christensen, that posts newspaper and scholarly articles concerned with tax avoidance and the global movement of capital.

Transnationale.org (www.transnationale.org/anglais/dossiers/finance/paradis.htm) is the tax haven and offshore banking-focused web page of an international organisation broadly devoted to consumer information and protection.

References

Armey, D. (2000) letter from US House of Representatives Majority Leader Dick Armey to US Treasury Secretary Lawrence Summers, available at www.tax-news.com/asp/res/st_offshorefuture_28_09_00.html.

Barnes, C. (2000) 'Narcotics: driven from Latin America, smugglers eye Caribbean', *Inter Press Service*, 26 January, available at web.lexis-nexis.com/.

Bromley, R. and Bromley, R. (1988) *South American Development. A Geographical Introduction*, 2nd edition, Cambridge University Press, Cambridge and New York.

CARICOM (2001) 'The Nausau Declaration on International Terrorism: The CARICOM Response', 11 October, available at: www.state.gov/coalition/cr/ddr/13795pf.htm [accessed January 2003].

Cobb, S. (2002) 'Offshore financial services and the Internet: creating confidence in the use of cyberspace?' paper presented at the annual meeting of the Association of American Geographers, Los Angeles, March.

Corbridge, S. (1993) *Debt and Development*, Blackwell, Oxford.

Daniels, P. W. (1993) *Service Industries in the World Economy*, Blackwell, Oxford.

Delude, J. (2000) 'Las Vegas of the Internet: Americans flock to Costa Rica to set up online casinos', *The San Francisco Chronicle*, 20 July, A12.

Economist (2000) 'Small states, big money', *The Economist*, 23–9 September.

Evans, R. (1993) 'Banking on the black economy', *Geographical Magazine*, September, 40–2.

Executive Time Magazine (2001) 'Caricom hits hard at G7's harmful tax report', *Executive Time Magazine: Journal of Business & Technology: Caribbean Edition*, 3, available at www.angelfire.com/journal/executivetime/.

FATF (Financial Action Task Force on Money Laundering) (2001) *Progress Reporting on Non-Cooperative Countries and Territories*, OECD (Organisation for Economic Co-operation and Development), Paris.

Ferguson, J. (1997) *Eastern Caribbean in Focus: A Guide to the People, Politics and Culture*, Latin American Bureau, London.

Gwynne, R., Klak, T. and Shaw, D. (2003) *Alternative Capitalisms: Geographies of 'Emerging Regions'*, Edward Arnold, London, and Oxford University Press, New York.

Hampton, M. and Christensen, J. (2002) 'Offshore pariahs? Small island economies, tax havens and the reconfiguration of global finance', *World Development*, **30**, 9, 1657–73.

Harvey, D. (1989) *The Condition of Postmodernity*, Blackwell, Oxford.

Illeris, S. (1996) *The Service Economy: A Geographical Approach*, John Wiley, Chichester, UK, and New York.

James, J. (1999) *Globalization, Information Technology and Development*, St Martin's Press, New York.

Klak, T. (1998a) Is the neoliberal industrial export model working? An assessment from the eastern Caribbean', *European Review of Latin American and Caribbean Studies*, **65**, 5 & 67–90.

Klak, T. (ed.) (1998b) *Globalization and Neoliberalism: The Caribbean Context*, Rowman & Littlefield, Lanham, Maryland.

Lakeway International Equities Ltd (1998) *What is Offshore Investing*? available at www.lakewayinternational.com/whatisoffshore.html.

Lashmar, P. and Manneh, M. (2001) 'Grenada puts clamp on banks', *The Independent*, 18 March, Business Section, 3, available at web.lexis-nexis.com/.

Lee, R. (2001) 'Grenada pushes on in e-commerce drive', *Tax-News. Com*, **22**, June.

Martin, A. (2001) 'A sure thing', *Harper's Magazine*, **April**, 96.

Masters, C., Ramsay, M. and Larsen, L. (2001) 'A clear and present danger', *Four Corners, Australian Broadcasting Corporation Investigative Journalism*, **14**, May.

Mullings, B. (1995) 'Telecommunications restructuring and the development of export information processing services in Jamaica', in Dunn, H. (ed.) *Globalization, Communications, and Caribbean Identity*, Ian Randle, Kingston, Jamaica, pp. 174–91.

Mullings, B. (1998) 'Jamaica's information processing industry: neoliberal niche or structural limitation?' in Klak, T. (ed.) *Globalization and Neoliberalism: The Caribbean Context*, Rowman & Littlefield, Lanham, Maryland, 135–54.

New York Times (2001) 'Finances of terror', editorial *New York Times*, 24 September.

Pantojas-Garcia, E. (2001) 'Trade liberalization and peripheral postindustrialization in the Caribbean', *Latin American Politics and Society* (formerly the *Journal of Inter-American Studies and World Affairs*), **43**, 57–77.

Panton, S. (2003) 'E-commerce no no's', *Jamaica Gleaner*, 5 February, D4.

Roberts, S. M. (1994) 'Fictitious capital, fictitious spaces? The geography of offshore financial flows', in Corbridge, S., Martin, R. and Thrift, N. (eds) *Money, Power and Space*, Blackwell, Oxford, pp. 91–115.

Roberts, S. M. (1995) 'Small place, big money: the Cayman Islands and the international financial system', *Economic Geography* **71**, 3, 237–56.

Robinson, M. (2001) 'The multilaterals' campaign against offshore', *Tax-news.com* **01**, January.

Sassen, S. (2000) *Cities in a World Economy*, 2nd edition, Pine Forge Press, Thousand Oaks, California, and London.

Selya, R. M. (1999) 'Taiwan as a service economy', in Bryson, J., Henry, N., Keeble, D. and Martin, R. (eds) *The Economic Geography Reader: Producing and Consuming Global Capitalism*, John Wiley, Chichester and New York, pp. 247–51.

Stroud, J. (1996) 'Porn calls move offshore; parents are shocked by costs', *St. Louis Post-Dispatch*, 18 February.

Taylor, P. J., Hoyler, M., Walker, D. R. F. and Szegner, M. J. (2001) 'A new mapping of the world for the new millennium', *Research Bulletin, Globalization and World Cities Study Group and Network* (forthcoming in *The Geographical Journal*), **30**, available at www.lboro.ac.uk/gawc/rb/rb30.html#f2.

UNCTAD (United Nations Conference on Trade and Development) (2000) *Building Confidence: Electronic Commerce and Development*, United Nations, New York.

Wade, R. (1996) 'Japan, the World Bank, and the art of paradigm maintenance: the East Asian miracle in political perspective', *New Left Review*, **217**, 3–36.

Wade, R. (2001) 'Winners and Losers', *The Economist*, 26 April.

Warf, B. (1999) 'Telecommunications and the changing geographies of knowledge transmission in the late 20th century', in Bryson, J., Henry, N., Keeble, D. and Martin, R. (eds) *The Economic Geography Reader: Producing and Consuming Global Capitalism*, Wiley, Chichester and New York, pp. 57–63.

Warf, B. (2001) 'Global dimensions of U.S. legal services', *The Professional Geographer*, **53**, 3, 398–406.

Chapter 10

GLOBALISATION AND THE CARIBBEAN

Introducing the uneven nature of globalisation

Scarcely uttered before the 1990s, *globalisation* has suddenly become one of the most common terms in our daily lexicon. The term is regularly invoked in corporate, political and academic depictions of worldwide trends, including those shaping the Caribbean. Despite the term's ubiquity, it is unusual to find a clear definition of globalisation or to see it interrogated against evidence. Too often, globalisation is presented as if it were some autonomous, pervasive and uncontrollable force. In the face of such an overwhelming force, people seem to have little choice but to adapt and make do. An additional problem with most discussions of globalisation is that they often imply that what is emerging is an inclusive and integrated 'global village' (cf. Klak 1998a). This chapter breaks with everyday usage of the term. Rather than presenting globalisation as its own force, and subsuming the Caribbean under the common and typically over-generalised notions of globalisation, this chapter takes a more place- and issue-specific approach. The chapter will first provide an overview of the general nature of globalisation before getting to the specific case of the Caribbean.

By 'globalisation' we refer to increases in the geographical scale, volume and velocity of transnational interactions (cf. Held *et al.* 1999). At the forefront of these growing trans-border relations are often profit-seeking corporations, but the transnational flows and influences are multi-dimensional. In addition to the economic aspects of globalisation, we identify five other dimensions: technological, political, cultural, ideological and environmental. Other chapters of this book have touched on several of these dimensions of globalisation. Some of the key components of each of these globalisation dimensions, and important studies of them in both the global and the Caribbean contexts, are as follows:

- *economic* – the growing role of TNCs and, in particular, in organising globally networked production systems; growing gaps between the beneficiaries of globalisation and the global majority (Gereffi and Korzeniewicz 1994; Gwynne *et al.* 2003);
- *technological* – the increased importance of electronic telecommunications and air travel (see Chapter 8; Castells 2000);

- *political* – neo-liberal free-market policies; trade treaties such as the US 807 tax code, which encouraged maquiladoras to locate in the Caribbean region; treaties that extend CBI and give Caribbean exporters access to US markets in exchange for greater access to Caribbean markets for US exporters; NAFTA and the movement towards hemisphere-wide free trade through the Free Trade Association of the Americas or FTAA (see Chapter 8; Klak 1998a; Ugarteche 2000; Hall 2001; Ramsaran 2002);
- *cultural* – the emerging global culture of mass consumption; worldwide TNC name-brand marketing; global (mainly US) TV through the widespread dissemination of cable; the transnational impacts of music, food and festivals such as Trinidad's Carnival (see Figure 10.1 and Chapter 2; Dunn 1995; Brysk 2000; Hall and Benn 2000; Scher 2003);
- *ideological* – TINA (There Is No Alternative to free-market capitalism, Margaret Thatcher's famous utterance from more than two decades ago); the collapse of the state socialist bloc and of Latin American and Caribbean socialist experiments (Chapter 7; Gwynne and Kay 2004);
- *environmental* – the increasing transnational scale of ecological damage such as atmospheric pollution and warming, sea-level rise, the degradation of fisheries, coral reefs and freshwater supplies, and deforestation; both the institutional and grassroots responses to these problems are also crossing international borders (see Chapter 1; NRC 1999; Lee *et al.* 2000; Goodbody and Thomas-Hope 2002).

This chapter will elaborate on many of these aspects of globalisation while focusing primarily on the economic facets – Caribbean and extra-regional corporations, industrialisation and trade. Throughout the discussion, we stress that current economic and political transformations must be specified at the local level to understand their impacts on people and places. By taking a more place-specific approach to globalisation, it becomes clear that current economic and political trends are not really globalised but are instead highly uneven geographically as well as socially. Our main point is therefore that globalisation is highly uneven spatially and socially, and that this unevenness occurs across a range of considerations such as control, access, advantage and impacts. These specificities are crucial for understanding globalisation impacts in the Caribbean.

Many scholars have examined trends associated with economic globalisation but have come to very different conclusions about their significance and, in particular, how much the role and powers of the state have declined relative to TNCs. One of the most comprehensive reviews of this literature (Held *et al.* 1999) finds that the interpretations of the changes fall into three groups: the strong or hyperglobalist thesis, the weak or sceptical thesis, and the nuanced or transformationalist thesis.

- *Hyperglobalist thesis* – according to this view, globalisation represents a new epoch of human history in which traditional nation-states have become

Figure 10.1 Cable TV and Caribbean culture

TV CHANNEL LISTING

Variety

02 TBS
03 TV Guide
04 Bulletin Board
05 Reggae Sun
06 LOVE TV
07 TVJ
08 Discovery
09 CVM
10 TNT
11 USA

Children

12 Disney
13 Toon Disney
14 Cartoon
15 Nickelodeon

Variety

16 National Geographic
17 PBS
18 Sci-fi
19 UPN
20 Court TV

Adult

21 _*Extasy (scrambled)_

News

22 CNN
23 CNNHn
24 CNN/Fn/Int
25 CNBC

Networks

26 Fox
27 CBS
28 ABC
29 NBC

Movie

30 Lifetime Movies
31 Western
32 HBO
33 Cinemax
34 Showtime
35 TMC
36 STARZ
37 Encore
38 AMC
39 Encore Action

Sports

40 ESPN
41 ESPN2
42 Speedvision
43 TNN
44 Fox Sports World
45 Msgn
46 GOLF
47 Outdoor

Sport news

48 CNN/SI
49 ESPN News

Music

50 MTV
51 MTV2
52 VH1
53 BET
54 BET on *Jazz*

Religious

55 3ABN
56 TBN
57 EWTN

Variety

58 Wisdom
59 Health
60 Oxygen
61 Lifetime
62 Family
63 Comedy
64 E! Entertainment
65 WPIX
66 HGTV
67 DIY
68 Food Net
69 TV Land
70 Game Show
71 WGN
72 Tech TV
73 TLC

International

74 BBC *America*
75 BBC *World*
76 CCTV-4
77 Sony TV
78 Univision
79 TV5 (French)
80 LBC
81 DW
82 Zee TV

Variety

83 Travel
84 Weather
85 CSpan
86 Animal Planet
87 A&E
88 History
89 Arts
90 Value Vision
91 Fox
92 HBO Comedy
93 Solid Gold Oldies *(audio)*
94 Lite Jazz *(audio)*

Local

95 Community Channel
96 *Data Carrier*
97 *Data Carrier*
98 Hype TV
99 Cable View

Variety

100 A1 TV
101 Fox
102 CBS
103 ABC
104 NBC
105 PBS
106 Goodlife
107 WE *(Romance)*
108 Bravo
109 MSNBC
110 IFC
111 Real TV
112 FXM
113 *Test Channel*
114 TCM
115 *Test Channel*
116 BET Movies
117 Previews
118 *Test Channel*
119 *Test Channel*
120 *Test Channel*

In recent years, many more Caribbean homes have obtained cable television hookups, whether through legal or, for some, illegal distribution systems. As this listing of channels in Jamaica suggests, the United States is the main source of programming, and this is having a significant cultural impact on the region.

ill-suited for an emerging global economy. This thesis emphasises the economic logic underlying globalisation and argues that it 'denationalises' economies through the establishment of transnational networks of production, trade and finance. Within this thesis are two contrasting interpretations. One is the pro-neo-liberal version, which celebrates the dominance of the market principle over state power. The other is a neo-Marxist view, which sees contemporary globalisation as the triumph of oppressive global capitalism. The geographical implications of the hyperglobalist thesis include new patterns of regional winners and losers. In the creation of these new world geographies, the pro-neo-liberals stress the advantages of global competition. Some spaces within a country may be made worse off as a result of such competition, but other spaces will have a comparative advantage in producing certain goods for global markets. Those holding pro-neo-liberal views tend to see all countries (rich and poor) benefiting from globalisation although everywhere will experience significant restructuring. In contrast, the neo-Marxists believe that global capitalism creates and reinforces inequalities both between and within countries.

- *Sceptical thesis* – sceptics argue that many of the fundamental features and empirical manifestations of global capitalism today remain as they were decades and even centuries ago. Using statistical evidence of world flows of trade, investment and labour from the nineteenth century, sceptics contend that contemporary levels of economic interdependence are by no means historically unprecedented. The sceptics think that 'true' globalisation requires a fully integrated world economy, which remains a long way off. One crucial economic factor of the world economy, labour, remains relatively immobile, particularly compared with capital. Sceptics emphasise the enduring power of national governments to regulate international economic activity. Thus, they regard the early twenty-first century as indicating only heightened levels of internationalisation. Geographically speaking, sceptics see globalisation as increasingly marginalising regions such as the Caribbean that are part of the world periphery. Globalisation provides economic growth for the core economies of the global North, and certain industrialising or natural resource-rich countries of the semi-periphery. However, a whole series of economic and political factors retard the economic growth in peripheral regions such as the Caribbean.
- *Transformationalist thesis* – this interpretation sees globalisation as a powerful transformational force which is responsible for a massive 'shake-out' of societies, economies, institutions of governance, and the world order. The direction of this 'shake-out' remains uncertain, since globalisation is seen as an essentially contingent historical process replete with contradictions. Contemporary processes of globalisation are historically unprecedented such that all governments and societies need to adjust to a world in which there is no longer a clear distinction between international

and domestic, external and internal affairs. The geographical aspects of this thesis emphasise the continuation of global divergence – increasing inequalities between and within countries.

All three theses regard capitalism as having entered a distinctly 'globalising' phase, though the nature and outcomes are much in dispute. What is interesting geographically is that most perspectives (apart from the pro-neo-liberal one) do not see global convergence (that is, less inequality between and within countries) resulting from globalisation. Growing inequalities appear to be the result of production, trade and finance becoming increasingly transnational. The transformationalist thesis sees some (but by no means all) countries, regions, communities and households benefiting from being more closely linked to the fortunes of the global economy, while many others will suffer. Divergence and unevenness therefore have become interwoven with globalisation. As a result, some argue that 'globalisation' (a word that implies convergence) can be a misleading term for the unfolding process. David Harvey suggests that we jettison 'globalisation' entirely and replace it with the 'uneven spatio-temporal development of capitalism' (Harvey 1995). That is a mouthful, but it may point in a fruitful direction. Similarly, corporate manoeuvring on the world stage leads Hilbourne Watson to this stark yet useful characterisation of the situation: 'Globalisation is capitalism in the age of electronics!' (Watson 2001). Many others that have studied the growing power of TNCs have labelled the current era as primarily one of 'corporate globalisation'. These are all improvements over the overgeneralised and undefined ways that globalisation is often presented.

Across the various interpretations of contemporary globalisation is a recognition that its technological underpinning is time–space compression. By this we mean that many places are becoming closer together in travel or communication time or costs (Knox and Marston 2004). Although such convergence has been occurring over centuries, the pace has quickened markedly in recent decades. To illustrate, consider three examples of cost reductions using 1990 as a benchmark: shipping costs have fallen by more than two-thirds since 1920; airline transport costs have fallen by more than 60 per cent since 1960; and the price of international telephone calls has fallen by 90 per cent since 1970 (Girvan 1999). But this is just the beginning. Satellites and the Internet have virtually eliminated the 'friction of distance' for communications. Messages, prices, television images and other information are transferred as quickly and cheaply across 10 miles as 10,000 (Harvey 1989).

Note, however, that access to these time–space compressing technologies is highly uneven. Time–space compression has opened the possibility for potentially new links between different parts of the world, and potentially new geographical configurations of the global economy. Computer-based technological advances have made it possible for certain types of production, capital flows, communication and decision making to become transnational in scope. Activities previously confined geographically can now be instantly coordinated across great distances and national boundaries. That time–space compression is made possible by technological breakthroughs should not be interpreted to mean that it plays out in politically or socially neutral ways. On the contrary, the capacity for, and the interest in, taking advantage of time–space compression is distributed highly unequally. In particular, corporate and individual capital emanating from core countries takes greatest advantage of time–space compression.

To capture the notion of economic actors operating on a world scale, we suggest replacing the more familiar term 'the multinational corporation' (MNC) with 'the transnational corporation' (TNC) (Dicken 1998). The evidence of global corporate power is clear. Although the world's 200 largest TNCs employ just 0.3 per cent of the total population, they now represent 28 per cent of global economic activity (Lietaer 2001). Relatedly, there is also increasing income inequality, as corporate owners and investors profit from TNC activity. In 1998, for example, the world's three richest people collectively had assets worth $156 billion. These were Bill Gates and Paul Allen of Microsoft (not coincidentally, a computer software giant judged monopolistic by the US Supreme Court), and Warren Buffett of the global insurance and investment firm Berkshire Hathaway. Their wealth is as much as the combined GDPs of the world's 43 poorest countries, home to 600 million people (UNDP 1999). The concentration of economic power just described is related to the advantages and powers that time–space compression delivers to those able to tap into it. TNCs coordinate production across national boundaries, benefit from differences in resources and regulations between the states involved, and switch activities (for example, labour deployment, transfer pricing) between states in pursuit of the firm's interests (see Box 9.1 on Enron's methods; Dicken 1998).

A key point to take from the above paragraphs is that TNCs (as well as individual capital), through their capacity to exploit time–space compression, have accrued certain advantages *vis-à-vis* national political actors, whose powers derive from immobile jurisdictions. Indeed, this imbalance has many impacts on the relationship between globalised economic actors and localised political actors. Globalised economic actors are now better able to reduce state regulatory oversight, and to move capital in and out of jurisdictions as they see fit. States are less able to tax mobile capital, and are therefore forced to downsize, which further weakens their authority. In turn, global capital has become involved in a greater range of activities, from the privatisation of state assets to a range of offshore activities made lucrative by the reduction in state oversight and regulation. Labour too is less mobile than capital and is therefore at a growing disadvantage.

What is the relevance of contemporary globalisation, as we have described it, for the Caribbean region? Time–space compression is highly uneven across global regions. Access and participation are highly correlated with a country's position in the world-system hierarchy. Core countries, especially their global cities (Taylor *et al.* 1999), are much more connected than peripheral regions. Concerns about airline transportation accessibility loom large in the Caribbean and are discussed later in the chapter. Furthermore, the roles that actors and interests from different regions play in transnational interactions are also hierarchically arranged. Core countries play most of the high-end, high-value and decision-making roles, whereas peripheral countries are involved mainly at the low end, concerning low-value and labour-intensive activities.

This chapter therefore primarily focuses on the way that advanced electronics-based communications and computer technologies have influenced activities in the Caribbean. These activities certainly include a range of economic sectors, but also the related policy responses. The four main themes are:

- the globalised origins and history of the Caribbean region in the plantation economies;
- new global divisions of labour, and mainly addressing the region's export-oriented apparel factories;
- the mass media, especially the print media, and air transportation;
- Caribbean transnational corporations (TNCs) and how they are faring in the globalisation era.

These globalisation issues are united by the fact that they involve transnational capital flows in and out of the Caribbean. Many of the issues discussed here have behind them an impetus on the part of transnational capital to invest in the Caribbean region so as to avoid some of the regulations and restrictions in the legal codes of core countries. The regulations and restrictions pertain to employee wages and benefits, and rights, finance and various profit-making activities. At the same time as transnational capital goes offshore to avoid these restrictions, it remains bound to those core countries because they represent the world's largest consumer markets for the products of that capital. This world-system hierarchical relationship holds true over a wide time frame and for a wide range of products. It can be applied to the slave plantations producing sugar for Western Europe, the factories employing low-wage labour to sew garments for the United States, and the Internet firms providing offshore tax havens or gambling opportunities for US clients, which were discussed in Chapters 8 and 9.

Globalisation and the Caribbean past (sixteenth to twentieth centuries)

For Caribbean people, all the talk today about the uniqueness of the present era of globalisation is cause to chuckle. They know their histories, and realise they have been through many earlier rounds of globalisation. The Caribbean region is shaped now, as it has been under previous phases of capitalism, by the ideas and actions of outside investors and political leaders. This chapter therefore looks to the Caribbean's past experience with globalisation for clues to its present and future prospects.

Historically, the Caribbean is perhaps the most globalised world region. Since the 1500s it has been controlled by outside powers, based economically on imported labour, cleared to create monocultural landscapes of sugar cane, bananas or other crops, and reliant on the import of virtually everything else needed to sustain local populations (Richardson 1992; Mandle 1996). Caribbean sugar cane processing was industrialised before European production was. Because all dimensions of Caribbean society were exogenously constructed and transplanted there, the culture of Afro-Caribbean people from the outset has been detached from its historical and geographical roots, and therefore modern and global. In other words, the organisation of Caribbean societies and their associated production systems have for half a millennium been devoted to distant markets and profit-making demands. During centuries of colonial rule, Caribbean culture emerged as a unique and distinctive result of the myriad contributions of a wide range of African and Asian traditions, West Indian adaptations, and many foreign influences introduced by Europeans or picked up during circular migration for work in other parts of the hemisphere and in Europe (Richardson 1992; Mintz 1996).

For the Caribbean, therefore, current 'globalising' trends represent yet another round of powerful external influences for a region historically shaped by exogenous decisions and events. The Caribbean's historical global integration, modernisation and industrialisation underlie the region's current abject dependency within the world system. The importance of path dependency in world systems theorising leads to the expectation that the Caribbean's historical dependency is an ongoing situation. The historical relationship between the Caribbean and core regions established a pattern that is difficult to break out of today. The Caribbean region is now mostly independent from Europe politically, but it is still reeling under the historical legacies of dependency on outside authorities, suppliers, markets and geopolitical agendas, as the World Trade Organization's banana rulings aptly illustrate (Payne and Sutton 2001). The Caribbean's extreme trade dependency, primarily involving the trade imbalances with the United States shown by the data in Table 8.3, is an important current manifestation of the continuity of external dependency. Now that the entire world has entered the present era of (US-dominated) globalisation, the Caribbean's long-term predicament offers a chronicle of the impacts of exposure to many previous rounds of transformation of global capitalism, and valuable lessons for other economically peripheral regions.

Globalisation and the Caribbean today

As discussed in the first section of this chapter, one distinctive aspect of the global economy today is the integration of production across space and, more importantly, across national borders. The production of a particular consumer good often involves a host of contributors and intermediaries from several world regions. Take the example of a relatively simple item such as a dress shirt. Before it is displayed for sale in a department store in the United States, it may have circumnavigated the globe. A garment begun as cotton grown and processed into fibre in India or China may well have been shipped by companies flying the flag of Liberia, Malta or the Bahamas to factories owned by South Koreans but located in the Dominican Republic, where female sewers are reliable, inexpensive and without union representation. In fact, this Asia–Caribbean–US triangle is one of the world's most important inter-regional production networks (Bonacich 1994; Pantojas 2001).

Economic geographers and sociologists have labelled and conceptualised these multi-step labour and production processes. They are defined alternatively as extended commodity chains (to emphasise the networks of people, firms and states behind the products), or as vertically disintegrated production (to emphasise the segments or nodes of production, which are often discrete in terms of knowledge, control and ownership) (Gereffi and Korzeniewicz 1994; Johnston *et al.* 1994).

As we continue to assess the distinctiveness of the current round of global restructuring, it is important to note, as the globalisation sceptics have emphasised, that a complex division of labour and production across international borders is not itself new (Hopkins and Wallerstein 1994). What *is* new is the number of commodities involved, their contributions to global trade, the accelerated speed of transactions, and the reduced costs of movement. Note further that the commodities conceptualised in this way include not only things conventionally manufactured, but also much more. In the current era of globalisation, they encompass an array of services from tourism, other entertainment such as music and festivals, agro-industries and medicine, to computers, data processing, offshore banking and e-commerce. The diverse list of globalised commodities points to the fact that divisions of labour and production are also shifting rapidly in terms of where, how and what. Any production advantages or investor interests in the Caribbean are likely to be highly ephemeral. Caribbean policy making becomes especially difficult in this context, and the experience of the workers in export-oriented activities becomes highly volatile, as the examples in this chapter illustrate.

For now, we should return to the above example of the global voyage of a conventional commodity, a shirt. It is highly relevant to the Caribbean's recent experience with globalisation. More than any other industry, apparel factories have relocated to the Caribbean (and Central America) during recent decades. This foreign investment has been the principal response to policy efforts to find new productive roles in evolving global commodity chains. Indeed, the 'export market niche' strategy of neo-liberal development is all about identifying specialised roles in current global trade (Klak 1998a, 1998b). Unfortunately, as apparel exemplifies and as Chapter 8 discussed, the Caribbean has been able to contribute to the new global economy primarily at the low end of the value-added continuum. At the same time, global consumer culture has much more thoroughly infiltrated the region. The result is strong Caribbean domestic demand for imported consumer goods, mainly from the USA, and huge negative trade balances. Even the region's most industrialised country, Trinidad and Tobago, faces this problem. Its exports of petroleum products, steel and other manufactured goods to the Caribbean, Central America and Mexico give it a huge intra-regional trade surplus. But this is more

Box 10.1: A globalisation whodunit

In 1995 and 1996 at least 88 fevered Haitian children under the age of six died mysteriously after their parents gave them a sweet-tasting medicine. After the children weakened from an initial dose of the medicine, their parents, thinking the original fever was worsening, gave them more of the medicine. The children first lost their ability to walk, stand and crawl, and then died of kidney failure. Most were from CiteSoleil, the hemisphere's worst shanty town located near the outskirts of Port-Au-Prince. An international investigation by the World Health Organization, the Pan American Health Organization, the US Food and Drug Administration and the US Center for Disease Control eventually discovered the cause. The medicine, which was supposed to be sweetened with glycerin, was actually laced with similar-tasting diethelene glycol, a toxic but common ingredient in antifreeze, solvents and lacquers. Parents had unknowingly poisoned their own children. The investigation traced the poison to Haiti's largest pharmaceutical company, Pharval, which had purchased barrels labelled as glycerin from the Vos shipping company in Rotterdam, the Netherlands. Vos had purchased the mislabelled toxin from a German firm, Helm AG. Helm in turn had bought it from SinoChem (China National Chemicals Import and Export Corporation), a state-owned firm and the world's 307th largest TNC. SinoChem was itself supplied by another Chinese firm, but the investigation has been unable to determine which one. The trail has reached a dead end. Neither the US nor the Chinese governments has publicised the name of the firm that originally produced and mislabelled the substance. US Congresswoman Nancy Pelosi's explanation for the non-disclosure is that 'the only consideration that is important in Washington DC is the money that is involved in the US–China trade'. Because the production costs of the toxin are less than half that of glycerin, the mislabelling may well have been motivated by profit (Beck 1997; Public Citizen 1997).

than offset by a continuing trade deficit with the United States, totaling $1.3 billion in 1998 (CARICOM 2000).

For many of those imported consumer goods, it is difficult to track their entire production paths because they change hands so often (Box 10.1). Under the veil of anonymity, consumer standards and safety can be compromised. But that assumes that there actually are standards to uphold. With commodity chains increasingly fragmented and extended across many national territories, individuals may behave ethically and safely within the confines of their legal jurisdiction and/or their narrow knowledge of those affected, but then cause harm down the line. Legal experts are grappling with a growing number of such cases of questionable transnational accountability. Too often those negatively affected by such transnational labyrinths are poor people. They are least aware of the globalised nature of trade and most vulnerable to any 'cost saving' that compromises safety.

Global commodity chains, global consumerism and Caribbean apparel factories

When the women who work in the Caribbean apparel factories learn how much the clothes they sew sell for in the United States, they invariably totter in disbelief. Their surprise comes from a simple comparison of the relatively high prices the consumer products fetch abroad and their own wage rates, which are often below subsistence level. Ironically, Caribbean wage rates are relatively high compared with competing locations for the same factories such as Indonesia or China. The workers' shock points to two telling features of the commodity chains that locate their labour-intensive assembly operations in the Caribbean and Central America. We will next examine each of these features for their lessons concerning the Caribbean's insertion into the capitalist world system in the era of electronic globalisation.

The first feature is the class- and region-based differences in knowledge of, and access to, the mechanisms and technologies of time–space compression that underlie this contemporary form of production. There is now an extreme contrast between those that direct the global commodity chains and practically everyone else. On the one hand, the factory owners are mostly from core or semiperipheral countries, and are globally oriented and aware. On the other hand, the direct producers in the capitalist global economy have a very narrow understanding and awareness of the global commodity chains in which they participate. The factory workers' limited vision extends to their products' consumers in core countries, and to most other people living in the Caribbean. People in the USA shopping at Target, K mart or Wal-Mart know little about the conditions behind the products. Similarly, few people in the Caribbean outside those working in the factories even know that their island exports products to Disney, Sears or Victoria's Secret, let alone how sales prices compare with wage rates.

The second feature is the decoupling of production and consumption. In the heyday of industrial capitalism in core countries, workers earned enough to buy the products of their own labour, and that of other factories. Workers' earnings were, in Marxist terms, enough to reproduce their own labour power. For the export-oriented factories of the Caribbean and many other peripheral countries, that basic principle of industrial capitalism no longer holds. Box 10.2 explains how women working in garment factories manage to survive on such paltry wages.

From 1990 to 1994, the hyper-exploitation of female factory workers illustrated in Box 10.2 was the basis for substantial growth in CARICOM's manufacturing exports to the United States. Exports grew from $283 million to $508 million over those four years. However, for many female factory workers in St Lucia and elsewhere in the Caribbean, the bad situation described in Box 10.2 has only got worse. Despite the fact that wages are below subsistence level, they are nonetheless considered high compared with many other countries outside the core that are competing for the same investment. Since 1 January 1994 and the implementation of NAFTA, foreign owners have closed Caribbean factories and relocated

Box 10.2: How to survive where wages are below subsistence level

For the typical female factory worker in the Caribbean, the pay level is less than what is required to meet their basic needs. These women are forced to resort to a wide range of additional and creative survival strategies. The example of St Lucia helps to reveal how the factory workers and their families survive (Figure 10.2). For many St Lucian women these strategies include some combination of wage work, informal sector activities, homework, backyard gardening, food provision by relatives and neighbours, monthly support payments from one or more fathers of the children in the household, and rural living to minimise the cost of living. The factory pay rates are not high enough to allow many of the workers to live in an urban area, where public services and other amenities are better but rents are higher and subsistence food production is not possible.

Figure 10.2 Typical monthly wage and expenses (in US$) for female workers in export-oriented factories in St Lucia

1. **Wages:** $148; plus $74 child support (if father provides it)
2. **Expenses:**
    a. rent for housing in the countryside (cheaper than in town): $37
    b. utilities: $37
    c. minibus transportation to and from work (EC$1.50 per ride): $22
    d. other costs (these must be covered by the $126 of remaining monthly income, or the equivalent of about $4.20 per day):
        i. food:
            (1) supermarket prices are approximately 64% higher than in the U.S. for comparable products
            (2) women must obtain as much food as possible from subsistence sources (grow own, get donations from family and neighbours, exchange, barter)
        ii. other children's needs including school supplies, familial health care, clothing, household needs, other transportation costs, entertainment, savings ...

Data source based primarily on figures provided by Theo Van Katwyk, Managing Director of National Glove, a division of Maine Brand Manufacturing, for workers in his factory; reconciled with data from other factories and the St Lucia National Development Corporation in September 1997; food prices in supermarkets in Castries and Oxford Ohio were compared by the author in November 1997; US$1 = EC$2.67.

Source: Klak (1998b)

them to Mexico in large numbers to take advantage of lower wage rates and better access to the North American markets (Pantojas-Garcia 2001). Factory owners have found even lower wage rates in China, Indonesia and other parts of East Asia. Since 1994, CARICOM's manufacturing exports to the USA have fallen every year. They stood at $422 million in 1998, the most recent available year (CARICOM 2000). Many Caribbean factory workers that had underpaid jobs now have no jobs at all.

Both of the features of the global commodity chains described above demonstrate how third world producers and first world consumers are now disconnected. They are separated by distance, borders and ignorance of each other's reality and the commodity chains that link them. Their ignorance helps to perpetuate what is otherwise an unstable and unsustainable, let alone unethical, form of global production and consumption (Hartwick 1998). Compared with the impersonal nature of most components of contemporary mass consumption, Caribbean tourism is an unusual economic sector whereby citizens of the global North visit peripheral countries. But most tourists see little of the islands they visit, because they are mostly confined to cruise ships or to all-inclusive resorts (Chapter 11).

The mass media and communication linkages in the wider Caribbean

In an era of dramatic time–space compression, good intra-regional communication and transportation linkages are essential for Caribbean development. In the light of this concern, the following discussion focuses primarily on the printed news media and commercial airline traffic. The intent is to illustrate some of the broad patterns of information flows, media politics, connectivity, external influence and local expression. We have already noted that the widespread availability of cable TV stations from the United States has provided an additional avenue through which the Caribbean is heavily influenced by US culture (Figure 10.1). While US news broadcasts, soap operas, comedy and drama series, and movies dominate television in the English-speaking Caribbean, local TV stations struggle to provide serious local programming. Similar points about regional vulnerability and external influences could be made through the examples of radio, film, Internet access, shipping and other forms of communication and connectivity (see Dunn 1995).

The Caribbean's only region-wide newspaper is the international edition of the *Miami Herald*. That the Caribbean itself has no region-wide daily newspaper comes as little surprise given its geography and history. It is a less industrialised region of small islands separated by vast swaths of sea. The region is also divided by language, culture, national identity and, last but not least, mercantile capitalists-cum-publishers who protect their newspapers' home markets.

None of these obstacles is assuaged by the intra-regional transportation limitations left over from the 'hub-and-spoke' trade and communication patterns of the colonial powers. For the last century, the United States has considered the Caribbean part of its backyard (Kenworthy 1995). Miami, Florida, is the wider Caribbean's regional capital (Portes and Stepick 1993). Consider the commercial airline linkages that are fundamental to integrating and fostering a regional agenda through CARICOM. The US cities of Miami and, secondarily, San Juan, Puerto Rico, are more accessible (in terms of travel time, trip frequency and cost) to virtually any Caribbean country than they are to each other (Negrón-Muntaner and Grosfoguel 1997). Try to arrange some flights through an Internet travel service to see for yourself. You will find that ticket prices within the Caribbean are often several times higher than for travelling to and from Miami. Many Internet travel services do not even offer round-trip tickets between such CARICOM members' capital cities as Paramaribo, Suriname and Port-au-Prince, Haiti. And if you want to get to Kingston, Jamaica, from Port-au-Prince, add 1000 miles to what would be a 300-mile direct trip because you are routed through Miami.

Concerns about these intra-Caribbean transport limitations prompted the ACS in 1999 to launch a programme called 'Uniting the Caribbean by Air and Sea'. The programme seeks to establish a region-wide transportation policy and to have resources shared among regional carriers (Pantojas-Garcia 2001). Meanwhile, several developments in the air transport situation are not encouraging. Many government-supported Caribbean regional airlines have lost money. During the 1990s, Air Jamaica stopped serving the southern Caribbean after trying it for only a few months, while several regional airlines have folded. The casualties include Carib Express (it flew for only 15 months), Air Caribbean and, most recently, Guyana 2000 Airlines. Those few Caribbean locations with direct flights from Miami on US carriers in recent years, such as San Juan, Antigua, Grenada and St Lucia, have had to pay for that service with $1.5 million or more in annual subsidies to the airlines (American and TWA, both owned by AMR Corp.).

The US war on terrorism, and especially its war against Iraq in 2003, only made the economically precarious conditions facing the region's airlines worse. Air Jamaica, after surviving a US$90 million loss in 2002, saw reservations fall by nearly 40 per cent during the first two weeks of the US assault. This decline forced Air Jamaica to reduce flights to some US cities it regularly services, thereby worsening the airline's decline in passengers (Jacobs 2003). The other two surviving and smaller Caribbean airlines, BWIA and LIAT, are in even worse shape. The already financially strapped governments of the region needed to provide multi-million dollar bailouts to keep them afloat during the first half of 2003, and the two airlines laid off employees and agreed to merge in April 2003 to help them through the crisis. As Jean Holder, secretary-general of the Caribbean Tourism Organisation, described the situation: 'It is something of a miracle that LIAT, BWIA and Air Jamaica, which have been teetering on the brink, can remain in business' (*Financial Times Investor* 2003). These events vividly demonstrate how the Caribbean region's enduring external dependency leads to economic crises triggered by faraway events that have nothing directly to do with the Caribbean. Furthermore, the impacts of dependency appear to be accentuated and more immediate in the era of globalisation.

The 'hub-and-spoke' pattern, which applies to the flights in the Caribbean by the major international airlines, carries over to the print news media. As one of the major US dailies, the *Miami Herald* has had access to the capital, transportation and communication linkages, and electronic technologies to overcome the networking obstacles that are endemic to the Caribbean. The newspaper has for decades printed English and Spanish editions, and has distributed them via the commercial aircraft bound from Miami to the island capitals. In the late 1990s it began transmitting its daily paper electronically to the Caribbean. The *Herald* went first by satellite, and now goes more cheaply by the Internet, to the presses of major country-based papers. Newspapers in eight Latin American and Caribbean territories (the Bahamas, Cancun (Mexico), Curaçao/Bonaire, the Dominican Republic, Jamaica, Panama and St Maarten) print and circulate the *Herald* locally. The *Herald* is further distributed on a daily basis to proximate islands by passenger boat. Note that the *Miami Herald* has been able to extend and deepen its Caribbean regional presence despite the recent decline across the US newspaper industry (including the *Herald*) owing primarily to falling advertising revenues.

The regional impact of the *Herald* should not be exaggerated, but its availability across the Caribbean is unparalleled and its editorial voice is without competitors at that regional scale. Regarding Caribbean politics, the *Herald* emanates from an environment in south Florida that has been tolerant only of views that are fully anti-Castro and anti-communist, and which favour isolating Cuba. As a *Human Rights Watch/Americas* report on metropolitan Miami remarked on the subject of Cuba, 'only a narrow range of speech is acceptable, and views that go beyond those boundaries may be dangerous to the speaker' (*Human Rights Watch/Americas* 1994). Right-wing terrorists in Miami have for years threatened or punished media voices daring to express other views. Jim Mullin has identified 68 acts of violence by Miami's right-wing Cuban community between 1968 and 2000 (Mullin 2000). As an example of the group's ideology, The *Miami International Edition* on 2 June 2001 carried a photo of Cuban American singer Gloria Estefan wearing a tee-shirt saying 'CUBA B.C.' (Before Castro). Asked why she created the tee-shirt, Estefan replied, 'I wanted to celebrate the Cuba when everybody was happy'. Research provides an alternative view of Cuba in the 1940s and 1950s by documenting the prostitution, corruption, US mafia activities, poverty, and gender and racial inequalities (Pérez 1988; Cirules 1993; Segre *et al.* 1997).

Similar to the *Miami Herald* in terms of their conservative politics and their perspective emanating from the core are US weekly news magazines. The international editions of such magazines as *Time* and *Newsweek* have for long been sold at Caribbean newsstands and been read by higher-income Caribbean people. The region-wide presence of conservative news sources from the USA influences how Caribbean people see the world and themselves.

The local print media are largely confined to individual islands. Caribbean newspapers demonstrate some qualities that are valuable to open political debate and expression. Much of the print news media in many European ex-colonies is modelled after European traditions of journalism. The Caribbean is well represented by the more politically conscious news media traditions that come out of the liberal democratic societies of Europe. Representing that tradition are the *Trinidad Express* and the *Trinidad Guardian*, and two Bajan newspapers, the *Nation* and the *Barbados Advocate* (see the following section on 'Caribbean TNCs' regarding ownership of some of these newspapers). The *Advocate* has been a long-time supporter of the social democratic Barbados Labour Party (BLP). Both islands have literacy rates over 97 per cent, well-developed journalistic traditions and significant readerships. There are also some quality print media in the non-independent Caribbean. French dailies are distributed and read in Guadeloupe and Martinique, for example.

As is the case for many cities in the USA and the UK, several Caribbean countries have their own afternoon newspaper tabloids that are widely read. Some of these tabloids are notorious for their photos of nude women and their sensationalist stories. Perhaps less well known, however, is that many Caribbean tabloids also present politically informative points of view. For example, the *Bomb* in Trinidad has always been a successful outlet for political critiques and exposés of political wrongdoing. The *Star* in St Lucia also has a long tradition as a watchdog and oppositional newspaper.

The Caribbean is also well represented by long-time conservative institutions such as Jamaica's *Daily Gleaner*, founded in 1834. The clearest and most decisive impact of the *Gleaner*'s conservative leanings was its 'ideological war', in the words of one of the paper's senior contributors (John 1999), against the Democratic Socialist regime of Michael Manley in the 1970s. While the CIA's destabilising interventions against Manley are better known (Blum 1986), the *Gleaner*'s consistent portrayal of a society out of control were no less significant. The paper's daily and graphic presentation of chaos and violence during the final years of the Manley government helped to ensure a huge electoral win for the conservative Jamaican Labour Party in 1980 (Meeks 2000).

Caribbean TNCs

How are the major indigenous firms of the Caribbean faring in the era of electronic globalisation? Are they competitive in the fast-paced and fluid international environment? How committed are Caribbean TNCs to regional development? The region can boast of many diversified firms with deep historical roots that offer an array of quality products, ranging from foods and beverages to chemicals, construction materials and professional services. The growing openness of borders and regions, however, suggests that these firms be measured by wider international standards. Some comparisons are therefore made with major US and global TNCs.

Of course Caribbean economies and firms are small by global standards, but the scale of difference is worth considering as the hemisphere and the world move towards greater openness, more all-encompassing free-trade agreements and head-to-head competition. As Jamaica's ambassador to the US, Richard Bernal, pointed out in US Congressional testimony over FTAA, 'The seven largest

US companies each have sales revenue larger than the combined GDP of the 21 Caribbean and Central American countries' (Bernal 1999). The Caribbean's largest firm in terms of sales is Neal and Massey of Trinidad and Tobago. It is highly diversified and owns operations in more countries (14) than any Caribbean firm besides Goddard (discussed later). Neal and Massey's *annual sales* are little more than General Motors' sales *per day* (Table 10.1).

Similarly, the USA's largest employer, Wal-Mart, with a staff of 675,000, is 100 times bigger the Caribbean's larger employer, Lascelles deMercado, with 6,800 workers. Lascelles deMercado is a Jamaican firm involved in liquor, sugar cane, drugs, insurance and car sales. The family traces back to Spanish immigration to the island in the eighteenth century. A size disparity of 100–150/1 continues down the list of major US and Caribbean corporations (Table 10.1).

Comparisons can also be made with major global TNCs. The 500th largest TNC in the *Fortune Magazine* annual list is The Limited, a US clothing retailer. It had assets of US$4 billion and sales of US$9.7 billion in 1999. No Caribbean firm approaches this size. With these proportions in mind, we briefly profile a few of the other top Caribbean firms that are involved in manufacturing, drawing from a range of sources, most notably *Executive Time Magazine* (2001a). All but two of the top 20 firms are headquartered in Trinidad and Tobago, Jamaica or Barbados.

The Ansa McAl conglomerate from Trinidad and Tobago ranks second in the region in both sales and employment (Table 10.1). If measured in terms of its assets of more than US$500 million, Ansa McAl ranks first. It has been one of the region's most aggressive firms in terms of acquiring assets in other countries, particularly Barbados. Perhaps best known for Carib Beer, Ansa McAl also produces bottles, chemicals and building materials, invests in banking, insurance and real estate, sells cars, imports and exports a range of products, and publishes the *Trinidad Guardian* and *Barbados Advocate* daily newspapers.

The Grace Kennedy conglomerate of Jamaica ranks third in the region in sales and eighth in employment. It specialises in food production, distribution and finance, and has been judged as one of the Caribbean's most consistent performers over the decades. Barbados Shipping and Trading is that country's largest corporation, and the region's seventh and ninth largest in sales and employment, respectively. It specialises in retailing, distribution, manufacturing, maintenance services and insurance. Goddard of Barbados ranks 11th in sales and seventh in employment. Goddard is an especially diversified company involved in catering, importing and exporting, food and beverages, car rental and real estate. It owns firms in more Caribbean and Latin American countries (18) than any other Caribbean company. In 1999, Goddard had $210 million in assets, while its sales grew to $354 million.

The largest firm in Puerto Rico, Popular Inc., is considerably larger than any entity in the English-speaking Caribbean. Its sales revenues were $2.2 billion in 1999. Popular Inc. is a diversified financial firm involved in banking, mortgages, insurance, leasing and investing. It has offices in San Juan, Caracas, Miami and Santo Domingo, and clients in 12 Latin American countries. Still, Popular Inc.

Table 10.1 The 20 largest Caribbean corporations

Sales revenues (in millions of $)	Corporation	Rank	Employees	Corporation	Rank
164,068.9	General Motors Corp.	#1 in US for comparison	675,000	Wal-Mart Stores, Inc.	#1 in US for comparison
500.0	Neal & Massey (T&T)	1	6,800	Lascelles Demercado (Ja)	1
326.3	Ansa McAl Ltd (T&T)	2	6,000	Neal & Massey (T&T)	2
321.4	Grace Kennedy & Co. (Ja)	3	5,000	Jamaica Producers Group (Ja)	3
293.7	NCB Group (Ja)	4	4,500	telecom of Jamaica (Ja)	4
250.0	telecom of Jamaica (Ja)	5	4,435	Ansa McAl Ltd (T&T)	5
243.5	Jamaica Producers Group (Ja)	6	4,355	NCB Group (Ja)	6
229.0	Barbados Shipping & Trading (B'dos)	7	2,783	Goddard Enterprises (B'dos)	7
184.6	Atlantic Tele-Network (St Croix)	8	2,700	Grace Kennedy (Ja)	8
169.2	Trinidad & Tobago Unit Trust Corp. (T&T)	9	2,500	Barbados Shipping & Trading (B'dos)	9
165.8	Bank of Nova Scotia (Ja)	10	2,323	Royal Bank of T&T (T&T)	10
159.0	Goddard Enterprises (B'dos)	11	2,161	Republic Bank Ltd (T&T)	11
138.4	Lascelles Demercado (Ja)	12	1,591	Bank of Nova Scotia (Ja)	12
128.4	Republic Bank Ltd (T&T)	13	1,572	Banks DIH Ltd (B'dos)	13
117.9	Desnoes & Geddes (Ja)	14	1,500	Desnoes & Geddes (Ja)	14
108.9	Jamaica Broilers Group (Ja)	15	1,300	Life of Jamaica Ltd (Ja)	15
91.2	Barbados External Telecom (B'dos)	16	1,100	Atlantic Network Inc. (St Croix)	16
86.7	Barbados Light & Power (B'dos)	17	1,100	Demerara Distillers Ltd (Guyana)	17
80.0	Little Switzerland, Inc. (St Thomas)	18	1,022	First Citizens Bank Ltd (T&T)	18
62.43	Carib. Cement Co. (Ja)	19	1,000	Bank of Commerce (T&T)	19
58.54	Jamaica Flour Mills Ltd (Ja)	20	884	Bank of Nova Scotia (T&T)	20
28,818.0	Citicorp	#20 in US for comparison	141,700	Dayton Hudson Corp.	#20 in US for comparison

Data are for the English-speaking Caribbean in 1996; key: T&T = Trinidad and Tobago; Ja = Jamaica; B'dos = Barbados
Source: Bernal (1999)

does not approach the size of the world's 500 largest firms. It is about a quarter of the size of the 500th ranked firm, The Limited. In fact, if the 300 largest Puerto Rican firms were considered as one, in aggregate they would rank just behind the world's 206th largest firm, the drug and bathroom products firm Bristol-Myers Squibb, with $20.22 billion in revenue (Lozano 2000).

Caribbean TNCs are indeed small compared with the global Fortune 500 firms, but unlike the global giants, their capital investment portfolio tends to stay within the region. Caribbean TNCs therefore contribute more positively than global firms to regional development. In contrast, global TNCs tend to expropriate the capital and profits from non-core regions in a more 'counter-development' dependent relationship. While development economists commonly bemoan the lack of domestic capital in the global South, domestic/regional TNCs such as those of the Caribbean are providing much-needed local control of capital investments in agricultural industries, manufacturing, services and commerce.

We leave the last words on the future prospects of Caribbean TNCs to CARICOM. As regional economic borders become increasingly porous, and as the need arises for Caribbean firms to be competitive beyond the region, CARICOM sees the only hope for the future for Caribbean TNCs through regional integration of production, which has thus far proved elusive:

> The challenge facing the Region remains the imperative of implementing programmes to make its productive sectors, both in goods and services, internationally competitive. This requires a fundamental paradigm shift in the modes of business organization. The trends in CARICOM's trade with the EU is indicative of the stagnant or declining relative position of the Region in the EU market both in terms of growth and importance. No single CARICOM country has the resources – human, technological, financial, or otherwise – to effectively respond (CARICOM 2000).

Conclusion and policy implications

As has been the case throughout Caribbean history, the region is now largely shaped by political and economic structures established for the benefit of outside interests or local elites. As with previous rounds of globalisation, most policies within the region are reactions to outside stimuli, rather than proactive attempts to promote sustained development. The potential for developmental success through such reactive tactics is greatly hampered in the current era of electronic globalisation. The world is growing more unequal at all levels (Wade 2001), and it is harder for peripheral regions such as the Caribbean to make relative gains. Much of the capital that the Caribbean has recently been able to attract is fleeing regulation. Such capital is unlikely to provide a basis for the region's sustained development. The exodus of apparel factories over recent years in pursuit of lower wages and better market access should provide a broader lesson about relying on the various types of mobile capital discussed in this chapter and the previous one on offshore services.

The most promising ventures for regional development are likely to be found among those with strong local roots and commitments. These might be found amongst Caribbean TNCs, and perhaps the international success of the Jamaica-based Sandals group of resorts could serve as one model to which regional firms can aspire (see Chapter 11 on tourism). Regional development might also be fostered through the newly emerging non-traditional exporters from the region, local media outlets, and locally run e-commerce operations. This is not to romanticise local firms. They need to be scrutinised carefully and supported selectively by state policy based on their potential longer-term contributions, their commitment to collaborating with policy makers and other local firms with forward and backward linkages, and their interest is contributing to meeting national and especially CARICOM developmental goals. This kind of targeted, strategic state support follows the East Asian development model. It runs counter to the dominant neo-liberal way preferred by the World Bank and the United States, which treats all firms equally (Wade 1996). Policies that 'treat all firms the same' spread scarce public resources too thin and waste subsidies on footloose or unscrupulous investors. This leaves inadequate resources to target and support the most promising firms (Klak 1998b). On this neo-liberal 'level playing field', bigger, more mobile and more internationally experienced foreign firms dominate (Klak 1995). The Caribbean has enough negative experience coddling footloose and exploitative firms while neglecting the most promising ones. It is time to try a different development model.

What does the future of corporate globalisation hold for the Caribbean? A western hemisphere region-wide free-trade zone looms on the horizon. But this involves a negotiation process on very uneven grounds. The hemispheric power disparities are hard to exaggerate. The United States is but one of 35 sovereign states in the hemisphere, but its GDP comprises 76 per cent of the total. CARICOM countries account for only 0.2 per cent of total hemispheric GDP. Adding in the six other small countries between Mexico and South America still does not reach 1 per cent of hemispheric GDP (Girvan 1999).

Unfortunately, the rules and regulations recently drafted for the FTAA (FTAA – Negotiating Group on Investment 2001) reproduce and extend the neo-liberal model with which the Caribbean has ample experience to seek to avoid. That document's chapter on investment greatly facilitates the trans-hemispheric mobility of capital and capitalists. International military incursions are allowed to secure and protect investments. 'Key personnel' (read: TNC management) are allowed to move freely throughout the hemisphere unencumbered by national discretion or even passports. Labour is not similarly mobilised. The document desires to outlaw any 'buy local' campaign. For their part, the small countries are seeking special short-term protections within the FTAA. The intent is to buffer the immediate and devastating impacts of head-on competition with foreign TNCs. But how many concessions the small countries can obtain remains in doubt.

Useful websites

The Ansa McAl (www.ttol.co.tt/ansamcal/) conglomerate from Trinidad and Tobago is the region's second largest firm and perhaps the most diversified; it produces beer, bottles, chemicals and building materials, invests in banking, insurance and real estate, sells cars, imports and exports a range of products, and publishes regional newspapers.

New Internationalist (www.newint.org/) is a monthly news magazine critical of top-down economic globalisation and sympathetic to building capacities to resist and define local priorities at the grassroots level.

The United Nations Economic Commission for Latin America and the Caribbean (ECLAC) (2002) *Globalisation and Development*, available at www.eclac.cl/cgi-bin/getProd.asp?xml=/publicaciones/xml/0/10030/P10030.xml&xsl=/tpl-i/p9f.xsl&base=/tpl-i/top-bottom.xsl, A major report providing an overview of globalisation's economic impacts and restructuring patterns, and the relative successes, failures and continuing vulnerabilities of Latin American and Caribbean states.

The World Bank (www.worldbank.org/data/) collects and makes publicly available some of the best data on globalisation trends.

References

Barnes, C. (2000) 'Narcotics: driven from Latin America, smugglers eye Caribbean', *Inter Press Service*, 26 January.

Beck, S. (1997) 'US pressure mounts over poisonings', *South China Morning Post*, September 30.

Bernal, R. L. (1999) 'Free trade area of the Americas', testimony to House International Relations Subcommittee on International Economic Policy and Trade, Hearings on the Free Trade Area of the Americas (FTAA), 15 October.

Blum, W. (1986) *The CIA: a Forgotten History*, Zed Books, London.

Bonacich, E. (1994) *Global Production: The Apparel Industry in the Pacific Rim*, Temple University Press, Philadelphia.

Brysk, A. (2000) *From Tribal Village to Global Village: Indian Rights and International Relations in Latin America*, Stanford University Press, Stanford, California.

CARICOM (2000) *Caribbean Trade and Investment Report 2000: Dynamic Interface of Regionalism and Globalisation*, The Documentation Centre, Caribbean Community Secretariat, Georgetown, Guyana.

Castetls, M. (2000) The rise of the network society, 2nd edition, Blackwell, Oxford, UK, and Malden, Massachusetts.

Cirules, E. (1993) *El imperio de La Habana*, Casa de las Américas, Ciudad de La Habana, Cuba.

Dicken, P. (1998) *Global Shift: Transforming the World Economy*, 3rd edition, Guilford Press, New York.

Dunn, H. (ed.) (1995) *Globalisation, Communications and Caribbean Identity*, St Martin's Press, New York.

Economist (2000) 'Small states, big money', *The Economist*, 23–29 September.

Evans, R. (1993) 'Banking on the black economy', *Geographical Magazine*, **September**.

Executive Time Magazine (2001a) 'Corporate profiles (e-time wise index)', *Executive Time Magazine: Journal of Business & Technology: Caribbean Edition*, **3**.

Executive Time Magazine (2001b) 'Caricom hits hard at G7's Harmful Tax Report', *Executive Time Magazine: Journal of Business & Technology: Caribbean Edition*, **3**.

FATF (the Financial Action Task Force on Money Laundering) (2001) *Progress Reporting on Non-Cooperative Countries and Territories*, OECD (Organisation for Economic Co-operation and Development), Paris.

Ferguson, J. (1997) *Eastern Caribbean in Focus: A Guide to the People, Politics and Culture*, Latin American Bureau, London.

Financial Times Investor (2003) 'Caribbean carriers will merge to stay afloat', *Financial Times Investor*, **1 May**, available through lexisnexis.com.

FTAA – Negotiating Group on Investment (2001) 'Report of the negotiating group on investment to the trade negotiations committee', available at www.wtowatch.org/library/

Gereffi, G. and Korzeniewicz, M. (1994) *Commodity Chains and Global Capitalism*, Praeger, Westport, Connecticut.

Girvan, N. (1999) 'Globalisation and counter-globalisation: the Caribbean in the context of the south', paper presented at seminar at the University of the West Indies, Mona, entitled 'Globalisation: a strategic response from the south', 1–2 February.

Goodbody, I. and Thomas-Hope, E. (eds) (2002) *Natural Resource Management for Sustainable Development in the Caribbean*, Canoe Press, University of the West Indies, Kingston.

Gwynne, R., Klak, T. and Shaw, D. (2003) *Alternative Capitalisms: Geographies of 'Emerging Regions'*, Edward Arnold, London, and Oxford University Press, New York.

Gwynne, R. and Kay, C. (2004) *Latin American Transformed: Globalisation and Modernity*, 2nd edition, Edward Arnold, London.

Hall, K. (ed.) (2001) *The Caribbean Community: Beyond Survival*, Ian Randle Publishers, Kingston.

Hall, K. and Denis B. (eds) (2000) *Contending with Destiny: The Caribbean in the Twenty-first Century*, Ian Randle Publishers, Kingston.

Hartwick, E. (1998) 'Geographies of consumption: a commodity-chain approach', *Environment and Planning D, Society and Space*, **16**, 423–37.

Harvey, D. (1989) 'Time–space compression and the postmodern condition', in Harvey, D. (ed.) *The Condition of Postmodernity*, Blackwell, Oxford, UK, and Cambridge, Massachusetts.

Harvey, D. (1995) 'Globalisation in question', *Rethinking Marxism*, **8**.

Held, D., McGrew, A., Goldblatt, D. and Perraton, J. (1999) *Global Transformations: Politics, Economics, and Culture*, Polity Press, Cambridge.

Hopkins, T. K. and Wallerstein, I. (1994) 'Commodity chains in the capitalist world-economy prior to 1800', in Gereffi, G. and Korzeniewicz, M. (eds) *Commodity Chains and Global Capitalism*, Praeger, Westport, Connecticut.

Human Rights Watch/Americas (1994) Report on freedom of speech in metropolitan Miami, available at www.hrw.org/.

Jacobs, S. (2003) 'Caribbean leaders seek solutions to drop in tourism', *Associated Press*, 8 April.

John, G. R. (1999) 'Favourite son looks back on old lady', *Daily Gleaner*, 23 December.

Johnston, R. J., Gregory, D. and Smith, D. M. (1994) *The Dictionary of Human Geography*, 3rd edition, Blackwell, Oxford.

Kenworthy, E. (1995) *America/Américas: Myth in the Making of U.S. Policy Toward Latin America*, Pennsylvania State University Press, University Park.

Klak, T. (1995) 'A framework for studying Caribbean industrial policy', *Economic Geography*, **71**.

Klak, T. (ed.) (1998a) *Globalisation and Neoliberalism: The Caribbean Context*, Rowman & Littlefield, Lanham, Maryland.

Klak, T. (1998b) 'Is the neoliberal industrial export model working? An assessment from the eastern Caribbean', *European Review of Latin American and Caribbean Studies*, **65**.

Knox, P. and Marston, S. (2004) *Places and Regions in Global Context*, 3rd edition, Prentice Hall, Upper Saddle River, NJ.

Lakeway (1998) 'What is offshore investing?' Lakeway International Equities Ltd.

Lashmar, P. and Manneh, M. (2001) 'Grenada puts clamp on banks', *Independent*, 18 March.

Lee, K., Holland, A. and McNeill, D. (eds) (2000) *Global Sustainable Development in the Twenty-first Century*, Edinburgh University Press, Edinburgh.

Lietaer, B. (2001) *The Future of Money: A New Way to Create Wealth, Work and a Wiser World*, Century/Random House.

Lozano, D. I. (2000) 'Top 300 locally owned companies continue setting highs', *Caribbean Business*, 2 November.

Mandle, J. R. (1996) *Persistent Underdevelopment: Change and Economic Modernization in the West Indies*, Gordon and Breach, Amsterdam.

Martin, A. (2001) 'A sure thing', *Harper's Magazine*, April.

Meeks, B. (2000) *Narratives of Resistance: Jamaica, Trinidad, the Caribbean*, University of the West Indies Press, Mona, Jamaica.

Mintz, S. (1996) 'Enduring substances, trying theories: the Caribbean region as oikoumene', *The Journal of the Royal Anthropological Society*, **2**.

Mullin, J. (2000) 'The burden of a violent history', *Miami New Times*, available at www.miaminewtimes.com/issues/2000-04-20/mullin.html

Negrón-Muntaner, F. and Grosfoguel, R. (eds) (1997) *Puerto Rican Jam: Rethinking Colonial Nationalism*, University of Minnesota Press, Minneapolis.

NRC (National Research Council) Board on Sustainable Development, Policy Division (1999) *Our Common Journey: A Transition Toward Sustainability*, National Academy Press, National Academy of Sciences, Washington.

Pantojas-Garcia, E. (2001) 'Trade liberalization and peripheral postindustrialization in the Caribbean', *Latin American Politics and Society*, **43**.

Payne, A. and Sutton, P. (2001) *Charting Caribbean Development*, University Press of Florida, Gainesville.

Pérez, L. A. (1988) *Cuba: Between Reform and Revolution*, Oxford University Press, New York.

Portes, A. and Stepick, A. (1993) *City on the Edge: The Transformation of Miami*, University of California Press, Berkeley.

Possekel, A. K. (1996) 'Offshore financial centres in the Caribbean: potential and pitfalls', *Caribbean Geography*, **7**, 2.

Public Citizen (1997) Report on diethelene glycol poisoning in Haiti, available at www.citizen.org.

Ramsaran, R. (ed.) (2002) *Caribbean Survival and the Global Challenge*, Ian Randle Publishers, Kingston, and Lynn Rienner Publishers, Boulder, Colorado.

Richardson, B. (1992) *The Caribbean in the Wider World, 1492–1992: A Regional Geography*, Cambridge University Press, New York.

Roberts, S. (1994) 'Fictitious capital, fictitious spaces? The geography of offshore financial flows', in Corbridge, S., Martin, R. and Thrift, N. (eds) *Money, Power and Space*, Blackwell, Oxford.

Roberts, S. M. (1995) 'Small place, big money: the Cayman Islands and the international financial system', *Economic Geography*, **71**, 3.

Scher, P. (2003) *Carnival and the Formation of a Caribbean Transnation*, University Press of Florida, Gainesville.

Segre, R., Coyula, M. and Scarpaci, J. L. (1997) *Havana: Two Faces of the Antillean Metropolis*, John Wiley, New York.

Taylor, P. J., Walker, D. R. F. and Beaverstock, J. V. (1999) 'Introducing globalisation and world cities (GaWC): researching world city network formation', in Sassen, S. (ed.) *Global Cities – The Impact of Transnationalism and Telematics*, UNU/IAS Project Report.

The Light (of Anguilla) (nd) 'Governor accepts drug bust share cheque . . . Chief Minister comments!', *The Light (of Anguilla)*, **310**.

Ugarteche, O. (2000) *The False Dilemma: Globalisation: Opportunity or Threat?* Inter Pares, Ottawa, and Zed Books, London and New York.

UNDP (1999) *Human Development Report*, United Nations, New York.

Wade, R. (1996) 'Japan, the World Bank, and the art of paradigm maintenance: the East Asian miracle in political perspective', *New Left Review*, **217**.

Wade, R. (2001) 'Winners and losers', *The Economist*, 26 April.

Wallerstein, I. (2000) 'Globalisation or the age of transition? A long-term view of the trajectory of the world-system', *International Sociology*, **15**.

Watson, H. (2001) 'The Caribbean and Latin America in the crisis of neoliberal capitalism', unpublished paper available from author.

Chapter 11

TOURISM, ENVIRONMENT AND DEVELOPMENT

Introduction

The international tourist industry has developed into a complex, multi-functional and geographically dispersed system of services. Capital accumulation and profit dictate the organisation of this global industry by involved enterprises (and entrepreneurs), industries (owners, stock holders, workers), markets (hoteliers, retailers, travel agencies), transportation industries (airlines, cruise lines, taxi and transit services), producer services (information, communication, financial and 'back-office' services), and state agencies (tourism promotion industries, ministries of tourism, regulatory ministries) (Britton 1991). It is this multi-faceted and highly profitable international capitalist *ensemble* that interacts with the environment in a two-way system of direct and indirect feedbacks and relations. On the one hand, environmental resources provide one of the basic ingredients – the natural and/or people-made 'place' and island setting for the tourist to visit, enjoy, 'consume' and relax away from the rigours of the work regime 'at home'. The international destination in the Caribbean is a different, or exotic, resource to be produced and consumed 'on vacation', but is generally packaged, marketed and promoted in the major generating markets of North America and Europe.

This chapter traces the trials and tribulations of this complicated industry as Caribbean governments, people, institutions and agencies have struggled to balance the returns from it with the host of accompanying impacts, many of which are negative. First, the historical record is examined. Then, several geographers' and tourist scholars' models are presented and evaluated. Finally, the contemporary concerns with the industry's environmental and societal consequences are brought to the fore, because they constitute the gravest threats to the viability of this complex global industry. Sustainable tourism, ecotourism, alternative tourist niches, may be today's ideas, or ideals, to help the region's maturing tourist landscapes avoid the negative consequences of unregulated, mass tourism. However, the Caribbean's 'love–hate' relationship with tourism will not be easily resolved. There are no easy answers to managing and controlling this 'commodification of places'.

Caribbean tourism's problematic growth

The beginnings: quality niche markets

It was the Caribbean's balmy tropical climate that first made the islands of the region a tourist destination for the wealthy, or 'quality' visitors wishing to recuperate, relax and recreate among their upper-class equals. Colonial ties, by and large, dictated the venues. British royalty, members of the 'establishment' and their playmate entourages went to upmarket hotels and their own vacation homes in Barbados and Jamaica. The French went to Martinique, the Dutch to Curaçao. Many would spend lengthy sojourns in the islands. Americans 'wintered' in the West Indies, in the nearby islands of the Bahamas and Cuba, some by yacht, others by passenger steamship service.

Jamaica's first short-term tourists were associated with the development of the banana industry, travelling from Europe to the Caribbean and back on the regular banana boat service. A few Jamaican hotels were built to accommodate this fledgling group of visitors, but their financial state was scarcely ever secure and profitable. In contrast, Cuba and the Bahamas enjoyed regular steamship services from the United States, and their early entry into tourism spawned growth and considerable prosperity. Later, in the 1920s and 1930s, successful authors, playwrights, film stars and reclusive millionaires also sought out Caribbean/ West Indian havens – in Tobago, Jamaica, Nevis, for example; some for solitude, others for solicitous excitement.

After the turn of the (twentieth) century, a burgeoning wealthy middle class in Europe and North America sought alternative sights, places and tourist excitement in such tropical venues, and Cuba in particular proved especially attractive. By the 1920s, American desires for tropical exoticism had been fuelled by brochures and Hollywood characterisations. Their growing appetites for entertainment, excitement, recreation, romance and indulgence took them to Cuba in large numbers, effectively generating the first mass tourist flow into the region. For nearly four decades, Havana and Cuban hospitality beckoned American tourists. Some, like the Mafia and other underground city bosses, gave this Caribbean metropolis a seamy, criminal identity. On the other hand, the island should be remembered more for its capacities to charm and welcome wealthy, middle-class Americans, and elite South Americans and Spanish seeking a luxurious and naughty 'Riviera in the American Tropics', than as a harbour and haven for the likes of Al Capone (Schwartz 1997):

> A trip to Cuba, a little rum and rumba, were movies-come-true for throngs of bankers, lawyers, industrialists, teachers, sales clerks and housewives who boarded steamships bound for Havana (Schwartz 1997:14).

Elsewhere, mass tourism in the Caribbean was, by and large, a post-1950 phenomenon. Chiefly responsible for this was the advent of regular non-stop jet services from major developed countries like Britain, Canada, the United States and Holland in the 1960s. Caribbean governments, international agencies and the private sector considered promoting tourism as a development strategy, in part because of the troubles plantation agriculture was beginning to face, and in part because the success of the quality luxury markets seemed to offer a place for an enlarged, less exclusive tourism (Issa 1959; Blake and Carrington 1975). Now, in addition to climatic ambience, the sun and the sea were promoted as part of the Caribbean's exotic tourism image.

The post-1950s: embracing mass tourism

The first Caribbean countries to be major tourist destinations were islands with adequate public services, a good-sized airport, and strong commercial or political ties with a wealthy 'mother country'. Cuba's importance as a major destination for the United States market declined rapidly after Castro's takeover of the island in 1959. Its hotels and casinos lost business from 1960 onwards due to the ostracism of Cuba by the US administration. Consequently, the Bahamas, Jamaica and Puerto Rico prospered as alternatives for North Americans.

Since the 1940s, a modest tourist industry, catering to exclusive hotel and cruise ship clientele, had grown in Puerto Rico, mainly in and around San Juan. By the 1970s, American visitors to the island exceeded a million, and there were growing numbers of cruise ship visitors, as well as US military and returning Riquenos swelling the temporary visiting volumes. For British vacationers, Antigua, Barbados and Jamaica were their preferred 'islands in the sun'. Barbados also profited from its longstanding mercantile relationship with Canada and welcomed that Commonwealth country's wealthier 'snowbirds'. On its own in the North Atlantic, Bermuda profited from proximity to the eastern seaboard of the United States, and welcomed a mixed market of quality and less wealthy visitors seeking exclusive and socially secure vacations.

In the early 1960s, the rapid economic growth enjoyed by an ever-expanding middle class in Europe and North America had brought them considerable increases in disposable incomes, making exotic 'intercontinental' and 'international' vacations a more viable possibility. The advent of jet-service airlines increased the flexibility for short trips as well as lowering air transportation costs, providing the more exclusive steamship services with competition. The marketing and promotion of overseas travel by an increasingly efficient travel industry in North America, as well as in Europe, led to diversification as well as expansion of Caribbean tourism.

Motivated by these changes in organisation and transportation, the 1960s became a decade of *optimism*, with the more adventurous, repeat traveller to the Caribbean choosing new or different exotic hideaways. Many, but not all, Caribbean governments became convinced that the rapid expansion of tourist demand for 'sun, sand and sea' augured a prosperous future for their resource-limited economies. They were, however, unsure about this new venture, and let themselves

Plate 11.1 The 'platinum' coast of Barbados, 1974
Source: Dennis Conway

be directed by outside industry 'experts', rather than taking control or regulating the industry themselves.

In addition to giving away territorial control of coasts and islands to wealthy outsiders, they allowed exclusive tourist enclaves to be established. Mill Reef Resort, a private retreat in Antigua financed by a group of US millionaires (developed relatively early in 1949), was one notorious example. Paradise Island, off Nassau in the Bahamas (with major development taking off in 1961), was another (Pattullo 1996). Promising island locations and even some small dependencies were sought by tourist promoters, or entrepreneurs, to provide exclusive resorts for the 'fabulously, rich and famous'. Luxury tourist enclaves were appropriated in the Grenadines, island dependencies of St Vincent; the island of Mustique was literally bought by a foreign company, Mustique Company Ltd; and a 99-year lease for another, Palm Island, was agreed upon for a total rental of a mere US$99. An early tenant of a Mustique property was Princess Margaret, Queen Elizabeth's sister. A more recent celebrity tenant is Richard Branson, CEO of Virgin Industries. The exclusiveness of hideaway resorts remains a very high-priced commodity to this day.

The existing stock of accommodation was inadequate. They might have been locally owned, but most were relatively small, modest in terms of services, and the standards of upkeep varied. Driven by perceived imperatives to provide comfortable and attractive accommodation to the anticipated influx of wealthy Americans, Canadians and Europeans, island governments or colonial administrations provided

Plate 11.2 The 'new' Hilton Hotel and Carlisle Bay, Barbados, from the air, 1968
Source: Dennis Conway

generous financial incentives to attract foreign investors willing to develop and manage hotels and related tourist projects. Hotel chains like Hilton and Holiday Inn were invited to provide these modern tourist accommodations, and the resulting investment scramble in the late 1960s created a proliferation of foreign-owned and foreign-managed hotels (Plate 11.2). Some were successful, or after the inevitable teething troubles associated with start-ups in the Caribbean – labour problems, service breakdowns, construction and repair problems – managed to establish themselves and gain a profitable foothold in the industry. Many, however, were inadequately financed, poorly designed and inefficiently operated. They changed hands often (after bankruptcy, or financial troubles), and sometimes were closed down altogether by the international consortium, millionaire financier or financial group that had primarily operated a hotel to take advantage of the subsidisation, or to launder illicit money (Block 1990).

The 1970s: reality strikes, criticism rages – reassessments are necessary

In the 1970s, Caribbean tourist industry fortunes waxed and waned, although overall the decade was one of continued growth, averaging 7 per cent annually (CTRC 1983). More importantly, however, the 1970s was a decade of *realisation* and *reassessment*. The optimism of the 1960s was countered by academic criticism, by development planners' concerns for the costs of infrastructure development and improvement, by the industry's reappraisal of its investment returns, by the increasing costs of hotel and resort management and by the dawning realisation that factors beyond the control of the host countries were dictating preferences and tourist travel patterns. The Caribbean's longstanding dependence on external forces was to be repeated in tourism's case, and the dramatic fluctuations and downturns in the global economy that occurred in the 1970s only exacerbated the region's problems with this fickle and complex international industry.

In academic circles, whether in the Caribbean, North America or Europe, criticism was vehement and passionate on the numerous negative impacts of mass tourism on the socio-cultural values of the hapless and despoiled society. Tourists were alien intruders, wealthy, white, privileged outsiders and dubious ambassadors of a wasteful and extravagant society. What did these tourist invaders offer the Caribbean or other underdeveloped regions? Different and often warped values – swinging lifestyles, alien metropolitan values, disruptive religious practices, extreme racial attitudes, incompatible mores of sexual freedom, alternative work ethics and conspicuous, lavish consumption patterns – in sum, tourists were an alien patronising presence whose main concern was non-productive self-indulgence and hedonism (Hiller 1976, 1979a; Erisman 1983; Husbands 1983).

The 1980s: rationalisation and redefinition – better management and informed analysis

If the 1970s was a decade for realisation, reflection and reassessment, the 1980s was one of *rationalisation* and *redefinition*. Although tourism fortunes in the early years of the decade were severely impacted by the global recession, the later half saw an upturn after 1983, which peaked in 1987 and flattened out to the end of the decade. These later years suffered from hurricane impacts – Gilbert in 1986 and Hugo in 1989 – and the weakening positions of two major North American airlines, Eastern and Pan American. Despite these setbacks, the decade as a whole registered an increase in stopover visitors, from 6 million in 1980 to 10 million by 1990. Globally, the region's share of world tourism increased from 2.1 per cent in 1980 to 2.3 per cent by 1990.

In large part this increase was due to the rapid expansion of tourism in the Dominican Republic, where international capital in alliance with local entrepreneurs took advantage of the government's *laissez-faire* stance and invested heavily in resort development. On the other hand, in Haiti, a decade of political instability

Plate 11.3 *Cunard Princess* in St George's Harbour, 1984
Source: Dennis Conway

and fears that AIDS originated on the island kept most American visitors away and fundamentally disrupted the tourist industry. Elsewhere, Grenada's tourist industry was similarly affected by political (and military) upheavals, though it gradually rebounded after the island's return to normalcy (see Chapter 9 for details of Grenada's 'Revo' and subsequent US-led invasion). In addition, the Caribbean's importance as the major region for a rapidly expanding North American cruise ship industry took off in the 1980s, with passenger arrivals increasing from 3.3 million in 1980 to 6.3 million in 1991 (Mather and Todd 1993) (Plate 11.3).

In the region's more developed tourist islands, however, government and private sector approaches to this fluctuating and fickle industry matured and deepened. The Caribbean Tourism Research and Development Centre, later to become the Caribbean Tourism Organisation (CTO), grew to have a membership of 29 Caribbean countries and a host of non-government members – the Association of Caribbean Universities and Research Institutes, Caribbean Hotel Association, the New York-based Caribbean Tourism Association, the British-based Eastern Caribbean Tourist Association, and Christian Action for Development in the Caribbean. From its inception, CTRC conducted fundamental studies and analyses of Caribbean tourism with a view to reducing the social costs and increasing the benefits to the region's economy. Later, as CTO, the centre assumed more responsibility for promotion, marketing and proactive planning (Holder 1979).

Experience and research, moreover, provided leaders with a clearer view of the approach necessary to integrate tourism with development objectives. Most Caribbean government leaders now accept that tourism is and will continue to be an essential ingredient in their countries' development trajectories. International financial institutions like the World Bank, the Inter-American Development Bank, the Canadian International Development Agency and the region's own Caribbean Development Bank (CBD) have also become less hesitant to support tourism as development. The CBD views the growth of a balanced and financially healthy regional tourist industry as one major likely contributor to increased employment, and its more positive attitude reinforces that of the Caribbean governments themselves (CDCC 1987).

Most governments of islands with a developed tourist industry therefore promoted tourism diversification and sought ways to widen markets beyond the traditional North American and European ones. This promotional effort was undertaken in concert with local hotel associations, the international chains, airlines, travel agencies and operators. Regional cooperation rather than regional competition was the organisational message. The Barbadian and Jamaican governments both recognised the need for more creative 'destinational development'. In attempts to recapture their shares of the repeat tourist market, they encouraged diversification of attractions and (together with their private sector) promoted a wider range of entertainment options, water sports, sightseeing and dining. This 1980s concern was with creation of a more distinctive Caribbean product that better reflected their island heritages and cultural distinctiveness, while keeping their 'sun, sand and sea' image alive and vibrant.

Indeed, with most islands developing a tourism sector by the 1980s, the region as a whole offered a wide variety of destinations in terms of the attractions offered and the diverse cultures to experience. Coastal tourism prevailed as the regional staple, but visitors had a wide range of environments to experience: the 'cloud forests' of Martinique, the rugged terrain and tropical rain forest of Dominica, the volcanic peaks of the Pitons in St Lucia, for example. Historical and cultural attractions were also included in the tourist itinerary: archeological sites became tourist attractions in Montserrat and Antigua, and plantation houses were converted into hotels in several islands, for example Nevis, Martinique and Barbados. The capital cities' colonial architecture was now viewed as part of the tourist landscape, and gentrification of historic buildings and the landscaping of inner city parks and avenues put on the agenda (Thomas 1991).

Although the Caribbean had started with a fairly small stock of tourist accommodations, mostly comprising small, privately owned family hotels and guest houses, by the 1980s the region's diversity of offerings had widened quite considerably. In addition to the existing local stock, there were small boutique inns and plantation houses catering to European long-staying visitors in places such as Nevis and Anguilla, as well as Barbados and Jamaica. There was also an enormous increase in the number of new resort hotels built on beachfront property in underdeveloped coastal areas, often outside existing tourist concentrations. Then there were condominium developments in St Maarten, the Cayman Islands and the US Virgin Islands, more budget hotels were added to the existing stock on the south coast of Barbados, and new high-rise, mass market hotels proliferated in St Maarten and Aruba (Mather and Todd 1993).

All-inclusive resorts, modelled after the Club Mediterranean concept, made their entrance into the Caribbean mix in the 1980s, with two notable Jamaican entrepreneurial ventures being Butch Stewart's Sandals and John Issa's SuperClubs (Pattulo 1996). International hotel corporations also began to return to the Caribbean, after their pull-out in the 1970s, and North American chains – Hilton, Holiday Inn, Marriott and Sheraton – sought to establish themselves in the main urban centres of the region's more developed islands, which were better served by their domestic carriers – Kingston, Jamaica; San Juan, Puerto Rico; Santo Domingo, the Dominican Republic; Nassau, the Bahamas; even Bridgetown, Barbados (Mather and Todd 1993).

Mass tourism was evolving into a more diversified set of market opportunities, and the 1980s heralded calls for Caribbean tourist officials and operators to respond to this 'new tourism' (Poon 1989). Environmental awareness, the conservation of nature, the fascination to learn about one's cultural roots, growing interest in others' histories, cultures, indigenous peoples' rights were ideas gradually coming to the fore in the 1980s. Changing population demographics in the major markets of North America and Europe were altering tourist orientations, and the 'maturing' host sectors responding to these new special needs and interests needed to create an even more differentiated tourist product. Yuppies, senior citizens and single women travellers were all categories of tourists seeking and expecting a different experience from the mass production of sand, sea and sun in a modernised, undifferentiated Caribbean resort. They were looking for real, authentic, natural, healthy and activity-oriented experiences, as befits their maturity, their life-styles and their self-perceptions (Poon 1989). These 'new tourists' were indeed similar in profile and in expectations to the earliest 'wealthy visitors' to the Caribbean – royalty, the landed gentry, the 'upmarket' colonial few – but now they represented a substantial (much larger) proportion of the tourist population. The low-density tourist environments that those early sojourners preferred had been transformed to denser, less regulated offerings in many Caribbean venues to cater for mass tourism.

Today's 'new tourism' has a variety of new faces, sport tourism, marina development and yachting, and scuba diving among them. Other alternative tourist activities, such as culture tourism's music festivals, sporting tournaments and associated professional association conventions, are low-impact/low-penetration models that avoid the problems of mass tourist impact (Conway 1993). Providing for such clients is likely to realise more returns than mass tourism, even though it can be criticised as pandering to the upper-class tastes of the international wealthy.

The 1990s: tourism 'comes of age' – environmental concerns become critical

The early 1990s were not good for the Caribbean tourist industry, with the Gulf War and recession-like conditions in all of the world's major generating markets – North America, Europe and the Pacific Rim – significantly affecting overseas tourism worldwide. Two major American airline carriers – Eastern and Pan American – folded in 1991 and 1992 respectively. Regional carriers such as BWIA and LIAT were struggling, and only American Airlines and British Airways continued to prosper from their monopolistic control of air access to the Caribbean. Islands that had built a dependence on the now-recalcitrant North American market suffered the most. Eventually, however, the North American situation brightened, and by the end of the decade the European market was again buoyant. Towards the end of the decade, stay-over numbers were on the increase again at an average annual rate of 5.5 per cent, the industry was expanding again, so that in 1998 the total of stay-over visitors comprised nearly 20 million. In an even more extreme fashion, cruise ship passenger visits, which had been 4.5 million in 1990, had risen to 12 million by 1998.

Tourism was now the major economic sector in more and more Caribbean islands, and by the end of the century every Caribbean destination had a recognisable tourism sector, even Trinidad and Tobago (Table 11.1). As measured by visitor expenditures as a proportion of island economic activity (GDP), the small islands of Aruba, Turks and Caicos, and the Cayman Islands were heavily dominated by tourism, in excess of 58 per cent. These, together with other micro-states such as St Maarten and the US and British Virgin Islands, also had the highest visitation proportions, with ratios exceeding 7,000 arrivals per thousand residents in 1998. The mature, high-density destinations of Barbados, Antigua and Barbuda, and the Bahamas had medium-range expenditure proportions, with tourism's contribution being between 35 and 45 per cent. Many others, Jamaica and several of the Windward Islands, had visitor expenditure proportions in the 20–30 per cent range. Cuba's visitor expenditure proportion had risen to 9 per cent, Haiti's was back up to 18 per cent and the Dominican Republic's had risen to 13 per cent. Only Suriname's tourist industry remained stagnant and not very significant (Table 11.1). In terms of visitor volumes, it was the very smallest islands in the northern Caribbean – the US and British Virgin Islands, St Maarten, Turks and Caicos – and Aruba in the south that experienced the heaviest relative influxes of stay-over visitors.

The 1990s was, significantly, the decade in which Castro's Cuban regime turned the clock back and re-embraced tourism as a development strategy and hard currency earner. Losing economic support from the Soviet Union at the beginning of the decade, the island weathered the Special Period of hardships brought on by this loss and a hardened effort to continue the tightening of commercial sanctions by the United States administration and Congress (as manifest in the Helms–Burton Act), by encouraging tourism. Tourist dollars were sought by promotion

Table 11.1 Visitor expenditures in the Caribbean as a percentage of gross domestic product, 1991/95–2000

Destination country	Average expenditure (% of GDP) 1991	1995	2000	Tourist (stay-over) arrivals (000s)
Cuba	2.05	7.41	11.22	1,774
Dominican Republic	n.a.	11.78	14.51	2,973
Haiti	3.03	19.31	n.a.	141
Puerto Rico	n.a.	4.32	3.78	3,341
Jamaica	36.57	24.88	21.26	1,323
Trinidad and Tobago	1.97	1.47	3.59	398
Barbados	31.78	39.01	33.01	545
Guyana	10.08	15.13	14.58	105
Grenada	22.74	29.90	27.28	129
St Vincent and the Grenadines	30.01	18.38	32.94	73
St Lucia	45.92	57.14	63.75	270
Dominica	18.62	18.04	28.18	70
Antigua and Barbuda	87.47	59.47	63.36	237
St Kitts and Nevis	47.77	33.49	28.09	73
Montserrat	21.24	38.57	43.17	10
Belize	26.05	13.14	16.44	196
The Bahamas	38.59	39.36	44.03	1,596
Bermuda	33.61	27.60	20.25	328
Curaçao	16.67	n.a.	n.a.	191
Aruba	41.66	35.93	32.39	721
Suriname	0.56	6.98	7.64	58
Guadeloupe	n.a.	n.a.	n.a.	807
Martinique	6.88	8.51	n.a.	526
Anguilla	65.53	79.38	83.06	44
Virgin Islands (British)	76.61	76.76	48.97	281
Cayman Islands	36.83	76.18	n.a.	354
Saba / St Eustatius	n.a.	n.a.	n.a.	9
St Maarten / Ste Martin	n.a.	n.a.	n.a.	432
Turks and Caicos	63.66	64.12	n.a.	151
Virgin Islands (US)	46.55	n.a.	56.74	607
			Total	17,763

Sources: CTO (2001), Tables 61 and 71

of a revived Havana and its cabaret entertainment attractions (including sex tourism, unfortunately (O'Connell Davidson 1996)). The Cuban government also invited European capital to help in the expansion of tourist resorts, and Spanish, Italian and European consortia were quick to respond. As a consequence, by the end of the decade, European tourists were a healthy market for Cuba's tourist industry.

Box 11.1: Sex tourism: the Caribbean's other 'S'

Early practices of sex tourism: female (and male) prostitution

- Female prostitution, with visiting military – sailors in particular – as their clients, has a long history under colonialism.
- The Dutch islands of Aruba and Curaçao attempted institutional control of their brothels and sex workers, predominantly for health reasons.
- Prostitution in Cuba became an integral aspect of that island's flourishing tourism image during the 1950s, in which black and mulatto women provided sexual labour to predominantly white North American men, and thereby reinforced age-old social conventions of exoticism and eroticism. The Cuban revolution then changed the situation dramatically, though not completely, by closing brothels and rehabilitating the approximately 150,000 working women. Some women, now working independently, continued to cruise the hotels of Havana to ply their trade among male visitors, but it was not until the 1990s, 'the Special Period', that sex tourism again became a common means for impoverished young Cuban women and men to seek dollars in sex work.

Sex tourism diversifies

- In the late 1970s, studies of sex tourism in Barbados began to investigate the 'beach boy' phenomenon, in which young black men 'hustled' for material compensation for the social and sexual favours they rendered to visiting white women. Young black men hung around the beaches, offering companionship and a 'good time' to visiting single white female tourists. Commonly, sex was one of the offerings, and it was expected that the woman would provide some monetary or material compensation for such services. Trips abroad to the visitor's home country might even enter into the agreement.
- Illegal trafficking and sex slavery is not commonplace in the Caribbean, except in one island territory, the Dominican Republic. By 1996, estimates of around 50,000 Dominican women working locally, for both tourists and local men, and approximately another 50,000 working in prostitution abroad (mainly in North America), were being cited, with poverty and illegal trafficking identified as major causes.
- Child prostitution is a recent concern among researchers, in large part because of the increasingly younger age of girls/women and boys/men being co-opted into sex work. Conditions of coercion, violence, domestic abuse and poverty among the young in the poorest households, or on the streets, has led to the growth of this variant of sex tourism internationally.

(Box continued)

It is not, known, however, how broad-spread this socially destructive practice is in the Caribbean, although studies in Guyana, the Dominican Republic and Cuba have acknowledged its emergence. Not to be overlooked in this disturbing development is the role that male fantasies for sex with a 'racialised other' play in the emergence of child prostitution as a new exoticism in the tourism sector.

- A recent revisionist view of sex tourism provides a more interesting perspective: namely the view that male and female sex work, particularly forms of 'hustling', represent a strategy to counter their powerless position in the existing social orders and racialised, class power and dominance of Caribbean post-colonial societies. It is a strategy that allows young men and women a form of freedom from oppressive and exploitative national (and global) political economic relations that keep them in poorly paid work or poverty in the islands, and positions them to gain access to a life that takes them away from their miserable social situation. For some, this might entail migration abroad with their 'lover'; a trip that might not be expected to be permanent, but at least it widens their horizons and offers them 'a chance'.

Source: Kempadoo (1999)

In similar fashion to the 1980s, the fluctuating fortunes of the industry during the 1990s highlighted problem areas that prompted action from Caribbean governments and private sectors. Modernised transportation and infrastructure needs were viewed as essential, and international agencies were now more ready to finance major upgrades of island infrastructure – airport development and redevelopment, port and cruise ship facilities, and road network improvements.

Most importantly, environmental issues were thrust into the limelight, in large part as a consequence of the Rio Earth Summit of 1992, Agenda 21, and the United Nation's Programme of Action for the Sustainable Development of Small Island Developing States. The latter advocated a plan for Sustainable Tourism Development in Small Island Developing States in which the principles of tourism and environmental management were spelled out. It espoused actions to be promoted at the national, regional and international levels in order to ensure the viability of the tourism sector and its harmonious development with the cultural and natural endowments of small island developing states (POA/SIDS 1994).

The Caribbean Tourism Organisation had an equally broad perspective on how to ensure a sustainable tourist industry in the Caribbean. Bound by an underlying principle that tourism should be for the economic and social benefit of Caribbean people, CTO advocated the following goals for tourism in the 1990s: increased profitability, competitive rates (and fares) of air access from major generating markets, maintenance of product quality, improved local acceptance for the industry, building a competitive local workforce, and strengthening links between tourism and other economic sectors (Youngman 1998). Then, in 1998, the CTO board widened these previous objectives to give high priority to environmental and socio-cultural considerations by endorsing a sustainable tourism development strategy and plan of action for the Caribbean. In particular, the development of community-based tourism development, in which local communities and stakeholders involve themselves in stimulating and managing their own national and local tourism development, was advocated as a means for ensuring the industry's sustainability (CTO 1998).

Theoretical models of Caribbean tourism

Dependency models

Early critics

Time and time again, international tourism has been criticised as an alien, dominating industry plaguing the Caribbean (Turner and Ash 1975; Barry *et al.* 1984). Perhaps the pursuit of leisure is so easy to ridicule, or perhaps depictions of tourism as escapism, hedonism and the like make its defence unseemly. Whatever the reason, mass tourism was seen as a vehicle for the continued dependency and subordination of the Caribbean (periphery) to the whims and capricious wishes of the affluent in core countries, North America especially (Hiller 1976, 1979a; Bugnicourt 1977; de Kadt 1979; Britton 1980; Erisman 1983). Latin American dependency theory was invariably evoked to support these critical assessments.

'Plantation' tourism

Early geographical evaluations of Caribbean tourism noted the similarities between the structural contexts of the region's dominant plantation society and its economic dependent relations, and that of mass tourism. Alluding to the situation as one of continuity, rather than evolution or change, the plantation's 'conditioning' of society was seen as ideal for the introduction, or penetration, of international tourism. Similarities included the dominant role of external capital in investment, ownership of resources and management of the region's agro-industry/industrial bases (Weaver 1988).

Three successive stages of cultural landscape changes were envisaged in this 'plantation tourism landscape' model, based upon the relationship between agriculture and tourism in Antigua, and the almost total substitution of the former by the latter:

1. pre-tourism (1632–1949) – agriculture is the dominant component of the economy, and tourism is negligible in both relative and absolute terms;

2. transition (1950–1969) – agriculture and tourism temporarily coexist as the two most important economic sectors;
3. tourism dominant (1970 to present) – tourism stands alone as the dominant economic component of the Antiguan economy.

Highly generalised spatial profiles of land use were depicted as primary, secondary and tertiary 'spaces' ranked according to the degree of dominance of the tourism industry in the landscape. A 'non-tourist space' category was the fourth land use. A historical analysis of Antigua's transformation in the 1950–1970 period demonstrated the close linkages between the plantation system's ownership of island resources and the ownership and appropriation of the landscape for the tourist industry (Weaver 1988). As others found, there was a 'geography' to the transformation of Antigua's tourism-impacted landscape, which Lorah (1995) would later broaden to detail the unsustainable nature of coastal zone pressures of land use and (mis)management practices in that same island (also see de Albuquerque 1991; McElroy and de Albuquerque 1997).

Models of urban tourism

Tourism's effect on transitions of the urban landscape in Caribbean islands was also modelled, using Antigua as a case in point (Weaver 1993b). In particular, the emergence of a specialised commercial zone adjacent to cruise ship docks in Caribbean ports, and the resultant uneven development of tourism activities beyond that specialised zone, during the mature phase of tourism dominance was seen as a distinctive infrastructural impact (see also Chapter 7). The model depicts a series of concentric circles that indicate declining tourism influence with distance from a specialised tourist zone close to the disembarkation point of cruise ships. Exceptions to the distance–decay pattern include resort strips along the coast beyond the port city's limits, as well as nodal attractions and accommodations in the main urban area. Of particular note in this model is the zone immediately inland from the specialised zone, the transitional CBD, in which tourists and locals are likely to have the most contact, invoking the spectre of cross-cultural conflicts, mismatched consumption bundles, and the like. Tourism spaces come to dominate small island landscapes, either through direct impacts or indirectly as infrastructure and transportation developments also become part of the built environment and link the higher-density urban retailing and historic districts with the coastal tourist strips and stay-over visitor accommodation landscapes. The more inaccessible interiors are spared urban or peri-urban transformation in this model, however.

Antigua was also the island that Thomas (1991) examined to depict a further urban landscape transition, which appeared to emerge in the more recent stages of tourism domination: no less than a 'gentrification of paradise'. This critical appraisal adds more explanatory substance to our understanding of tourism-induced urban transformations, because it stresses the importance of the emergence of new classes – a black business elite (with considerable political power) and a white expatriate entrepreneurial class (with considerable economic power) – to fill the power vacuum left by the old plantocracy. Neither have entrenched historical positions in Caribbean island society, but both appear to have vested interests that ally themselves to the old creole plantocracy, and all three have been involved in the gentrification of St Johns, Antigua, and in the political control of the city. Thomas goes further, however, suggesting that these new class allegiances are appropriating urban space and 'Disneyfying' it for tourist consumption. This is a more sophisticated assessment of tourism's influence on urban renewal and transformation in the insular Caribbean, and it remains a troubling reminder that tourism's impacts are not always externally induced and directed. The internal elites and local interest groups – an alliance of merchant capital, international finance and insurance, the old creole plantocracy – appear to be enthusiastic participants in the promotion and expansion of mass tourism, since they profit the most from the transformation of an 'amenable' built environment designed for unbridled consumption and tourist enjoyment (Harvey 1985, 1996; Thomas 1991).

Tourism life-cycle models

The destination life-cycle model

One of the most commonly used models of the evolution of tourist resort areas is Butler's (1980), which uses a logistic curve (emulating the product life-cycle) to illustrate the waxing and waning influences of tourism on resort landscapes (Figure 11.1). The cyclical evolution of resort areas is expected to be caused by numerous factors: changes in the preference and needs of visitors, the deterioration and possible replacement of tourist accommodation stock and infrastructure, and the transformation (and likely deterioration) of the area's original natural and cultural landscapes. In some cases, the vestiges of the original attractive context might continue, but they may be utilised for different purposes, or be supplanted by imported attractions, and by new 'consumerscapes' built to attract the latest waves of tourists.

The pattern of evolution is expected to cycle through a series of stages, proceeding slowly at first, then growing rapidly, to reach a stable, saturated stage before subsequently declining. The five stages are illustrated in Figure 11.1.

First, the *exploration stage* is characterised by small numbers of tourists, the area's attractions are the natural and historical/cultural landscapes, and their sojourning is relatively insignificant to the social life of the area's residents. As numbers increase, some local residents participate in the *involvement stage* and build small hotels and guest houses as a domestic-owned tourist industry develops. Contacts between visitor and resident are high at this stage, because the visitors are welcomed into a domestically generated tourist space that they share with residents. As the stage progresses, however, tourism becomes an emerging local industry,

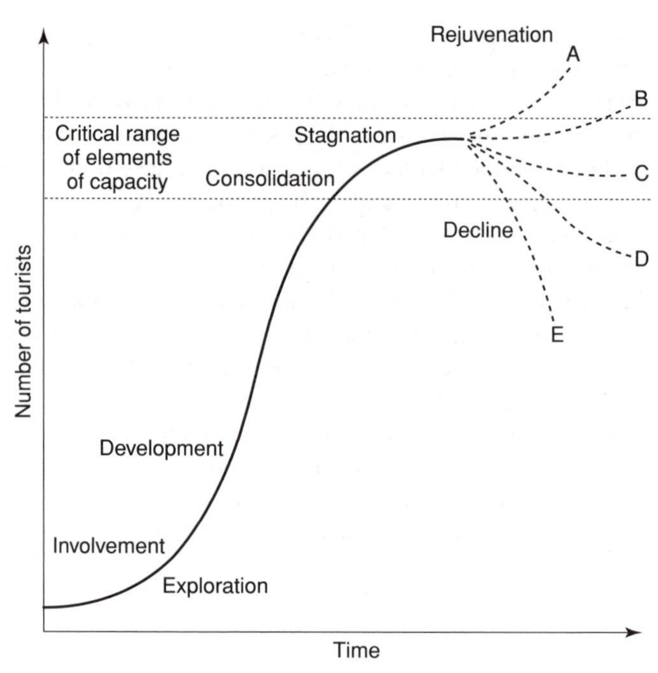

Figure 11.1 Hypothetical evolution of a tourist area

a basic market area for visitors can be defined, and a high-tourist season becomes an anticipated feature of social life. The first pressures on local governments for improving infrastructure and facilities come at this stage. Eventually (inevitably?), the *development stage* acquires a well-defined tourist market image, shaped in large part by advertising and tourist agency marketing in major generating markets, and rapid growth of the industry occurs. As this stage evolves, local involvement in, and control of, tourist development declines, while international influence increases, especially in the accommodation sector. Natural and cultural landscapes are marketed specifically to conform to an appropriate image. Often they can be supplemented by imported attractions to reinforce the area's image, while at the same time making it more comfortable and exclusively accessible to the visitor.

As the *consolidation stage* is reached, the rates of increase in tourists decline, although absolute visiting volumes may still increase, and the visitor-to-local ratio grows and changes at the expense of the latter. By this time, a major part of the area's economy is tied to tourism, the degree of international/external penetration is at its highest, and the tourism-related built environment is at or near capacity. These consolidated resort areas are no longer places for tourist–local interactions; they are more and more, tourist-dominated 'consumerscapes', where local needs and wants, if they do not mesh with the tourists, are subordinated. Resort cities will now have recreational business districts, and depending upon the maturity of the industry, the older accommodation stock will be ageing, and deemed second-rate or undesirable.

The *stagnation stage* represents the peak of the tourist area's evolution, where facilities are at capacity, visiting numbers reach their maximum, and the area is beset by numerous environmental, social and economic problems. The area may have a well-established image, but fashion has moved on, so that there is a heavier reliance on repeat visiting and convention/business tourism to maintain occupancy levels. Natural and cultural attractions have been superseded by imported 'artificial' attractions, and the resort's image is more 'created' than genuine. If there is any new tourist development, it is geographically separate from these stagnating resort areas, and the old areas may eventually suffer from over-building, over-capacity and over-extension, leading to a *declining stage*. On the other hand, *rejuvenation* might occur if the area's attractions can be fashioned, or repackaged, to appeal to a new set of tourist clients. Capturing a new set of tourists with casinos, or with packages offering licentious entertainment, and lax attitudes towards unfettered hedonism, might help rejuvenate these maturing, ageing resort areas. The Butler model appears to be somewhat fatalistic in its assessment of this waning of resort areas, but it remains a useful framework for others' more optimistic, or less pessimistic, attempts to gauge tourism's changing influences on Caribbean island landscapes.

A three-stage model of tourism styles

A more recent categorisation of Caribbean islands' tourism styles is based on a simplified version of Butler's product life-cycle model and it postulates three clearly identifiable stages of evolution and maturation:

1. *emergence*, or initial discovery;
2. *transition* to rapid expansion;
3. *mass-market maturity*, as defined by visitor saturation (de Albuquerque and McElroy 1992; McElroy and de Albuquerque 1991).

Each stage is associated with its own respective growth trajectory; first slow *emergence*, then rapid *transition*, followed by growth stagnation during the *maturity* stage.

To empirically substantiate this model, the late-1980s demographic, socio-economic and tourist market patterns of 23 small islands were examined, and islands were assigned to three relatively distinct subgroups (Figure 11.2). Stage III islands have a mass-market, mature destination style characterised by high visitor volume, high-density mass marketing, well-developed infrastructure and considerable hotel accommodation. Stage II islands are in transition style, characterised by rapid visitor growth, burgeoning hotel and infrastructure construction, and high seasonality of tourism patterns. Stage I islands are those with an emerging, low-density, long-staying style with limited infrastructure and hotel capacities.

According to this assessment of Caribbean tourism styles, the mature stage III islands include the six oldest and earliest, most tourist-dominated destinations – Aruba, the Bahamas, Barbados, Bermuda, St Maarten and the US Virgin

Figure 11.2 Caribbean small-island tourism stages and styles

Stage I (emerging)		Stage II (intermediate)		Stage III (mature)	
Low density	Retirement		*Substyles*	High density	
Long-staying	Nature tourism	Rapid growth	Fishing	Mass market	Shopping
West Indian	Small hotels	Europeans	Sailing	Short-staying	Gambling
Winter residence	Local control	High seasonality	Diving	N. Americans	Conventions
				Slow growth	Large hotels
				Bermuda	
				Bahamas	
				US Virgin Islands	
				Barbados	
				Aruba	
				Curaçao	
				St Maarten	
		Antigua			
		Martinique			
		Guadeloupe			
		Caymans			
		British Virgin Islands			
		Turks/Caicos			
		Bonaire			
		St Lucia			
		Anguilla			
		St Kitts/Nevis			
		Grenada			
St Vincent					
St Eustatius					
Saba					
Montserrat					
Dominica					

Source: De Albequerque and McElroy (1991, 1992)

Islands. At the other end of the scale, five emerging, low-density stage I islands are identified – Montserrat, Saba, St Eustatius, Dominica, and St Vincent and the Grenadines. The remaining 11 islands examined are intermediate stage II destinations still in transition in the late 1980s. They include four Windward Islands (Grenada, Guadeloupe, Martinique and St Lucia), four Leeward Islands (Anguilla, Antigua and Barbuda, the British Virgin Islands, and St Kitts and Nevis), two northern island clusters (the Cayman Islands, Turks and Caicos), and the remaining member of the Netherlands Antilles (Bonaire).

De Alburquerque and McElroy then pursue other characteristics of these island clusters and categorise each stage with distinctive tourism profiles (Figure 11.2). Stage III mature destinations are high-density, mass-market venues, predominated by short-staying North Americans and experiencing low rates of expansion. Hotel capacities are large, but fluctuating, and declining occupancy rates are troublesome. Like Butler's stagnation stage, attempts to diversify and rejuvenate the tourist bundle have led these mature islands to develop duty-free shopping, open themselves to gambling and casino development, and seek new visitors from convention and business markets.

Stage I emerging destinations, on the other hand, are still low-density, long-staying venues, with regional Caribbean visiting being important, and some retirement and winter residential tourism persisting because of the underdeveloped nature of mass tourism, to date. They still have a vibrant, locally owned small hotel and guest house sector, nature tourism favours their tropical, unspoilt landscapes, and exclusive visitors still return to these hideaways, precisely because of their underdeveloped state. The stage II destinations of the late 1980s were the emerging (stage I) destinations of the 1960s and 1970s, and the range of tourism development and penetration varies widely among this largest intermediary group.

A tourism penetration index

Examining tourism's effects on 20 small Caribbean islands, McElroy and de Albuquerque (1996) refined their Tourism Styles model further by constructing a composite tourism penetration index. Utilising three variables – per capita visitor spending (economic effect), daily visitor densities per 1,000 resident population (social effect), and hotel rooms per square kilometre (socio-environmental effect) – a cluster analysis ranked 20 island destinations in an attempt to recreate three sub-groups of tourism penetration (Figure 11.3). According to these quantitative rankings, the most penetrated, high-density islands comprise a group topped by St Maarten, with the Cayman Islands, Aruba, Bermuda and the US Virgin Islands now mass-market venues (with the British Virgin Islands appearing to be a statistical anomaly in this top-ranked group).

At the bottom of the ranking are three relatively underdeveloped Windward Islands (Dominica, St Vincent and the Grenadines, and Grenada), all low-density destinations, with limited infrastructure, limited airline access and small-scale, locally owned accommodation stock. A further quantitative attempt to categorise

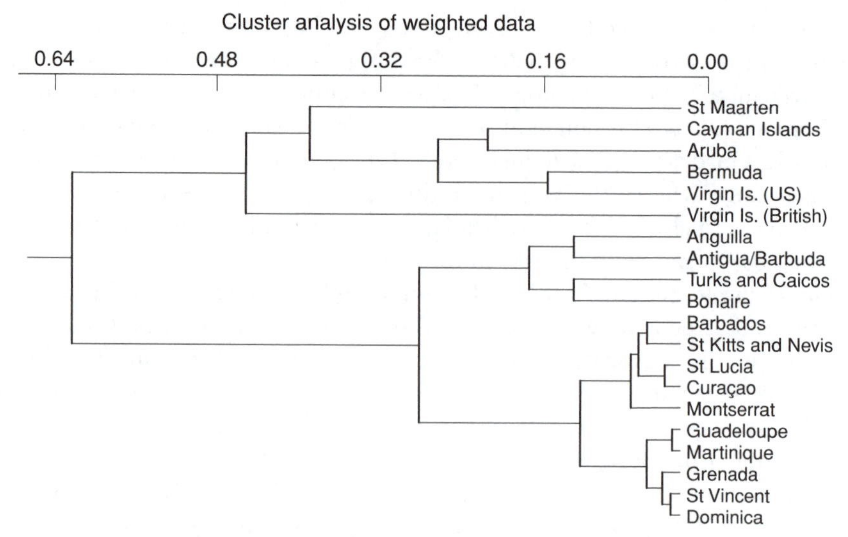

Figure 11.3 Tourism penetration index
Source: McElroy and de Albuquerque (1996)

the intermediate, transition group derived a further grouping of islands into four medium–high penetrated destinations (Anguilla, Antigua and Barbuda, the Turks and Caicos, and Bonaire), four medium–low penetrated destinations (Barbados, St Kitts and Nevis, St Lucia and Curaçao), and a least penetrated group comprising the aforementioned Windward Islands (Dominica, St Vincent and the Grenadines, and Grenada) together with their neighbouring French DOMs, Guadeloupe and Martinique.

Tourism and environment – 'commodified' relationships

Theoretical insights into the 'commodification of places'

In today's globalising capitalist world, leisure, vacationing and tourist visiting are behaviours and contexts which appear to have four organising characteristics: privatisation, individuation, commercialisation and pacification (Rojek 1985). There is the recognition of the individual as distinct from the community, which has not only facilitated the differentiation and specialisation of leisure pursuits but has also co-opted people and institutionalised specific cultural practices compatible with social control and prevailing class relations (Lefebvre 1976; Rojek 1985). Importantly, leisure activities have become increasingly 'commodified', with international tourism organised by a complex of institutions in sending and receiving markets with the express purpose of creating, coordinating, regulating and distributing 'exchange values' rather than 'use values' (Britton 1991).

Given this logic of a 'culture of consumption', the value to the individual consumer of the travel mode, cultural events and associated goods packaged through the tourism production process, lies in the quality and quantity of the experience promised and symbolically 'purchased'. The leisure/tourism commodity becomes an alternative to the alienated and stressful work regime individuals are experiencing; a sought-for escapism and means to an end for the individual, albeit short-lived and vicariously valued. It becomes the purchase of a different life-style; a statement of taste and demonstration of the possession of 'cultural and symbolic capital'; a bestowal of status; an invigoration of the body; an uplifting of the spirit; a broadening of the mind; and a confirmation or challenge of attitudes (Featherstone 1990; Britton 1991).

For the Caribbean tourist industry there is a need to provide tourists with views, and visions, in accordance with their perceptions of the islands' people, their scenery, sights, places and culturally different behaviours, all in a commercial and institutionalised 'package' constructed to satisfy the tourist demand for a leisure-time interlude and alternative escape. Catering to tourists from the metropolitan and highly urbanised cultures of North America and Europe, as the Caribbean does, means that the tourism experience offered is typically not authentic, but constructed, commodified and socially comfortable for the individual tourist or group of individuals; i.e. their social power and status is not threatened or undermined (Nettleford 1990; Britton 1991). Infrastructure developments must be familiarly metropolitan; exclusionary practices and hedonist pursuits must be tolerated, if not locally sanctioned (Karch and Dann 1981); tourists must be welcomed, islanders must be civil, even when tourist behaviours flout local cultural norms; and the tourism industry has to be socially sanctioned locally, nationally and internationally (Conway 1983a).

All the while, there is a 'commodification of place' that accompanies the commodification of leisure. Certain places and sites – with their landscapes, social practices, buildings, residents, symbols and meaning – achieve the status of tourist sites because of their physical, social, cultural and commercial attributes. Internationally, the small islands of the Caribbean are highly favoured in this regard (Hudson 1986). And, despite the local traditions and heritages that have tended to endow island people with communal rights and opportunities, tourism brings a commodification of 'leisure space' in such small islands that alienates as well as appropriates. Commodification assumes two generic forms: (a) the legal recognition, or transfer, of commercial property rights involving ownership or lease of the site itself – a building, recreation land or beach, or (b) where the attraction cannot be appropriated directly, the inclusion of the tourist experience or geographical attributes of the place turns it into a saleable commodity (a tour, the ambience of a hotel in close proximity to a significant site, visual souvenir potential, or symbolic image with recognisable connotations). With both forms, such commodified places provide opportunities for the generation of capital (rents and profits) by virtue of their special qualities and status and the captured proprietary assets that emanate from their promotion and exploitation (Britton 1991).

The Caribbean tourist industry has three main ways to create and promote tourist attractions. It can take advantage of already existing cultural attractions or 'curiosities', by co-opting them for the purposes of accumulation into tourist products: tours to historic sights and major cultural spectacles, for example. It can create its own attractions, such as all-inclusive recreation resorts, theme parks or cruise ship 'adventures(?)'. Tourism can also be co-opted into other commercial ventures in order to enhance the market profile, commercial returns or social legitimisation of such ventures. Duty-free shopping centres, festival markets, airport and seaport expansion and modernisation come to mind as examples of these cooperative 'productions of paradise'.

The commodification of these tourist facilities and attractions occurs in two ways. First, the industry contrives to impart (additional, enhanced) meaning to its places by associating them with the attributes of the non-commercially created (public goods) attractions. Of necessity, this also means creating and promoting meaningful images of specific places and sites. Second, non-commercially created attractions or non-tourism derived commercial attractions – such as shopping centres – take on (new and other) meanings by being associated with or incorporated into the tourism 'bundle'. Local authority is undermined, local empowerment is difficult to sustain, and local environments are changed for ever. International tourism offers economic returns to those participating – albeit short- and medium-term returns – and this reality too often preoccupies the minds and concerns of Caribbean governments, at the expense of other considerations. In so many ways, therefore, international tourism constructs as it commodifies, alienates as it appropriates, and dominates as it penetrates.

Despite the critical analyses that have exposed the ecological and socio-cultural limits to growth of international tourism in both the fragile and more robust island ecosystems of the Caribbean (McAfee 1991; McElroy and de Albuquerque 1991, 1996, 1998; Pattullo 1996), it is invariably the *economic growth potential* of this international industry that is always offered as a future panacea for Caribbean development. Yet, questions of tourism's ecological sustainability remain as important caveats to claims of the industry's promise for the Caribbean.

Sustainable tourism in the Caribbean

A model of new tourism for the Caribbean

To help develop a sustainable tourist industry in the Caribbean, Poon (1989, 1993) advocated several changes: putting the environment first; making tourism a leading sector; strengthening marketing and distribution channels; building a dynamic private sector; and seeking to be globally competitive rather than intra-regionally. Such unbridled enthusiasm for the benign direct effects and supra-national influences of global capitalism must be questioned, however. These forces are dominant globally, certainly. Nevertheless, Caribbean decision makers, in the public and private sectors alike, must retain a healthy scepticism of the effects of such global external forces. Most crucial in this respect must be the progressive and proactive involvement of the state in its 'stewardship role', to maintain each country's landscapes, coastal zones and marine habitats, as well as cultural heritages. There must be progressive co-management and local democratic economic decision making, involving not only those within the tourist sector, broadly defined, but also local communities, public–private coalitions of local authorities, NGOs, artisans, citizens, even children. If, 'capital-by-invitation' is nothing more than today's variant of the failed 'industry-by-invitation' model that haunted Caribbean industrialisation experiences (see Chapter 9), then *scepticism* should be the watchword.

Furthermore, the unknown, under-explored, and most probably under-valued dimensions of a sustainable tourism industry might very well be the industry's capacities to incubate and spawn a wider set of multiplier effects than previously envisioned. These should go beyond the essential linkages that tourism must engender between local agriculture, local micro-businesses (formal and informal), local producer services, even local manufacturing–customising of imported technologies and commodities (both product lines in need of customising for tropical/ Caribbean island uses). The wider set involves tourism's place in transnational systems, in inter-regional associations and cooperative ventures (rather than competitive scenarios), and tourism's partnership with other sectoral expansions, especially communication services development and customising. The state's role in aiding and facilitating co-management initiatives, where tourism planning, environmental conservation and marine resources management and local, democratic economic decision making are joint undertakings, treated as a merged bundle of policy mandates and practices, is crucial for future success. The long-term survival of Caribbean tourism is dependent upon environmentally conscious plans of action.

Expanding upon this warning, McElroy and de Albuquerque (1991) identify five dimensions to the specific contours of such structural disequilibrium in small island societies of the Caribbean, embedded, as they have become, in the open tourist economy–closed environment interface. First, the large size of the international tourist economy interacting upon the delicate, tightly bound, insular environment generally produces wide income disparities and resource imbalances (also see Bellar 1987). Sustainable land and marine uses are replaced by services and non-productive real estate speculation. Formally mixed-use or pristine coastal and upland landscapes are commodified as tourist 'leisure spaces'. Second, tourism penetration and growing dominance clashes with the cultural modes of island life to produce alien residuals, which overload the absorptive capacity of an island's fragile ecology. Common overloads occur with solid waste disposal and unsightly garbage sinks and landfills, with sewage discharges and toxic pollution of surface water bodies, rivers and near-offshore reefs, with groundwater contamination and depletion, and with physical changes of the coastal zones and attendant habitat damages (IRF 1996). Third, seasonality in tourism demand contributes to system overrun and stresses the environment at peak times. Consequently, repeated cycles of high and low peaks of tourist activity generate local concerns for the slack period and hotel and accommodation under-utilisation, while their commensurate declines in retailing activity and services leads to pressures for year-round tourism. Fourth, there appears a common preoccupation among island policy makers and tourist industry promoters with increasing the number of tourist arrivals instead of focusing on net visitor expenditures and the 'tourist multiplier effect'. International airline interests also fuel such perspectives (Conway and Jemilio 1991). Fifth, to satisfy the overhead and profit criteria of international hotel, cruise ship, airline and tour operators, the tourist sector has its own built-in, long-run propensity to expand visitor densities up to, and beyond, the social and ecological limits of any given island destination. These five dimensions cumulatively institutionalise an environmentally incompatible growth process leading to a high-density, high-consumption style of mass tourism that is unsustainable as well as dysfunctional (McElroy and de Albuquerque 1991).

It would appear to be crucial for those islands not already enmeshed in high-density/high-impact styles of mass tourism to seek low-density/low-impact alternatives. Among Caribbean islands, the smallest in resource area and most threatened in terms of their fragile resource base need to avoid the expansion path (see Briguglio *et al.* 1996a, 1996b). Those in the Caribbean already too far along might well need to rethink their strategy and refurbish their tourism image if they are not to descend into the type of 'tourism death spiral' that two of the US Virgin Islands – St Croix and St Thomas – are experiencing, due in large part to the substantial environmental degradation the islands have suffered (IRF 1996).

Ecotourism: can a sustainable model of nature tourism be developed?

Conservationists and wildlife preservationists began promoting the idea of ecotourism as a way to both protect wildlife and wilderness, and preserve indigenous peoples' ways of life, while allowing limited tourist access to such frontier

Box 11.2: Ecotourism

Defining ecotourism

Ecotourism is travel to relatively undisturbed natural areas for study, enjoyment or volunteer assistance. It is travel that concerns itself with the flora, fauna, geophysical landscapes and ecosystems of an area as well as the people (caretakers and stakeholders) who live nearby, their needs, their culture and their relationship to the land. It views natural areas both as 'home for all of us' in a global sense ('eco' meaning home), but 'home to nearby local residents' specifically. It is envisioned as a tool for both conservation and sustainable development, especially in areas where local people are asked to forgo the consumption/production use of land and forest resources for their sustenance.

(Box continued)

Contexts

Remote, exotic frontiers, with cultural minorities, natural ecosystems and bio-diverse habitats. Marine and coastal ecosystems as well as land-based ecosystems are sites for this alternative tourism.

Principles

- Entails a type of use that minimises negative impacts to the environment and to local people.
- Increases the awareness and understanding of an area's natural and cultural systems and the involvement of visitors in issues affecting those systems.
- Contributes to the management and conservation of legally protected and other natural areas.
- Maximises the early and long-term participation of local people in the decision-making process that determines the kind and amount of tourism that should occur.
- Directs economic and other benefits to local people that complement rather than replace traditional practices (farming, fishing, forest foraging, social systems, etc.).
- Provides special opportunities for local people and nature tourism employees to visit natural areas and learn more about the fauna, flora, ecosystems and habitats that other visitors come to see.

Conceptual elements

1. Ecotourism activities should have the minimum of impact on the natural environment, using all possible mechanisms to avoid these impacts.
2. Ecotourism activities should attempt to educate both the visitors and the local populations about the natural area they are visiting, as well as broader concepts of environmental conservation.
3. Ecotourism activities should benefit local communities and regions, both economically and culturally. This means assuring a maximum, and if possible equitable, monetary benefit, as well as attempting to minimise the cultural impact.

4 Ecotourism activities should contribute to local and national conservation efforts, both economically and in any other way possible. Ecotourism as an economic activity cannot exist without those natural areas on which it depends.

5 Planning a partnership between local peoples, tourism operators, national park authorities, conservationist groups (NGOs) and other involved government officials is the key to obtaining long-term success of ecotourism in particular, and sustainable development in general.

regions or remote interiors (Boo 1990). Ecotourism invokes strong conservationist ethics, and nature tourism is a logical extension, where wild, natural and even human-modified landscapes are distinctive, exotic, rich and colourful, and enticing to photographers, naturalists, outdoor enthusiasts and the like (Whelan 1991). Both versions of low-density style tourism, together with their urban fellow traveller – heritage tourism – anticipate that the participants in these alternative tourism packages would be, generally, self-selected culturally sensitive visitors, reasonably well educated and wealthy, who would understand the ecological limits they were under while enjoying the outdoor and truly local experience. Not only would the cultural differences between visitor and the visited be muted, but the smaller number of wealthier visitors would bring sufficient influxes of money to offset their significantly reduced numerical size and its curtailment of mass tourist dollars, or pounds.

An essential aspect of ecotourism and nature tourism is the active involvement of local stakeholders and communities in the management of the tourist projects, but this has been found easier to conceptualise than implement. The continuing conflict between stakeholders in the Soufrière Marine Management Area in St Lucia, despite the project's early promise, is a noteworthy caution in this respect (Trist 1999). On the other hand, islands such as Dominica, Grenada, St Vincent and the Grenadines, and the British Virgin Islands, which have not embraced mass tourism and are still in a stage I situation, might very well develop profitable nature tourism ventures that are both economically profitable and environmentally friendly. Ecotourism, as a local, 'bottom-up' model of new tourism, is certainly a promising avenue to pursue, as islands seek to diversify their tourism offerings. Unfortunately for many Caribbean islands, their coastal zones are already past redemption.

Environmental degradation of coastal zones

Caribbean coastal zones have undergone transformation since early colonial times, as Watts (1987) has so clearly demonstrated. Forest clearing was rapid, soil quality deterioration resulted from intensive plantation cultivation, siltation occurred, and offshore marine zones were the recipients of the effluent and sewage discharges from coastal settlements. Natural hazards – hurricanes, floods and volcanic eruptions – also contributed to environmental change. Coastal pollution has accompanied mining, industrialisation, urbanisation and infrastructure development, because there has been inadequate monitoring, a lack of assessment systems, and an overall lack of oversight of waste-disposal practices.

Tourism, however, has grown to become a major direct contributor to sewage and solid waste pollution of island marine resources in virtually every Caribbean island, and where islands have matured to stage III, high-density tourism, the industry is the prime contributor to coastal erosion and sedimentation from construction-related activities. High-density tourism also has indirect effects. Much of the oil and fertiliser/pesticide pollution of coastal waters can be attributed to tourism-related activities – with yachts, ferries and cruise ships dumping their bilge-water and waste into Caribbean waters, and with 'modern' lawn-care practices on golf courses, second home complexes and condominium resorts being sources of nitrate and phosphorus contamination during storm run-offs (IRF 1996).

The Island Resources Foundation's 1996 study of *Tourism and Coastal Resources Degradation in the Wider Caribbean* identifies a comprehensive list of tourism's direct and indirect impacts and demonstrates the extent of the industry's destructive record of environmental degradation. Displacement of traditional uses and users is pervasive, and major physical changes such as construction of causeways, jetties, groynes, piers and wharves, dredging and spoil disposal are widespread. Also, coastal and marine habitat damage is widespread, caused by sand removal, dune destruction, destruction of mangroves and salt ponds, and anchoring damage to coral reefs and sea grass beds. Problems of solid waste disposal are especially critical in the smaller islands and dependencies, which face difficulties siting disposal facilities and are too small to support economically viable recycling programmes. Toxic chemical pollution and nitrification from surface water run-off plague Caribbean tourism coastal zones, and when added to herbicide, pesticide and fertiliser treatments in plantation agriculture and resort landscaping, the accumulations in near-shore marine waters can be destructive to marine life and water quality.

Sewage discharges are generally believed to be the most serious land-based pollutant of the Caribbean's coastal waters, and non-existent or poorly designed sewerage systems in coastal hotels and resorts contribute to this problem (Archer 1985; Gladfelter and Ogden 1994). Other physical impacts related to tourism activities include groundwater contamination, construction-related sediment load increases in rivers and estuaries, sand mining, scarring of mountain faces, wilful deforestation of the island's tropical forest ecosystems, embayment filling, and poorly designed transportation, building and resort facilities with little aesthetic value and inappropriate architecture.

The loss of non-renewable resources is a major negative impact of tourism's penetration. Sand mining and removal, when combined with regional sea-level rise (global warming's contribution to the equation) and increased rates of coastal erosion caused by human and natural factors can deplete beaches, alter their profiles and degrade their environmental value (Cambers 1996). In Cuba, sand mining for construction purposes has degraded some of the island's famous beaches east of Havana, including the vaunted Varadero beach. Southern beaches such as Caimito, Rosario, Mayabeque, Cajio and Guanimar have been degraded by discharges from nearby sugar *Centrales*, and three cattle-breeding feedlots have created massive discharges of pollutant excrement (IRF 1996).

The over-harvesting of renewable marine resources is also tourism-connected, but local evaluation of depleted stocks of such resources as 'sea eggs' (sea urchins), mangrove oysters, sea grass beds and red mangrove are subordinate to tourist perceptions of shortages of 'steak fish' (dolphin/mahi-mahi), Caribbean *langoustine*, garoupa and conch. A recent study by the World Resources Institute (1998) argues that two-thirds of Caribbean coral reefs are in jeopardy, with the reefs of Jamaica,

Plate 11.4 Inter-island schooners in Carenage Harbour, Bridgetown, Barbados, 1968
Source: Dennis Conway

Barbados, Dominica, Haiti, the Dominican Republic and Puerto Rico under high potential threat because of over-fishing and pollution.

Tourism activity also has direct effects on the coastal zone's human environment, its local residents and socio-cultural environment, because the multiple use that results invariably generates cultural conflicts, socio-economic conflicts and conflictual power relationships, if not outright aggression and antisocial behaviour – insults, stand-offs and put-downs. Displacement of fishers and other traditional water users, conflicts between fishing communities, diving and sailing tourist groups and companies, conflicts over local versus tourist use of public spaces, and especially public access to beaches are commonplace in Caribbean tourism destination areas (see Conway and Lorah 1995; Trist 1999 and Pugh and Potter 2001 for disputes between local stakeholders – fishermen, divers, yachtsmen, hoteliers – in the Soufrière Marine Management District in St Lucia).

Rarely do local, traditional uses of coastal resources compete, and triumph, over tourist-prompted uses, except perhaps in Barbados. In this island, as elsewhere in Antigua and Jamaica, at the beginning of their mass-tourism experience, some established hotels sought exclusive rights of access to their beachfront. This policy did not gain support from the Barbados government, however, since it had

Box 11.3: Tourism's negative side-effects

Hedonism and cultural imperialism

The litany of opinion on tourism's disastrous effects on Caribbean societal and cultural mores appears endless and, in many ways, is persuasive (Karch and Dann 1981). The following are criticisms levelled by Caribbean commentators:

- Enclave tourism and the tourist havens – the hotels, nightclubs, duty-free commercial concession zones – exclude the local population or at best tolerate the local 'colour'.
- The tourist spreads 'cultural imperialism' and as an outsider in pursuit of hedonistic delights – invariably a member of metropolitan society with a different standard of values, behaviour and expectations – contributes to alienation of the local culture.
- The employment structure of the tourist industry perpetuates the Caribbean's plantation and slavery legacy, a colonial class division wherein white foreigners control and manage while black natives work and serve.
- The juxtaposition of the rich 'white' tourist with the poor and usually 'black' local is viewed as an obvious cultural conflict, and consequently tourism is blamed for everything from vandalism to increased crimes of violence, social unrest and even child abuse.
- The cultural diffusion of modern decadent ways via the tourist 'demonstration effect' leads to a debasement of local morality.
- Prostitution is encouraged, sex tourism flourishes as a result, and sexually transmitted diseases (and now AIDS) are freely transmitted, with little vestiges of governmental control (O'Connell Davidson 1996; Howe and Cobley 2000). Drug trafficking, gambling and casino activities are accommodated in some Caribbean countries over the objections of concerned residents (Maingot 1994).

to respond to popular outcry, not to mention civil disobedience and even calypso satire to maintain 'Windows to the Sea' for local Bajans (Hutt 1980). Such enclave tourism would eventually re-emerge in the Caribbean, with the welcoming of all-inclusives to such islands as Nevis (the Four Seasons), St Lucia (Sandals) and Antigua (Sea Clubs), and they certainly restrict local access to their facilities, including their beachfronts. However, in the 1990s, local acceptance of such exclusive practices by all-inclusives appears to be given begrudgingly, perhaps because their management has learned to employ local labour, purchase local produce and use local businesses to help run the resort's many activities, services and infrastructure maintenance (Timms 1999).

Socio-environmental costs revisited

So vehement and persistent has been the outcry against these socio-cultural disruptions of tourism that international agencies, development experts and even Caribbean governments have long held ambivalent opinions about its place in future development scenarios. Tourism is not the only 'modernising' outside influence, however. Continuing urbanisation, industrial diversification, return migration and international circulation are common transitions in all but the micro-dependencies – the Grenadines, Guadeloupe's dependencies, the outer islands of the Bahamas, for example. In other urbanised, modernising island societies, European and/or North American cultural penetration is considerable, deep and enduring. Today's globalising world is more open than ever before, and the *open* societies of the Caribbean, with or without tourism, will experience many of these alien importations. The tourist industry's concern with crime and visitor harassment is one issue of contemporary importance and sensitivity.

Illegal global industries and their links to tourism

Griffith and Munroe (1995) and Maingot (1994) remind us of the growing significance of 'underground/informal' activities in the Caribbean region. These require further elaboration, especially given our concern for environmentally sensitive tourism development, because wilful, corruptible decision making can so easily replace, or subordinate, rational democratic forms of conduct – the sort always hoped for in Caribbean 'civil society'. Being in the 'backyard of the United States' does not favour the region's micro-states in this era of increasing complexity and complicity.

The threats to Caribbean social and national security by the current international regime of corruption and violence unfortunately appear to be an integral part of the globalising system today (Maingot 1994). The 1970s and 1980s witnessed an array of violent challenges to legitimate authority that seemed to find fertile ground in the area's long-held vulnerability to traditional and bureaucratic corrupt practices, while adding new forms to the volatile pot. There was the growth of offshore banking and its attendant money-laundering potential. The growth of the region's importance as a transshipment region for Colombian drugs on their way to the insatiable appetites of the United States was another violent reality. The often tacit involvement of the region's own criminal sector in gun running, drug running, blockade avoiding, and illicit fraudulent financial schemes, not to mention offshore banking in its legitimised forms, is also an important macro-structural obstacle currently at play in the region (Richardson 1992; Griffith and Munroe 1995; Pattulo 1996). Notably, Maingot (1994) warned of the legitimisation of a new form of 'market corruption': a corruption among officials that suits the logic of today's cut-throat marketplace, with the corrupt official regarding his office as a business, where income is to be maximised, or the risks decided, knowing the costs of sanctions and the pay-offs justify the risk (Heidenheimer 1970).

Where are the Caribbean paradises where such market corruption and international financial intrigue happens? Well, there is the Bahamas (Block 1990, 1994), with a recent past of underground money-laundering schemes and speculative tourist resort investment schemes, where such celebrated, or infamous, people as Howard Hughes, Roberto Vesco and later Carlos Lehder (drug smuggler of Norman's Cay fame), who was a major player among the Medellin drug cartel, played and enjoyed Bahamian hospitality. The Pindling government's complicity in drug trafficking was eventually exposed by the 1984 Commission of Inquiry, but the Bahamian political regime did not founder.

St Maarten/St Martin was another island where international financiers, US crime syndicates and Bahamian internationalists focused their investment efforts. Some projects failed, while others were refinanced and transferred between international hotel chains (Block 1994). Maingot (1994) also commented on the Mafia's presence in St Maarten, and the international drug control agencies indict St Maarten as an island where large shipments of cocaine transit through to Puerto Rico and the Netherlands, and where offshore banking facilities, casino/resort complexes, high-volume American tourism in Aruba and St Maarten, and stable currencies make the islands of the Netherlands Antilles attractive to money-laundering organisations and their operations (INCSR 1996).

Proximity to the United States poses additional problems for the orderly management of Caribbean tourism resources, in part because of the region's historical linkages with its all-powerful northern neighbour and the hegemony US business (legal and illegal) has enjoyed in this incorporated, offshore 'backyard'. Although examples of US FBI and CIA, state-sponsored 'clientelism' with organised crime, violence and murder are drawn from operations in the Dominican Republic (Block 1990), the threat that they may eventually reach other Caribbean islands is ever present. Similarly, the threat posed by the growth of New York city/New Jersey Mafia syndicates' waste disposal network and its penetration of the Caribbean into Jamaica (Block 1990, 1994) is another environmental disaster waiting to happen. The globalisation dynamic that is enveloping the insular Caribbean deserves extra vigilance. Never before has the need for a softer, lower-impact, locally managed and locally produced tourism style been so imperative.

Conclusion: Caribbean tourism's promise, problems and constraints

Tourism remains one of the Caribbean region's only 'comparative advantage' in today's neo-liberal, globalising era, but the consequences and impacts of this 'difficult-to-regulate' industry have been problematic. In simple terms, one decisive issue revolves around the *quality versus quantity* argument. Mass tourism, the unregulated expansion of hotel and tourist facility building, urban over-building and the wholesale 'commodification' of island environments has had largely negative consequences. On the other hand, the diversity of 'new tourism' with its

Table 11.2 Cruise tourism: recent trends, 1996–2000

Country	Population in 2002	Cruise ship calls 1996	Cruise ship calls 2000	Cruise passenger arrivals (000s) 1996	Cruise passenger arrivals (000s) 2000
Antigua and Barbuda	67,400	212	331	270.5	429.4
Bahamas	300,500	1,844	1,892	1,687.1	2,512.6
Barbados	276,600	528	485	510.0	533.3
Belize	263,000	n.a.	70	0.2	58.1
Bermuda	64,000	147	165	181.7	209.7
Cayman Islands	36,300	525	605	800.3	1,030.9
Dominica	70,200	290	287	193.5	239.8
Grenada	89,200	393	360	267.0	180.3
Guadeloupe	435,700	429	249	589.5	392.3
Jamaica	2,680,000	466	504	658.2	907.6
Martinique	422,300	409	288	408.4	286.2
Puerto Rico	3,958,000	746	698	1,025.1	1,301.9
St Kitts and Nevis	38,700	320	343	85.8	164.1
St Lucia	160,100	297	389	182.2	443.6
St Martin/ St Maarten	67,764	473	492	657.4	868.3
Trinidad and Tobago	1,163,700	n.a.	n.a.	46.4	82.2
Virgin Islands (British)	21,300	203	230	159.6	188.5
Virgin Islands (US)	123,500	917	1,014	1,316.4	1,768.4

Source: CTO 2001

focus on niche markets, alternative tourists and their higher ratios of expenditure per tourist is more promising, even though it is likely to be harder to manage and implement successfully.

The rapid increase of the cruise ship industry in the Caribbean region, on the other hand, is a looming spectre that requires careful management (Table 11.2). Cruise ships disgorge 'day-tripper' passengers in volumes that stretch local capacities yet provide modest returns to island economies. Bermuda's restrictions on cruise ship visits is a model to contemplate, though it is likely to meet stiff opposition from the extremely powerful corporations that have interests in cruiseship developments – Disney/AOL/Time-Warner, for example. Another external interest that wields virtual monopoly power and is likely to threaten regional self-reliance is the airline industry. Regional carriers such as BWIA and LIAT are relatively weak, whereas the monopolistic control of American Airlines/American Eagle over the United States market, and of British Airways over the British market, will ensure that the promotion of tourist 'quantity' takes precedence over 'quality' (or smaller passenger volumes).

Towards sustainability – the ecological imperative for regulating and managing tourism

This chapter has fully documented the fluctuating fortunes of tourism, and of Caribbean countries that have, to varying degrees, embraced the industry. Environmental protection is even more crucial a policy imperative than it was at the industry's inception. The Caribbean's environment is *the* crucial resource that sustains the tourist industry. Its sustenance is as important for the future of Caribbean societies as material prosperity, wealth creation and welfare provision is for Caribbean people. Enlightened but well-integrated planning and policy making, which *regulates and manages* tourism so that the industry's destructive consequences are reduced, or mitigated, is a necessity (also see Wilkinson 1997). Environmental conservation, preservation and sustenance of the Caribbean's fragile marine and terrestrial ecosystems is also a necessity (see Lorah 1995; IRF 1996).

There have been international calls for proper planning approaches to ensure balanced and sustainable tourism development (Inskeep 1987; Wilkinson 1997), and suitable policies for implementing an appropriate set of prescribed planning measures (OECD 1980; UNEP 1982; POA/SIDS 1994). Among these latter measures are elements from the broader class of environmental and development policies derived from Agenda 21, which have been adapted to tourism. Central concerns are (a) control of tourism growth away from environmentally sensitive areas; (b) restrictions imposed on the types, extent and intensity of activities permitted in an area; (c) proper management of residuals generated by tourism; and (d) minimisation of conflicts between tourism and competing land uses (Briassoulis and Van der Straaten 1992).

The record of progress, however, is scarcely laudable. Implicated in the lack of local initiative and the general reluctance of Caribbean governments to be proactive in their management styles and concerns with mass tourism is the nature of the international industry itself: its international multi-functional character, its global reach and its 'commodification of places'. The Caribbean's 'love–hate' relationship with tourism will not be easily resolved. There will need to be a concerted effort by concerned citizenry – the Caribbean people – to influence their governments and colonial administrations to have the political will and strength of their own convictions to put longer-term environmental and ecological conservation and preservation priorities ahead of the short-term economic goals that currently drive their policy making and decision making. The continued sustainable health of Caribbean tourism demands this long-term commitment; the people's future depends upon it.

References

Archer, E. D. (1985) 'Emerging environmental problems in a tourist zone: the case of Barbados', *Caribbean Geography*, **2**, 1, 45–55.

Barry, B., Wood, T. and Preusch, D. (1984) *The Other Side of Paradise: Foreign Control in the Caribbean*, Grove Press, New York.

Beller, W. S. (1987) *Sustainable Development and Environmental Management of Small Islands*, NTIS Springfield, Virginia, PB. 88-100029.

Blake, B. W. and Carrington, E. (1975) 'Tourism as a vehicle for Caribbean economic development', paper presented at the Regional Seminar on Tourism and its Effects, Nassau, November.

Block, A. A. (1990) *Masters of Paradise: Organized Crime and the Internal Revenue Service in the Bahamas*, Transaction, New York and London.

Block, A. A. (1994) *Space, Time and Organized Crime*, Transaction Books, New Brunswick and London.

Boo, E. (1990) *Ecotourism: The Potentials and the Pitfalls*, World Wildlife Fund, Washington.

Briassoulis, H. and Van der Straaten, J. (1992) *Tourism and the Environment: Regional, Economic and Policy Issues*, Kluwer, Dordrecht, Boston and London.

Briguglio, L., Archer, B., Jafari, J. and Wall, G. (1996a) *Sustainable Tourism in Islands and Small States: Issues and Policies*, Volume 1, Cassell and Pinter, London.

Briguglio, L., Archer, B., Jafari, J. and Wall, G. (1996b) *Sustainable Tourism in Islands and Small States: Case Studies*, Volume 1I, Cassell and Pinter, London.

Britton, R. A. (1980) 'Shortcomings of third world tourism'. in Vogeler, I. and de Souza, A. (eds) *Dialectics of Third World Development*, Allanheld, Osmun, Montclair, NJ.

Britton, S. (1991) 'Tourism, capital and place: towards a critical geography of tourism', *Environment and Planning D: Society and Space*, **9**, 451–78.

Bugnicourt, J. (1977) 'The new colonialism: tourism with no return: part 1', *Development Forum*, **5**.

Butler, R. W. (1980) 'The concept of a tourist area cycle of evolution: implications for management of resources', *Canadian Geographer*, **24**, 1, 5–12.

Cambers, G. (1996) *1995 Annual Report: COSALC I: Beach and Coastal Stability in the Lesser Antilles*, UNESCO and University of Puerto Rico, Sea Grant College Program: Environment and Development in Coastal Regions and Small Islands.

Conway, D. (1983a) *Tourism and Caribbean Development*, Hanover, NH: University Field Staff International Report, **27**, p. 12.

Conway, D. (1993) 'The new tourism in the Caribbean: reappraising market segmentation', in Gayle, D. J. and Goodrich, J. N. (eds) *Tourism Marketing and Management in the Caribbean*, Routledge, London and New York, pp. 167–77.

Conway, D. and Jemilio, J. (1991) 'Tourism, air service provision and patterns of Caribbean airline offer', *Social and Economic Studies*, **40**, 2, 1–45.

Conway, D. and Lorah, P. (1995) 'Environmental protection policies in Caribbean small islands: some St Lucia examples', *Caribbean Geography*, **6**, 1, 16–27.

CTO (1998) *Caribbean Tourism Statistical Report, 1998*, Caribbean Tourism Organization, St Michael, Barbados.

CTO (2001) *Caribbean Tourism Statistical Report, 2001*, Caribbean Tourism Organization, St Michael, Barbados.

CTRC (1983) *Caribbean Tourism Statistical Report, 1983*, Caribbean Tourism Research and Development Centre, Christ Church, Barbados.

de Albuquerque, K. (1991), 'Conflicting claims on the Antiguan coastal resources: the case of Mckinnon's and Jolly Hill Salt Ponds', in Girvan, N. and Simmons, D. (eds), *Caribbean Ecology and Economics*, Caribbean Conservation Association, St Michael, Barbados, pp. 195–205.

de Albuquerque, K. and McElroy, J. (1992) 'Caribbean tourism styles and sustainable strategies', *Environmental Management*, **16**, 5, 619–32.

de Kadt, E. (1979) *Tourism: Passport to Development? Perspectives on the Social and Cultural Effects in Developing Countries*, Oxford University Press, New York.

Erisman, H. M. (1983) 'Tourism and cultural dependency in the West Indies', *Annals of Tourism Research*, **10**, 3, 337–61.

Featherstone, M. (1990) 'Perspectives on Consumer Culture', *Sociology*, **24**, 1, 5–22.

Gladfelter, E. H. and Ogden, J. C. (1994) 'The status of relevant scientific capabilities and knowledge of land-based marine pollution problems in the Caribbean', discussion paper, Regional control of land-based marine pollution: lessons from other regions of the wider Caribbean project, Woods Hole Oceanographic Institute Marine Policy Center, Woods Hole, Massachusetts, with the US Environmental Protection Agency, Office of International Affairs.

Griffith, I. L. and Munroe, T. (1995) 'Drugs and democracy in the Caribbean', *The Journal of Commonwealth and Comparative Politics*, **33**, 3, 357–76.

Harvey, D. (1985) *The Urbanization of Capital*, Vols I & II, Johns Hopkins University Press, Baltimore.

Harvey, D. (1996) 'Possible urban worlds', *Justice, Nature and the Geography of Difference*. Blackwell, London, pp. 403–38.

Heidenheimer, A. T. (1970) *Political Corruption*, Holt, Rinehart & Winston, New York.

Hiller, H. L. (1976) 'Escapism, penetration and response: industrial tourism and the Caribbean', *Caribbean Studies*, **16**, 2.

Hiller, H. L. (1979) 'Tourism: development or dependence?' in Millet, R. and Will, W. M. (eds) *The Restless Caribbean: Changing Patterns of International Relations*, Praeger, New York.

Holder, J. S. (1979) *Caribbean Tourism: Policies and Impacts*, Caribbean Tourism Research Centre, Christ Church, Barbados.

Howe, G. and Cobley, A. (2000) *The Caribbean AIDS Epidemic*, University of the West Indies Press, Mona, Jamaica.

Hudson, B. J. (1986) 'Landscape as resource for national development: a Caribbean view', *Geography*, **71**, 2, 116–21.

Hutt, M. B. (1980) *Windows to the Sea*, Caribbean Conference of Churches and Cedar Press, Bridgetown, Barbados.

INCSR (1996) *International Narcotics Control Strategy Report, 1996*, Department of State, Bureau for International Narcotics and Law Enforcement Affairs, Washington, March.

Inskeep, E. (1987) 'Environmental planning for tourism', *Annals of Tourism Research*, **14**, 1, 118–35.

IRF (1996) *Tourism and Coastal Resources Degradation in the Wider Caribbean*, Island Resources Foundation, St Thomas, USVI.

Issa, A. (1959) *A Survey of the Tourist Potential of the Eastern Caribbean: Observations Made During a Tour of the Area, 16 May–3 June, 1959*. Federal Government of the West Indies, Kingston, Jamaica.

Karch, C. A. and Dann, G. H. S. (1981) 'Close encounters of the third world', *Human Relations*, **34**, 4, 249–68.

Kempadoo, K. (1999) *Sun, Sex, and Gold: Tourism and Sex Work in the Caribbean*, Rowman & Littlefield, Lanham, Maryland.

Lefebvre, H. (1976) *The Survival of Capitalism: Reproduction and Relations of Production*, Allison and Busby, London.

Lorah, P. (1995) 'An unsustainable path: tourism's vulnerability to environmental decline in Antigua, *Caribbean Geography*, **6**, 1, 28–39.

Maingot, A. P. (1994) *The United States and the Caribbean: Challenges of an Asymmetrical Relationship*, Westview Press, Boulder, Colorado, and San Francisco.

Mather, S. and Todd, G. (1993) *Tourism in the Caribbean*, Economist Intelligence Unit, London Special Report, 455.

McAfee, K. (1991) *Storm Signals: Structural Adjustment and Development Alternatives in the Caribbean*, South End Press and Oxfam America, Boston.

McElroy, J. and de Albuquerque, K. (1991) 'Tourism styles and policy responses in the open economy-closed environment context', In Girvan, N. P. and Simmons, D. (eds) *Caribbean Ecology and Economics*, Caribbean Conservation Association, Barbados, pp. 143–65.

McElroy, J. and de Albuquerque, K. (1996) 'Sustainable alternatives to insular mass tourism: recent theory and practice', in Briguglio, L., Archer, B., Jafari, J. and Wall, G. (eds) *Sustainable Tourism in Islands and Small States: Issues and Policies*, Pinter, London and New York, pp. 47–60.

McElroy, J. and de Albuquerque, K. (1998) 'Tourism penetration index in small Caribbean islands', *Annals of Tourism Research*, **25**, 1, 145–68.

Nettleford, R. (1990) 'Heritage tourism and the myth of paradise', *Caribbean Review*, **16**, 3 & 4, 8–9.

O'Connell Davidson, J. (1996) 'Sex tourism in Cuba', *Race and Class*, **38**, 1, 39–48.

OECD (1980) *The Impact of Tourism on the Environment*, OECD, Paris.

Pattulo, P. (1996) *Last Resorts: The Cost of Tourism in the Caribbean*, Cassell and the Latin American Bureau, London.

POA/SIDS (1994) *Report of the Global Conference on the Sustainable Development of Small Island Developing States*, Bridgetown, Barbados, 25 April–6 May, United Nations, Sales No. 94.I.18, Chapter 1, Resolution 1, Annex II.

Poon, A. (1989) 'New approaches to tourism', *Caribbean Contact*, **March**, 8–9.

Poon, A. (1993) *Tourism, Technology and Competitive Strategies*, CAB International, Wallingford, Oxford.

Pugh, J. and Potter, R. B. (2001) 'The changing face of coastal zone management in Soufriere, St Lucia', *Geography*, **86**, 3, 247–60.

Richardson, B. C. (1992) *The Caribbean in the Wider World, 1492–1992*, Cambridge University Press, Cambridge.

Rojek, C. (1985) *Capitalism and Leisure Theory*, Tavistock, Andover, Hampshire.

Schwartz, R. (1997) *Pleasure Island: Tourism and Temptation in Cuba*, University of Nebraska Press, Lincoln and London.

Thomas, G. A. (1991) 'The gentrification of paradise: St. John's Antigua', *Urban Geography*, **12**, 5, 469–87.

Timms, B. (1999) 'Linkages between domestic agriculture and the hotel sector in St Lucia', unpublished masters thesis, Department of Geography, Indiana University Press, Bloomington.

Trist, C. (1999) 'Recreating ocean space: recreational consumption and representation of the Caribbean marine environment, *The Professional Geographer*, **51**, 3, 376–87.

Turner, D. and Ash, J. (1975) *The Golden Hordes: International Tourism and the Pleasure Periphery*, Constable, London.

UNEP (1982) 'Tourism', chapter 14 in *The World Environment, 1972–1982*, Tycooly International Pubs, Dublin.

Watts, D. (1997) *The West Indies: Patterns of Development, Culture and Environmental Change since 1492*, Cambridge University Press, Cambridge.

Weaver, D. B. (1988) 'The evolution of a "plantation" tourism landscape on the Caribbean island of Antigua', *Tijdschrift voor Economische en Sociale Geographie*, **79**, 5, 313–19.

Weaver, D. B. (1993) 'Model of urban tourism for small Caribbean islands', *Geographical Review*, **83**(2), pp. 134–40.

Whelen, T. (1991) *Nature Tourism: Managing for the Environment*, Island Press, Washington and Covelo, California.

Wilkinson, P. F. (1997) *Tourism Policy and Planning: Case Studies from the Commonwealth Caribbean*, Cognizant Communication Corporation, New York, Sydney and Tokyo.

World Resources Institute (1998) *Reefs at Risk: A Map-based Indicator of Threats to the World's Coral Reefs*, World Resources Institute, Washington.

Youngman, M. (1998) 'Tourism and the environment: sustaining the sector', *Caribbean Handbook, 1998/1999*, FT Caribbean (BVI), Tortola, British Virgin Islands.

Chapter 12

POLITICAL REALITIES IN THE CARIBBEAN

Introduction

The region's islands (and mainland enclaves) have prospered and declined through a period of some 500 years of externally dominated incorporation into a succession of metropolitan empires and domains. Early on, there were European 'encounters', which consolidated patterns of commercial wealth in some islands, plantation colonialism and sugar cane prosperity in many others, even buccaneering and smuggling in others. Colonial regimes imported and imposed European class values and systems on island societies, with local *creole* cultures suppressed or subordinated to lower-class status. Slavery thoroughly changed the demographic profiles of island populations, turning them from white-dominated to Afro-Caribbean majorities (see Chapter 2). However, under colonialism European class stratification upheld a racial continuum in its societal hierarchy, the usual socio-political structure being a white-to-black class stratification (see Chapter 5). From top to bottom, came the powerful 'white' elites, then 'brown', mixed-race middle classes and the black powerless masses as the majority underclass. It would take over three centuries, eventual decolonisation and several decades of societal transformation thereafter before this rigid racial/class hierarchy would be challenged and modified to any appreciable degree (see Chapter 5).

In the twentieth century, however, there has been a widely differing record of changes in political system interactions and evolutions in specific locales: island states, island clusters and Caribbean mainland enclave societies. Especially after the Second World War, the colonial- and metropolitan-propelled agendas of many Caribbean regimes came under challenge. Returning soldiers, socially conscious intellectuals and professionals, union activists and nationalists, compared the hardships of island homes with the relatively advanced states of European mother countries and found them wanting. This prompted social unrest and motivated political independence movements. Political mobilisation and popular anti-colonialist platforms swept many territories to political independence, though some remained tied to their 'mother country'.

Politically, strategically and economically, successive United States administrations in the twentieth century, often in partnership with vigorous US corporate private enterprise, gradually extended their hegemony throughout the region. Defining events in the subordination of British influence in the hemisphere and the dominance of US political and economic interests were the 1895 Venezuela/British Guiana border dispute over Essequibo province, and the Spanish–American war of 1899–1902. In the former confrontation, the USA invoked the 'Monroe Doctrine' and the British diplomatic retreat signalled its hegemonic replacement by the USA. To start the Spanish–American War the *USS Maine* was sabotaged (by CIA agents, it now appears), the US Congress declared war, the marines invaded Cuba, and at the cessation of hostilities the USA took possession of Cuba, Puerto Rico and the Philippines. Soon afterwards, the Commonwealth of Puerto Rico and the US Virgin Islands would become US colonies in the Caribbean.

From that time to the present, US administrations and the military establishment have viewed the Caribbean as a primary defence zone. Anti-communist, Cold War politics ruled geopolitical stances. Dictators could be supported by US administrations (and the occasional marine invasion, or two) if they declared themselves opposed to socialism and communism. The Panama Canal's strategic importance for US commerce and the region's importance for the transit of oil through its straits also added to the region's identity as a sea where the national security interest of the US must be protected. It was to be expected therefore, that after the Second World War and the subsequent decolonisation of Caribbean territories, British hegemonic influence and interest would decline. France and the French Caribbean, on the other hand, held on firmly (Burton and Reno 1995).

There were Caribbean challenges to this neighbourly aggression, however: one in particular being substantial. Cuba, under Fidel Castro's popular regime, and strongly supported by the USSR until its collapse in 1990/91, actively and successfully campaigned on behalf of its sovereignty and independence. Cuba remains defiant and independent to this day, and perhaps only the remnants of a Cold War mentality among US administrative officials, Miami's anti-Castro, Cuban-American institutions and Washington caucus keep the America–Cuba stand-off alive and vehement (Stubbs 1989; Azicri 2000; Calvo and Declercq 2000). Grenada's socialist Marxist 'New Jewel Revolution', on the other hand, lasted only a few years: from 1979 to 1983. President Johnson resorted to 'gunboat diplomacy' when he authorised the 1965 Marine invasion of the Dominican Republic to quell a popular socialist succession to governmental power in that island client state. First, President Clinton in the mid-1990s, then George W. Bush in 2004, repeated this tactic when political crisis after crisis followed President Aristides' unstable and capricious career as titular head of that troubled island. Of consequence in this decision to intervene was Haiti's uncomfortable proximity to the US mainland, not to mention the media's representation of Haitian boat people as a 'flood of refugees', and the perceived need for decisive action by the US executive branch. In short, the Caribbean region's geopolitical situation is clear, being under US surveillance and dominance, or as President Reagan personified it; 'our backyard Mediterranean Sea' (Conway 1983a, 1983b).

This is not to say that US geopolitical hegemony in the hemisphere has not undergone transformation in most recent times. Indeed, the world recession of the early 1980s, the massive indebtedness of Latin American countries (among them several Caribbean unfortunates – Jamaica, Haiti and the Dominican Republic), and the subsequent structural adjustment programmes (SAPs) of the world's international financial institutions – the World Bank and the International Monetary Fund – has brought the hemisphere (and the world) into today's neo-liberal era of ever-freer scenarios for global finance, commerce and production (see Chapters 5, 8 and 10). United States 'domestic' capital interests and political administrations are now becoming subordinate clients to the hegemony of international corporate capital and international financial capital. Today's forces of globalisation and neo-liberalism are ushering in an era of international 'fictitious capital', is one commentator's perception (Roberts 1994), 'arrogant capital' is another's condemnation (Phillips 1994). By the mid-1990s, the World Trade Organization (WTO) – completely dominated (and directed) by business interests of the G-8 cadre and the US–EU–Japan 'triad' – had become the chief global institution for negotiating, codifying and enforcing neo-liberal discipline in inter-state economic relationships and conflicts. A triumphalist ideology, some would label it religious faith (Cox 1999), was embraced with a fervour that is only now being questioned in the light of the 1998 Asian meltdown, the bursting of the 'dot-com' bubble in the world's stock markets and the recent Enron scandal in the United States (Conway and Heynen 2002). Globalisation is fostering, and is fostered by, neo-liberal modernisation and expansion in US commercial, financial and strategic relations with its hemisphere, as witness the Bush administration's promotion of FTAA in the early years of the twenty-first century, on the heels of the Clinton administration's promotion of NAFTA in the last decade of the twentieth century. Perhaps now more than ever, Caribbean political-economic relations are more often subordinate to wider hemisphere and global issues, and are treated as incidental, minor irritants in the greater capitalist schemes afoot (Bryan 1995; Girvan 1999).

This chapter details the diversity in political regimes, dwelling on the paths of post-colonial manoeuvring for 'independence' and assessing the political and economic consequences of the chosen routes, whether radical, conservative, revisionist or authoritarian. The series of attempts at regional associations and regional political institution building – from the ill-fated West Indies Federation to CARIFTA, CARICOM and most recently the Association of Caribbean States (ACS) – are also featured. Today, however, US hegemony embraces the whole region and the onslaught of global forces and neo-liberalism, which favour corporate and G-8 interests over those of the global periphery, threaten to fundamentally undermine such regional interest groupings. Caribbean micro-states have become even smaller geopolitical players in today's globalising, neo-liberal system (Thorndike 1988; Bakan *et al.* 1993; Aponte Garcia and Gautier Mayoral 1995; Sutton 1995; Dookeran 1996). Yet the people of the region continue to struggle to adjust to this new set of externally directed, dependent relations, as

Plate 12.1 Queen's Royal College, Port of Spain, Trinidad, 1991
Source: Dennis Conway

they have to earlier external regimes. Problems common to many less developed countries persist in the region: structural problems of high unemployment, especially among young men; unequal access to housing and land; gender inequalities; and the common dearth of livelihood opportunities in rural sectors, in the crowded shanty towns of the region's urbanised areas, and in the marginal situations of the region's poorer masses. There are no outright successes, but post-colonial progress is in evidence. There has been considerable advancement of social agendas and, despite their small size and geopolitical vulnerabilities, some Caribbean micro-states – like Barbados and Anguilla – have managed themselves remarkably well (Blackman 1991; Connell 1993). The situation is not hopeless, is not beyond redemption, and not beyond despair (Nettleford 1989).

The region's educated have always been at the forefront of opposition to the 'forces of imperialism', whatever their guise. The educational systems of many of the ex-colonial islands have remained their treasured asset, and these reservoirs of human and social capital certainly offer hope and potential for the future. Notably, Caribbean women have made considerable strides in their quest to participate in local politics, and this movement towards gender equity is to be lauded. Most significantly in today's globalising world, the islands' migration diasporas constitute a huge transnational resource reservoir of people, capital, information and material resources that the Caribbean can little afford to ignore. Many Caribbean political regimes have come to expect involvement of their overseas supporters in North America and elsewhere in the region, and dual nationality is a commonly accepted Caribbean practice. Remittance flows from overseas migrant communities are as valued as direct financial aid and assistance from such neighbours as the United States, Canada and Britain, in large part because they are both substantial and more reliable than the latter's overseas aid efforts (Payne 1998). Even the modest amounts of foreign aid and assistance by the richer countries of Europe and North America that previously found their way to the Caribbean have dwindled, and today's greater reliance on foreign direct investment has not helped the region's prospects. The Caribbean, it appears, will have to forge its own path, and find new political ways to make its concerns known and make itself heard more persuasively. More of this later, however. Let us start with the post-Second World War experiences of decolonisation and the emergence of post-colonial regimes.

Caribbean post-colonial regimes: many divergent political paths

The Caribbean has a colonial past that is clearly at the root of much of the political diversity in the region. The political institutions found in these colonial and post-colonial societies are extremely diverse. Westminster-style parliamentary democracies hold sway in Commonwealth Caribbean countries. Although most are politically independent, the vestiges of British administrative practices persist. The remaining British dependencies in the Caribbean follow colonial traditions of self-governance, with British Foreign Office and ministerial oversight dictating foreign and military affairs in such islands as Bermuda, the Cayman Islands, the Turks and Caicos Islands, and the British Virgin Islands.

The French Départements D'Outre Mer of Guadeloupe and Martinique have assemblies and all the metropolitan institutions expected of an overseas French department. Indeed, they are now part of the European Union, and benefit from EU subsidies and support programmes. The Netherlands Antilles islands of Aruba, Bonaire, Curaçao and St Maarten are self-governing colonies with the Dutch government and the EU being the major suppliers of aid and assistance to these micro-state colonies. Puerto Rico and the US Virgin Islands, as commonwealth dependencies of the United States, mirror their mother country in their congressional institutions. The Dominican Republic's neo-colonial experience with that neighbouring 'big brother' has also led to US-style government practices. The former are US colonies, however, and benefit from mainland/federal welfare programmes. The Dominican Republic and its people, on the other hand, have come to rely on immigration to the United States and the remittance flows that this movement has stimulated as much as formal aid and assistance from Washington, or US corporate investment.

The Cuban political regime, since Fidel Castro's takeover and prompt ostracism by the Kennedy administration of the United States, has followed socialist, Marxist-Leninist lines of central command and control. Consequently, Cuba has a 'popular democracy' form of government modelled after the Soviet Union, but given its own distinctive Caribbean island character through the nearly 40 years of its survival and evolution. Even after the collapse of the Soviet Union and the withdrawal of its financial and trade support, Castro's Cuba has managed to sustain its socialist regime through the Special Period of hardship to the present day, with only some relaxation of national authoritarian control in the economic sectors of tourism and trade. Haiti, on the other hand, has long suffered, both as an independent nation in the authoritarian grip of elite cabals and dictators and as a neo-colonial tributary of the United States. The latest two decades of political unrest have not led to any marked improvement in the lives of this island's people, who grow more desperate with the decline in civil authority and the growth of drug Mafia influence.

The 1960s: paths towards post-colonialism

In most of the Caribbean island territories, 300 years of colonial, plantation exploitation left enduring legacies of subordination and inferiority, so that the growth of the political independence movements in the post-Second World War era was a desperate struggle against considerable odds. The alliance of sugar and merchant capital to continue these classes' dominance of island economies was one ever-present structural force. The emergence of 'reconstituted peasantries', as a small farming class always at the margins of agrarian society where plantation labour still offered proletarian wages, was another. On the other hand, there was the growth in political awareness and activism among a proletarian working class as labour struggles in the estates, oil fields in the case of Trinidad and in the services fomented labour organisation (Boland 2001).

Urban life-styles gained sway and as more education opportunities were provided for colonial youth, middle-class intellectuals and radical commentators found their voices and dissent and activism became part of the political scene. Ideological enfranchisement prospered, in part helped by international socialist movements, and further cemented in the West Indian Regiment's experiences of two World Wars. Alliances of a professional, *creole* middle class with the masses brought on struggles for constitutional reform and a widened franchise, thereby fomenting another societal transformation of note. Some would argue this alliance undermined the prospects for real progressive societal transformation, however, while it successfully fought for, and won, political independence in several Commonwealth Caribbean countries – Jamaica, Trinidad and Tobago, Barbados, Guyana, Antigua and Barbuda (Sunshine 1985; Beckles 1990). The Haitian revolution of 1799 and 'independence' in 1804 was given its due as a significant historical benchmark, made more so by Caribbean intellectuals' writing and political activism in the 1950s and 1960s (for example, C. L. R. James (1963)). Another benchmark and defining political 'moment' was Cuba's revolution in 1959, the consolidation of Fidel Castro's regime, the ensuing escalation of Cold War rhetoric between Castro and successive US administrations, Cuba's insistent embracing of a Caribbean identity, and its outspoken activism on behalf of third world and non-aligned grievances against the United States administration's policies, aggressive tactics and hegemonic assumptions. As a Caribbean post-colonial independence movement, Cuba's socialism, its non-aligned political stance, its defiant opposition to US government hegemony, were to be constant reminders and prompts for socialist activists in the region.

The region's economic profile in the post-Second World War era inherited the structural limitations of the plantation past. The region's depressed state was an accumulation of decades of hardship, of declining fortunes in the plantations, of not so benign colonial neglect, and of the structural limitations of many an island's once profitable but now economically stagnating and too small plantation-mercantile economy. The Caribbean's plantation systems might seek to diversify their export staple crop, growing bananas or citrus fruits instead of sugar in the smaller volcanic islands, substituting coffee for cocoa in the larger ones, but such adjustments to agrarian production did little to bring back the profitable days of the previous centuries.

The various colonial office responses to the plight of their Caribbean outposts were similarly pessimistic. European mother countries perceived their West Indian small island territories as geographical problem areas, with overpopulation and small arable land areas making an untenable equation. They held Eurocentric environmental determinist views that suggested development in such tropical climates was virtually impossible and accompanied this with a racist view that denied industrialisation as an option for 'such people' (Blood 1958; Kuznets 1960). In the British West Indies, assessment of viable projects was to be guided by Colonial Office permission, and often such projects were merely adequate, scarcely ever forward-looking or expansive. For example, some of the smallest Windward Islands like St Vincent and Grenada were not deemed 'developable' and airport facilities with the most meagre and limited capabilities were finally built, often reluctantly. The same was argued with respect to upgrading port facilities and communications networks and developing health care facilities. Simply providing the most basic facilities with the most limited technological equipment was considered 'more than adequate' (O'Loughlin 1962). In the French West Indies, the departmentalisation of the islands in 1946 heralded the erosion of the traditional economic base (agriculture, fishing, craft industries) and its replacement by a service economy completely reliant on imported French manufactured goods. Indeed, more than a century of building self-reliance and autonomy was supplanted by a new dependence on French/European consumer goods and services, and a heavy reliance on French government financial support (Burton and Reno 1995).

The post-colonial path was difficult from the start, with the new political leaders suffering from Fanon's (1967, 1968) 'post-colonial mentalities' of inferiority and subordinate class consciousness. The nationalist model(s) of economic development these leaders adopted were more often than not outgrowths of the postwar colonial development strategies, constrained by the same set of limiting visions that they inherited from their colonial mother administrations, development advisers, foreign engineers, agricultural extension offices, teachers, doctors, health practitioners, legal experts and metropolitan commercial interests. Domestic capital stocks were also limited. Merchant capital had prevailed by the nineteenth century and allied with plantation capital these *comprador elite* were not at all interested in productive investments on behalf of the national interest (Ambursley and Cohen 1983). Commonly, therefore, political independence found popular, democratically elected political administrations responsible for physical and social infrastructure expenditures, while private commercial capital prospered on its own terms. As it turned out, it would take 30 years of 'independence' in some of the more successful islands – Barbados and Trinidad and Tobago, for example – for this divisive obstacle to be overcome, or at least made democratically feasible (Manning 1973; Beckles 1990; Ryan 1991).

There were, however, several aspects of this struggle for political independence that reflected progressive social change. The main pressures against the injustice of the old colonial regimes came from the popular masses – the workers, unionised and organised, peasantry, and the unemployed. Mass mobilisation legitimised the political climate as never before, a broader franchise was realised, there was more local autonomy, better pay and better working conditions, better prices for agricultural goods, and more security of living standards. Unfortunately, the very allegiance between the new professional class and the masses that had enlarged and enlivened the mass movements that posed a real challenge to the old elite structures, and to the colonial authorities, proved to be these movements' weak point. Time and time again, the new leaders of the independence movements, who were either drawn from the intermediate professional classes, or who betrayed their class roots once in power, chose to *depoliticise* the masses rather than further build upon their support and reverse the neo-colonial ordering of institutional power and authority (Thomas 1988; Beckles 1990).

The 1970s and 1980s: experimenting with alternative political models

Even by the beginning of the 1970s, the nationalist agendas for economic development and social progress were in conflict, with many of the early promises broken and progressive platforms under modification. Yet, there was a plethora of paths followed, spanning the ideological spectrum. Cuba's state-socialist, central-command, peoples' democratic model was the earliest radical alternative, and Fidel Castro's successful management of this island's development path situates Cuba's communist experiment as the most extreme variant. Also left-leaning was Michael Manley's leadership of the People's National Party (PNP) and its democratic socialist regime in Jamaica in the 1970s. On the other hand, Forbes Burnham's Co-operative Socialist Republic of Guyana, though espousing a people's socialist agenda, was more influenced by racial patrimony and bias than a consistent ideological path during the 1970s through to the mid-1980s. Grenada's 'Peaceful Revolution' brought the New Jewel Movement to power in 1979 with an agenda that also espoused socialist goals, but US opposition and internal dissent brought this relatively short-lived experiment to an abrupt and bloody end (Conway 1983a, 1983b).

Attempting a middle way, Trinidad and Tobago's Prime Minister Eric Williams and his People's National Movement (PNM) party embarked on a state capitalist development path, which had as its centrepiece the purchase of foreign-owned assets and industries. More centrist and pragmatic in their ideological orientations, Barbados' leading political parties, the Democractic Labour Party

Plate 12.2 Maurice Bishop released from house arrest, October 1983
Source: Dennis Conway

Plate 12.3 PRG armoured cars on their way to arrest Maurice Bishop, October 1983
Source: Dennis Conway

(DLP) and the Barbados Labour Party (BLP), and their leaders, Errol Barrow (DLP), Grantley Adams (BLP) and Tom Adams (BLP), chose to pursue a less adventurous and more managerial path of modernisation, tourism development and economic diversification. Indeed, this island's relative 'success' demonstrates lessons others might well follow. On the other hand, the post-colonial governments and colonial administrations of several other islands in the Windward and Leeward groups – Antigua and Barbuda, St Kitts and Nevis, St Lucia and St Vincent and the Grenadines (governments), and Anguilla, Montserrat and the British Virgin Islands (administrations) – chose *cautious continuity* over such drastic alternative paths. Few continued their plantation-dominated dependent state, with most embarking on tourism-led development that some have rather harshly characterised as a substitute form of 'plantation dependency' (Weaver 1988) (see Chapter 11 on tourism for a more in-depth treatment of this notion).

Finally, and ideologically driven by the neo-classical economic orthodoxy of the arguments of (St Lucian born and Nobel Laureate) Sir Arthur Lewis, Puerto Rico's republican government enthusiastically followed an industrialisation strategy that sought 'modernisation at all costs' by embracing externally directed capitalisation of the island's economy. This highly varied democratic spectrum was not the total picture of the Caribbean's political landscape, however. Haiti suffered under the 'Papa Doc' and 'Baby Doc' Duvaliers' authoritarian dictatorship, which mainly survived because of its allegiance to the United States in that country's anti-communist policy, and its domestic tactics of suppression of any kind of political opposition. The country's stagnation and ensuing degradation of its agrarian landscape was in large part a consequence of political myopia and neglect. The Dominican Republic had a republican assembly as its representative government, but time and time again, US external influence interfered in domestic affairs, anti-communism doctrines played highly significant roles in this government's decision making, and Gulf and Western's corporate interests were protected at the expense of the Dominican people.

The political regimes varied, and the genesis of each nation's or territory's development path was invariably due to a combination of internal and external factors, as well as regional influences. Indeed, the 'models' varied, but the lessons each offered were scarcely obvious in the regime failures, the policy disappointments and the, often unrealistic, goals. As it turned out, the 1970s was not an era where peripheral capitalist, or socialist, models of development were going to be given the time to mature, evolve and become refined or restructured according to each regime's longer-term objectives or social goals. International, macrostructural events interceded to such an extent that the countries of the Caribbean region faced an economic development 'crisis', which was in part a growing 'debt crisis', partially due to regime fallibilities, partially dependency-related, and partially an outcome of the region's continuing peripheral and marginal relations with metropoles. There was the unravelling of the Bretton Woods agreement initiated when President Nixon took the US dollar off the gold standard, and made more disorderly by the growth of eurodollar markets. Then there were the two OPEC oil price shocks in 1974–1975 and again in 1978, which posed such energy cost burdens to render many national accounts in default. Also, the ensuing heavy indebtedness heralded the intrusion of the IMF in the fiscal affairs of many Caribbean countries, Jamaica especially.

Cuba, isolated by United States administrative *fiat*, sought neo-colonial support from the Soviet Union to avoid its version of the 1970s crisis. Puerto Rico's industrialisation progress faltered, but its people's answer was migration and circulation to the US mainland, or greater reliance on the federally mandated social welfare net. The Dominican Republic's openness to multinational corporate involvement (Wiarda and Kryzanek 1992) did not spare this country from structural malaise and downturns of national performance, which also prompted the beginning of a large-scale exodus to America, or New York city, often by way of neighbouring Puerto Rico (Hendricks 1974; Morrison and Sinkin 1982).

Dependent territories, the French DOMs, British West Indian associated states, its small island colonies, and Netherlands Antilles groupings perhaps fared the best. The annual subsidies flowing from European exchequers, the metropolitan infrastructural investments, and the aid and assistance of overseas agencies might not have been excessive, but they did lend support to these Caribbean territories during this restructuring crisis. While the emigration avenue was not open for British West Indian colonials after 1962, migration and circulation to the French and Netherlands metropoles served to provide these Caribbean 'commonwealth citizens' with some overseas opportunities. For the newly independent nations of the remainder of the Caribbean – countries like Jamaica, Trinidad and Tobago, Guyana, Barbados, and Antigua and Barbuda – their nationalist experiments foundered, or stumbled, by the beginning of the 1980s.

Alternatives across the Caribbean political spectrum

In the following section, the wide spectrum of political regimes and their development paths during the 1970s and 1980s are given more regional specificity. First, radical paths are presented, followed by more conventional paths, and finally colonial and neo-colonial paths are covered.

Radical 'socialist' paths

Guyana's cooperative socialism

One alternative path was Guyana's 'cooperative socialism', which came into being in 1970, four years after British Guiana gained political independence. It was promoted by Forbes Burnham's People's National Congress (PNC) government on a wave of militant, anti-colonial popular support. Complaints over Reynolds Alumina, Alcoa and Alcan ownership of the bauxite industry also figured in the conflict. Long-held traditions of trade unionism and anti-colonial fervour were common among both the Indo- and Afro-Guyanese sub-populations, so that the PNC adopted a political platform that espoused popular socialist rhetoric while the party and clients pursued the more self-serving agenda of legitimising the state's (i.e. their) ascendance to become a national bourgeois class (Thomas 1983). The principles of cooperative socialism were quite explicit, if impractical to implement. There was to be a tri-sector national economy under private, state and cooperative control, with the cooperative sector eventually becoming the dominant sector. There was to be an expansion of state ownership through the nationalisation of foreign assets. There was a declaration that from henceforth a strategy of national development would be embraced and a programme to 'feed, clothe and house' the nation would be implemented. Finally, the PNC as the nation's ruling party was to be the vanguard party with ultimate authority, even over the state. Unfortunately, political propaganda and rhetoric were more in evidence than the actual delivery of promises in the PNC's national development programme to deliver basic goods. It was not too long before the Burnham regime's record of corruption, nepotism and patronage excesses, and its gratuitous diversions of state revenues, became legendary in the region, if not the butt of region cartoonists in regional newspapers and cause for widespread disrespect among the region's political leaders.

There were 'grassroots' alternatives, as people and communities struggled with day-to-day survival. The rapid growth of the informal sector, locally operated as a 'parallel economy', was one response. It was common for the remote rural communities that were not PNC-patronage recipients to revert to subsistence practices, plus petty commodity production, so dealing in the parallel economy served their purposes. Migration across the border to seek jobs in Suriname was another strategy, but it was not until the late 1980s that international emigration to the United States began in earnest. Prompting this exodus was the growing need for 'hard currency' (especially US dollars) to trade in the parallel economy that had flourished. Through the decade of the 1980s, Guyana continued to stagnate and regress. The PNC continued its authoritarian practices, effectively disenfranchising the masses, and the IMF-structural adjustment programme 'conditionalities' did not alter the situation for the better.

Quite abruptly, Forbes Burnham's death during a surgical operation in Havana in 1985 signified the end of Guyana's 'shame'. Desmond Hoyte continued to lead a PNC government but his was a more conciliatory position. Eventually, 1992 saw the end of nearly 28 years of PNC rule. The opposition party, the People's Progressive Party (PPP), was voted into office and the much-awaited return of normalcy, democracy, civility and national self-respect for Guyanese, both at home and abroad, arrived. The ten years of relative 'normalcy' that followed was, tragically, disrupted by racial politics and a social divisiveness that has Guyanese in two political camps – the Indo-Guyanese and Afro-Guyanese. This social schism has not helped return the country to its former (1960s) self-reliant and promising beginning. Rather, the country is politically divided, racial politics are

Box 12.1: The Caribbean's charismatic leaders

A widely held image of Caribbean people sees them as flamboyant, effervescent and self-assured. Many being widely travelled, thoroughly modern and well educated, they are often characterised as 'a highly sophisticated political people' (Lewis 1968: 26). It is generally acknowledged that the region has produced world leaders in many diverse fields: Nobel Prize winners in economics and literature, reggae maestros, calypso kings, steel band extravaganzas, and world-class athletes, cricketers and baseball players. In the realm of politics, regional and international, the Caribbean has yielded statesmen of renowned calibre and political finesse, many known as much for their charisma, political astuteness and charm as their populism and popularity. Specifically, it was during the heady days of decolonialism and the mobilisation of the people against the colonial '*massa*' that a succession of charismatic and popular political leaders leapt to the fore:

- Fidel Castro, of course, was one of these populist leaders, with his leadership stretching beyond 40 years, through to today. Castro's charismatic authority is based on his personal qualities, as well as those attributed to him by his loyal public, and despite many changes in the country's situation, as well as in the maintenance of a balance between institutional power and personal power, Fidel Castro's regime remains as strong and popular as ever.
- Eric Williams, an academic historian by training, led his Trinidadians through a successful anti-colonial struggle, and built a party around his personality and authority, which stayed in power for 28 years. He moved ideologically from left-leaning populism to centrist positions and eventually to relatively conservative politics in these 20 or more years, and attempted to chart a 'third way' for independent Trinidad and Tobago.
- Michael Manley was the son of a former prime minister of Jamaica – Norman Manley – and he was unquestionably one of the two most charismatic political leaders to emerge, the other being Maurice Bishop. Manley wrote books, charmed audiences, absolutely flummoxed interrogating journalists, and generally charmed and won over everyone he met.
- Maurice Bishop was one of the New Jewel Movement leaders (the other being Bernard Coard) who were thrust into leadership roles in 1969, when their 'Peaceful Revo' overthrew the corrupt, but nevertheless constitutionally legitimate government of Eric Gairy in Grenada. Maurice charmed everyone, he enjoyed adulation, and won over the people with his charisma and good looks. Eric Gairy before him had also enjoyed his place as a charismatic leader and spokesman for the poor man, but his alter ego, and superstitious side, as well as his intolerance of criticism, eventually led to his overthrow.

(Box continued)

- Errol Barrow may not have been a charismatic leader in the mould of a Manley or Bishop, but he certainly promoted populism and social democracy as the foundation of his Democratic Labour Party's base, and his polemic debates with Sir Grantley Adams, the leader of the Barbados Labour Party, situated Errol among these other more charismatic luminaries. Though espousing populism, Barrow's pragmatism prevailed, and the political course Barbados took was never overly volatile, or ideologically shrill.
- Forbes Burnham, and his long-time political foe in British Guiana and then independent Guyana, Cheddi Jagan, were both charismatic leaders of their respective ethnic/racial groups; the former a champion of the Afro-Guyanese, the latter leader and spokesperson for the Indo-Guyanese. Both were consummate speech-makers, both gained the loyalty of their people, but Guyana in the end suffered from the ethnic divisions they created and fostered, and a promising independent Caribbean country was torn apart, run down and underdeveloped as they fought and postured on their political platforms.

Source: Alalhar, (2001)

at their sharpest and most destructive, and the ongoing 'hard times' of the Guyanese people has only encouraged more emigrants to flee, postponed returnee plans, and made rebuilding of the country that Forbes Burnham ruined more difficult a task.

Jamaica's democratic socialism

Another alternative path was Jamaica's 'democratic socialism' devised by Prime Minister Michael Manley as his political party, the People's National Party, embraced socialist teachings. Manley's (1974) book *The Politics of Change* served as the blueprint for social transformation. In it, he advanced some of C. L. R. James' (1969) ideas for participatory democracy. He believed in the necessity for a 'politics of popular participation' and generally adopted Nyerere's (1969) critical position towards colonialism, capitalism and imperialism, i.e. attitudes and psychological barriers need to be overcome, and self-confidence and national pride instilled before there will be success in productive enterprise, in local independence, or in progressive development. Like the Guyanese premier, Michael Manley took on the foreign-owned bauxite corporations – Alcoa and Alcan – although he did not go so far as to attempt the outright nationalisation of their assets. Instead, Manley's Jamaican government introduced a production levy set at 7.5 per cent of the selling price of alumina, replacing the previous method that strongly favoured these multinational mining corporations.

The bauxite levy controversy, international exposure and acclaim, and the Jamaican leader's eloquence, thrust Michael Manley onto the world stage as a spokesperson for the 'New International Economic Order' in the Non-Aligned Movement, Socialist International, and on behalf of the third world in general. Cuba's Fidel Castro and Michael Manley both made challenging speeches on behalf of the Caribbean and Latin American oppressed peoples, but the PNC's willingness to entertain closer political and economic ties with Cuba evoked suspicion and hostility from the United States' State Department. A series of destabilisation efforts were put into effect, including CIA activities, media pressure on tourist visiting and loan denials (Thomas 1988). Although there were economic repercussions, and an escalation in social tensions, the Jamaican people's responses in the 1976 elections were predictably partisan and nationalistic. Michael Manley's PNP was returned to power with a landslide vote, with the charismatic leader going so far as securing Rastafarian support, but 1977 was to see the premier's demise.

After confronting the IMF's draconian conditionality package with his 'we are not for sale' declaration, within three months Prime Minister Michael Manley had to about-face on his polemic stance and accept an IMF credit line of US$74 million. This political humbling also meant Jamaica had to abide by the 'humiliating' conditions of the IMF's structural adjustment programme (SAP): begin a programme of anti-labour measures; hold wages constant; agree to a 40 per cent devaluation of the Jamaican dollar; and undertake a series of cutbacks in public spending (Latin American Bureau 1983). For the next two years, Michael Manley and the PNP abandoned their populist agenda, attempted to regain the political high ground via pragmatism and generally lost credibility among the electorate. It was no surprise that Edward Seaga and the opposition Jamaican Labour Party was swept into power on an electoral landslide of their own in the 1980 elections (Ambursley 1983).

Prime Minister Seaga took great pains to cultivate a friendship with the new Republican President of the United States, Ronald Reagan, and offered Jamaica as a worthy partner and recipient of development assistance under the soon-to-be-launched Caribbean Basin Initiative. Since then, successive governments have struggled under IMF conditionality strictures. Despite programmes to foster higher levels of foreign direct investment, and other neo-liberal plans and policies, the island has failed to become Seaga and Reagan's open-economy, export-oriented 'model of success', and Jamaica remains the troubled island it has been since political independence (Klak 1995).

Grenada's New Jewel Movement – 'Revo'

A third alternative path was attempted by the New Jewel Movement in Grenada, whose accession to power in 1979 in a bloodless coup started their 'Peaceful Revolution' and gave this small (45-person-strong) group of intellectuals and radical leaders an opportunity to chart a 'non-capitalist path' towards development

(Jacobs and Jacobs 1980). Grenada's People's Revolutionary Government (PRG) promoted a socialist programme that was optimistic as well as idealistic. Several objectives were framed to thoroughly redevelop the island's economy: (1) Point Salins international airport was to be constructed to handle wide-bodied jets as the much-needed infrastructure for a restructured, locally owned tourism industry; (2) growth of a mixed economy with three major institutional bases – state, cooperative and private – was to be encouraged, with the state playing the leading role; (3) the quality of life of citizens was to be improved through a comprehensive programme aimed at upgrading social services and ensuring basic needs were met; and (4) overseas trade and the portfolio of foreign aid and assistance were to be diversified, with the PRG particularly courting assistance and linkages with COMECON countries and Cuba, and improving South–South cooperation (Thomas 1988).

Bernard Coard as Minister of Finance certainly managed to diversify the PRG's assistance portfolio, and Maurice Bishop, the Prime Minister, befriended Fidel Castro and Cuban aid and assistance was very much in evidence in revolutionary Grenada. Unfortunately, the international acclaim that Bishop garnered championing the anti-imperialist cause on behalf of the Non-Aligned Movement was often made at the expense of the Reagan administration. Bishop's rhetoric, like Michael Manley's, was answered by US State Department reaction and displeasure. Bishop's principled stance(s) might have been championed by the US Congressional Black Caucus, but the Republican US administration was anything but amused. In the end, the 'Revo' lasted only four years before internal strife within the ruling People's Revolutionary Government – notably a military coup and assassination of Maurice Bishop and other followers – provided an opportunity for the US military and President Reagan's administration to coordinate the invasion and occupation of the Windward 'Spice Isle' in mid-October 1983 (for details of this 'gunboat action' see Conway 1983a, 1983b). After a few months of military occupation, Grenada was gradually readmitted back into the fold, its airport was finished, its tourist facilities opened to foreign finance, and its national economy was to be open, export-oriented and foreign-capital dominated. A conservative, middle-of-the-road path was to be followed, and the nationalist/socialist fervour of the 'Peaceful Revo' was suppressed, shelved or left behind by the majority in the island.

A middle way: radical rhetoric allied to conservative economics

Trinidad and Tobago's state capitalism

There were political paths that embraced neo-classical economic views for national planning and industrialisation, a favoured model coming from one of the region's favourite sons and Nobel Laureate, St Lucia-born Sir Arthur Lewis. Puerto Rican development was to take its great step forward with 'Operation Bootstrap', with the island's republican government's unabashed encouragement of footloose industries to its shores. 'Industry-by-Invitation' was the more derogatory label used by the radical economists at the University of the West Indies, the new world group, to personify this encouragement of international corporate penetration. Puerto Rico provided a regional showcase for this model, yet its experience was not one to emulate (Ruth *et al.* 1994). One Commonwealth Caribbean island that attempted a variant of this industrialisation model was Trinidad and Tobago. Trinidad was blessed with oil and gas resources, both on- and offshore, or so the politicians wished it and duly spent thoughtlessly. Its alternative path was labelled 'state capitalism', or as their Prime Minister Eric Williams called it, a 'middle path for development' (Sandoval 1983; Conway 1984).

Using OPEC-derived windfall gains from the island's oil and gas exports, from 1975 onwards, Prime Minister Eric Williams' People's National Movement (PNM) party embarked on a comprehensive modernisation scheme to buy outright most foreign-owned assets and fully industrialise the Trinidadian economy. Five main thrusts were at the fulcrum of this state capitalist model: (1) the utilisation of oil revenues to create large-scale resource-intensive export industries, especially energy-intensive export industries such as fertilisers and iron and steel production; (2) the negotiation of state and transnational corporation joint ventures; (3) the accompanying development of import-substituting manufacturing and assembly plant operations with export and domestic market capabilities; (4) the use of the oil boom surpluses to upgrade basic infrastructure; and (5) the disbursement of government revenues into welfare funds, plus massive, yearly transfers and subsidies to bolster state-owned industries that were not profitable, though they had large patronage-fuelled payrolls. 'Capitalism Gone Mad' was the Mighty Sparrows' satirical calypso that perceptively summed up the situation. By the end of the oil bonanza in 1983, Trinidad's economic engine had

Plate 12.4 The 'Red House' of Parliament, Woodford Square, Port of Spain, Trinidad and Tobago, 1991: site of the Muslimeen armed uprising
Source: Dennis Conway

overheated, the glut of revenues had not realised a much-needed diversification, and the exchequer was once again reliant on foreign indebtedness (Conway 1984; Thomas 1988; Ryan 1991; St Cyr 1993).

Restructuring, currency devaluation, increased impoverishment and long-felt hardship among the ranks of the unemployed, the poor and the less fortunate was the downside of Trinidad's experiences through the 1980s. The country's elite did not suffer tremendously, however. There were eventual changes of government after such a long reign by the PNP, but there was little change in development strategy or ideology. Rather, racial politics dominated the swings of regime (Ryan 1988, 1991). There was an armed coup in 1990 by a Muslimeen army under Imam Abu Bakr, which attempted – unsuccessfully as it turned out – to unseat the reigning party in power, the NAR under its Tobagonian Prime Minister A. N. R. Robinson (*Trinidad Express* 1990). This disrupted but did not dramatically change politics as usual in Trinidad. Successive governments attempted to govern and set policies and domestic agendas, but the private sector was given a relatively free hand, and private (domestic and foreign) capital determined investment priorities and infrastructural developments.

Corporate activities in Trinidad rebounded eventually, and with the merger of Jamaica's Geddes Grant Corporation with Trinidad's Neal and Massey Group in 1992, the region gained its first major multinational corporation (Gayle 1995). Other joint-venture projects with oil and gas multinationals were brought on line, and Trinidad's resource extracting sector rebounded to bring a new wave of external capital into the island. The Trinidad and Tobago government also attempted to develop a tourism industry, promoting Tobago as an upmarket resort for the European wealthy, though whether such economic growth ambitions and plans will bring about the sought-for economic diversification of the island's economy is yet to be seen (St Cyr 1993). Some have cautioned that Trinidad's multiethnic diversity, and the embedded political economic conflicts such cultural divisiveness causes, will inevitably limit the country's development potential (Deosoran 1996).

Pragmatic conservatism

Barbados' pragmatic conservatism

On the other hand, the political path taken by Barbados after gaining political independence in 1966 might best be characterised as 'conservative *and* pragmatic'. Five dimensions characterised the island's progress, which some suggest may be the region's success story (Blackman 1982, 1991; Worrell 1987). First, the two major parties, the Barbados Labour Party (BLP) and the Democratic Labour Party (DLP), have both ruled as residing governments in this parliamentary democracy, exchanging positions as 'opposition' and 'ruling party' in accordance with the popular vote. Both hold to parliamentary codes of civility, both view the promotion of political and social stability as a vital necessity, and both agree on high standards of democratic tolerance. Both are committed to maintaining law and order, respect for the judiciary and the police force, and both hold to traditions of 'social democratic ideals', built upon British and European Fabianism and models of the welfare state. Second, the economic policies successive governments have pursued since independence are non-ideological – being technocratic, pragmatic and rational. Barbadian economic policy making has never been framed in confrontational posturing, and its conservatism has never been flaunted, since the essence of the island's low-key approach has been cultivated as the most pragmatic approach for its governments to practice.

Third, every government, BLP or DLP, whether led by Sir Grantley Adams, Tom Adams, Erskine Sandiford, Errol Barrow or Owen Arthur, has, by and large, accepted the colonial mandate and more nationalist strategies, with their outward orientation and their heavy reliance on foreign capital investment. Each government's goal was to seek to make the programs they inherited more effective, so that continuity was maintained even as regimes changed. Three major developments can be identified in this respect: (1) greater stress being placed on the importance of management for successful economic development and fiscal planning; (2) providing for the systematic development of institutions that will improve labour skills at all levels within the country; and (3) the deliberate cultivation of an outlook and a management style that constantly tries to anticipate global economic changes, particularly in the North American and European markets that the country might exploit in the tourism, export-processing and offshore financial services sectors. Fourth, there were measures introduced to facilitate the inflow of foreign capital, and both political parties have maintained close working relationships with the Central Bank. Additionally, as part of this open-economy policy, successive governments have encouraged local industry to be export-oriented, particularly within the regional CARICOM market. Fifth, all governments have retained ongoing programmes which were designed to upgrade the island's physical infrastructure and improve basic services. All governments have been committed to a common purpose to invest in education, health and welfare services, to respect and defend the island's national heritage, and to undertake the responsibility to use the most applicable forms of institutional control and comprehensive planned use of the island's physical landscape, its coastal zone environment, and its human and natural resources – to wit, the island's quality of life (Blackman 1982; Thomas 1988).

Small island pragmatism and external dependency

Like Barbados, Antigua embraced mass tourism as its development path, deliberately turning away from plantation sugar production. Tourism came to dominate the island's economy, and hotel growth along its many coastal inlets and bays continued apace, but the successive Bird regimes rarely paid much attention to environmental degradation, mangrove depletion and the coastal zone's ecological health (Lorah 1985). Other small islands, like St Lucia, St Vincent and the Grenadines, Dominica, St Kitts and Nevis, on the other hand, hung on to their bilateral trade agreements, which favoured their plantation exports – mainly bananas. They also attempted to enlarge their tourist sector and attract offshore export-oriented assembly manufacturing. As another alternative, some of the smallest British colonies, such as the Cayman Islands, Turks and Caicos, and Bermuda, looked to 'offshore' international finance for their salvation (Thorndike 1988; Connell 1993). For all Caribbean islands, tourism provided their national accounts with some revenues, while the performances in other sectors did not improve. Despite most diversification experiments, and despite external and regional development programmes that tried their utmost to bring about productive changes in small farming, innovations in agricultural enterprises, and introduce forest farming, aquaculture and non-traditional export crops as alternatives, many of the small islands' external earnings remained highly dependent on export staple crops like bananas, sugar, coffee and citrus. For these small islands, the 1980s were to merge with the early 1990s as a period of stagnation, or of little expansion, without any promising signs that their development paths would be easier to chart (Worrell 1987).

Neo-colonial regimes

Although most of the Caribbean island nations had sought political independence, colonial ties and neo-colonialism persisted. The neo-colonial situations are only summarised here, since they present comparable models to the other alternatives examined earlier, rather than models to be advocated. On the other hand, McElroy and de Albuquerque (1996) have recently presented an interesting and somewhat suggestive case, that colonial dependence has been advantageous for some micro-states, especially those seeking to attract offshore financial operations: for example, the Cayman Islands, Turks and Caicos, and Bermuda. For other Caribbean islands, like Puerto Rico, the Dominican Republic and Cuba, the disappointing development experiences under neo-colonial dominance are all too apparent (Sunshine 1985). Again, in the interests of brevity, treatment of every Caribbean territory experiencing neo-colonialism is not attempted. For example, the political and economic turbulence on the island of Hispaniola and its two nation-states, the Dominican Republic and Haiti, is not given comparative coverage, although Haiti's extremes are (also see Fass 1990; Lundahl 1979, 1983, 1992 for more in-depth treatment of Haiti's dilemma).

Puerto Rico's modernisation and 'industry-by-invitation' regime

Puerto Rico's 'Operation Bootstrap' started in 1947 when US industrial corporations were invited to open their factories on the island, with a package of incentives to subsidise their investments, including tax-free repatriation of profits, freedom from US income taxes, and promises of low-wage, non-union labour. In its earlier years, the industrialisation experience was claimed to be an 'economic miracle', with average annual growth rates of GDP of 6 per cent in the 1950s, 5 per cent in the 1960s and 4 per cent in the 1970s. Outside foreign investment, mostly from US sources, increased from $1.4 billion in 1970 to $24 billion in 1979. Puerto Rico also had the highest per capita level of imports from the USA and 34 per cent of total US foreign direct investment. After an earlier enticement of labour-intensive light industrial enterprises, like shoes, glassware and clothing (many leaving after their 15 years tax-shelter time was up), petroleum refining and petrochemical production came to predominate. After 1973, the refining sector slumped, but pharmaceuticals stayed on, exacting a severe price in terms of pollution and offshore environmental degradation (Meyn 1996). In addition, such rapid industrialisation came with its geopolitical price: namely greater dependency on foreign capital and greater dependency on US subsidies and welfare assistance. Paralleling the successive waves of invited industry have been excesses in nuclear power plant siting, with toxic spillage and the poisoning of land and water eventually igniting popular local resistance (Berman Santana 1996). Mass circulation between island and mainland has been the accompanying human response to this rapid and destructive pattern of industrial transformation of the Puerto Rican economy, and the structural unemployment that still haunts the island is a constant reminder of Operation Bootstrap's myopic vision (Sunshine 1985).

French Départements D'Outre Mer: modernisation and metropolitan dependency

Modernisation of the infrastructure and of the leading sectors of the island's economic bases of the French Départements D'Outre Mer (DOMs) – Martinique and Guadeloupe – benefited tremendously from metropolitan subsidies and from preferential access of their tropical produce to the European market. These dependencies, however, maintained metropolitan levels of consumerism and the import bills could only be matched by transfers from (French) central government coffers. European tourism and Euro-port, maritime transportation industries are now flourishing developments in Guadeloupe, while plantation agriculture in Martinique has been somewhat successful in developing its export capabilities of rum and liqueurs. With national deputies serving their overseas departments in Paris, island politics is directed by the mayors and councils of the island's capital cities, so it is scarcely surprising that the infrastructure of these administrative capitals is well developed and well financed. The urbanised cores of the islands are highly developed, while the rural hinterlands have become depopulated beyond the beachfront regions that foster tourist development. Even more European than before, these DOMs are extensions of Europe in the Caribbean, but now the social class tensions that separate islanders and Europeans and occasionally lead to outbreaks of racial violence have replaced the earlier colonial class divisions of *bekés* and ex-plantation workers. French and European ideas and identities compete with island national identities of *negritude* and appear to be overpowering them (Burton and Reno 1995).

The Dominican Republic's US corporate dependency

The Dominican Republic, however, experienced Gulf & Western's dominance in its sugar industry, without the longer-term returns that would translate into domestic sector growth and expansion. When Gulf & Western abruptly pulled out in 1978, and sold its considerable holdings, this perhaps symbolically signalled that the Balaguer government's attempts to replicate its own version of Brazil's economic miracle had foundered: export revenues were in decline, debt obligations were increasing, and civil unrest mounting (Wiarda and Kryzanek 1992). To counter this, and to take advantage of US proximity and the globalisation of the region, enclave tourism has been embraced here as an alternative to failed industrialisation strategies. Export-processing zones have been built to entice producer services, data processing and industrial assembly development, and domestic agricultural policies have been implemented to serve the growing tourist industry. Perhaps most important in the restructuring of the Dominican Republic economy in recent times has been the growth of linkages between the emigrant communities in New York city and family and communities on the island. Remittance flows are now substantial, a sizeable (and growing) proportion of these repatriated funds are finding their way into small business ventures, and the commercial ties between the island and the mainland are now of value for both political and socio-cultural exchanges. Dominican political hopefuls regularly campaign among these US-based communities, overseas Dominicans vote at home, and the bi-national relations are deeply embedded and transnational in character.

Haiti's problematic transitional regimes: from dictators to presidents?

Haiti, long exhausted by the exploitative and autocratic regimes of the Duvaliers and their elite clients, found US aid fomenting and entrenching greater and greater dependency than the political economic situation would have reflected (Fass 1990). Migration, and environmental degradation of arable land resources, forest cover and river basin systems, as well as widespread corruption and authoritarian inhumanity, characterised this unfortunate people's lives and struggles for existence. By the early 1980s, appreciable numbers of Haitian 'illegal immigrants' had begun making their way across the Caribbean to the Bahamas and the US mainland. The US media personified this 'flood of refugees' as a consequence of the prevalent corruption and inequities in Haiti, and Jean-Claude Duvalier's luxurious excesses were cause for self-righteous criticism. Anti-regime conspiracies were hatched among Haitian army officers and widespread outbreaks of popular protest occurred. Duvalier, sensing the inevitable, resigned in 1986 and departed Haiti for a life of luxurious exile in his Mediterranean villa (Kretchick *et al.* 2003).

Duvalier's departure hardly solved Haiti's political crisis, however. Jean-Claude gave way to a military *junta* led by Lieutenant General Henri Namphy. Then, to create a semblance of legitimacy, the *junta* orchestrated the election of an émigré, Professor Leslie Manigat, who lasted only five months in the presidency before Namphy claimed the office for himself in June 1988. Namphy, in turn, lasted only three months before Prosper Avril replaced him. Avril served over a year before yielding to an interim presidency, which was followed in 1990 by the internationally supervised democratic election of President Jean-Bertrand Aristide. Aristide had been a relatively obscure priest at St Jean Bosco church in the impoverished community of La Saline, but emerged as a national figure in 1986 because of his outspoken criticism of the Duvalier regime. Aristide's election, while reflective of popular support for a charismatic priest, scarcely signified a change in Haiti's political culture. As an outspoken advocate for the poor and the powerless, Aristide used Catholic and Voodoo theology to promote his/their views on the severe societal inequities that dominated island politics. As president, Aristide lacked political experience, possessing neither the tact nor pragmatism to bring about a social consensus. Indeed, his inflammatory rhetoric had quite the opposite effect, playing to the masses but alienating the middle and upper classes.

With Aristide ousted by a military coup on 30 September 1991, the new crisis now more directly involved the United States. American disquiet, again fuelled in large part by the well-publicised flotilla of 'boat people' bound for Florida, put Haitian politics back into the international spotlight. The US intervened militarily, operating under a UN mandate to re-establish social order and prevent the island from slipping into anarchy. Aristide spent three years in exile in the United States, but then was returned to power in 1994 under US auspices. Unfortunately, a deepening cycle of corruption, and significantly, the withdrawal of international/US aid in 1997, saw the increasing penetration of drug Mafias with drug trafficking operating quite openly in this impoverished and desperate nation, to the detriment of civil society and stability. A flawed election, the blatant corrupt practices of the Aristide administration, an overall lack of progress and a general decline in living standards for all but those Aristide supporters and clients eventually prompted another internal crisis and civil war in 2004. Like Baby Doc before him, Aristide was 'encouraged' to go into exile, and the US administration of George W. Bush has once again been left with the island's future in its hands. Emigration has been one answer for many. For those staying behind, day-to-day survival is the order of the day. Political authority has rarely represented the rural majority, whether it was meted out by a self-declared dictator, occupying army administration or elected official. Today, the continued cycle of corruption and, significantly, the withdrawal of international/US aid in 1997, has seen the increasing penetration of drug Mafias with drug trafficking operating quite openly in this impoverished and desperate nation, to the detriment of civil society and stability.

Cuban dependency on the Soviet Union

Cuba's growing dependence upon trade and commerce, and financial support from the Soviet Union and Eastern Europe's COMECON countries, also reflected the country's plight under a neo-colonial regime. Yet, in distinct contrast to the experience of several of the incorporated Caribbean peripheral states (the exceptions being the French DOMs and Netherlands Antilles), Cuban social transformations were considerable and deep, social welfare services were equitably administered, and *microbrigade* housing performances were impressive. Despite the entrenchment of Castro's central, command-control style of government and the growth in power and privilege of the Communist Party on the island, the Cuban masses enjoyed a comprehensive education and welfare safety net that was considerably better than their counterparts in the region's other neo-colonial territories, particularly Haiti and the Dominican Republic (Mathey 1997). This was to change, however, with the demise of the Soviet Union, the withdrawal of its essential financial and commercial support and the onset of the 'Special Period' of hardship and increased isolation (enforced by US administration embargos) starting in 1990 (Plate 12.5).

To counter the embargo and the loss of Soviet external support, from 1990 onwards the Cuban government embarked on a plan to deal with these difficulties in a relatively short time period. After the first years of severe shortages and rationing during the Special Period – 1990 to 1994 – the island's food production programme was directed towards supplying domestic needs and environmentally friendly methods of crop treatment; herbicide and pest control were experimentally adopted. Tourism and export industries were encouraged to bring much-needed hard currency to the island, and improved connections with Spain, Europe and Latin America helped rebuild the tourist industry to its current healthy state. Research was accelerated in export-oriented industries that promised new commercial possibilities, such as biotechnology, pharmaceuticals and medical equipment, and steps were taken to save energy and/or develop new local sources of energy. Despite the US-imposed embargo, the continual flow of remittances from émigré Cuban-Americans brought much-needed financial support to many Cuban families, helping them to meet their basic needs. Tourism, remittances and related commercial micro-enterprises have 'dollarised' the local economy in the largest cities – Havana, Santiago de Cuba – causing class divisions and distinctions to re-emerge. At the same time, the growth of Cuba's tourism industry, and its 'success' in helping the Cuban government weather the Special Period, not only ameliorated the harsh conditions that food shortages, petroleum shortages and low-wage regimes had brought, it also helped save the Revolution, and sustained Castro's socialist government. Of course, a not inconsequential force contributing to the regime's 40-plus years of authority has been the personality and nationalistic charisma of Fidel Castro himself.

The 1980s and 1990s: the ascendancy of neo-liberalism

The decade of the 1980s was greeted by the Reagan administration's counter-inflationary strategies to bring the US domestic economy into order using draconian 'supply-side' policies and fiscal measures. The small island economies of the Caribbean, along with the rest of the world, were dragged into a global

Plate 12.5 Che Guevara mural in Havana, 1992, in need of repair during the 'Special Period'
Source: Dennis Conway

recession, almost irrespective of their national economic performance to that date. Many countries' exchequers were saddled with onerous debt burdens. Region-wide tourism volumes dropped, international commodity prices plummeted (except for marijuana), and the recession brought island unemployment volumes to unprecedented high levels. For many Caribbean (and Latin American and African) economies, the 1980s was a decade-long 'depression' of continuing hardship, global indebtedness, and externally imposed restructuring imperatives.

In addition to the external shocks that rained down on Caribbean economies, there were longstanding internal (dis)economies that made these distorted economies doubly, trebly vulnerable to growing imbalances in their national accounts:

reliance on external financing for too-high import bills,[1] weak economic management in public and private sectors, and the high costs of maintaining the social welfare net (Bakan *et al.* 1993). But it was the peripheral vulnerability of these economies to the world recession that were most significant: the second oil shock of 1978 and 1979 dramatically increasing energy costs, the slackening of demand and declines in prices of export commodities, and the strictures on discretionary money in North America and Europe, which reduced tourist volumes, shrank tourist spending and forced bankruptcy among fledgling tourism entrepreneurs in the region. Global foreign direct investment was in retreat during the 1980–1983 recession, and the Caribbean peripheral economies ranked low on any priority list for diversions of the shrinking pool of international capital (FDI).

The global competitiveness of Caribbean export staples and mineral resources continued to decline in relation to other regions of the South (the 'restructuring' competitors in Latin America, Africa and Asia). There was increased protectionism on the part of Northern countries, especially the United States, with bilaterally negotiated quotas for Caribbean commodities reduced and renegotiations for increases in foreign market shares in East and South-east Asia disadvantaging Caribbean commodities, such as coffee and other tropical products. East Asian and Asian trade relations were to be strengthened, and the competitiveness of the small Caribbean economies was by and large disparaged, viewed as having limited potential in the globalising marketplace that promised much for an emergent global commercial order. The 'debt trap', as some critics have labelled the IMF/World Bank 'conditionalities', scarcely provided a lending hand to the struggling Caribbean states and their peoples (Latin American Bureau 1983).

The 1990s to today: Caribbean micro-states under neo-liberalism's discipline

External forces and macro-structural imperatives direct and determine each country's ministries' policies and practices, scarcely providing the necessary leeway for regional, communal actions, and local practices. Bilateral negotiations through CBI, CARIBCAN and the EU's Lomé agreements dominated until a few years ago (Sutton 1995). Multilateralism has always 'played second fiddle' to bilateralism in the United States' dealings with its Caribbean neighbours. The track record of recent US administrations' overall neglect of Caribbean affairs, unless they were indirectly or directly involved with protecting US domestic national security and/or promoting commercial ventures, is scarcely a 'good neighbour' policy. The often highhandedness of US administrative initiatives, such as the 'ship rider agreement' (which authorised the US Coast Guard to not be overly concerned with sovereignty issues when searching vessels in the Caribbean Sea for drug shipments), or the clumsy manner in which the INS deportation of criminal immigrant aliens back to Caribbean countries was carried out, did little to assure the Caribbean people that the US government respected their national rights of autonomy and sovereignty. Even so, GATT rhetoric insists market protection needs to be swept away and global commerce opened further. Regionally, IMF and World Bank 'conditionalities' and neo-liberal free-market ideologies reign supreme, leaving little room for regional political cooperation or for the Caribbean Development Bank's multilateral lending programmes (Hardy 1995). The crisis faced by Caribbean finance ministries, economists, development planners, industrial development corporations, even the region's multinational corporations, is deeply structural, deeply global and highly problematic. It has always been so for these micro-states, ever since their incorporation into the earliest imperial domains in the fifteenth and sixteenth centuries. And, it is today, in the first decade of the new millennium. Today, more than ever, regional political allegiances offer a way forward, which FTAA and other externally dominated neo-liberal frameworks do not!

From the West Indies Federation to CARIFTA/CARICOM and the Association of Caribbean States (ACS): post-colonial attempts at regional integration

One longstanding idea or 'seminal truth', according to Gordon Lewis (1968), has been for Caribbean countries to join in a federation and seek economic and political 'regional integration'. Indeed, some regional configurations were suggested under colonial auspices. For example, the 1897 British Royal Commission's proposed a federation between Barbados and the Windward Islands, and the 1938 Caribbean Labour Congress supported integration that included the Bahamas, Belize and British Guiana. The Moyne Royal Commission in 1938, on the other hand, viewed such integration as a far-off ideal (Thomas 1983).

The West Indian Federation: 1958–1961

The first serious attempt at such a regional grouping was the British government's coordination of the West Indian Federation of ten of its colonial island territories in 1958. With its centre of government in Trinidad and ministers selected from throughout the territories, this federation governmental experiment was never able to overcome national rivalries. It lasted less than four years, with Sir Alexander Bustamante's nationalist message in favour of Jamaica's independence bringing it down in 1961. Regionalism was viewed sympathetically, however. Eminent regional economists such as William Demas argued long and consistently on behalf of the advantages of regional integration for small, disadvantaged territories (Demas 1965; Chernick 1978). Despite the success of ensuing national movements in the 1960s of the federation's major players, namely Jamaica, Trinidad and Tobago, Guyana and Barbados, others were first directed towards other federal arrangements by the British Colonial office but then eventually sought their own independent enfranchisement. Eric Gairy's Grenada declared independence in 1974, after flirting with integration ideas with Trinidad and Tobago.

Then, five of the British Commonwealth's 'associated states' – Dominica, St Lucia, St Vincent and the Grenadines, Antigua and Barbuda, and St Kitts, Nevis and Anguilla – all sought and won political independence between 1978 and 1983. Subsequently, Anguilla petitioned to rejoin the British colonial family and, in 1996, Nevis petitioned (unsuccessfully) to sever its ties with St Kitts (McElroy and de Albuquerque 1996). Elsewhere, among other European colonial groupings, the Netherlands maintains its widely scattered federation of colonial island territories and the French Départements D'Outre Mer have evolved to become Caribbean extensions of a unified European Union, albeit a Francophone extension (Sutton 1995).

From CARIFTA to CARICOM

The collapse of the federation might have promoted political independence for island territories, but it did not remove the integrationist ideas from the development debate. Heads of government among the Commonwealth Caribbean family of independent nations held annual meetings, if nothing else to keep open the dialogue for greater unity (CARICOM 1981). There were after all several regional interests to foster, or to collectively 'own': the University of the West Indies was one, the highly successful West Indies cricket team another (Beckles and Stoddart 1995). So, in December 1965, the heads of government of Antigua, Barbados and Guyana collectively agreed to form a three-country free-trade agreement, CARIFTA, and three years later in 1968, this Caribbean Integration and Free Trade Agreement was expanded to incorporate all ten West Indies Federation members plus Guyana. CARIFTA admitted Belize in 1974 and the Bahamas in 1983.

In 1973, the Treaty of Chaguaramas formalised CARICOM as a common market to replace CARIFTA, heralding a strengthening of the integrationist quest. CARICOM was to be based on three principles: (1) a common market, based on free trade within the group, a common external tariff, a commitment to the progressive removal of non-tariff barriers to trade, harmonised fiscal incentives and free intra-regional movements of capital (notably omitted was a similar commitment to negotiate for the free movement of labour within CARICOM); (2) establishment of areas of cooperation and the formation of inter-ministerial committees to collaborate in areas of health, education, science and technology, energy, mining and natural resources management, agricultural development, food, and industry, and cooperation in the development of communications and transportation infrastructure; and (3) the coordination of foreign policies (Thomas 1983).

In 1989, CARICOM member states agreed to advance beyond a common market towards more comprehensive integration. In 1991, the main areas of emphasis in the creation of a single market and economy were agreed. These included the completion of the arrangements of the free internal movement of goods of regional origin; mechanisms for the free movement of services, capital and labour; and the greater harmonisation of laws and regulations affecting commerce, including customs laws and procedures, intellectual property, competition policy, corporate taxation, dumping and subsidisation. CARICOM's membership in 2001 stood at 15 states and territories, 13 of which are independent, with Montserrat being an overseas territory of the United Kingdom, and the two non-Commonwealth countries joining later being Suriname (in 1996) and Haiti (2000). Three associate members of the Caribbean Community (CARICOM) are the British Virgin Islands, the Turks and Caicos Islands and Anguilla. In addition, the Caribbean Community has negotiated reciprocal trade agreements with Colombia, Venezuela and the Dominican Republic. Officially, Cuba is still not a member of the Caribbean Community, but member countries have often come to the defence of its government and upheld the principle that as a Caribbean country its interests should be included in broader regional deliberations.

The Association of Caribbean States

The latest effort at regional integration is the establishment of the Association of Caribbean States (ACS).The Association of Caribbean States is an organisation for consultation, cooperation and concerted action among the countries of the Greater Caribbean. The convention establishing the ACS was signed on 24 July 1994 in Cartagena de Indias, Colombia, with the aim of promoting consultation, cooperation and concerted action among all the countries of the Caribbean, comprising 25 member states and three associate members. Eight other non-independent Caribbean countries are eligible for associate membership. Its headquarters are in Port of Spain, Trinidad and Tobago. ACS members span the Caribbean Basin and include South and Central American countries as well as those from the wider Caribbean: Antigua and Barbuda, the Bahamas, Barbados, Belize, Colombia, Costa Rica, Cuba, Dominica, the Dominican Republic, El Salvador, Grenada, Guatemala, Guyana, Haiti, Honduras, Jamaica, Mexico, Nicaragua, Panama, St Kitts and Nevis, St Lucia, St Vincent and the Grenadines, Suriname, Trinidad and Tobago, Venezuela, Aruba, France, and the Netherlands Antilles. The CARICOM Secretariat, the Latin American Economic System (SELA), the Central American Integration System (SICA) and the Permanent Secretariat of the General Agreement on Central American Economic Integration (SIECA) were declared Founding Observers of the ACS in 1996. The United Nations Economic Commission for Latin America and the Caribbean (ECLAC) and the Caribbean Tourism Organisation (CTO) were admitted as Founding Observers in 2000 and 2001 respectively.

The objectives of the ACS are enshrined in the convention and are based on the following: the strengthening of the regional cooperation and integration process, with a view to creating an enhanced economic space in the region; preserving the environmental integrity of the Caribbean Sea, which is regarded as the common patrimony of the peoples of the region; and promoting the sustainable development of the Greater Caribbean. Its current focal areas are trade, transport, sustainable tourism and natural disasters, and to date three summits have been held. The inaugural summit was held in Trinidad and Tobago in 1995. The second ACS summit was held in the Dominican Republic in 1999 to establish the sustainable tourism zone of the Caribbean and the third was held in Margarita, Venezuela, in 2001, where plans of action were laid out concerning the special and differential treatment of small islands and on the sustainable tourism zone of the Caribbean. The ACS remains a fledgling organisation with a modest budget and little policy-making authority, but as a forum for regional integrationist ideas it serves a useful political purpose. Together with CARICOM, the ACS provides opportunities for a unified regional voice to be heard.

Conclusion

The political diversity in the contemporary Caribbean remains one of its defining features. One of the world's few remaining socialist/communist regimes, Fidel Castro's Cuba, is still intact, despite 40 years of antagonism and Cold War political manoeuvring between the United States and Cuban administrations. There are French, British and Dutch colonial dependencies, still adhering to the political institutional models of their European mother countries: social democratic regimes with European-style parliaments, assemblies, polders, civil services and ministerial hierarchies. The Westminster-style of government favoured among Commonwealth Caribbean independent nation-states may have come under scrutiny for its inflexibility, and its 'first-past-the-post' dictum may have been constitutionally challenged by the Trinidad and Tobago 18–18 tie between the two major parties these last three years, but it remains the favoured model elsewhere in the region – Barbados, Antigua, Dominica and the rest. Republican institutional governments modelled after the United States' Presidential and Congressional structure also find their place in Puerto Rico, the US Virgin Islands and the Dominican Republic, albeit with some regional modifications according to the much smaller size of these states or colonies.

Only Haiti still struggles with its lack of political institutional stability and efficiency, and its lack of social democratic traditions. And, as recent events show, Haiti's democratic future is anything but certain. Elsewhere, democratic regimes are the Caribbean standard – the people's voices and representatives; be it the one-party 'popular democracy' of Cuba, the US-influenced 'economic democracy' of Puerto Rico, or the European-style 'social democracies' of the British, French and Dutch ex-colonies and dependencies. Political participation in fair and free elections is enjoyed by the majority of Caribbean people, informed political comment and criticism is a valued social and cultural trait in these island societies. Small island politics has its closeness, its personalities, its absurdness, and its colourful character – its music, calypso, satire, resistance traditions. Sometimes it has violent turns, as in the case of Jamaican politics in recent times. Nevertheless, democracy is not so easily compromised, even as senseless violence and criminal activity appear to be growing as destructive forces in the regional political scene. External political forces, external financial dealings and corporate penetration, external illegal commerce and external criminal activity – money laundering, drug, small arms and people smuggling, racketeering and the like – certainly threaten the autonomy of the region's political realms (Desch *et al.* 1998), but the peoples' resilience is ever-present. The political stability of the Caribbean is not under threat; the hard-fought democratic freedoms are not so easily undermined, Caribbean political institutions have endured and enjoy the majority of the people's involvement and enfranchisement.

Note

1. The IMF and international financial institutions (Banks such as Barclays International, Citicorp, Chase Manhattan, and Canadian National) are implicated in this indebtedness, because they were only too willing to increase debt ceilings, offer new loans to already indebted treasuries, and generally encourage Caribbean finance ministers 'to live beyond their means'.

References

Allahar, A. (2001) *Caribbean Charisma: Reflections on Leadership, Legitimacy and Populist Politics*, Ian Randle and Lynn Reiner, Kingston, Jamaica, London and Boulder, Colorado.

Ambursley, F. (1983) 'Jamaica: from Michael Manley to Edward Seaga', in Ambursley, F. and Cohen, R. (eds) *Crisis in the Caribbean*, Heinemann, London, pp. 72–104.

Ambursley, F. and Cohen, R. (1983) *Crisis in the Caribbean*, Heinemann, London.

Aponte Garcia, M. and Gautier Mayoral, C. (1995) *Postintegration Development in the Caribbean*, Social Science Research Center, Rio Pedras, Puerto Rico.

Azicri, M. (2000) *Cuba Today and Tomorrow: Reinventing Socialism*, University of Florida Press, Gainesville.

Bakan, A. B., Cox, D. and Leys, C. (1993) *Imperial Power and Regional Trade: the Caribbean Basin Initiative*, Wilfred Laurier University Press, Waterloo, Ontario.

Beckles, H. M. (1990) *A History of Barbados: From Amerindian Settlement to Nation State*, Cambridge University Press, Cambridge.

Beckles, H. M. and Stoddart, B. (1995) *Liberation Cricket: West Indies Cricket Culture*, Ian Randle, Kingston, Jamaica.

Berman Santana, D. (1996) *Kicking off the Bootstraps: Environment, Development and Community Power in Puerto Rico*, University of Arizona Press, Tucson.

Boland O. N. (2001) *The Politics of Labour in the British Caribbean*, Ian Randle, James Currey and Markus Wiener, Kingston, Oxford and Princeton.

Blackman, C. N. (1982) *The Practice of Persuasion*, Cedar Press, Bridgetown, Barbados.

Blackman, C. N. (1991) 'The economic management of small island developing countries', *Caribbean Affairs*, **4**, 4, 1–12.

Blood, Sir H. (1958) *The Smaller Territories: Problems and Future*, Conservative Political Centre, London.

Bryan, A. T. (1995) *The Caribbean: New Dynamics in Trade and Political Economy*, Transaction, New Brunswick, NJ, and London.

Burton, R. D. and Reno, F. (1995) *French and West Indian: Martinique, Guadeloupe and French Guiana Today*, University Press of Virginia and Macmillan Caribbean, Charlottesville and London.

Calvo, H. and Declercq, K. (2000) *The Cuban Exile Movement: Dissidents or Mercenaries?* Ocean Press, Melbourne and New York.

CARICOM (1981) *The Caribbean Community in the 1980s*, report by a Group of Experts to the Caribbean Common Market Council of Ministers, Caribbean Community Secretariat, Georgetown, Guyana.

Chernick, S. (1978) *The Commonwealth Caribbean: the Integration Experience*, World Bank and Johns Hopkins University Press, Baltimore.

Connell, J. (1993) 'Island microstates: development, autonomy and the ties that bind', in Lockhart, D. G., Drakakis-Smith, D. and Schembri, J. (1993) *The Development Process in Small Island States*, Routledge, London and New York, pp. 117–47.

Conway, D. (1983a) *Grenada–United States Relations Part I, 1979–1983: A Prelude to Invasion*, University Field Staff International Report, 39, Hanover, NH.

Conway, D. (1983b) *Grenada–United States Relations Part II, October 12–27, 1983: Sixteen Days that Shook the Caribbean*, University Field Staff International Report, 40, Hanover, NH.

Conway, D. (1984) *Trinidad's Mismatched Expectations: Planning and Development Review*, University Field Staff International Report, 26, Hanover, NH, p. 12.

Conway, D. and Heynen, N. C. (2004) 'Introduction: the ascendency of globalization and neoliberalism', in Conway, D. and Heynen, N. C. (eds) *Globalization's Dimensions: Destructive, Disciplinary and Contradictory*, Rowman & Littlefield.

Conway, D. and Heynen, N. (2002) 'What is globalization? A political (economy) time–space schematic as explanation', unpublished manuscript, Department of Geography, Indiana University, Bloomington.

Cox, H. (1999) 'The market as god: living in the new dispensation', *Atlantic Monthly*, **March**, 18–23.

Demas, W. G. (1965) *The Economics of Development in Small Countries with Special Reference to the Caribbean*, McGill University Press, Montreal, Canada.

Deosaran, R. (1996) 'Political management of conflict in a multicultural society', in Dookeran, W. C. (ed.) *Choices and Change: Reflections on the Caribbean*, Inter-American Development Bank, Washington, pp. 137–49.

Desch, M. C., Dominguez, J. I. and Serbin, A. (1998) *From Pirates to Drug-lords: the Post-Cold War Caribbean Security Environment*. State University of New York Press, Albany, NY.

Dookeran, W. C. (1996) *Choices and Change: Reflections on the Caribbean*, Inter-American Development Bank, Washington.

Fanon, F. (1967) *Wretched of the Earth*, Penguin Books, Harmondsworth.

Fanon, F. (1968) *Black Skins, White Masks*, Penguin Books, Harmondsworth.

Fass, S. M. (1990) *Political Economy of Haiti: The Drama of Survival*, Transaction, New Brunswick, NJ.

Gayle, D. J. (1995) 'The evolving Caribbean business environment', in Bryan, A. T. (ed.) *The Caribbean: New Dynamics in Trade and Political Economy*, Transaction, New Brunswick, NJ and London, pp. 135–56.

Girvan, N. (1999) 'Globalisation and counter-globalisation: the Caribbean in the context of the South', paper prepared for the International Seminar on Globalization: A Strategic Response from the South, University of the West Indies, Mona, Jamaica.

Hardy, C. (1995) *The Caribbean Development Bank*, Lynne Reiner and The North-South Institute, Boulder, Colorado, and Ottawa, Canada.

Hendricks, G. (1974) *The Dominican Diaspora*, Teachers College Press, New York.

Jacobs, W. R. and Jacobs, B. I. (1980) *Grenada: The Route to Revolution*, Casa de las Americas, Havana, Cuba.

James, C. L. R. (1963) *The Black Jacobins: Toussaint L'Ouverture and the San Domingo Revolution*, Vintage Books, New York, originally published 1938.

James, C. L. R. (1969) *State Capitalism and World Revolution*, Detroit: Facing Reality.

Klak, T. (1995) 'A framework for studying Caribbean industrial policy', *Economic Geography*, **71**, 3, 297–317.

Kuznets, S. (1960) 'Economic growth in small nations', in Robinson, E. (ed.) *Economic Consequences of the Size of Nations*, Macmillan, London.

Latin American Bureau (1983) *The Poverty Brokers: The IMF and Latin America*, Latin American Bureau, London.

Lewis, G. K. (1968) *The Growth of the Modern West Indies*, MacGibbon and Kee, London.

Lorah, P. (1995) 'An unsustainable path: tourism's vulnerability to environmental decline in Antigua', *Caribbean Geography*, **6**, 1, 28–39.

Lundahl, M. (1979) *Peasants and Poverty: A Study of Haiti*, St Martin's Press, New York.

Lundahl, M. (1983) *The Haitian Economy*, St Martin's Press, New York.

Lundahl, M. (1992) *Politics or Markets: Essays on Haitian Underdevelopment*, Routledge, London and New York.

Manley, M. (1974) *The Politics of Change: a Jamaican Testament*, Andre Deutsch, London.

Manning, F. E. (1973) *Black Clubs in Bermuda: Ethnography of a Play World*, Cornell University Press, Ithaca, NY.

Mathey, K. (1997) 'Self-help housing strategies in Cuba: an alternative to conventional wisdom', in Potter, R. B. and Conway, D. (eds) *Self-Help Housing, the Poor and the State in the Caribbean*, University of Tennessee Press, Knoxville, and University of the West Indies Press, Kingston, Jamaica, pp. 164–87.

McElroy, J. L. and de Albuquerque, K. (1996) 'The social and economic propensity for political dependence in the insular Caribbean', *Social and Economic Studies*, **44**, 2 & 3, 167–93.

Meyn, M. (1996) 'Puerto Rico's energy fix', in Collinson, H. (ed.) *Green Guerillas: Environmental Conflicts and Initiatives in Latin America and the Caribbean*, Latin American Bureau, London.

Morrison, T. K. and Sinkin, R. (1982) 'International migration in the Dominican Republic: implications for development planning', *International Migration Review*, **16**, 4, 819–36.

Nettleford, R. (1989) 'Caribbean crisis and challenges to year 2000', *Caribbean Quarterly*, **35**, 1 & 2, 6–16.

Nyerere, J. (1969) *Freedom and Socialism*, Oxford University Press, Dar es Salaam, Tanzania.

O'Loughlin, C. (1962) 'Economic problems of the smaller West Indian islands', *Social and Economic Studies*, **11**, 1, 44–56.

Payne, D. W. (1998) 'Emerging voices: the West Indian, Dominican and Haitian diasporas in the United States', policy papers on the Americas, Volume IX, Study 11, CSIS Americas Program.

Phillips, K. (1994) *Arrogant Capital: Washington, Wall Street and the Frustration of American Politics*, Little, Brown and Co., Boston and New York.

Roberts, S. M. (1994) 'Fictitious capital, fictitious spaces? The geography of offshore financial flows', in Corbridge, S., Martin, R. and Thrift, N. (eds) *Money, Power and Space*, Blackwell, Oxford, pp. 88–120.

Ruth, R., Grusky, S. and Rodriguez, J. (1994) 'Puerto Rico and the Caribbean Basin Initiative: uneasy interdependence', in Hilbourne, A. and Watson, H. (eds) *The Caribbean in the Global Political Economy*, Lynn Reiner and Ian Randle, Boulder, Colorado, London and Kingston, Jamaica, pp. 207–21.

Ryan, S. (1988) *Trinidad and Tobago: The Independence Experience, 1962–1987*, University of the West Indies, Institute of Social and Economic Research, St Augustine, Trinidad and Tobago.

Ryan, S. (1991) *Social and Occupational Stratification in Contemporary Trinidad and Tobago*, University of the West Indies, Institute of Social and Economic Research, St Augustine, Trinidad and Tobago.

Sandoval, J. M. (1983) 'State capitalism in a petroleum-based economy: the case of Trinidad and Tobago', in Ambursley, F. and Cohen, R. (eds) *Crisis in the Caribbean*, Heinemann, London, pp. 247–68.

St Cyr, E. (1993) 'The theory of Caribbean-type economy', in Lalta, S. and Freckleton, M. (eds) *Caribbean Economic Development: The First Generation*, Ian Randle, Kingston, Jamaica, pp. 8–16.

Stubbs, J. (1989) *Cuba: The Test of Time*, Latin American Bureau, London.

Sunshine, C. A. (1985) *The Caribbean: Survival, Struggle and Sovereignty*, EPICA, Washington.

Sutton, P. (1995) 'The "New Europe" and the Caribbean', *European Review of Latin American and Caribbean Studies*, **59**, December, 37–57.

Thomas, C. Y. (1983) 'State capitalism in Guyana: an assessment of Burnham's co-operative socialist republic', in Ambursley, F. and Cohen, R. (eds) *Crisis in the Caribbean*, Heinemann, London, pp. 27–48.

Thomas C. Y. (1988) *The Poor and the Powerless: Economic Policy and Change in the Caribbean*, Monthly Review Press, New York.

Thorndike, T. (1988) *No End to Empire*, Department of International Relations and Politics, Stafford Polytechnic and Foreign and Commonwealth Office, London.

Trinidad Express (1990) *Daily Express: Trinidad under Siege: the Muslimeen Uprising – 6 Days of Terror*, Trinidad Express Newspapers Ltd, Port of Spain, Trinidad and Tobago.

Weaver, D. B. (1988) 'The evolution of a "plantation" tourism landscape on the Caribbean Island of Antigua', *Trjdschrift voor Economische en Sociale Geographie*, **79**, 5, 313–19.

Wiarda, H. J. and Kryzanek, M. J. (1992) *The Dominican Republic: A Caribbean Crucible*, 2nd edition, Westview Press, Boulder, Colorado.

Worrell, D. (1987) *Small Island Economies: Structure and Performance in the English Caribbean since 1970*, Praeger, New York.

INDEX